physica status solidi (b)

AKADEMIE-VERLAG · BERLIN

VOLUME 62 · NUMBER 2 · 1974

THEORIA CUM PRAXI

Classification Scheme

1. Structure of Crystalline Solids
    1.1 Perfectly Periodic Structure
    1.2 Solid-State Phase Transformations
    1.3 Alloys. Metallurgy
    1.4 Microstructure (Magnetic Domains See 18; Ferroelectric Domains See 14.4.1)
    1.5 Films
    1.6 Surfaces
2. Non-Crystalline State
3. Crystal Growth
4. Bonding Properties
5. Mössbauer Spectroscopy
6. Lattice Dynamics. Phonons
7. Acoustic Properties
8. Thermal Properties
9. Diffusion
10. Defect Properties (Irradiation Defects See 11)
    10.1 Metals
    10.2 Non-Metals
11. Irradiation Effects (X-Ray Diffraction Investigations See 1 and 10)
12. Mechanical Properties (Plastic Deformations See 10)
    12.1 Metals
    12.2 Non-Metals
13. Electron States
    13.1 Band Structure
    13.2 Fermi Surfaces
    13.3 Surface and Interface States
    13.4 Impurity and Defect States
    13.5 Elementary Excitations (Phonons See 6)
        13.5.1 Excitons
        13.5.2 Plasmons
        13.5.3 Polarons
        13.5.4 Magnons
14. Electrical Properties. Transport Phenomena
    14.1 Metals. Semi-Metals
    14.2 Superconductivity. Superconducting Materials and Devices
    14.3 Semiconductors
        14.3.1 Films
        14.3.2 Surfaces and Interfaces
        14.3.3 Devices. Junctions (Contact Problems See 14.3.4)
        14.3.4 High-Field Phenomena, Space-Charge Effects, Inhomogeneities, Injected Carriers (Electroluminescence See 20.3; Junctions See 14.3.3)
    14.4 Dielectrics
        14.4.1 Ferroelectrics
15. Thermoelectric and Thermomagnetic Properties
16. Photoconductivity. Photovoltaic Effects
17. Emission of Electrons and Ions
    17.1 Field Emission Microscope Investigations
18. Magnetic Properties
    18.1 Paramagnetic Properties
    18.2 Ferromagnetic Properties
        18.2.1 Ferromagnetic Films
    18.3 Ferrimagnetic Properties
    18.4 Antiferromagnetic Properties

(Continued on cover three)

physica status solidi (b)

basic research

Volume 62 · Number 2 · Pages 335 to 728, K63 to K124, and A9 to A16

April 1, 1974

PSSB 62(2) 335—728, K63—K124, A9—A16 (1974)

AKADEMIE-VERLAG · BERLIN

Subscriptions and orders for single copies should be addressed to
AKADEMIE-VERLAG, 108 Berlin, Leipziger Straße 3 – 4 or to Buchhandlung KUNST UND WISSEN, Erich Bieber, 7 Stuttgart 1, Wilhelmstr. 4 – 6 or to Deutsche Buch-Export und -Import GmbH, 701 Leipzig, Postschließfach 160

Editorial Note:

"physica status solidi (b)" undertakes that an original paper *accepted for publication* before the 8th of any month will be published within 50 days of this date unless the author requests a postponement. In special cases there may be some delay between *receipt* and *acceptance* of a paper due to the review and, if necessary, revision of the paper.

Schriftleiter und verantwortlich für den Inhalt: Professor Dr. Dr. h. c. P. Görlich, 102 Berlin, Neue Schönhauser Str. 20 bzw. 69 Jena, Humboldtstr. 26. Chefredakteur: Dr. H.-J. Hänsch. Redaktionskollegium: Prof. Dr. E. Gutsche, Dr. H.-J. Hänsch, Dr. H. Lange, Dr. S. Oberländer. Anschrift der Schriftleitung: 102 Berlin, Neue Schönhauser Str. 20, Fernruf: 4223380. Verlag: Akademie-Verlag, 108 Berlin, Leipziger Str. 3—4, Fernruf: 220441, Telex-Nr. 112020, Postscheckkonto: Berlin 35021. — Die Zeitschrift „physica status solidi (b)" erscheint jeweils am 1. des Monats. Bezugspreis eines Bandes 140,— M; Sonderpreis für die DDR 120,— M. Bestellnummer dieses Bandes: 1068/62. Jeder Band enthält zwei Hefte. Gesamtherstellung: VEB Druckerei „Thomas Müntzer" Bad Langensalza.
Veröffentlicht unter der Lizenznummer 1310 des Presseamtes beim Vorsitzenden des Ministerrates der Deutschen Demokratischen Republik.
Printed in the German Democratic Republic.

Contents

Short Notes

Pre-printed Titles

Systematic List

The Author Index of Volume 62 Begins on Page 729

(It will be delivered together with Volume 63, Number 1.)

Original Papers

phys. stat. sol. (b) **62**, 343 (1974)

Subject classification: 13.1 and 20.1; 2; 22.7

Groupe de Spectroscopie des électrons[1]), *Faculté des Sciences, Reims*

Optical Constants and Electronic Structure of Crystalline and Amorphous As_2S_3 in the 3 to 35 eV Range[2])

By

J. Perrin, J. Cazaux, and P. Soukiassian

By Kramers-Kronig analysis, the complex dielectric constant tensor of monocrystalline As_2S_3 is deduced from energy losses in the 3 to 35 eV range suffered by a 40 keV electron beam. The weak optical anisotropy observed in this spectral range is explained and the results are discussed in terms of a valence bonding model with two groups of valence bands. Upon the amorphous to crystalline transition it is the optical transitions originating in the resonance bonds which are only influenced.

A partir du spectre des pertes d'énergie comprises entre 3 et 35 eV et subies par un faisceau d'électrons de 40 keV nous déduisons, à l'aide d'une analyse de Kramers et Kronig, les trois composantes du tenseur de la constante diélectrique de As_2S_3 monocristallin. On rend compte de la faible anisotropie optique observée dans ce domaine spectral et nous interprétons les résultats en termes de transitions faisant intervenir deux groupes de bandes de valence. Le passage de l'état cristallisé à l'état amorphe influence uniquement les liaisons de résonance.

1. Introduction

Optical properties of binary compounds of the form $A_2^{V}B_3^{VI}$ have received attention in the past [1 to 4], but due to the lack of optical polarizers above the LiF cut-off and except the use of synchrotron radiation, complete determination of the dielectric constant tensor is optically limited up to 14 eV.

Besides, for lamellar crystals like As_2S_3, we cannot deduce at first sight, from reflectance measurements at oblique incidence the optical constants along the principal axis perpendicular to the layers.

So, to overcome these difficulties, we have applied the electron energy loss technique to extend up to 35 eV along the three principal axis the complex dielectric constant of As_2S_3 which was previously determined, using reflectance experiments, by Zallen et al. [4] (limited up to 14 eV along *a*- and *c*-axis only).

From experimental results, we have observed that optical anisotropy, in this energy range, is of MoS_2 type and weaker than those of graphite and boron nitride.

Applying the ε_2 sum rule, we suggest a band structure scheme following Kastner's model [5] for semiconductors containing group VI elements in two-fold coordination.

We also compare the optical properties of As_2S_3 in crystalline and amorphous form.

---

1) Laboratoire de Microscopie et Spectroscopie électronique, Equipe associèe au C.N.R.S. n⁰ 376.

2) This work partially covers the 3th cycle thesis of J. Perrin.

2. Experimental Procedures

2.1 Crystal structure and sample preparations

Crystalline As_2S_3 is monoclinic ($a = 11.48$ Å; $b = 9.58$ Å, $c = 4.23$ Å) of space group C_{2h}^5 with 20 atoms, i.e., 4 molecules/unit cell. The atoms are arranged in a layer structure with layers $\perp \boldsymbol{b}$ and weak As–S bonding between layers and, therefore, the crystals exhibit natural cleavage $\perp \boldsymbol{b}$ [6].

Thus films used were prepared by successive cleavage, with an adhesive tape, of a single crystal of natural orpiment (mineral name of As_2S_3). After dissolution of adhesive tape, films were self-supported on electron microscope grids.

The structure of the sample was checked by electron diffraction patterns: the a- and c-axis were lying in the plane of the crystals. Amorphous films of As_2S_3 were obtained by evaporation of the same basic orpiment crystal onto room temperature NaCl substrate and controlled by electron diffraction.

2.2 Electron energy loss experiments

2.2.1 Description of the method

The general procedure follows the method applied by Zeppenfeld [7] to uniaxial lamellar crystals: At normal incidence and for a scattering angle θ of about 7×10^{-4} rad, the impulse $\hbar \boldsymbol{k}$ transferred to the crystal by incoming electrons (40 keV energy range) and determined from momentum conservation lies in the sample plane. If, by an azimuthal rotation of the sample, we put successively a-axis and c-axis in the scattering plane, we get energy loss spectra related to these two directions which are proportional to $-\text{Im } 1/(\tilde{\varepsilon} \parallel \boldsymbol{a})$ and $-\text{Im } 1/(\tilde{\varepsilon} \parallel \boldsymbol{c})$. As in the isotropic case [8] a Kramers-Kronig analysis allows us to get first $\text{Re } 1/(\tilde{\varepsilon} \parallel \boldsymbol{a})$ and $\text{Re } 1/(\tilde{\varepsilon} \parallel \boldsymbol{c})$ and the complex dielectric constant $\tilde{\varepsilon} \parallel \boldsymbol{a}$ and $\tilde{\varepsilon} \parallel \boldsymbol{c}$ in the spectral range observed (here 3 to 35 eV).

To have $\tilde{\varepsilon} \parallel \boldsymbol{b}$, it is necessary to perform experiments at oblique incidence ($\alpha = 45°$) and with the axis $\boldsymbol{a}$ ($\boldsymbol{c}$) in the scattering plane. As in the uniaxial case, we deduced, from energy loss spectra obtained for two symmetrical scattering angles $\theta+$ and $\theta-$ ($\theta+ = \theta- \approx 5 \times 10^{-4}$ rad), the corresponding energy loss function: $\text{Im } 1/\tilde{\varepsilon}(\omega, \theta+)$ and $\text{Im } 1/\tilde{\varepsilon}(\omega, \theta-)$; then, by a Kramers-Kronig analysis involving these two functions we get $\text{Re } 1/\tilde{\varepsilon}(\omega, \gamma)$ (where γ is the angle between the impulse $\hbar \boldsymbol{k}$ and the b-axis — see Fig. 1).

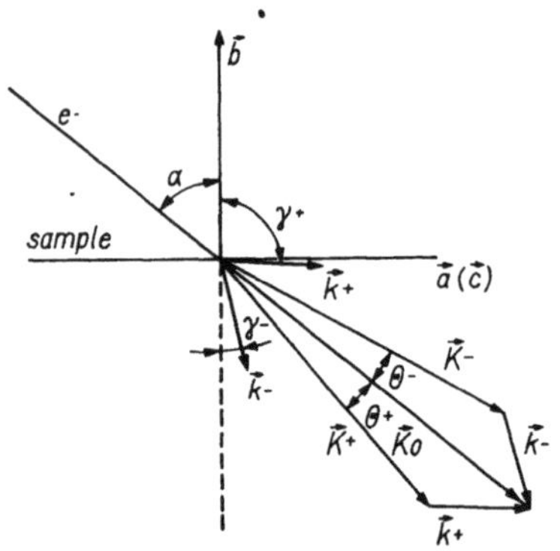

Fig. 1. Experimental arrangement

Finally the required value is given by the following expression:

$$\tilde{\varepsilon} \parallel \boldsymbol{b}(\omega) = \frac{\tilde{\varepsilon}(\omega, \gamma) - \tilde{\varepsilon} \parallel \boldsymbol{a}(\omega) \sin^2 \gamma}{\cos^2 \gamma}$$

putting in the ($\varepsilon \parallel \boldsymbol{a}$) result got at normal incidence.

2.2.2 Experimental conditions

Electron energy loss experiments were performed in an electron diffraction apparatus with a Möllenstedt type velocity analyzer [9] having an energy resolution of about 0.2 eV. The angular divergence of the incoming electrons is better than 10^{-4} rad and gives a knowledge of γ angles with a precision better than 5° in the whole spectral range. As we shall see later, it was not necessary to take the energy spread of the incident beam (about 1 eV) in the transformation of the observed loss spectra into the corresponding energy loss function to have a good agreement between the dielectric constant deduced from optical measurements and the same quantity deduced from energy loss experiments: this is due to the fact that spectra of this type of covalent materials show broad peaks without fine structures. At last, as a consequence of the narrow angular divergence, the scattered electron flux is about 10^3 electrons/s and such a low intensity was recorded with the help of a silicon barrier detector which has a low noise and good quantum efficiency for electrons in the 50 keV range.

3. Results

3.1 Energy loss function

Starting from the observed loss peak and to get the true energy loss function corresponding to single volume losses, it is often necessary to eliminate effects due to Cerenkov radiation, surface plasmons, and multiple scattering [8].

In such substances as As_2S_3 where ε_1 reaches values as high as 20 in the 3 to 8 eV energy range and with incoming electrons of 40 keV ($v \approx 0.4\, c$) one has to expect Cerenkov radiation emission; but the corresponding electron energy loss [8, 19] is concentrated in a scattering angle (5×10^{-5} rad) which is smaller than the scattering angle chosen for our experiments (5 to 7×10^{-4} rad), so we have not taken this effect into account.

In the same way, we have verified "a posteriori" that surface effects can be neglected in our observed spectra: The surface plasmon intensity ($\propto -\text{Im}\, 1/(\tilde{\varepsilon} + 1)$) is maximum in an energy range which corresponds to $\varepsilon_1(\omega) = -1$, so, when their contribution is important in the measured spectra, we have to find, after a Kramers-Kronig analysis, additional singularities at this energy in the calculated dielectric function. So, due to the smooth variation of ε_1 and ε_2 at around 14 eV ($\varepsilon_1 \approx -1$) we can say that surface effects do not affect our experimental results.

On the other hand, we have taken into account multiple loss effects and the $1/k^2$ dependence of the observed peak with respect to the true energy loss function [8]. So, after application of the corresponding ($\theta^2 + \theta_E^2$) correction, we have normalized the observed energy loss spectra, putting the following values of $\varepsilon_1(0)$ in the near infrared found in the literature [10] ($\varepsilon \parallel \boldsymbol{a} = 9.6$, $\varepsilon \parallel \boldsymbol{c} = 6.8$,

$\varepsilon(\text{am}) = 6.7$ at 0.1 eV) in the relation

$$\operatorname{Re} \frac{1}{\varepsilon(0)} - 1 = \frac{2}{\pi} \int_0^\infty \operatorname{Im} \frac{1}{\varepsilon(\omega)} \frac{d\omega}{\omega}.$$

The deduced functions $-\operatorname{Im} 1/(\varepsilon \parallel \boldsymbol{a})$, $-\operatorname{Im} 1/(\varepsilon \parallel \boldsymbol{b})$, $-\operatorname{Im} 1(\varepsilon \parallel \boldsymbol{c})$, and $-\operatorname{Im} 1/\varepsilon(\text{am})$ are shown in Fig. 2.

The energy loss spectra for crystalline material show a strong maximum at 20 eV (19.8 eV $\parallel \boldsymbol{a}$, 20 eV $\parallel \boldsymbol{b}$, and 19.9 eV $\parallel \boldsymbol{c}$) and weaker structure (more or less pronounced with the polarization direction) around 7 to 8 eV. These spectra which look like those observed in MoS_2 or graphite are typical of solids involving two groups of interband transitions [11].

For the amorphous form of As_2S_3, the main difference with the crystalline form is the shift toward lower energies of the main peak (19.2 eV); a shift which is due to the difference in the free electron plasma energy $(\hbar\omega_p(\text{cryst}) = 18$ eV, $\hbar\omega_p(\text{am}) = 17.4$ eV$)$ as a consequence of density differences ($d_{\text{am}} = 3.2\ d_{\text{cryst}} = 3.43$). On the other hand, one has to expect a very weak anisotropy of the optical constants in the energy range 10 to 35 eV as a consequence of the similarities observed between energy loss spectra corresponding to the three main directions of polarization.

3.2 Optical constants

The optical constants ε_1 and ε_2 deduced from energy loss functions are presented in Fig. 3 (3a: $\parallel \boldsymbol{a}$, 3b: $\parallel \boldsymbol{b}$, 3c: $\parallel \boldsymbol{c}$, 3d: am) and can be compared (dotted line) with the partial optical results given by Zallen et al [4]. Agreements between them is rather good and here, it is difficult to say if differences, when they occur, are due to experimental errors (in the energy interval 3 to 10 eV optical measurements are probably more accurate than energy loss ones), or if they take a more fundamental aspect: one gets $\tilde{\varepsilon}_T(k, \omega)$ from reflectance experiments ($k \approx 0$) and $\tilde{\varepsilon}_L(k', \omega)$ from energy loss spectra [12, 13]. Nevertheless, independently of the experimental process, the main features of $\varepsilon_2(\omega)$ for crystalline As_2S_3 are the same: a sharp peak, whose position (from 3 to 3.7 eV) is related to the polarization direction, is followed by a weaker bump at around 9 eV and the trough between them is more or less pronounced with respect to the principal axis concerned (this trough is specially clearly defined at 7 eV for $\varepsilon_2 \parallel \boldsymbol{b}$).

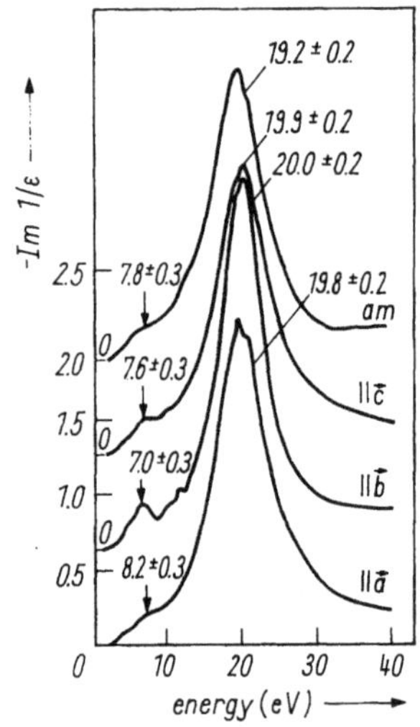

Fig. 2. Energy loss function of crystalline ($\parallel \boldsymbol{a}$, $\parallel \boldsymbol{b}$, $\parallel \boldsymbol{c}$) and amorphous As_2S_3

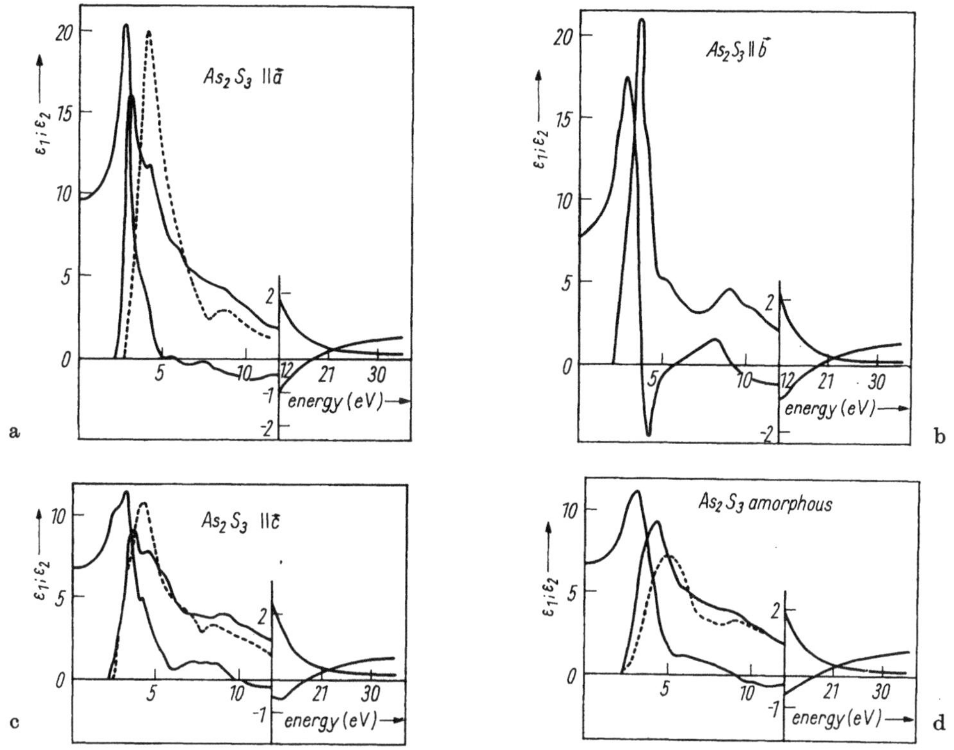

Fig. 3. Dielectric constant tensor of crystalline As_2S_3; a) $\varepsilon \parallel \boldsymbol{a}$, b) $\varepsilon \parallel \boldsymbol{b}$, c) $\varepsilon \parallel \boldsymbol{c}$, and d) dielectric constant of amorphous As_2S_3; (dotted line: ε_2 from partial optical results of Zallen et al. [4])

We also see that the anisotropy acts more on the amplitude of the peaks than on their energetic position and that no other singularity appears between 10 and 35 eV. The main effect, going from crystalline to amorphous form, is an intensity decrease of the first maximum and a shift towards an higher energy value (4.2 eV).

In the area of fundamental absorption (from 2.3 to 3.5 eV), our experimental results were enough accurate to verify the linear variation of $(\hbar\omega\alpha)^{1/2}$ with respect to $\hbar\omega$ (we find 980 cm^{-1} $eV^{-1/2}$ for the proportionality coefficient (the slope) to compare with Felty and Myers' optical results ([14]: 850 cm^{-1} $eV^{-1/2}$) for amorphous As_2S_3 and to find no linear variation for crystalline As_2S_3 ($\parallel \boldsymbol{b}$).

4. Discussion

4.1 Band structure scheme

The electron energy band structure of As_2S_3 is not yet known but it is certainly complicated (by its 28 valence electrons per molecule) so, taking into account our energy resolution (0.2 eV), it is out of our purpose to discuss $\varepsilon_2(\omega)$ in terms of interband transitions at some critical points of the Brillouin zone; but nevertheless, starting from $\varepsilon_2(\omega)$ curves and the corresponding ε_2 sum rule, we are able, with some chemical bond arguments, to draw the general outline of this

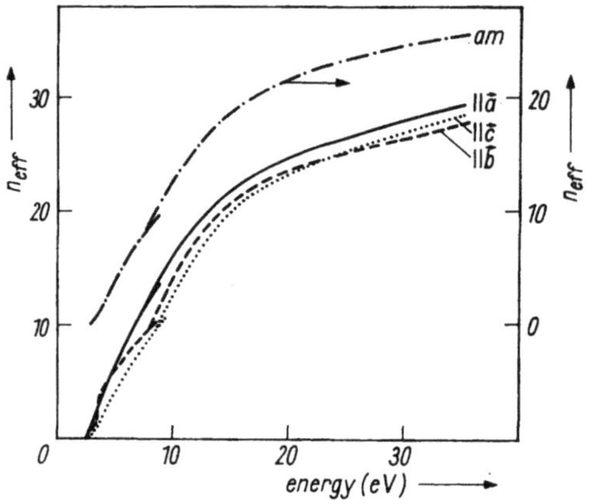

Fig. 4. n_{eff} vs. E from ε_2 sum rule for crystalline and amorphous As_2S_3

band structure. The electronic states of such a covalent solid may be considered, to a first order, to be a broadened superposition of molecular orbital states of the constituent bond. In $A_2^VB_3^{VI}$ materials we have twelve electrons per molecule to insure covalent bonding (σ states) between A and B atoms and we find the corresponding anti-bonding band (σ^* states) higher in the energy scale; between these two bands (one empty: σ^*, one full: σ), we have to place the unshared electrons (lone pair bands).

In this picture, the first peak of $\varepsilon_2(\omega)$ (at 3.4 eV) corresponds to electronic transitions between these lone pair band and the conduction band (σ^*), while the second peak has to be related with electronic transitions from σ band to σ^* band. To find the number of electrons suffering such transitions we have applied the ε_2 sum rule:

$$n_{eff}(E) = \frac{2}{\pi}\,\frac{m\varepsilon_0}{\hbar^2 N e^2}\int\limits_0^{E_0} \varepsilon_2(E')\,E'\,\mathrm{d}E'\,,$$

but the upper limit E_0 has to be taken at around 15 eV (above this energy, ε_2 is smoothly decreasing without any structure); so, as we can see in Fig. 4, the corresponding curve gives a value (20 electrons/molecule at 15 eV) in agreement with those (22 electrons/molecule) deduced from Kastner's model [5] (for semiconductors which contain group VI elements in twofold coordination).

So, in such a model, the bonding bands have to include twelve electrons per molecule and the lone pair bands ten unshared electrons: this last value can be clearly seen at the inflection point in the n_{eff} curves and the corresponding energy ($\approx$8 eV) represents the frontier between the two groups of interband transitions. (Due to a partial overlapping of the two valence bands, these inflection points are less marked than in graphite case [7].)

Such a band structure scheme was also proposed by Zallen et al. [4] but with a number of electrons (16 per molecule) populating the lone pair band, this number includes $3s^2$ of the chalcogen atom.

As experimental sensitivity is good enough to detect 3d transitions of gallium in GaS at around 20 eV [13] and from lack of singularities in the $\varepsilon_2(\omega)$ curve in the 15 to 25 eV energy range, we conclude that interband transitions of the $3s^2$ electron state of the sulphur atom are probably forbidden, or very weak.

We have to notice that such a situation exists too for Sb_2Se_3 [3] and MoS_2 type materials [15, 11].

4.2 Optical anisotropy

In spite of the other highly anisotropic physical properties, crystalline As_2S_3 presents in the 5 to 30 eV range optical anisotropy which is of the same order as that of transition metal dichalcogenides [16] and considerably smaller than anisotropy of other layered crystals like graphite [8] or boron nitride [17].

Without any considerations about selection rules associated with the polarization direction, this effect can be easily understood if we observe that in BN or graphite, bonds have the same directions as principal axis or principal planes in the lattice and that such a situation does not hold in MX_2 or As_2S_3 crystals where coordination units in the basis form trigonal prisms or octahedra. Optical properties in the vacuum ultraviolet are governed by covalent bonding between atoms in the basis rather than symmetries of the space lattice (which only give a little correction).

Analogies observed between crystalline and amorphous results (in the amorphous form only short-range order is conserved) square with such an hypothesis.

4.3 Amorphous material

We have seen in the ε_2 curves that only the first peak was affected ($E < 7$ to 8 eV) when we go from crystalline to amorphous As_2S_3, so we can conclude that the lack of long-range order only affects electronic transitions between the weakly bonding resonance electron bands and the conduction bands.

To give a more quantitative explanation of such an effect, it is possible between many theories related to amorphous semiconductors, to assign the increase in energy of the first group of interband transitions to a decrease of the mean interatomic spacing: such an explanation was given, for instance, with the use of the phenomenological theory of Philips, by Brodsky and Stiles [18] for amorphous GeTe, but such an explanation requires the introduction of some cracklines in the film structure to explain the apparent contradiction between a decrease of the density and a decrease of the interatomic spacing. So, in the present state of things, we are not able to conclude and further experiments are necessary.

Acknowledgements

We wish to thank Dr. Cervelle (Laboratoire de Minéralogie et Cristallographie — Université Paris VI) for supplying natural orpiments and we are indebted to Dr. Kunstreich for suggestion about the use of the electron detector.

References

[1] D. L. Greenaway and G. Harbeke, Optical Properties and Band Structure, Pergamon Press London 1970, 2nd Ed.

[2] V. V. Sobolev, S. D. Shutov, Yu. V. Popov, and S. N. Shestatskii, phys. stat. sol. **30**, 349 (1968).
S. D. Shutov, V. V. Sobolev, Yu. V. Popov, and S. N. Shestatskii, phys. stat. sol. **31**, K23 (1969).

[3] J. C. Shaffer, B. van Pelt, C. Wood, J. Freeouf, K. Murase, and J. W. Osmun, phys. stat. sol. (b) **54**, 511 (1972).

[4] R. Zallen, R. E. Drews, R. L. Emerald, and M. L. Slade, Phys. Rev. Letters **26**, 1564 (1971); Solid State Commun. **10**, 293 (1972); and private communication.

[5] M. KASTNER, Phys. Rev. Letters **28**, 355 (1972).
[6] N. MORIMOTO, Mineral J. (Sapporo) **1**, 160, 1954.
[7] K. ZEPPENFELD, Dissertation Universität Hamburg, 1969.
[8] J. DANIELS, C. FESTENBERG, H. RAETHER, and K. ZEPPENFELD, Springer Tracts in Modern Physics **54**, 77 (1970).
[9] J. CAZAUX, Thesis Paris, 1970, A.O. 54.90;
R. VILANOVE, Thesis Paris, 1971, A.O. 58.85;
R. VILANOVE, C. R. Acad. Sci. (France) **271**, B 1101 (1970);
J. CAZAUX, J. Microscopie **8**, 637 (1969).
[10] R. ZALLEN, M. L. SLADE, and A. T. WARD, Phys. Rev. B **3**, 4257 (1971).
[11] J. CAZAUX, Optics Commun. **3**, 221 (1971).
[12] P. BAMMES, R. KLUCKER, E. E. KOCH, and T. TUOMI, phys. stat. sol. (b) **49**, 561 (1972).
[13] J. CAZAUX, J. PERRIN, and P. SOUKIASSIAN, Colloque sur les propriétés optiques des semi-conducteurs à grande bande interdite, Montpellier, 5—7 septembre 1973; J. Phys. Suppl. C3 **35**, 225 (1974).
[14] E. J. FELTY and M. B. MYERS, Private communication, in: N. F. MOTT and E. A. DAVIS, Electronic Processes in Non-Crystalline Materials, Clarendon Press, Oxford 1971.
[15] W. Y. LIANG and S. L. CUNDY, Phil. Mag. **19**, 1031 (1969).
[16] K. ZEPPENFELD, Optics Commun. **1**, 377 (1970).
[17] R. VILANOVE, C. R. Acad. Sci. (France) **272**, B 1066 (1971).
[18] H. H. BRODSKY and P. J. STILLES, Phys. Rev. Letters **25**, 798 (1970).
[19] E. KRÖGER, Z. Phys. **216**, 115 (1968).

(Received December 28, 1973)

phys. stat. sol. (b) **62**, 351 (1974)

Subject classification: 13.1; 22.2.1; 22.2.3

A. F. Ioffe Physico-Technical Institute, Academy of Sciences of the USSR, Leningrad

Optical Effective Mass of Holes in p-Type $A^{III}B^{V}$ Compounds

By

A. S. Filipchenko, D. N. Nasledov, and L. N. Radaikina

A comparative study is made of the concentration dependences of the optical effective masses of holes in a series of p-type $A^{III}B^{V}$ compounds. It is shown that in all compounds studied the optical effective mass of holes begins to increase under high concentrations of carriers. A qualitative explanation is given of such increase. The comparison of the concentration dependences of m_p for various compounds shows that the valence bands are similar in various $A^{III}B^{V}$ compounds. The structure of the valence bands in $A^{III}B^{V}$ compounds is substantially different from the structure of the conductivity bands. Therefore in theoretical calculations of different effects it is necessary to distinguish quite distinctly the dependences on the density-of-states effective mass and on the conductivity effective mass, i.e. to be very cautious in the use of the quasiisotropic band approximation.

Es werden die Konzentrationsabhängigkeiten der optischen Massen der Löcher in einer Reihe von p-leitenden $A^{III}B^{V}$-Verbindungen verglichen. Es wurde gefunden, daß in allen untersuchten Verbindungen die optischen effektiven Massen der Löcher bei hohen Ladungsträgerkonzentrationen ansteigen. Eine qualitative Erklärung dieses Anstiegs wird angegeben. Der Vergleich der Konzentrationsabhängigkeiten von m_p für verschiedene Verbindungen zeigt, daß die Valenzbänder in den verschiedenen $A^{III}B^{V}$-Verbindungen ähnlich sind. Die Struktur der Valenzbänder in den $A^{III}B^{V}$-Verbindungen ist grundsätzlich von der Struktur der Leitungsbänder verschieden. Darum muß bei theoretischen Berechnungen verschiedener Effekte genau zwischen der effektiven Masse der Zustandsdichte und der effektiven Masse der Leitfähigkeit unterschieden werden, d. h. man muß sehr sorgfältig beim Benutzen der quasiisotropen Bandnäherung sein.

1. Introduction

For many years the problem of the valence band in $A^{III}B^{V}$ compounds has been attracting the attention of the researchers throughout the world. Nevertheless there are considerable discrepancies in the interpretation of the various effects in p-type $A^{III}B^{V}$ compounds. One of the principal causes of these discrepancies is the quite arbitrary choice of values of the hole effective mass. In spite of a considerable amount of work on this subject, the question of the magnitude of the hole effective mass in these compounds remains unsolved up to now. Thus, for p-InSb one can find in the literature a whole series of values m_p from 0.1 to 0.6 m_0. Much the same can be said concerning the other compounds of $A^{III}B^{V}$ type.

The most studied of these compounds is indium antimonide. Analysing the whole amount of literature data concerning the hole effective mass in p-InSb one can divide them into three groups.

1.1 Effective mass values of holes obtained from different transport effects

This group comprises the majority of published works. They are giving quite different values of m_p from 0.1 to 0.6 m_0. The literature values of m_p in p-InSb obtained from investigations of the transport effects are presented in

Table 1

p (cm^{-3})	m_p/m_0	method	ref.	note
—	0.1	thermo-e.m.f.	[3]	100—700 °K
—	0.18	thermo-e.m.f.	[4]	—
—	0.18	thermo-e.m.f.	[5]	333 °K
—	0.21	thermomagnetic effect	[6]	313 °K
—	0.24	thermomagnetic effect	[6]	283 °K
—	0.29	thermomagnetic effect	[6]	263 °K
—	0.42	thermo-e.m.f. under pressure	[7]	room temp.
6.7×10^{15}	0.13	thermo-e.m.f.	[8]	160—200 °K
10^{16}—10^{18}	0.35	thermo-e.m.f.	[9]	—
—	0.25	thermomagnetic effect	[10]	—
—	0.6	mobility	[11]	290 °K
10^{17}	0.15	mobility	[12]	100—300 °K
10^{12}—10^{19}	0.4	mobility	[13]	77, 300 °K
—	0.34	mobility	[14]	—
—	0.19	Hall effect	[15]	—
—	0.18	mobility	[16]	—
10^{14}—10^{19}	0.35	mobility	[17]	82 °K
—	0.25	mobility	[18]	—
10^{14}—10^{15}	0.17	mobility	[19]	1.3 °K
10^{13}—10^{17}	0.55	mobility and $\Delta\varrho/\varrho$	[20]	80—500 °K
—	0.5	mobility	[21]	—
—	0.5	mobility	[22]	—

Table 1. But we know that the method of calculating the values of the carrier effective mass from the transport effects is in itself extremely inaccurate. One must know the mechanism of the current carrier scattering, the exact values of many parameters of the material, and these questions are most often containing considerable difficulties. In case of p-$A^{III}B^{V}$ these difficulties are aggravated by the introduction of the relaxation time into the theoretical calculations of the transport effects. This makes the calculations not strict, as it is known [1] that the valence band in $A^{III}B^{V}$ compounds is not isotropic. The necessity of an exact knowledge of the carrier scattering mechanism is eliminated if the investigations of thermomagnetic effects are carried out in strong magnetic fields. But it is very difficult to attain the condition of a strong field in p-$A^{III}B^{V}$ compounds. Besides, these experiments are giving values of the density-of-states effective mass (in case of a simple isotropic band). It is not clear what would be obtained in case of a non-isotropic valence band.

1.2 Effective mass values obtained from optical investigations

The optical investigations (plasma reflection, Faraday effect, etc.) are a much more exact method of determining the effective mass values. As a rule the optical investigations are giving the values of the conduction effective mass [2]. Table 2 presents the literature values of m_p in p-InSb obtained by the optical method. It should be noted that here also one can find considerable divergencies between the values of the effective mass though not as large as in Table 1.

Table 2

p (cm^{-3})	m_p/m_0	method	ref.	note
—	0.4	absorption	[23]	—
1.9×10^{19}	0.2	IR reflection	[24]	290 °K
6.1×10^{18}	0.18	IR reflection	[26]	100, 300 °K
8.5×10^{18}	0.18			
1.5×10^{19}	0.18			
2.25×10^{19}	0.20			
5.1×10^{19}	0.23			
—	0.44	magneto-optical effect	[27]	[111] 20 °K
—	0.42			[110]
—	0.32			[100]
$1.8 \times 10^{18}-9 \times 10^{19}$	0.13–0.47	IR reflection	[28]	290 °K
—	0.19	magneto-optical effect	[25]	[111]
—	0.18			[110]
—	0.16			[100]

Table 3

p (cm^{-3})	m_p/m_0	ref.	note
5×10^{14}	0.18	[29]	2.2 °K
	0.20		
—	0.3	[33]	—
10^{14}	0.34	[30]	[100] 10–65 °K
	0.45		[111]
	0.42		[110]
—	0.3	[31]	77 °K
—	0.35		10–20 °K
—	0.2; 0.45; 0.85		weak peaks
$10^{14}-10^{15}$	0.35	[32]	[100] 4.2 °K
	0.45		[111]
	0.42		[110]
5×10^{14}	0.016–0.467	[34]	0–175000 Oe, 4 °K
—	0.17	[35]	[100] 1.8 °K
—	0.16		[110]
—	0.31		
—	0.15		[111]
—	0.34		[110] 10–65 °K
—	0.42		[110]
—	0.45		[111]

1.3 Hole effective mass values obtained from cyclotron resonance

The cyclotron resonance is perhaps the most accurate method of determining the values of the effective mass. There exists in the literature a considerable amount of works investigating the cyclotron resonance in p-InSb (see Table 3). But here also we find no unity of opinions. One must note the work published by Button et al. [34] who studied the cyclotron resonance in p-InSb in a wide interval of magnetic fields (up to 175 Oe), i.e. actually in the Landau subbands. As a result they obtained a series of m_p-values ranging from 0.016 to 0.46 m_0.

Besides there is in the literature a certain number of works attempting to calculate the theoretical values of the effective mass of holes in p-InSb [36 to 39].

The other p-type $A^{III}B^{V}$ compounds have been considerably less studied, but here also we find rather vast series of m_p-values (Table 4). It should be noted that there are in the literature almost no data on the concentration dependences of the effective masses of holes.

Table 4

compound	m_p/m_0	method	ref.	note
InAs	0.4	absorption	[40]	"at a reasonable distance from the band edge"
InAs	0.41	absorption	[41]	
InAs	0.33	thermo-e.m.f.	[5]	500 to 800 °K
GaSb	0.39	mobility	[42]	—
GaSb	$0.38 > m_p/m_0 > > 0.56$	mobility	[43]	—
GaSb	0.23	IR reflection	[44]	—
GaSb	0.35	thermo-e.m.f. and mobility	[45]	10^{17} to 10^{19} cm^{-3}, 300 °K
GaSb	0.23	IR reflection	[46]	—
GaSb	0.38	cyclotron resonance	[47]	[110]
	0.36			
	0.36			[111]
	0.26			[100]
GaSb	0.5	thermo-e.m.f.	[48]	—
GaAs	0.68	absorption	[49]	—
GaAs	0.38	magnetoab- sorption and Faraday effect	[50]	10^{15}—10^{17} cm^{-3}
GaAs	1.1	thermo-e.m.f.	[51]	10^{17}—10^{20} cm^{-3}
GaAs	0.35	Faraday effect	[52]	5.4×10^{16}—7.7×10^{19} cm^{-3}
	1.0	mobility and N-E effect		
GaAs	0.28	IR reflection	[53]	4.6×10^{19}
	0.38			6.7×10^{19}
	0.44			1.8×10^{20} cm^{-3}

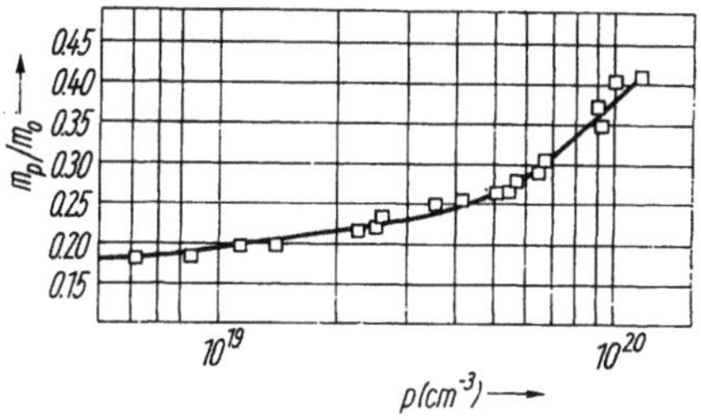

Fig. 1. Concentration dependence of the optical effective mass of holes for p-InSb

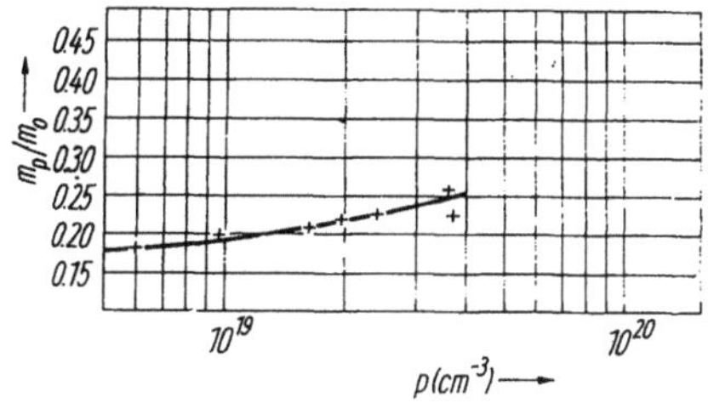

Fig. 2. The same as in Fig. 1 for p-InAs

In the present paper we are studying the concentration dependences of the optical effective mass of holes in some p-type $A^{III}B^V$ compounds, namely in InSb, InAs, GaSb, and GaAs. The investigations were carried out in an as wide as possible interval of concentrations. We investigated the spectral dependences of the reflection coefficient at room temperature. The values of the effective masses were calculated by the method described in [54].

2. Results of Measurements

Fig. 1 shows the concentration dependence of the optical effective mass of holes in p-InSb which we obtained in [55]. It can be seen that m_p remains constant ($\approx 0.18\ m_0$) up to hole concentrations 1 to 2×10^{19} cm^{-3} and begins to rise after it, obtaining the value $\approx 0.4\ m_0$ at $p = 1.5 \times 10^{20}$ cm^{-3}.

Fig. 2, 3, and 4 present the concentration dependences m_p for InAs, GaSb, and GaAs, respectively, obtained from the spectral dependences of the reflection coefficient.

The effective mass of holes in p-InAs crystals (Fig. 2) behaves in the same way as in p-InSb crystals. To our regret we had no crystals with $p > 4 \times 10^{19}$ cm^{-3} and could not trace the behaviour of m_p in crystals with larger hole concentrations.

In p-GaSb crystals (Fig. 3) m_p remains constant ($\approx 0.19\ m_0$) up to concentrations $\approx 6 \times 10^{19}$ cm^{-3}, slightly rising only after it.

In p-GaAs crystals (Fig. 4) m_p remains constant ($\approx 0.20\ m_0$) up to $p \approx (7 \text{ to } 8) \times 10^{19}$ cm^{-3} and rises after it, attaining the value $0.36\ m_0$ at $p = 1.5 \times 10^{20}$ cm^{-3}.

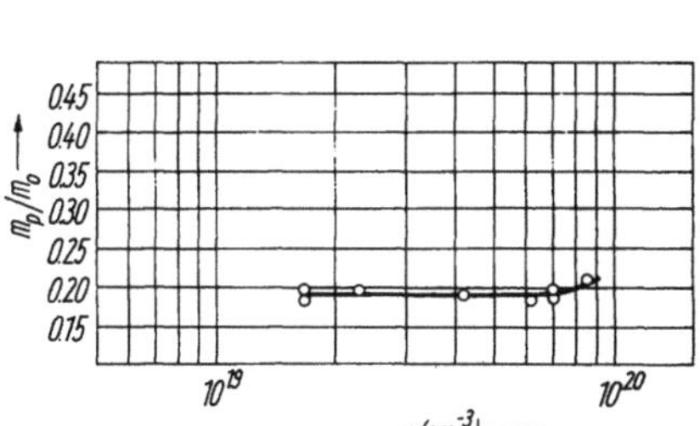

Fig. 3. The same as in Fig. 1 for p-GaSb

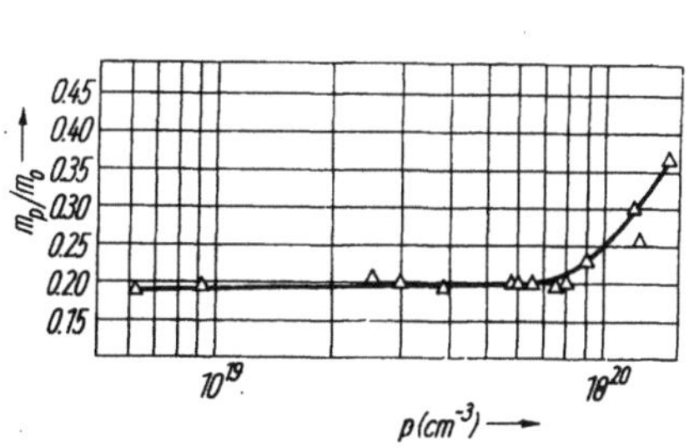

Fig. 4. The same as in Fig. 1 for p-GaAs

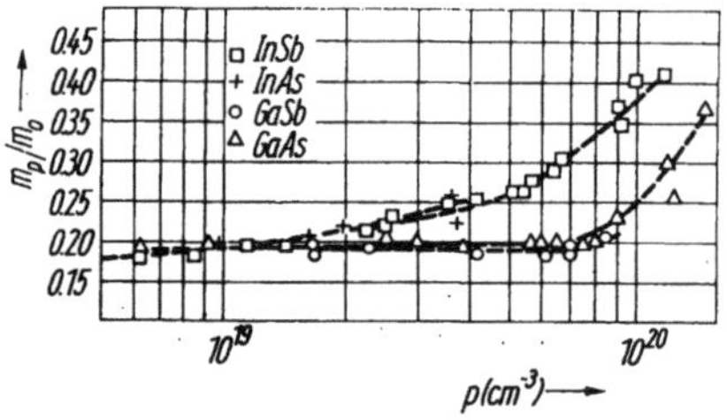

Fig. 5. Comparison of experimental concentration dependences of the optical effective mass of holes for all the investigated $A^{III}B^{V}$ compounds

To make the things more obvious all the concentration dependences are brought together on Fig. 5. It can be seen that up to concentrations $\approx 2 \times \times 10^{19}$ cm^{-3} in all investigated compounds the optical effective masses are practically the same. With increase of the hole quantity the concentration dependences m_p are drifting apart doing it in pairs: p-InSb with p-InAs and p-GaSb with p-GaAs.

So we can draw this first conclusion: as the differences between the optical effective masses of holes in the investigated $A^{III}B^{V}$ compounds are quite slight it may be supposed that the valence bands in these compounds also are similar.

3. Discussion

In our paper [55] we explained the concentration dependence of the optical effective mass of holes in p-InSb by transition from ellipsoidal isoenergetic surfaces to warped ones. In this connection we must examine in greater detail the results obtained in [35]. This work is developing the cyclotron resonance investigations carried out in [30]. The author of [35] studied the cyclotron resonance in p-InSb at carrier temperatures 10 to 65 °K and at 1.8 °K and obtained two sets of values of the cyclotron effective mass of holes. He explained these results by the difference in the structure of the valence band near its top and in its depth. Table 5 shows the experimental results [35] in comparison with the values of the cyclotron masses of holes theoretically calculated by Kołodziejczak [36] for the ellipsoidal (near the top of the valence band) and the warped (in the depth of the band) isoenergetic surfaces. It can be seen that the agreement between theoretical and experimental values m_c is quite satisfactory. Hence it follows that the theoretical calculations of Kołodziejczak are

Table 5

crystallographic direction	m^*/m_0				
	experiment [35] 10–65 °K	theor. calcul. for the warped surf. [36]	experiment [35] 1.8 °K	theoretical calculation [35]	theor. calcul. for the ellipsoidal surf. [36]
[100]	0.34	0.35	0.17	0.21	0.22
[110]	0.42	0.42	0.16	0.17	0.18
			0.31	0.30	0.30
[111]	0.45	0.45	0.15	0.15	0.16
			between 0.20 and 0.30	0.26	0.27

well correlated by experiment. In [55] we used just Kołodziejczak's calculations to explain the concentration dependence of m_p. His theoretical calculations were carried out for two limiting cases: the purely ellipsoidal and the purely warped isoenergetic surfaces. In experiments one has to deal mostly with a transition region of energies.

Thus for indium antimonide the concentration dependence of the optical effective mass of holes is well explained if we suppose that with increase of concentration we are passing from the region of ellipsoidal to the region of warped isoenergetic surfaces.

It may be supposed that the concentration dependences of the optical effective mass of holes in the other investigated $A^{III}B^{V}$ compounds also can be explained within the framework of this model, i.e. by transition from ellipsoidal to warped isoenergetic surfaces. A certain difference between the concentration dependences of m_p in different $A^{III}B^{V}$ compounds (Fig. 5) can be explained by the fact that the transition from ellipsoidal to warped isoenergetic surfaces takes place at different hole concentrations. It may be connected with the fact that in different compounds the energy in the valence band maxima surpasses the energy corresponding to the Brillouin zone centre in different degrees, i.e. that, roughly speaking, the purely ellipsoidal region has different energetic sizes. Besides, the density of states in the region of ellipsoidal isoenergetic surfaces, according to the theoretical calculations of Kołodziejczak and Zukotyński [38], is different in different compounds. For the investigated compounds the theoretically calculated values of the density-of-states effective mass m_{ds} are [38]:

for InSb $m_{ds} = 0.5732\ m_0$,

for InAs $m_{ds} = 0.6736\ m_0$,

for GaSb $m_{ds} = 0.6517\ m_0$,

for GaAs $m_{ds} = 0.7847\ m_0$.

As the dependences $m_p = f(p)$ for InSb and InAs are practically the same, one can say that the energy ε_0 (the surpass of energy in the valence band maxima) for InSb should be greater than for InAs. The same may be said concerning the pair GaSb–GaAs because m_{ds} in InAs is greater than in InSb and $m_{ds}\,(\mathrm{GaAs}) > m_{ds}(\mathrm{GaSb})$.

One of the principal conclusions drawn from the comparative investigations of the optical effective masses of holes in p-type $A^{III}B^{V}$ compounds carried out in the present work is that the differences between the valence bands in the investigated compounds are substantially smaller than between the corresponding conduction bands.

The exactness of the effective mass values determined from the spectral dependences of the reflection coefficient was examined in detail in [56]. We only can notice that in all compounds investigated in the present paper an appreciable increase of m_p-values begins at concentrations $\approx 2 \times 10^{19}\ \mathrm{cm}^{-3}$, i.e. when the degeneration is already taking place, and the method [54] of determining the values m_p from the reflection spectra is quite applicable. Therefore the increase of the optical effective mass of holes with the rise of concentration cannot be explained by an experimental error (as the authors of [28] tried to do).

It can be affirmed with a sufficient degree of certainty that there exists a dependence of the optical effective mass of holes (the conductivity mass) on concentration in all compounds of $A^{III}B^{V}$ type which are investigated in the present paper.

It also can be affirmed that the optical effective mass of holes (the conductivity mass) in these compounds is substantially smaller than the density-of-states mass. It is confirmed by investigations of thermo-magnetic effects, for instance in [51], and also by the theoretical calculations of Kołodziejczak and Zukotyński [38].

Thus, the analysis of literature data on both the experimental m_p-values and the theoretical calculations, as well as the results of our present work, lead to the conclusion that in investigating different physical phenomena in p-type $A^{III}B^{V}$ compounds one has to be very careful in choosing the hole effective mass values.

The structure of the valence bands in $A^{III}B^{V}$ compounds substantially differs from the structure of the conduction bands in which, as it is a rule considered, the density-of-states masses and the conductivity effective masses coincide ($m_{ds} = m_p$). We consider that in p-$A^{III}B^{V}$ compounds $m_{ds} \neq m_p$, the difference being quite substantial.

Hence it follows that in carrying out theoretical calculations of various effects it is necessary to distinguish quite distinctly the dependences on the density-of-states effective mass and on the conductivity effective mass, i.e. to be very cautious in the use of the quasi-isotropic band approximation.

References

[1] E. O. Kane, J. Phys. Chem. Solids **1**, 249 (1957).

[2] O. Madelung, Fizika poluprovodnikovikh soedinenii elementov III i V grup, Izd. Mir, Moscow 1967.

[3] J. Tauc and M. Matyas, Czech. J. Phys. **5**, 383 (1955).

[4] H. Welker and H. Weiss, Solid State Phys. **3**, 1 (1956).

[5] H. Weiss, Z. Naturf. **11a**, 131 (1956).

[6] M. Rodot, P. Duclas, F. Kover, and H. Rodot, C. R. Acad. Sci. (France) **242**, 2522 (1956).

[7] S. Porowski, A. Duracz, and S. Żukotyński, phys. stat. sol. **3**, 1555 (1963).

[8] H. P. R. Frederikse and E. V. Milczarek, Phys. Rev. **99**, 1889 (1955).

[9] J. Ginter and W. Szymańska, phys. stat. sol. **3**, 1389 (1963).

[10] M. Rodot, phys. stat. sol. **3**, 1751 (1963).

[11] C. Hilsum and A. Rose-Innes, Poluprovodniki tipa $A^{III}B^{V}$, Izd. int. Lit., Moskva 1963.

[12] R. G. Breckenridge, R. F. Blunt, W. R. Hosler, H. P. R. Frederikse, J. H. Becker, and W. Ovshinsky, Phys. Rev. **96**, 570 (1954).

[13] K. I. Vinogradova, V. V. Galavanov, and D. N. Nasledov, Fiz. tverd. Tela **4**, 1673 (1962).

[14] R. W. Cunningham, Bull. Amer. Phys. Soc. **10**, 386 (1965).

[15] R. F. Broom and A. Rose-Innes, Proc. Phys. Soc. **B69**, 1269 (1956).

[16] A. J. Strauss, J. appl. Phys. **30**, 559 (1959).

[17] V. A. Kakoshkin and D. N. Nasledov, Sbornik Fiz. Svoistva Poluprov., $A^{III}B^{V}$ i $A^{III}B^{VI}$, Izd. AN Azerb. SSR, Baku 1967 (p. 284).

[18] V. A. Kakoshkin, Izv. Akad. Nauk SSSR, Ser. fiz. **28**, 980 (1964).

[19] H. J. Hrostowski, F. J. Morin, T. H. Geballe, and G. H. Wheatley, Phys. Rev. **100**, 1972 (1955).

[20] A. Berkeliev and V. V. Galavanov, Sbornik Fiz. Svoistva Poluprov., $A^{III}B^{V}$ i $A^{III}B^{VI}$, Izd. AN Azerb. SSR, Baku 1967 (p. 229).

[21] HUAN CI-SHEN, U CZI-CAN, and TAN CHZUN-UN, Acta Phys. Sinica **20**, 418 (1964).
[22] C. HILSUM, Proc. Phys. Soc. **76**, 414 (1960).
[23] F. STERN, Proc. Internat. Conf. Semicond. Phys., Prague 1960 (p. 363).
[24] W. G. SPITZER and H. Y. FAN, Phys. Rev. **106**, 882 (1957).
[25] S. ZWERDLING, W. H. KLEINER, and J. P. THERIAULT, Proc. Internat. Conf. Semicond. Phys., Exeter 1962 (p. 455).
[26] A. S. FILIPCHENKO, G. G. KULIKOV, D. N. NASLEDOV, and YU. I. UKHANOV, phys. stat. sol. **17**, K113 (1966).
[27] C. R. PIDGEON and R. N. BRAUN, Phys. Rev. **146**, 575 (1966).
[28] G. P. SKORNIAKOV, I. M. TSIDILKOVSKI, T. P. CHUKINA, and V. A. KAKOSHKIN, Fiz. tverd. Tela **9**, 2625 (1967).
[29] G. DRESSELHAUS, A. F. KIP, C. KITTEL, and G. WAGONER, Phys. Rev. **98**, 556 (1955).
[30] D. M. BAGGYLEY, M. L. A. ROBINSON, and R. A. STRADLING, Phys. Letters A **6**, 143 (1963).
[31] H. T. TOVER and S. ASCARELLI, Proc. Internat. Conf. Semicond. Phys., Moscow 1968 (p. 348).
[32] YU. B. ERMOLOVICH, Trudy Kievskogo Politekh. Inst. **45**, 114 (1963).
[33] R. H. DEXTER and B. LAX, Phys. Rev. **99**, A635 (1955).
[34] K. J. BUTTON, B. LAX, and C. C. BRADLEY, Phys. Rev. Letters **21**, 350 (1968).
[35] M. L. A. ROBINSON, Phys. Rev. Letters **17**, 963 (1966).
[36] J. KOŁODZIEJCZAK, Proc. Internat. Conf. Semicond. Phys., Paris 1964 (p. 1147).
[37] J. KOŁODZIEJCZAK, S. ŻUKOTYŃSKI, and H. STRAMSKA, phys. stat. sol. **14**, 471 (1966).
[38] J. KOŁODZIEJCZAK and S. ŻUKOTYŃSKI, phys. stat. sol. **16**, K55 (1966).
[39] Y. OHMURA, J. Phys. Soc. Japan **21**, 1886 (1966); Japan. J. appl. Phys. **6**, 972 (1967).
[40] F. STERN and R. M. TALLEY, Phys. Rev. **106**, 882 (1957).
[41] F. STERN and F. MATOSSI, Phys. Rev. **111**, 472 (1958).
[42] H. N. LEIFER and W. C. DUNLAP, Phys. Rev. **95**, 51 (1955).
[43] S. ZWERDLING, B. LAX, K. J. BUTTON, and L. M. ROTH, J. Phys. Chem. Solids **9**, 320 (1959).
[44] A. K. RAMDAS and H. Y. FAN, Bull. Amer. Phys. Soc. **3**, 121 (1958).
[45] V. I. IVANOV-OMSKI, B. T. KOLOMIETS, and ZOU-HUAN, Fiz. tverd. Tela **4**, 383 (1962).
[46] W. M. BECKER, A. K. RAMDAS, and H. Y. FAN, J. appl. Phys. **32**, 2094 (1961).
[47] R. A. STRADLING, Phys. Letters A **20**, 217 (1966).
[48] J. T. EDMOND, R. F. BROOM, and F. A. CUNNELL, Proc. Rugly Conf., London 1957 (p. 109).
[49] T. S. MOSS and T. D. F. HAWKINS, Infrared Phys. **1**, 11 (1962).
[50] S. NARITA, M. KOBAYASHI, and N. KONKE, Proc. Internat. Conf. Semicond. Phys., Moscow 1968 (p. 369).
[51] O. V. EMELYANENKO, F. P. KESAMANLY, D. N. NASLEDOV, V. G. SIDOROV, and G. N. TALALAKIN, phys. stat. sol. 8, K155 (1965).
[52] F. P. KESAMANLY, E. E. KLOTINSH, D. N. NASLEDOV, and G. N. TALALAKIN, Izv. Akad. Nauk Latvinian SSR, Ser. fiz. **4**, 14 (1966).
[53] F. P. KESAMANLY, YU. I. MALTSEV, D. N. NASLEDOV, and YU. I. UKHANOV, phys. stat. sol. **13**, K119 (1966).
[54] A. A. KUKHARSKII and V. K. SUBASHIEV, Fiz. tverd. Tela 8, 753 (1966).
[55] L. P. BOLSHAKOV, A. S. FILIPCHENKO, D. N. NASLEDOV, and L. N. RADAIKINA, phys. stat. sol. **32**, K135 (1969).
[56] H. LYDEN, Phys. Rev. **134**, A1106 (1964).

(Received January 4, 1974)

phys. stat. sol. (b) **62**, 361 (1974)

Subject classification: 6 and 20.1; 22.5.4

Max-Planck-Institut für Festkörperforschung, Stuttgart

Long-Wavelength Optical Phonons in $K_{1-x}Tl_xCl$ and $K_{1-x}Tl_xBr$ Mixed Crystals

By

W. BAUHOFER and L. GENZEL

The $q \approx 0$ lattice vibrations of mixed crystals of $K_{1-x}Tl_xCl$ and $K_{1-x}Tl_xBr$ have been observed by far infrared reflectivity measurements. Contrary to the existing assumption, the system $K_{1-x}Tl_xCl$ clearly shows a mixed-mode behaviour similar to that of $K_{1-x}Rb_xI$. $K_{1-x}Tl_xBr$ is either a mixed- or a two-mode system. The experimental results are in good agreement with a modified version of the REI-model.

Die Gitterschwingungen bei $q \approx 0$ von $K_{1-x}Tl_xCl$- und $K_{1-x}Tl_xBr$-Mischkristallen wurden mittels Reflexionsmessungen im fernen Infrarot beobachtet. Im Gegensatz zu existierenden Annahmen zeigt das System $K_{1-x}Tl_xCl$ klar ein ähnliches Mischmodenverhalten wie $K_{1-x}Rb_xJ$. $K_{1-x}Tl_xBr$ ist entweder ein Misch- oder ein Zwei-Modensystem. Die experimentellen Ergebnisse befinden sich in guter Übereinstimmung mit einer modifizierten Version des REI-Modells.

1. Introduction

Since the work of Krüger et al. [1] no measurements of the zone-centre phonons of the mixed system $K_{1-x}Tl_xCl$ have been made. The frequencies of maximum reflectivity of [1], which were isolated by multiple reflections, lie somewhat higher than the actual ω_{TO}'s. This is particularly true for crystals with a broad reststrahl band such as TlCl. Krüger et al. found only one reflectivity maximum in all mixtures. Therefore, $K_{1-x}Tl_xCl$ was usually considered to be a one-mode system [2]. No X-ray measurements are mentioned in [1]. Hence, it is not known if their thallium rich samples were actually mixed crystals or merely clusters of the constituents.

We have investigated this system to check a modified version of the REI model [2] recently developed by Genzel et al. [3]. Following this model $K_{1-x}Tl_xCl$ should be a mixed mode system, while $K_{1-x}Tl_xBr$ should show a two-mode behaviour.

Recently, Müürsepp and Haav have described a method for producing solid solutions of $K_{1-x}Tl_xCl$ and $K_{1-x}Tl_xBr$ [4]. Powder mixtures of the pure end members are heated for 30 h at 410 °C under vacuum. After heating, the samples are rapidly cooled to room temperature. The resultant mixed crystals which are still in powder form turn out to be quite stable. The system $K_{1-x}Tl_xCl$ has a miscibility gap from $x = 0.65$ to $x = 0.95$. Up to $x = 0.65$ the mixed crystals have NaCl structure, between $x = 0.95$ and $x = 1$ they have CsCl structure. $K_{1-x}Tl_xBr$ mixed crystals can be formed with NaCl structure with x-values as high as 0.6. Unfortunately, it is impossible with this method to achieve any solubility of KBr in TlBr [4].

2. Experimental Methods

Following the method given in [4], we have made powder samples of the systems $K_{1-x}Tl_xCl$ and $K_{1-x}Tl_xBr$ with x-values of 0.05, 0.2, 0.5, 0.95 and 0.05, 0.2, and 0.5, respectively. After cooling the samples to room temperature they were investigated by X-ray diffraction measurements. The lattice constants of all mixtures are in good agreement with the values given in [4]. In addition to the strong reflections of the mixed crystals, the X-ray patterns show, for medium concentrations, weak reflections corresponding to the pure end members. However, this small amount of the constituents should not influence the reflectivity spectra of the mixed crystals.

For the far infrared reflectivity measurements the powder samples were pressed into pellets. The measurements were done at room temperature on a Polytec FIR-30 Fourier spectrometer in the 20 to 450 cm^{-1} range. Below 20 cm^{-1} the reflectivity is practically constant. The static dielectric constant was determined by filtering out all frequencies above 20 cm^{-1} and then measuring an average value of the reflectivity.

3. Results and Discussion

The reflectance spectra of the measured systems for various concentrations are plotted in Fig. 1. Both systems show two reflectivity maxima at medium concentrations. Therefore, they do not belong to systems with one-mode behaviour. For x-values smaller than 0.2 the low frequency reflectivity maximum gradually disappears. It is not possible, for these concentrations, to evaluate the frequencies of the weaker oscillator by reflectivity measurements. The values of ε_0 and ε_∞ of the pure end members used in the calculations were taken from [6].

3.1 $K_{1-x}Tl_xCl$

For two samples, $x = 0.2$ and $x = 0.5$, the transverse and longitudinal phonons were obtained by a Kramers-Kronig analysis of the reflectivity data. The

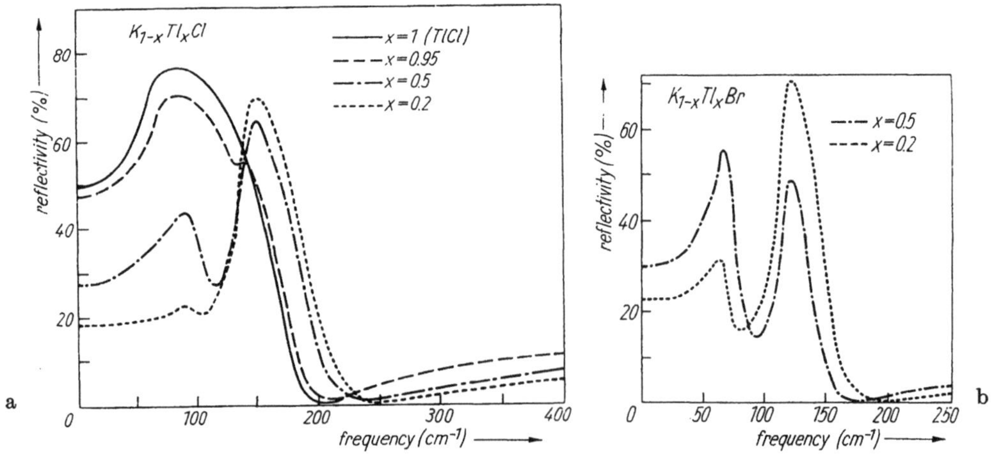

Fig. 1. a) Reflectivity spectra of $K_{1-x}Tl_xCl$; b) reflectivity spectra of $K_{1-x}Tl_xBr$

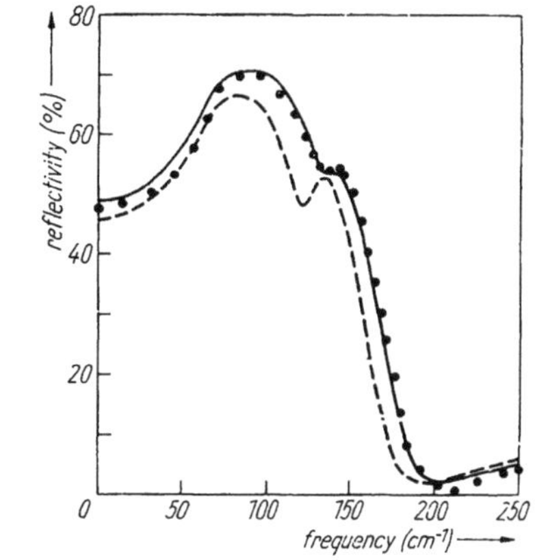

Fig. 2. Reflectivity spectrum of $K_{0.05}Tl_{0.95}Cl$ (dotted line). The solid line is calculated from a classical oscillator model, the broken curve is computed from the theory of [3] taking the damping constant of the oscillator fit

T-modes were taken as peaks in $\omega\varepsilon''$, where ε'' is the imaginary part of the dielectric function ε. The maxima of the imaginary part of ε^{-1} give the L-modes.

The Kramers-Kronig analysis for the sample with $x = 0.95$ yields only one strong peak in $\omega\varepsilon''$. A second T-mode is indicated by a shoulder at the high frequency side. A fit of the reflectance data to a classical oscillator dispersion gives this weak T-mode within ≈ 5 cm^{-1}. Fig. 2 shows the oscillator fit together with the reflectivity computed from the dielectric function given in [3]. Since no damping is introduced in this model, we took the damping constants of the oscillator fit. The longitudinal mode belonging to the weak oscillator was calculated from the generalized LST relation [5].

The transverse and longitudinal modes for various concentrations are plotted in Fig. 3a together with the theoretical curve calculated from the model of [3]. One sees clearly a mixed mode behaviour which is called case (b) in [3]. To date, the only other system showing such a mixed mode behaviour is $K_{1-x}Rb_xI$ [7]. The qualitative agreement is good. However, the theory does not give the right values for the impurity modes at $x = 0$ and $x = 1$. This point is discussed in detail in [3].

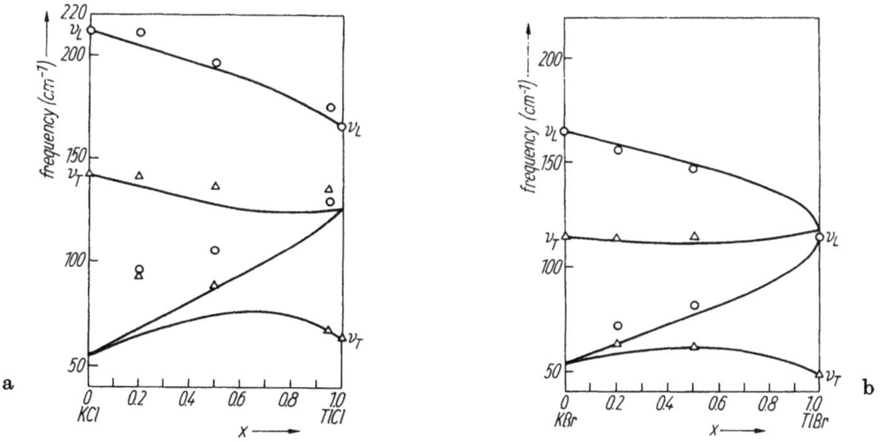

Fig. 3. a) Variation of the $q \approx 0$ optical phonons of $K_{1-x}Tl_xCl$ with composition. The open circles denote transverse modes, the triangles longitudinal modes. The solid line is calculated from the model of [3]; b) Variation of the phonon frequencies of $K_{1-x}Tl_xBr$ with composition. Open circles: transverse modes, triangles: longitudinal modes. The solid line is a calculation using the model of [3]

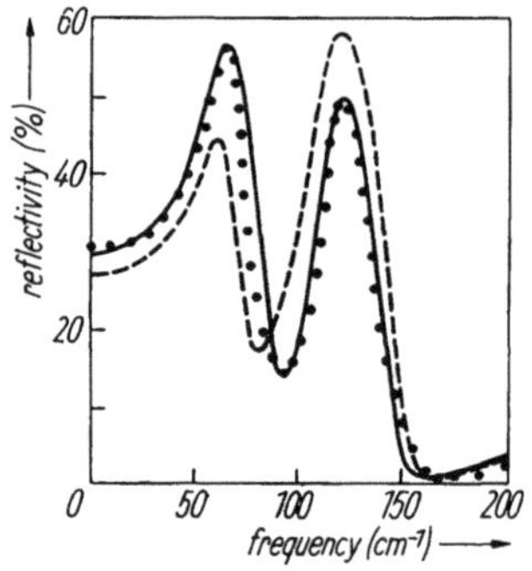

Fig. 4. Reflectivity spectrum of $K_{0.5}Tl_{0.5}Br$ (dotted line) The solid line is a classical oscillator fit; the broken curve is computed from the theory given in [3]

3.2 $K_{1-x}Tl_xBr$

For this mixed system, only samples with a thallium content of less than 60 mol% were available. Fig. 3b shows a comparison of the theoretical curve with the measured values of the T- and L-modes of two different mixtures. The agreement is remarkably good; this seems, however, to be somewhat fortuitous. Since we have no measurements on thallium rich samples, a clear distinction between mixed and two-mode behaviour is not possible. In Fig. 4, we have plotted an oscillator fit of the sample with $x = 0.5$ together with the reflectivity calculated from the dielectric function given in [3]. Again, we took the damping constant of the oscillator fit. The oscillator strength f of a mode with frequency ω_T is given to a good approximation by the relation

$$f = \frac{\varepsilon''_{max}\, \Gamma}{\omega_T}, \tag{1}$$

Table 1

Comparison of experimental oscillator strengths derived from a Kramers-Kronig analysis of the reflectivity data with calculated values using the theory of [3]

			oscillator strengths	
			low frequency mode	high frequency mode
$K_{1-x}Tl_xCl$				
	$x = 0.2$	exp.	0.42	2.93
		th.	0.83	3.07
	$x = 0.5$	exp.	5.83	2.39
		th.	3.54	3.3
$K_{1-x}Tl_xBr$				
	$x = 0.2$	exp.	2.61	2
		th.	1.03	2.6
	$x = 0.5$	exp.	6.31	1.21
		th.	4.2	2.14

where ε''_{max} denotes the maximum value and Γ the linewidth of the peak in ε'' belonging to this mode. In Table 1, the experimental oscillator strengths derived from a Kramers-Kronig analysis and using equation (1) are compared with the theoretical values which were computed using formula (19) of [3].

4. Conclusions

As predicted by the model presented in [3], the system $K_{1-x}Tl_xCl$ shows a mixed mode behaviour. $K_{1-x}Tl_xBr$ belongs either to the mixed or to the two-mode systems. This is also in agreement with the theory. Our experimental results show clearly that this relatively simple theory is able to describe the mode behaviour of many mixed systems without using any adjustable parameter. In addition, the model allows one to predict the concentrations needed to observe the respective types of mode behaviour in the reflectance spectra.

Acknowledgements

We wish to thank Prof. C. H. Perry and Prof. T. P. Martin for helpful discussions. Preparation of the samples by P. Senk and X-ray measurements by L. Viczian are also gratefully acknowledged.

References

[1] F. Krüger, O. Reinkober, and E. Koch-Holm, Ann. Phys. (Germany) **85**, 110 (1928).
[2] I. F. Chang and S. S. Mitra, Adv. Phys. **20**, 359 (1971).
[3] L. Genzel, T. P. Martin, and C. H. Perry, phys. stat. sol. (b), to be published.
[4] T. Müürsepp and A. Haav, phys. stat. sol. (a) **13**, K133 (1972).
[5] W. Cochran, Z. Krist. **112**, 465 (1959).
[6] R. P. Lowndes and D. H. Martin, Proc. Roy. Soc. A **308**, 473 (1969).
[7] J. H. Fertel and C. H. Perry, Phys. Rev. **184**, 874 (1969).

(Received December 12, 1973)

phys. stat. sol. (b) 62, 367 (1974)

Subject classification: 19; 13.1; 13.4; 22.1.1; 22.1.2

Institut für Angewandte Physik der Technischen Universität Clausthal

Electron Paramagnetic Resonance (EPR) of Phosphorus in Silicon-Rich Silicon–Germanium Alloys

By

H. VOLLMER and D. GEIST

g-values, hyperfine splittings, and line widths of the EPR-spectra of phosphorus substituted into $Si_{1-x}Ge_x$ alloys were investigated at 9.3 GHz and 1.8 K in the range $1 \geqq (1 - x) \geqq 0.84$ with phosphorus concentrations between 5×10^{15} and 10^{19} cm^{-3}. The peculiar variation of the measured quantities with the silicon concentration is discussed with regard to band structure investigations and to the individual surroundings of each phosphorus atom.

g-Faktoren, Hyperfeinstrukturaufspaltungen und Linienbreiten der EPR-Spektren von Phosphor in $Si_{1-x}Ge_x$-Legierungen wurden im Bereich $1 \geqq (1 - x) \geqq 0,84$ und bei Phosphorkonzentrationen zwischen 5×10^{15} und 10^{19} cm^{-3} bei $T = 1,8$ K im X-Band (9,3 GHz) untersucht. Die Abhängigkeit der gemessenen Größen von der Siliziumkonzentration wird anhand von Untersuchungen zur Bandstruktur und im Hinblick auf die lokale Umgebung jedes Phosphoratoms diskutiert.

1. Introduction

Comparison of the electron paramagnetic resonance spectra of the ground state of shallow donors in pure silicon [1] and pure germanium [2] crystals shows that g-value, hyperfine splitting, and line width are strongly influenced by the relevant electronic states of the host crystal. The measurements yielded information on the wave function of the impurity centers and on the band structure of the crystal (cf. [3]).

Because the changes in the band structure in going from pure silicon through the silicon–germanium alloys to pure germanium are believed to be known reliable, one would expect that the changes in the parameters of the EPR spectrum of phosphorus in Si–Ge alloys should be predictable. The measurements on phosphorus doped (5×10^{15} to 10^{19} cm^{-3}) $Si_{1-x}Ge_x$ alloys ($1 \geqq 1 - x \geqq 0.84$) to be presented here indicate that not only the overall composition given by the x-value but also the individual surroundings of each phosphorus atom must be considered.

2. Theory

2.1 Band structure

Bassani and Brust [4] have calculated the valence and the conduction bands of the alloy system by the pseudopotential method. According to them the position of only two bands, the conduction bands Λ_1 and Δ_2', is strongly depending on the composition of the alloy. With decreasing silicon content these two bands move down towards the top of the valence band. Correspondingly in silicon and silicon-rich alloys with $1 - x > 0.15$ the minimum of the lowest conduction band Δ_1 is lying along the cubic axes at $k_0 = 0.85 \cdot 2\pi/d$ (d cubic lattice constant).

For germanium-rich alloys $1 - x < 0.15$ and germanium on the other hand L_1 is lying lowest and responsible for the indirect gap.[1])

The most important evidence for the correctness of this picture is coming from measurements of the indirect gap by electrical [6, 7] and optical [8, 9] methods (intrinsic conductivity and optical absorption respectively). Furtheron magnetoresistance [10, 11] shows that the conduction band minima lie in the Δ-direction for $1 - x > 0.15$ and in the Λ-direction for $1 - x < 0.15$.

Effective mass values from measurements of the magnetic susceptibility [12 to 14] can be understood with the model [4]. Detailed optical reflection and electroreflection investigations [15 to 17] do agree with the calculations [4].

Therefore all quantities which are really influenced by the minimum of the conduction band should be similar to those in silicon for $1 - x > 0.15$ and similar to those in germanium for $1 - x < 0.15$.

With this classification the alloys investigated here ($1 - x \geqq 0.84$) lie on the silicon side and therefore the EPR spectra are similar to those in pure silicon.

2.2 Shallow donors

With the minimum of the conduction band lying at $\boldsymbol{k}_0$ in the Δ-direction the donor ground state wave function (singlet state A_1) is of the form [18, 19]

$$\Psi(\boldsymbol{r}) = \frac{1}{\sqrt{6}} \sum_{j=1}^{6} F^{(j)}(\boldsymbol{r})\, u^j(\boldsymbol{r}) \exp i\boldsymbol{k}_0^j\boldsymbol{r}\,, \tag{1}$$

where the sum runs over the six equivalent minima, $u^j(\boldsymbol{r}) \exp i\boldsymbol{k}_0^j\boldsymbol{r}$ is the Bloch function at the j-th minimum and $F(\boldsymbol{r})$ is the effective mass envelope function. The distance $\boldsymbol{r}$ is measured from the donor centre.

Neglecting the anisotropy of the effective masses the envelope function — adjusted to the observed activation energy ΔE_1 instead to the calculated energy $\Delta E_{\text{eff mass}}$ — outside a sphere with radius r_s is given by [20]

$$F(r) = C \exp\left(-\frac{r}{a^*}\right), \qquad (r > r_s) \tag{2}$$

where r_s depends on the atomic volume V_A

$$V_A = \frac{4\pi}{3}\, r_s^3 \tag{3}$$

and

$$a^* = \frac{\hbar^2 \varepsilon}{e_0^2 m^*} \left(\frac{\Delta E_{\text{eff mass}}}{\Delta E_1}\right)^{1/2} \tag{4}$$

(ε permittivity, m^* effective mass). For $r < r_s$ is assumed

$$F(r) = F(r_s)\,. \tag{5}$$

Normalising $\int |F(r)|^2\, d\tau = 1$ and using approximately equation (2) for the whole space [19] yields

$$C = (\pi a^{*3})^{-1/2}\,. \tag{6}$$

The conditions for the validity of this theory will be fulfilled better with a decreasing difference between the nuclear charges of the donor and the host

1) The self-consistent OPW calculations [5] are less consistent with the experimental results than the calculations [4]. This may be due to the assumptions used in the model [5]. We do not rely on this work.

lattice atoms [19]. Therefore in connection with this theory phosphorus as doping element is a correct choice.

3. Experimental Details

3.1 Equipment

The EPR measurements were performed at 9.3 GHz. The spectrometer allowed superheterodyne and straight detection. The magnetic field was modulated at 400 Hz.

The static magnetic field B_0 was measured with a nuclear-resonance magnetic field-meter. As the proton probe was outside the cavity the necessary correction was determined using a sample of phosphorus in pure silicon as a standard [1]. The latest value of the specific electron charge [21] was taken into account.

3.2 Samples

The samples were cut from Czochralski-grown crystals [22, 23] perpendicular to the growth direction. The crystals were pulled from quartz crucibles under very pure hydrogen or under vacuum. They were doped by Na_3PO_4 or $Ca_3(PO_4)_2$ added to the melt or by PBr_3 mixed with the hydrogen gas.

The samples were flat discs (diameter 6 to 9 mm, thickness 0.5 to 0.9 mm). Before making the EPR measurements they were etched with CP6 [24]. The investigated samples were polycrystalline with macroscopic monocrystalline regions. With decreasing silicon content these regions became smaller. Adjacent to each EPR sample in the crystals test samples were cut out for Hall-effect and conductivity measurements [22, 25]. From the room temperature values the phosphorus concentration of each EPR sample was determined by linear interpolation between the test samples. The silicon content was derived from Debye-Scherrer diagrams [26].

4. Experimental Results

4.1 Spectra

Fig. 1 and 2 show the measured EPR spectra of phosphorus in two different Si–Ge alloys with a small phosphorus concentration. Isolated donor atoms are

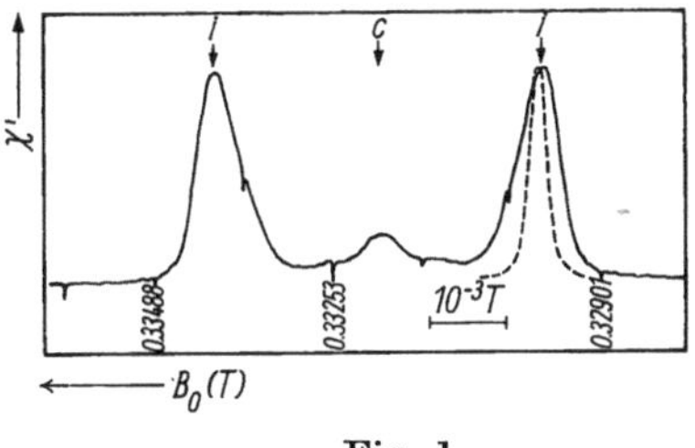

Fig. 1

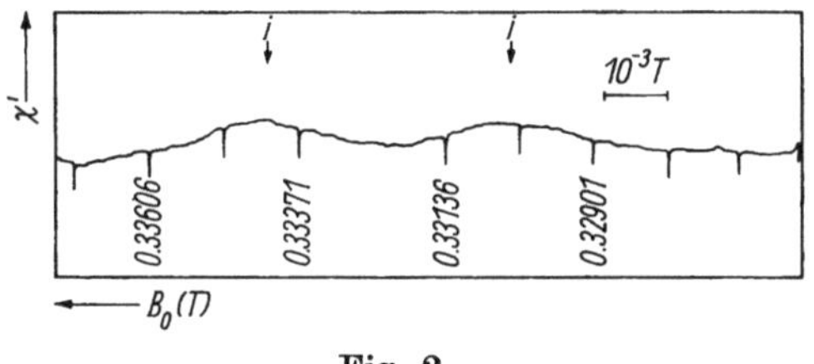

Fig. 2

Fig. 1. EPR spectrum of phosphorus in a Si–Ge alloy, dispersion signal, fast adiabatic passage. $1 - x = 0.987$; 6.2×10^{16} P cm^{-3}; $f = 9.2795$ GHz; $T = 1.8$ K. For comparison the low field line of P in pure Si is also drawn (dashed line)

Fig. 2. EPR spectrum of P in a Si–Ge alloy, dispersion signal, fast adiabatic passage. $1 - x = 0.893$; 1.2×10^{16} P cm^{-3}; $f = 9.2890$ GHz; $T = 1.8$ K

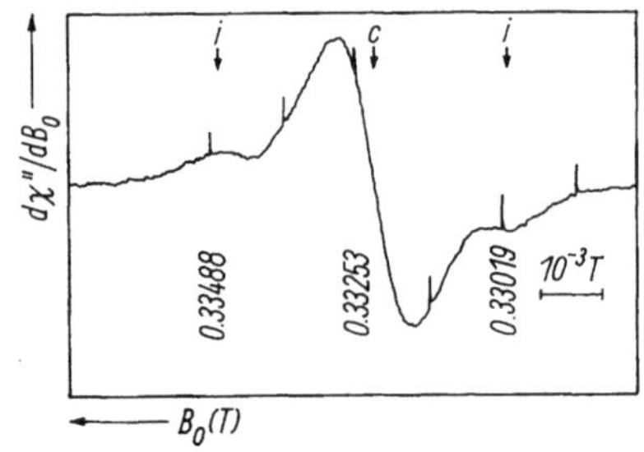

Fig. 3. EPR spectrum of P in a Si–Ge alloy, differentiated absorption line. $1 - x = 0.956$; 1.8×10^{17} P cm^{-3}; $f = 9.2184$ GHz; $T = 1.8$ K

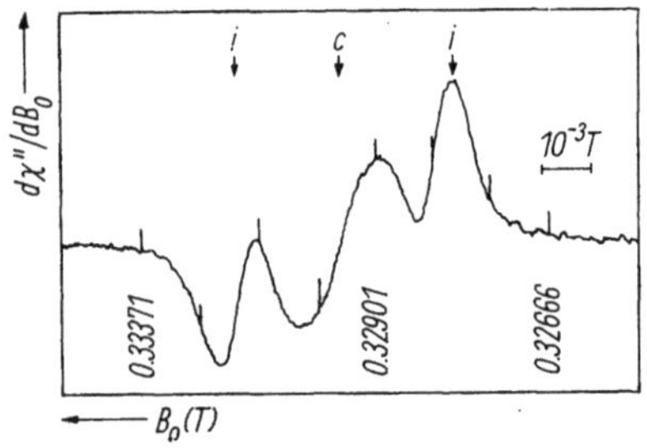

Fig. 4. EPR spectrum of P in a Si–Ge alloy, differentiated absorption line. $1 - x = 0.938$; 2.6×10^{17} P cm^{-3}; $f = 9.2887$ GHz; $T = 1.7$ K

essentially responsible for the two strong hyperfine lines (i in all figures) resulting from the isotropic hyperfine interaction of the donor electron with the phosphorus nucleus (spin quantum number $I = 1/2$).

The low field line of isolated phosphorus donors in pure silicon is shown for comparison in Fig. 1 (dashed curve). There are characteristic differences between the hyperfine lines (i) from pure silicon and from the alloys (cf. Fig. 1 and 2):

1. Each of the two lines in an alloy with $1 > 1 - x > 0.90$ is not symmetrical. But they are mirror-symmetrical (Fig. 1). For smaller values of $1 - x$ the asymmetry disappears (Fig. 2).

2. The lines become broader with decreasing silicon content.

3. The distance between the two lines i.e. the hyperfine splitting depends on the alloy composition.

Between and under these lines there are weaker lines from phosphorus clusters of different size (c in the figures). The lines result from interaction of neighbouring donor electrons according to the overlap of their wave functions [27]. The maximum of the central line g_c is shifted to lower fields with respect to the centre of the hyperfine lines g_a.

With increasing phosphorus concentration the intensity of the weak lines (c) increases, similar to pure silicon. Samples with an intermediate phosphorus concentration show spectra as in Fig. 3 and 4. The smaller the silicon content of the alloy the smaller is the phosphorus concentration where the lines of interacting donors appear. At high phosphorus concentration the spectrum consists only of one line (c) with Lorentzian shape (Fig. 5).

An influence of the magnetic field orientation against the crystal axes could not be observed.

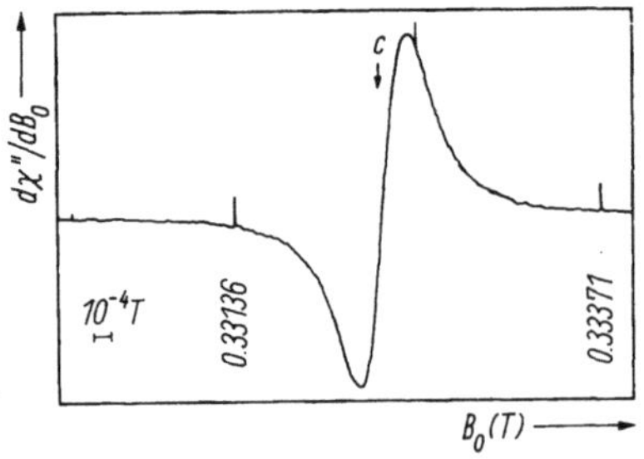

Fig. 5. EPR spectrum of P in a Si–Ge alloy, differentiated absorption line. $1 - x = 0.981$; 2.5×10^{18} P cm^{-3}; $f = 9.2894$ GHz; $T = 1.8$ K

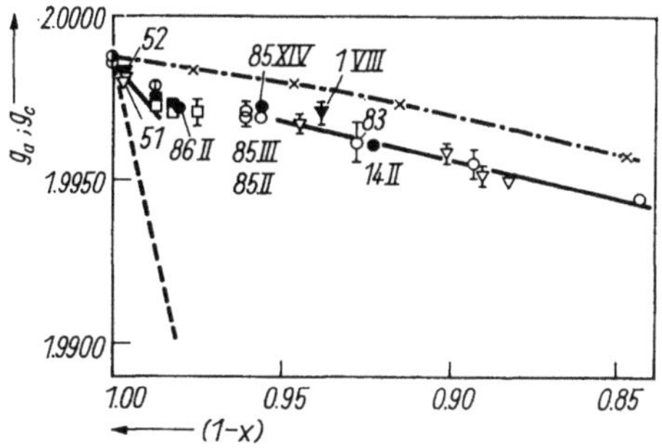

Fig. 6. g-values of P in Si–Ge alloys at $T = 1.8$ K. Open symbols: g_a-values of isolated donors, full symbols: g_c-values of P clusters. The crystals had been doped by Na_3PO_4 □, $Ca_3(PO_4)_2$ ▽, and PBr_3 ○. Dashed line: linear interpolation between pure Si and pure Ge. For comparison values taken from [1] (g_a ⦶ and g_c ◑) and from [28] (g_c ×) are drawn. The error bars correspond to the standard error of several measurements

4.2 g-values

Fig. 6 shows the variation of the g_a- and g_c-values with the silicon content of the samples. The g-values do not depend on the method by which the crystals have been doped with phosphorus.

4.2.1 Isolated donors

The g-factor g_a of isolated phosphorus atoms (open symbols in Fig. 6) at first decreases with decreasing silicon content down to about $1 - x = 0.988$ at a rate of

$$\Delta g = g_a - g_{Si} = -(0.12 \pm 0.02)\, x \qquad 1 \geqq 1 - x > 0.988\,. \tag{7}$$

Between $0.988 > 1 - x > 0.95$ the g-value is constant ($|\Delta g| < 0.01\, x$) and at smaller silicon concentration it decreases again according to

$$g_a = (1.9980 \pm 4\times10^{-4}) - (0.023 \pm 0.003)\, x\,, \qquad 0.95 > 1 - x > 0.84\,. \tag{8}$$

4.2.2 High phosphorus concentration

The g_c-value of the central line in highly doped samples (full symbols in Fig. 6) is lying close to the value g_a of the isolated donors. Fortunately for each of the samples with an intermediate phosphorus concentration both values could be determined simultaneously from a single registration (Fig. 3 and 4). From four samples

$$g_c(x) - g_a(x) = (3 \pm 1)\times10^{-4} \tag{9}$$

has been found, independent of the silicon content of the samples. Within experimental error equation (9) agrees with the value found in pure silicon [29].

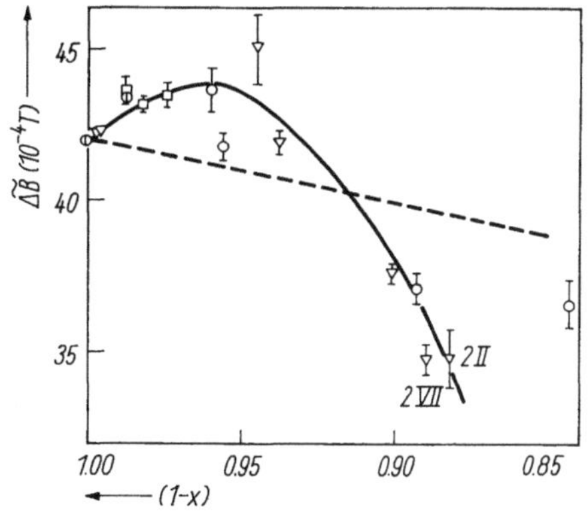

Fig. 7. Hyperfine splitting of P in Si–Ge alloys at 1.8 K. The full line is calculated according to Section 5.2, symbols see Fig. 6

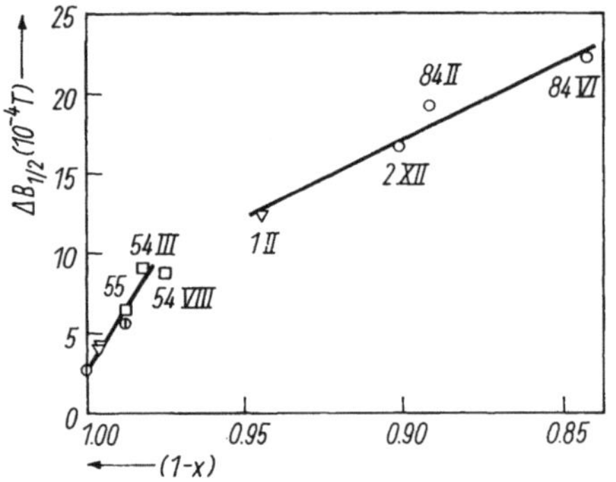

Fig. 8. Line width of one hyperfine line of P in Si–Ge alloys at 1.8 K, symbols see Fig. 6

4.3 Hyperfine splitting

With the silicon and germanium atoms distributed randomly on the lattice sites one expects local variations of $|\Psi(0)|^2$ and therefore different hyperfine splittings ΔB for the donors. Obviously the recorded spectra contain a weighted mean over all individual ΔB's. This distance $\widetilde{\Delta B}$ between the peaks of the hyperfine lines (i to i in Fig. 1 to 4) is taken as a measure for the hyperfine splitting (equation (12)). The dependence of $\widetilde{\Delta B}$ on the silicon concentration is drawn in Fig. 7. Again three regions are seen, approximately defined by $1 \geqq 1 - x > > 0.988$, $0.988 > 1 - x > 0.95$, and $0.95 > 1 - x > 0.84$ with an increasing, constant or decreasing value of $\widetilde{\Delta B}$. Once more the method of doping used is of no importance.

4.4 Line width

The total width at half power $\Delta B_{1/2}$ of one hyperfine line (Fig. 8) shows only two regions, a linear increase first at a rate of

$$\Delta B_{1/2} - \Delta B_{1/2}(\text{Si}) = (303 \pm 31)\, x \quad \text{for} \quad 1 \geqq 1 - x > 0.98 \tag{10}$$

and a second linear increase in a limited range

$$\Delta B_{1/2} = (7.2 \pm 0.9) + (100 \pm 8)\, x \quad \text{for} \quad 0.98 > 1 - x > 0.84\,. \tag{11}$$

For all measurements under fast passage conditions the microwave amplitude was low enough to yield a ratio line width to amplitude greater than 2×10^{-5} thus making superfluous any line width correction [1].

5. Discussion

5.1 g-values (Fig. 6)

In contrast to the large change of the g_a-value (isolated donors) which would be predicted by a simple linear interpolation between pure silicon and pure germanium the g-values depend only very little yet clearly measurable on the silicon content. This is reasonable in the silicon concentration range investigated here in which the band structure remains similar to that of pure silicon. Three effects have to be discussed.

a) After Liu's calculations [30] of the g-tensor of electrons in pure silicon the main contribution to the g-shift against the free electron value comes from the 2p-bands (core states) which are situated 81 eV below the top of the valence band. In germanium the 2p-bands are located yet much deeper. Between them and the valence band lie the 3s, 3p, and 3d bands. The spin–orbit matrix elements between the minimum of the conduction band Δ_1 and the band Δ_2' — the only one sensitive to the alloy composition (Section 2) — vanish by symmetry arguments. Therefore the core states will make the main contribution to the g-shift also in the silicon rich alloys. At present there are no calculations of the spin–orbit matrix elements between Δ_1 and core states available for germanium and the alloy system. Therefore a theoretical estimate of the g-shift in the Si–Ge alloys is not yet possible.

b) In addition to the changes in the bands strains in the lattice may cause a g-shift. Silicon and germanium have different atomic radii. In a tetrahedral surrounding they amount to $r_{Si} = 1.17 \times 10^{-8}$ cm and $r_{Ge} = 1.22 \times 10^{-8}$ cm [31]. Therefore local strains around the germanium atoms will be expected [32]. If donors are exposed to these strains, the symmetry will be distorted. This can alter the g-value in two ways:

1. The spin-orbit matrix elements between Δ_1 and Δ_2' will no longer vanish.
2. The six conduction band minima will no longer be equivalent [32], then equation (1) will fail.

c) The different sizes of the atoms involved may be the reason for the three-region-variation of the g_a- and g_c-values (Section 4.2). The atomic radius of phosphorus in tetrahedral symmetry is 1.10×10^{-8} cm [31]. So these small phosphorus atoms will seek regions with excess of the large germanium atoms reducing the strains there.

At low enough silicon concentration below about $1 - x < 0.95$ the local and the mean germanium concentration may be nearly the same. Going in the direction of a higher silicon concentration the phosphorus atoms will surround themselves with the largest number of germanium atoms useful for strain reduction. In this region the local germanium concentration will be higher than its mean value. With $[P] \ll [Ge]$, as it is always the case here, the local Ge concentration may remain nearly constant even if the mean Ge concentration becomes smaller $(0.95 < 1 - x < 0.99)$. For low enough germanium concentration this local concentration must finally decrease.

Thermodynamical calculations can only be helpful if one can ascertain that the crystals as grown are really in a thermodynamic equilibrium. This question is yet open.

A preferred insertion of donors at lattice distortions was recognized in germanium firstly by Wilson [2]. Also the EPR spectra of Mn^{2+} in $Cd_xHg_{1-x}Se$ [33] show that the Mn^{2+} ions are not distributed statistically in the lattice.

Umanskii et al. [34] conclude from diffuse X-ray scattering experiments that in their $Si_{0.5}$–$Ge_{0.5}$ alloys the germanium atoms themselves were not strictly distributed randomly. This may increase the effect on the g-values.

Here a short remark to the publication of Gverdtsiteli et al. [28] is adequate. Comparing their results on conduction electrons (central line, g_c) with the measurements of Feher [1] on isolated donors (hyperfine doublet, g_a) they conclude that the decrease of the g-value with decreasing silicon content is smaller for conduction electrons than for donors without interaction. This statement does not agree with the results of this work (Section 4.2.2). The difference in the g-values and in the rate of change for $1 - x > 0.95$ cannot be explained as a mistake in determining the silicon content of the alloys. To reconcile both sets of g-values, the necessary shift would exceed the limits of error $(\Delta x < 0.005)$. Remarkably the rate of change $-0.018\,x$ given by [28] is not too far from our value (equation (8)). On the whole the discrepancy may be caused by the properties of the different crystals, but it cannot be discussed further at this time on account of the lack of information on the degree of compensation and perhaps on other properties of the crystals.

5.2 Hyperfine splitting

There are two main influences of the alloy composition on the donor wave function which cause a change in the hyperfine splitting according to

$$\widetilde{\Delta B} \propto |\Psi(\boldsymbol{r} = 0)|^2 \propto |F(\boldsymbol{r} = 0)|^2 \, |u(\boldsymbol{r} = 0)|^2 \,. \tag{12}$$

1. The permittivity ε increases with decreasing silicon concentration. As the radius a^* is proportional to ε (equation (4)) the envelope function $F(r)$ broadens (equation (2)) and the value $\Psi(\boldsymbol{r} = 0)$ decreases resulting in a smaller hyperfine splitting.

2. The functions $u(\boldsymbol{r})$ are bunched to the lattice sites $\boldsymbol{r}_L$ more in germanium than in silicon because of the larger nuclear charge of germanium. This causes a larger hyperfine splitting in germanium [2].

An estimation of the hyperfine splitting was tried in the effective mass approximation (equation (1) to (6)). For $|u(\boldsymbol{r} = 0)|^2 = |u(\boldsymbol{r}_L)|^2 = \eta$ a linear interpolation between the values for pure silicon $\eta(\text{Si}) = 186$ [1] and pure germanium $\eta(\text{Ge}) = 1700$ [2] was taken.

$$\eta(x) = \eta(\text{Si}) + \{\eta(\text{Ge}) - \eta(\text{Si})\}\, x \,. \tag{13}$$

There was also assumed that $\eta(\text{Ge}, \text{L}_1) \approx \eta(\text{Ge}, k_0)$. The effective mass was assumed to be constant and equal to the value for pure silicon [13, 14]. For ΔE_1

$$\Delta E_1(x) = \Delta E_1(\text{Si}) - 0.170\, x \text{ eV} \tag{14}$$

was used.

With these assumptions the curve in Fig. 7 which is going through a maximum, was drawn. The adjustable free parameter in equation (14) has been chosen as 0.170 that the curve will meet the points near $1 - x = 0.89$. Two essential remarks should be made:

1. At $1 - x \approx 0.98$ the measured hyperfine splitting is larger than the curve indicates. This will be expected if at high enough silicon concentration the germanium atoms are not distributed randomly (Section 5.1).

2. The points at $1 - x = 0.956$, 0.945, and 0.843, deviate from the curve. From IR-transmission observations we know that these samples are not quite homogeneous.

5.3 Line width and form

The hyperfine doublet is inhomogeneously broadened due to the interaction of the donor electron with the magnetic moments of the ^{29}Si and ^{73}Ge nuclei. On account of the larger moment of ^{73}Ge the doublet line must be broader in the alloy than in pure silicon. But an unsymmetry cannot be expected because the assumption of a random distribution of the isotopes on the relevant lattice sites is adequate.

The measured doublet lines are a superposition of the EPR signals of donors with different surroundings. Therefore we have to discuss the consequences of the superposition of different g_a-values and different individual hyperfine splittings ΔB.

g_a-values: Because of the monotonous decrease of the g-values with decreasing silicon concentration a superposition of lines with different g-values will cause a broadening (I) or an asymmetry (II) or both but only in such a way that a simple shift of one component will bring it to coincidence with the second one. An asymmetry (II) would imply a distribution of different individual x_i-values not symmetrical to the mean value x.

Hyperfine splitting: For $1 - x \approx 0.97$ the hyperfine splitting measured by $\widetilde{\Delta B}$ is going through a flat maximum. So does ΔB (Section 4.3). Then the superposition of signals coming from donors with different surroundings in a sample with a mean silicon concentration in this range will lead to a mirror symmetry (III) of the two hyperfine components because ΔB cannot pass over its maximum value.

For mean concentrations $1 - x < 0.95$ ΔB decreases monotonously. A spread in the surroundings of the donors (symmetrical distribution of individual x_i-values around x expected) not too large will then result in a broadening (IV).

Now we can discuss the experimental results. The registrations show for small mean silicon content ($1 - x < 0.90$) symmetrically broadened hyperfine components and for higher concentrations mutual mirror symmetrical ones. This means that the spread of the hyperfine splitting is preferentially responsible for the lines hape according to the cases III and IV above.

For higher silicon content the line width shows a larger increase with decreasing silicon concentration than for $1 - x < 0.98$ (Fig. 8 and equation (10) and (11)). This also indicates that for $1 \geqq 1 - x > 0.98$ the local germanium concentration near the phosphorus sites exceeds its mean value x.

Because of the larger extension of the donor wave function in the alloys with decreasing Si–Ge ratio there is an overlap at smaller phosphorus concentrations than in pure silicon. Accordingly the lines from phosphorus clusters appear at a smaller phosphorus concentration with increasing germanium content.

6. Summary

Phosphorus doped $Si_{1-x}Ge_x$ alloys in the region $1 \geqq 1 - x \geqq 0.84$ were investigated with EPR at 9.3 GHz and 1.8 K. The spectra, although similar to those of phosphorus in pure silicon, reveal individual characteristic features.

The g-values of isolated phosphorus atoms (g_a in short) decrease only slightly with decreasing silicon concentration. The g-values of nonlocalized electrons (central line, g_c) differs by $(+3 \pm 1) \times 10^{-4}$ for all x from the value of localized ones (hyperfine doublet, g_a). So far assumptions of Gverdtsiteli et al. [28] could not be verified.

The decrease of the minimum phosphorus concentration necessary for the appearance of the central line with the decrease of the silicon concentration proves that the wave function of the donor ground state enlarges with decreasing silicon content (Section 5.3).

The variation of the hyperfine splitting going through a flat maximum at $1 - x = 0.97$ with the Si–Ge ratio could be explained on the basis of the effective mass theory.

The experimental results on the g-values, the hyperfine splittings and the line widths which depend in a peculiar manner (Fig. 6 to 8) on the silicon concentration in different x ranges may be explained by the (yet unproven) assumption that at values of $1 - x$ near 1 the phosphorus atoms are situated preferentially in regions with a locally enhanced germanium concentration (cf. Section 5.1 c).

That this effect can reduce lattice distortions is evident from the radii of the atoms involved.

Acknowledgements

We would like to thank K.-D. Hermbecker and J. Simon for growing the crystals, H. B. Koch for measuring the Hall effect and conductivity and D. Koch and R. Rösler for the liquid helium supply. Some helpful IR-photographs had been made by H. Behrens and P. Mossadegh-Pour. The experiments were efficiently supported by the Deutsche Forschungsgemeinschaft.

References

[1] G. FEHER, Phys. Rev. **114**, 1219 (1959).
[2] D. K. WILSON, Phys. Rev. **134**, A265 (1964).
[3] D. GEIST, Paramagnetische Elektronenresonanz in Halbleitern, in: Festkörperprobleme, Vol. 2, Ed. O. MADELUNG, Vieweg u. Sohn, Braunschweig 1962.
G. LANCASTER, Electron Spin Resonance in Semiconductors, Hilger and Watts, Ltd., London 1966.
[4] F. BASSANI and D. BRUST, Phys. Rev. **131**, 1524 (1963).
[5] D. J. STUKEL, Phys. Rev. B **3**, 3347 (1971).
[6] A. LEVITAS, Phys. Rev. **99**, 1810 (1955).
[7] G. BUSCH and O. VOGT, Helv. phys. Acta **33**, 437 (1960).
[8] S. M. CHRISTIAN and E. R. JOHNSON, Phys. Rev. **95**, 560 (1955).
[9] R. BRAUNSTEIN, A. R. MOORE, and F. HERMAN, Phys. Rev. **109**, 695 (1958).
[10] M. GLICKSMAN, Phys. Rev. **102**, 1496 (1956).
[11] M. GLICKSMAN and S. M. CHRISTIAN, Phys. Rev. **104**, 1278 (1956).
[12] G. BUSCH and O. VOGT, Helv. phys. Acta **33**, 889 (1960).
[13] D. GEIST, phys. stat. sol. (b) **46**, 283 (1971).
[14] D. GEIST and F. BALCK, phys. stat. sol. (b) **56**, 557 (1973).
[15] J. TAUC and A. ABRAHAM, J. Phys. Chem. Solids **20**, 190 (1961).
[16] E. SCHMIDT, Phys. Letters **21**, 640 (1966).
[17] J. S. KLINE, F. H. POLLAK, and M. CARDONA, Helv. phys. Acta **41**, 968 (1968).
[18] W. KOHN and J. M. LUTTINGER, Phys. Rev. **98**, 915 (1955).
[19] W. KOHN, Solid State Phys. **5**, 257 (1957).
[20] W. KOHN and J. M. LUTTINGER, Phys. Rev. **97**, 883 (1955).
[21] F. KOHLRAUSCH, Praktische Physik, Vol. 3, 6. Ed., B. G. Teubner Verlagsgesellschaft, Stuttgart 1968 (p. 168).
[22] K.-D. HERMBECKER, Diplomawork, Clausthal 1969.
[23] J. SIMON, Diplomawork, Clausthal 1970.
[24] P. J. HOLMES, Proc. Inst. Electr. Eng. **106**, T. B. Suppl. 10/18, 861 (1959).
[25] H. B. KOCH, Diplomawork, Clausthal 1972.
[26] J. P. DISMUKES, L. EKSTROM, and R. J. PAFF, J. phys. Chem. **68**, 3021 (1964).
[27] C. P. SLICHTER, Phys. Rev. **99**, 479 (1955).
[28] I. G. GVERDTSITELI, L. N. ALEKSANROV, and V. V. BOIKO, Fiz. tverd. Tela **9**, 169 (1967).
[29] H. KODERA, J. Phys. Soc. Japan Suppl. **21**, 578 (1966).
[30] L. LIU, Phys. Rev. **126**, 1317 (1962).
[31] LANDOLT-BÖRNSTEIN, Vol. 1, Part 4, 6. Ed., Springer-Verlag, Berlin/Göttingen/Heidelberg/New York 1950 (p. 521).
[32] D. K. WILSON and G. FEHER, Phys. Rev. **124**, 1068 (1961).
[33] K. LEIBLER, W. GIRIAT, Z. WILAMOWSKI, and R. IWANOWSKI, phys. stat. sol. (b) **47**, 405 (1971).
[34] Y. S. UMANSKII, V. L. PRILEPSKI, and S. S. GORELIK, Soviet Phys. — Solid State **7**, 2162 (1966).

(Received December 7, 1973)

phys. stat. sol. (b) 62, 377 (1974)

Subject classification: 14.2

I. Physikalisches Institut der Technischen Universität Berlin

Magnetische Bereichsstrukturen in Typ-I-Dünnschichtsupraleitern [1]

I. Der Zwischenzustand [2]

Von

U. KUNZE, B. LISCHKE und W. RODEWALD

Die magnetflußerfüllten Bereiche in dünnen supraleitenden Schichten (Pb, In) in senkrechten Magnetfeldern wurden durch die Dekoration mit Eisenteilchen sichtbar gemacht. Die Abhängigkeit dieser Zwischenzustandsstrukturen von der Schichtdicke wurde mit theoretischen Modellen verglichen. Unterhalb einer kritischen Schichtdicke von etwa $d_c \approx 0{,}1\ \mu m$ wird ein Übergang von den Strukturen des Zwischenzustandes zum zweidimensionalen Mischzustand beobachtet.

Patterns of the magnetic flux distribution in superconducting thin films (Pb, In) in perpendicular magnetic fields have been observed by decoration with small iron particles. The dependence of these intermediate state structures on the film thickness is compared with current theories. Below a critical thickness of about $d_c \approx 0.1\ \mu m$ a change from the intermediate state structures to the two-dimensional mixed state is observed.

1. Einleitung

Ein Supraleiter befindet sich im Zwischenzustand [1 bis 4], wenn supraleitende Bereiche neben magnetfelderfüllten normalleitenden Bereichen existieren [5 bis 10]. Der Übergang eines Supraleiters in diesen Zustand erfolgt durch ein äußeres, hinreichend hohes Magnetfeld.[3] In der vorliegenden Arbeit werden die magnetischen Bereichsstrukturen in Pb- und In-Schichten als Funktion ihrer Dicke systematisch untersucht. Das erfolgt mit der Essmann-Träuble-Technik [7], bei der die felderfüllten Bereiche an der Oberfläche des Supraleiters durch die Anlagerung feiner ferromagnetischer Teilchen direkt sichtbar gemacht werden. Dickere Proben wurden durch Walzen hergestellt, dünne Filme durch Aufdampfen hochreiner Materialien. Im senkrechten Magnetfeld wurden die Schichten auf 1,2 K abgekühlt und mit Fe-Teilchen dekoriert.

Diese Untersuchungen lieferten umfangreiche experimentelle Ergebnisse, welche die komplizierten Strukturen des Zwischenzustandes aufzeigen. Es wurde versucht, diese Strukturen mit vorhandenen Modellen zu vergleichen und deren Gültigkeitsbereiche aufzuzeigen.

Mit abnehmender Schichtdicke verringert sich der Abstand der magnetischen Bereiche, da dadurch die Streufeldenergie des Systems abnimmt. Allerdings muß für die Periodizität der felderfüllten normalleitenden und der feldfreien

---

[1]) Herrn Prof. Dr. H. Boersch zum 65. Geburtstag gewidmet.

[2]) Auszugsweise vorgetragen auf der „Internat. Conf. Magnetic Structures in Superconductors", 5.—8. 9. 1973, Argonne National Laboratory, Argonne, Illinois, USA.

[3]) Auch in stromdurchflossenen Drähten kann der Zwischenzustand durch einen überkritischen Strom erzeugt werden. Die dabei auftretenden Strukturen sind nicht Gegenstand der vorliegenden Untersuchungen.

supraleitenden Bereiche eine untere Grenze existieren, da die Bereiche nicht häufiger abwechseln können als es die magnetische Eindringtiefe λ bzw. die Kohärenzlänge ξ zulassen. Wegen der Quantisierung des Magnetflusses [11] werden die kleinst möglichen Bereiche dann gerade ein Flußquant $\Phi_0 = hc/2e$ enthalten. In sehr dünnen Schichten eines Typ-I-Supraleiters sollen nach Tinkham [12] tatsächlich Anordnungen einzelner Flußquanten existieren, die in dieser Arbeit direkt sichtbar gemacht werden konnten. In Analogie zu dem Mischzustand in Typ-II-Supraleitern [14 bis 16] werden diese Strukturen im folgenden als zweidimensionaler Mischzustand bezeichnet.

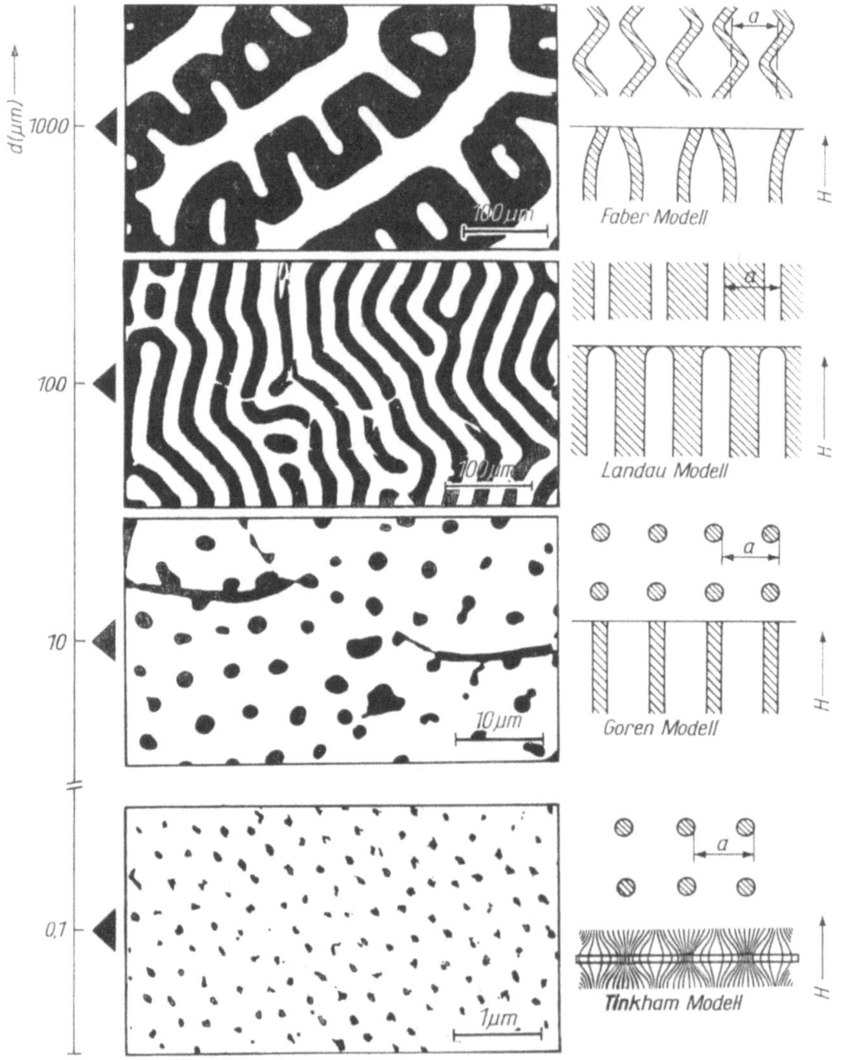

Fig. 1. Magnetische Bereichsstrukturen in Dünnschichtsupraleitern (dunkle Bereiche normalleitend) und Modelle zu ihrer theoretischen Beschreibung (schraffierte Bereiche normalleitend). Die Supraleiter (In und Pb) wurden bei 1,2 K in senkrechten Magnetfeldern $h = H/H_{cb} \approx 0{,}4$ untersucht

2. Der Zwischenzustand

2.1 *Qualitativer Vergleich der Bereichsstrukturen mit theoretischen Modellen*

Die im Zwischenzustand experimentell beobachteten Strukturen sind außerordentlich vielfältig. Nimmt man idealisierende Vereinfachungen in Kauf, so lassen sich die einfacheren Strukturen, zu deren Beschreibung theoretische Modelle existieren, in drei Gruppen unterteilen: Mäander-, Streifen- und Punktstrukturen (Fig. 1). Diese Modellstrukturen sind in den beiden untersuchten Materialien Pb und In unterschiedlich gut ausgeprägt. Daneben können andere und auch wesentlich kompliziertere Strukturen auftreten, so daß die angegebene Klassifizierung nur als eine Orientierungshilfe dienen kann. Die in den Figuren angegebenen Schichtdicken d markieren ungefähre Gültigkeitsbereiche, in denen charakteristische Strukturen überwiegend auftreten.

Die bei groben Bereichsmustern oberhalb der Probenoberfläche entstehenden Feldinhomogenitäten und die damit verbundene Streufeldenergie kann vermindert werden, wenn die Bereiche im Kristall zur Oberfläche hin aufspalten. Dabei beginnen sich die Wände zu falten. Dies soll nach Hubert [17] oberhalb einer charakteristischen Schichtdicke von $D_s \approx 800\,\Delta$ erfolgen, wobei Δ über die Oberflächenenergie σ (Energie pro Flächeneinheit) definiert ist:

$$\sigma = \frac{H_{cb}^2}{8\pi}\Delta \tag{1}$$

(H_{cb} kritische Feldstärke des massiven Supraleiters, $\Delta(\text{Pb}) = 630$ Å [18], $\Delta(\text{In}) = 3300$ Å [28]). Mit dem hier verwendeten Nachweisverfahren lassen sich nur die Strukturen an der Oberfläche des Kristalls, also nur die Faltung der Bereiche und die Ausbildung von Mäandern beurteilen. Diese treten tatsächlich oberhalb der charakteristischen Schichtdicke D_s von ca. 100 μm auf (Fig. 1), wobei die Form der Mäander durch das Faber-Modell [5] näherungsweise wiedergegeben wird.

Unterhalb dieses Schichtdickenbereiches treten Streifen- und Punktstrukturen auf, die qualitativ durch das Landau [19]- bzw. Goren-Modell [20, 21] beschrieben werden (Fig. 1). Dabei sind im allgemeinen Streifen- und Punktmuster auch nebeneinander zu beobachten, da nach Goren [20] ihr Energieunterschied gering ist und deshalb ihre Ausbildung von zufälligen Parametern abhängt. Nach der Theorie von Hubert [17] soll bei konstanter Schichtdicke für reduzierte Feldstärken $h = H/H_c < 0{,}2$ bevorzugt das Röhrenmodell nach Goren [20] auftreten. Das konnten wir durch unsere Experimente nicht eindeutig bestätigen. Mit abnehmender Schichtdicke wächst allerdings die Tendenz, Punktmuster bevorzugt auszubilden. Bei etwa 0,1 μm Schichtdicke bestehen die magnetischen Strukturen schließlich aus einzelnen Flußfäden mit je einem Flußquant Φ_0, deren Existenz in dünnen Schichten aus Typ-I-Supraleitern erstmals von Tinkham [12] postuliert worden war. (Man beachte die Änderung des Abbildungsmaßstabes in den magnetischen Strukturen der Fig. 1.)

2.2 *Quantitativer Vergleich der Bereichsstrukturen mit theoretischen Modellen*

Nach dem qualitativen Vergleich stellt sich die Frage, ob die Modelle auch quantitativ die Periodizitäten in den beobachteten Strukturen zu beschreiben vermögen. Zu ihrer Berechnung sei wegen der besonderen Einfachheit an das Landau-Modell [19] erinnert (Fig. 1). Die Zwischenzustandsstruktur besteht aus abwechselnd normal- und supraleitenden Regionen parallel zum äußeren Feld. An der Oberfläche ist eine Aufweitung der normalleitenden Streifen wegen der Stetigkeit der Feldkomponenten zu fordern [19]. Die vom Periodizitätsparameter a (Fig. 1) beeinflußten Energieterme des Systems setzen sich aus verschiedenen Anteilen zusammen. Zum Aufbau der Phasengrenze Normal–Supraleiter ist eine Grenzflächenenergie f_1 = Flächenelement $\times \sigma \times$ Anzahl der Grenzflächen $= dc\sigma(2b/a)$ notwendig (b Breite, d Dicke, c Länge der Schichten). Bezogen auf die Flächeneinheit der Oberfläche ergibt sich mit (1)

$$f_1 = \frac{H_{\mathrm{cb}}^2}{4\pi} \frac{\Delta d}{a}. \tag{2}$$

Die an den Oberflächen entstehenden Feldinhomogenitäten reichen bis in eine Tiefe in den Raum hinein, die etwa der Bereichsweite a entspricht. Die damit verbundene Streufeldenergie wird sich also zu a proportional erweisen. Auch die Aufweitung der normalleitenden Bereiche an der Oberfläche soll bis zu einer Tiefe a in den Supraleiter hineinreichen, so daß der damit verknüpfte Kondensationsenergieverlust proportional zu a ist. Bei der genaueren Berechnung des Kondensationsenergieverlustes und der Streufeldenergie treten elementar nicht mehr lösbare Integrale auf. Beide Terme lassen sich jedoch zu

$$f_{2,3} = a \frac{H_{\mathrm{cb}}^2}{4\pi} g(h) \tag{3}$$

zusammenfassen, wobei $g(h)$ eine tabellierte Funktion [22, 27] angibt.

Insgesamt folgt für die a-abhängigen Terme der Energie pro Oberflächenelement

$$f = \frac{H_{\mathrm{c}}^2}{4\pi} \left(\frac{\Delta d}{a} + a g(h) \right). \tag{4}$$

Aus der Minimalisierung dieser Gleichung bezüglich a folgt der Gleichgewichts-

Tabelle 1

Vergleich der experimentell beobachteten magnetischen Bereichsstrukturen in Typ-I-Supraleitern mit verschiedenen Zwischenzustands-Modellen (d Schichtdicke)

Experiment		Theorie	
Bereichsstruktur	Periodizitätsparameter a (µm)	Modell	Periodizitätsparameter a (µm)
Mäander ($d > 100$ µm)	130	Faber-Modell [5]	139
Lamellen (1 µm $< d <$ 100 µm)	28	Landau-Modell [19]	27
		Kuper-Modell [23]	8
		Landau-Verzweigungsmodell [24]	7
Punktstrukturen (1 µm $< d <$ 100 µm)	9	Goren-Modell [20]	9
		Andrew-Modell [25]	2

abstand der Streifen

$$a = \sqrt{\frac{\Delta d}{g(h)}}, \tag{5}$$

der sich gerade dann ergibt, wenn die Oberflächenenergie gleich der Streufeldenergie plus Kondensationsenergieverlust ist (Gleichung (4)). Mit abnehmender Schichtdicke d muß sich entsprechend (5) die Bereichsweite a vermindern, was sich sehr gut durch die in der Fig. 1 dargestellten Strukturen bestätigt. Allerdings soll der Zusammenhang von $a \sim d^{1/2}$ nur für dünne Proben $(d < D_s)$ gelten, da für dickere Schichten nach Hubert [17] $a \sim d^{2/3}$ erwartet wird $(d > D_s)$.

Der Gleichung (5) entsprechende Gleichungen lassen sich auch für die anderen Modelle erhalten. Der Periodizitätsparameter a wird demnach neben der Schichtdicke auch vom Material (Δ) und der magnetischen Feldstärke $(h = H/H_{cb})$ beeinflußt. Häufig hängt die Art der Strukturen auch von der magnetischen Vorgeschichte und der Materialbehandlung ab.

In zahlreichen, von uns an Pb und In durchgeführten Experimenten wurden zunächst die beobachteten Muster den entsprechenden Modellen zugeordnet (beispielsweise Streifen den Laminar-Modellen) und dann die gemessenen Periodizitäten mit den zu erwartenden Werten verglichen. Stellvertretend für diese Experimente sind einige Werte des Periodizitätsparameters a in der Tabelle 1 aufgeführt. In Bereichen, wo Mäander auftreten, können sie quantitativ durch das Faber-Modell [5] gut beschrieben werden. Streifenstrukturen werden am besten durch das Laminar-Modell nach Landau [19] und Punktstrukturen oberhalb von 1 μm Schichtdicke durch das Spot-Modell von Goren [20, 21] erfaßt. Die Branching-Modelle [24, 25] sollen nach Hubert erst oberhalb einer charakteristischen Schichtdicke Gültigkeit erhalten, wo allerdings im allgemeinen keine Punkt- oder Streifenstrukturen mehr zu beobachten sind. Allen bisher behandelten Modellen war die Voraussetzung gemeinsam, daß $a, d \gg \lambda, \xi$ sein sollte.

3. Übergang vom Zwischenzustand zum zweidimensionalen Mischzustand

Der Übergang eines Typ-I-Supraleiters vom Normal- in den Zwischenzustand erfolgt durch einen Phasenübergang 1. Art [20]. Die sich im Zwischenzustand ergebenden Bereichsstrukturen sind verhältnismäßig grob und enthalten eine Vielzahl von Flußquanten (Fig. 1). Der Übergang eines Typ-I-Supraleiters in den zweidimensionalen Mischzustand ist durch einen Phasenübergang 2. Art gekennzeichnet [12]. Er erfolgt bei sehr viel geringeren Schichtdicken, und die Bereichsstrukturen bestehen aus magnetischen Flußfäden mit geringer Distanz, die jeweils ein einzelnes Flußquant enthalten. Damit stellt sich abschließend die Frage, welche Kriterien eine bestimmte kritische Schichtdicke auszeichnen, bei der die Strukturen des Zwischenzustandes in die des zweidimensionalen Mischzustandes übergehen.

Die normalleitenden Lamellen des Landau-Modells [19] (Fig. 1) verbreitern sich trompetenförmig zur Oberfläche hin. Wegen der Vergrößerung des normalleitenden Bereiches muß die kritische Feldstärke $H_{c\perp}$ in der Nähe der Probenoberfläche kleiner als im Inneren sein. Der normalleitende Zustand einzelner Streifen erweist sich hier bei Feldern $H_{c\perp} < H_{cb}$ als stabil, was aus dem Einfluß der Oberflächenenergie bei gekrümmten Grenzflächen verstanden werden kann [20, 23]. Dieser sich in der Oberfläche des Supraleiters abspielende Effekt

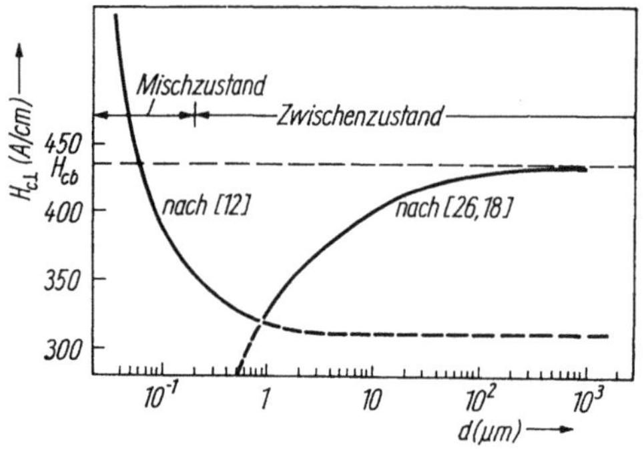

Fig. 2. Verlauf der kritischen magnetischen Feldstärke dünner Schichten nach Tinkham [12], Davies [26] und Cody und Miller [18]

muß sich bei dünnen Schichten immer stärker bemerkbar machen. Die Abnahme des kritischen Feldes mit sinkender Schichtdicke wurde von Davies [26] berechnet. Das Ergebnis stimmt mit Messungen von Cody und Miller [18] überein und ist in der Fig. 2 wiedergegeben.

Bei dünnen Schichten schließlich bestimmt ein gegenläufiger Effekt die kritische Feldstärke $H_{c\perp}$. Die Supraleitungselektronen werden an den Grenzflächen gestreut, wodurch sich die Materialeigenschaften des Supraleiters ändern und $H_{c\perp}$ mit weiter abnehmender Schichtdicke wieder ansteigt (Fig. 2). Dieser Effekt wurde von Tinkham [12] gefunden. Der Schnittpunkt beider Kurven der Fig. 2 zeichnet eine kritische Schichtdicke $d_c \approx 1\ \mu$m aus, unterhalb der ein Wechsel von der Zwischenzustandsstruktur zum zweidimensionalen Mischzustand erfolgen sollte.

Daß dieses Kriterium zu ungenau ist und in einem nachfolgenden II. Teil der Arbeit zu modifizieren sein wird, beweisen die experimentellen Ergebnisse. Sie liefern eine kritische Schichtdicke von $d_c \approx 0{,}1\ \mu$m. Unterhalb von diesem Wert können die Strukturen des zweidimensionalen Mischzustandes [13] beobachtet werden (Fig. 1).

4. Zusammenfassung

Durch die Untersuchung der Zwischenzustandsstrukturen in Pb- und In-Schichten als Funktion ihrer Dicke konnte gezeigt werden, daß die unterschiedlichen theoretischen Modelle jeweils Näherungen darstellen, die einen bestimmten Schichtdickenbereich quantitativ zu beschreiben vermögen. Bei sehr dünnen Schichten erfolgt ein Übergang der Strukturen in die des zweidimensionalen Mischzustandes, die in einer nachfolgenden Arbeit genauer beschrieben werden. Der zweidimensionale Mischzustand wurde dort an freitragenden, im Elektronenmikroskop durchstrahlbaren Schichten nachgewiesen, was die unmittelbare Untersuchung des Einflusses von Kristallbaufeldern auf die Verteilung der Flußquanten im Supraleiter ermöglicht.

Literatur

[1] P. G. de Gennes, Superconductivity of Metals and Alloys, W. A. Benjamin, Inc., New York 1966 (p. 40).
[2] C. G. Kuper, Introduction to the Theory of Superconductivity, Clarendon Press, Oxford 1968 (p. 78).
[3] R. D. Parks, Superconductivity, Vol. 2, M. Dekker, Inc., New York 1969 (p. 1235).
[4] W. Buckel, Supraleitung, Physik Verlag, Weinheim 1972 (p. 128).

[5] T. E. FABER, Proc. Roy. Soc. **A248**, 460 (1958).
[6] YU. V. SHARVIN, Soviet Phys. — J. exper. theor. Phys. **6**, 1031 (1958).
[7] H. TRÄUBLE und U. ESSMANN, phys. stat. sol. **18**, 813 (1966).
[8] H. KIRCHNER, Phys. Letters A **26**, 651 (1968); Siemens Forsch. Ber. **1**, 39 (1971).
[9] A. BODMER, U. ESSMANN und H. TRÄUBLE, phys. stat. sol. (a) **13**, 471 (1972).
[10] B. M. BALASHOVA und Y. SHARVIN, Soviet Phys. — J. exper. theor. Phys. **31**, 40 (1956).
[11] B. LISCHKE, Phys. Rev. Letters **22**, 1366 (1969).
[12] M. TINKHAM, Phys. Rev. **129**, 2413 (1963).
[13] H. BOERSCH, U. KUNZE, B. LISCHKE und W. RODEWALD, Phys. Letters A **44**, 273 (1973).
[14] A. A. ABRIKOSOV, Zh. eksper. teor. Fiz. **32**, 1442 (1957); Soviet Phys. — J. exper. theor. Phys. **5**, 1174 (1957).
[15] U. ESSMANN und H. TRÄUBLE, Phys. Letters A **24**, 526 (1967).
[16] B. LISCHKE und W. RODEWALD, Phys. Letters A **39**, 321 (1972).
[17] A. HUBERT, phys. stat. sol. **24**, 669 (1967).
[18] G. D. CODY und R. E. MILLER, Phys. Rev. **173**, 481 (1968); **173**, 494 (1968).
[19] L. D. LANDAU, Phys. Z. SU **11**, 129 (1937).
[20] R. N. GOREN, Report TR 2 (AD 703658), erhältlich von CFSTI, Springfield, Va. 22151 USA (1970).
[21] R. N. GOREN und M. TINKHAM, J. Low Temp. Phys. **5**, 465 (1971).
[22] E. M. LIFSCHITZ und J. W. SCHAPWINA, Dokl. Akad. Nauk SSSR **79**, 783 (1951).
[23] C. G. KUPER, Phil. Mag. **42**, 961 (1951).
[24] L. D. LANDAU, Nature **147**, 688 (1938); J. Phys. (USSR) **7**, 99 (1943).
[25] E. R. ANDREW, Proc. Roy. Soc. **A194**, 98 (1948).
[26] E. A. DAVIES, Proc. Roy. Soc. **A255**, 407 (1960).
[27] F. HAENSSLER und L. RINDERER, Helv. phys. Acta **40**, 659 (1967).
[28] R. P. HUEBENER und R. T. KAMPWIRTH, phys. stat. sol. (a) **13**, 255 (1972).

(Received November 28, 1973)

phys. stat. sol. (b) 62, 385 (1974)

Subject classification: 13.5.3 and 20.1; 22.4

Institut für Angewandte Festkörperphysik der Fraunhofer-Gesellschaft Freiburg i. Br. (a) and Fakultät für Physik der Universität Freiburg i. Br. (b)

Bound Small Polaron Optical Absorption in V^--Type Centres in MgO

By

O. F. SCHIRMER (a), P. KOIDL (a), and H. G. REIK (b)

A theory, based on the method of moments, is described which predicts exact lineshapes for the optical absorption of bound small polarons. This theory explains the position, large width, and high oscillator strength of the optical absorption of the V_{Al}^- centre in MgO and the polarisation of V^- in MgO, which has been observed under uniaxial stress.

Eine Theorie, beruhend auf der Methode der Momente, wird beschrieben, die exakt die Linienform der optischen Absorption gebundener kleiner Polaronen herleitet. Diese Theorie erklärt die Lage, große Breite und hohe Oszillatorenstärke der optischen Absorption von V_{Al}^- Zentren in MgO und die Polarisation von V^- in MgO, welche unter einachsigem Druck beobachtet wurden.

1. Theory

V^- type centres in alkaline earth oxides (a hole trapped at one of the O^{2-}-ions neighboring various negative defects [1]) show broad absorption bands [2 to 5] in the visible, the origin of which has so far not been understood [5]. With V^- [6] and V_{Al}^- [7] in MgO as examples (negative charge represented by a vacancy, isolated or associated with Al^{3+}, respectively), we will show that these bands can be explained if the distortion of the centres is taken into account.

Even in V^-, where the six O^{2-}-ions surrounding the vacancy are equivalent, the hole is localised at one ion at low teperatures and the centre shows tetragonal symmetry. This can be attributed to a distortion of the centre by a coupling of the hole to a T_{1u} phonon (coordinates Q_x, Q_y, Q_z), competing with the tunneling of the hole. This exchange ($J > 0$) between neighbouring O^{2-} p_z functions (Fig. 1a) leads to three O_h states [8] Γ_1^+, Γ_4^-, and Γ_3^+ separated by $4J$ and $2J$, respectively. If hole–T_{1u}–phonon interaction is introduced in a manner analogous to that described by Ham [9, 10], it can be shown that to each af the six $O^{2-}p_z$ functions there belongs a spherically symmetric oscillator, displaced in Q-space (Fig. 1b), with the potential energies (in the limit $J \to 0$)

$$E_{\pm x} = \frac{1}{2} \mu\omega_0^2((Q_x \mp 2E_{JT}/V)^2 + Q_y^2 + Q_z^2)$$

for the wells centred on the Q_x-axis and similarly for the four equivalent ones. E_{JT} is an abbreviation for $V^2/(2\mu\omega_0^2)$, where $\hbar\omega_0$ is the oscillator energy and V, the reduced matrix element[1]) describing the strength of hole–phonon coupling.

[1]) In principle, there are two reduced matrix elements, describing the coupling between the Γ_3^+ and Γ_4^- states and between the Γ_1^+ and Γ_4^- states. These, however, have to be chosen equal ($= V$), in order to describe complete localisation of the hole at one oxygen site, if the relevant coupling coefficients of Koster et al. [11] are used. See also [10] p. 48.

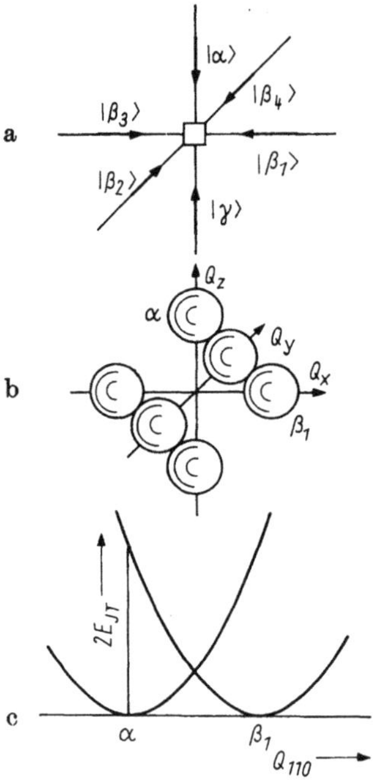

Fig. 1. a) The $O^{2-}p_z$ orbitals, which can capture a hole in the V^- centre. b) Configuration of the wells in the space of the coordinates of the T_{1u} mode. Contours of equal energy are sketched. c) E–Q diagram of the α and β well along their connecting direction

This hole–phonon interaction can be called a pseudo-Jahn-Teller effect [10]; with the same justification, the hole, accompanied by the lattice distortion, can be termed a small polaron [12], bound to the vacancy.

Anticipating later use, we call the six $O^{2-}p_z$ functions $|\alpha\rangle$, $|\gamma\rangle$ and $|\beta_i\rangle$ ($i = 1$ to 4), Fig. 1a. For simplification, we assume the $|\beta_i\rangle$ to be orthogonal, $\langle\beta_i|\beta_j\rangle = \delta_{ij}$. If the hole is in $|\alpha\rangle$, the eigenfunctions of the system, neglecting exchange, are Born-Oppenheimer products $|\alpha\rangle\,|\mu_\alpha\rangle$, where $|\mu_\alpha\rangle$ is an eigenfunction of the displaced oscillator and μ, the relevant set of quantum numbers. The analogous eigenfunctions for the other wells are $|\gamma\rangle\,|\mu_\gamma\rangle$ and $|\beta_i\rangle\,|\mu_i\rangle$.

In the following, we shall assume that the hole is captured in well α. For large $E_{JT}/\hbar\omega_0$, optical transitions will take place with fixed Q (Franck-Condon principle). The intersection of the potential sheets of two neighbouring wells, taken along their common [110] type direction, yields the simple E–Q relation of Fig. 1c. Since the energy of these two sheets changes equally for all directions perpendicular to [110], the transition energies depend only on this configuration coordinate for such a situation. For $E_{JT} \gg \hbar\omega_0$, transitions between oscillator levels of such potentials lead to a Gaussian, peaked at $2E_{JT}$, with second moment $2E_{JT}\,\hbar\omega_0$.

As it will turn out later (equation (4)) the integrated transition probability between two potential wells does not depend on the vibrational parts of the Born-Oppenheimer products. Such transitions are allowed by the admixture of neighbouring well functions $|\beta_j\rangle$ to $|\alpha\rangle$ via the exchange interaction:

$$|\alpha'\rangle = |\alpha\rangle - \frac{J}{\Delta E} \sum_i |\beta_i\rangle \,. \tag{1}$$

Analogously,

$$|\beta_i'\rangle = |\beta_i\rangle + \frac{J}{\Delta E} |\alpha\rangle \,. \tag{2}$$

The energy denominator is put equal to the average hole excitation to the β_i potential sheets, i.e., $\sim 2E_{JT}$. The dipole moment for a transition between sites α and β_i is then[2])[3])

$$\langle\alpha'|\, \boldsymbol{P}\, |\beta_i'\rangle = \frac{J}{(2E_{JT})}\, e\boldsymbol{l}, \;\; \boldsymbol{P} = -e\boldsymbol{x} \,. \tag{3}$$

[2]) Compare [8] for the analogous case of electric stabilisation.

[3]) After this derivation we shall call $|\alpha'\rangle$ and $|\beta_i'\rangle$ again $|\alpha\rangle$ and $|\beta_i\rangle$.

$\boldsymbol{l}$ is a vector connecting the two oxygen sites. It lies approximately along the [110] direction, depending on the actual distortion within the centre. The length of this vector can give the transition moment considerable magnitude, even if J/E_{JT} is small.

The absorption line shape will be determined by calculating its moments [13]. Defining Π as $\boldsymbol{P} \cdot \boldsymbol{e}$, where $\boldsymbol{e}$ is the polarisation vector of the light, the area of the shape function is

$$\langle E^0 \rangle = \mathrm{av}_a \sum_b \langle a|\Pi|b\rangle \langle b|\Pi^+|a\rangle \,,$$

where av_a means the thermal average over the states in well α; $|b\rangle$ are the exact excited states, related to the Born-Oppenheimer products $|\beta_i\nu_i\rangle$ by a unitary transformation. For later use, we also mention that the oscillator functions of the system, e.g. $|\nu_i\rangle$, if the hole is caught in $|\beta_i\rangle$, form a complete set: $\sum_{\nu_i} |\nu_i\rangle \langle\nu_i| = 1$. Because of the unitarity, we can replace $\sum_b |b\rangle\langle b|$ by $\sum_{\beta_i\nu_i} |\beta_i\nu_i\rangle\langle\beta_i\nu_i|$ and obtain

$$\begin{aligned} \langle E^0 \rangle &= \mathrm{av}_{\mu_\alpha} \sum_{i\nu_i} \langle \alpha\mu_\alpha|\Pi|\beta_i\nu_i\rangle \langle \beta_i\nu_i|\Pi^+|\alpha\mu_\alpha\rangle \\ &= \sum_i |\langle\alpha|\Pi|\beta_i\rangle|^2 \,\mathrm{av}_{\mu_\alpha} \sum_{\nu_i} |\langle\mu_\alpha|\nu_i\rangle|^2 = \sum_i |\langle\alpha|\Pi|\beta_i\rangle|^2 \,. \end{aligned} \tag{4}$$

The first moment reads

$$\langle E^1 \rangle = \frac{1}{\langle E^0 \rangle} \,\mathrm{av}_a \sum_b |\langle a|\Pi|b\rangle|^2 \,(E_b - E_a)\,,$$

where

$$E_b = \langle b|\mathcal{H}|b\rangle \quad \text{and} \quad E_a = \langle a|\mathcal{H}|a\rangle$$

$\mathcal{H} = \mathcal{H}_L + \mathcal{H}_J$ is the total Hamiltonian of the system, with $\mathcal{H}_L$, diagonal in the occupied well, having the spectrum of the spherical harmonic oscillator, and $\mathcal{H}_J$, the exchange interaction, being off-diagonal.

The lowest state, $|a\rangle$, is about $2E_{\mathrm{JT}}$ apart from the next higher states near the centre of the well, where the transitions occur. From the interpretation of the experimental data we infer that E_{JT} is large compared to J. We thus neglect any influence of $\mathcal{H}_J$ on the energy of $|a\rangle$ and keep only the first order modification of the wavefunction introduced in (1). Thus $\langle a|\,\mathcal{H}\,|a\rangle = E_{\mu_\alpha}$, the energy of the μ-th level of the oscillator centred at α. On the other hand, J enters in first order into the energy of the states b, since their component functions $|\beta_i\rangle$ are degenerate. The first moment becomes now:

$$\begin{aligned} \langle E^1 \rangle \langle E^0 \rangle &= \mathrm{av}_a \sum_b \langle a|\Pi|b\rangle \langle b|\,\mathcal{H} - E_{\mu_\alpha}|b\rangle \langle b|\Pi^+|a\rangle \\ &= \mathrm{av}_{\mu_\alpha} \sum_{ij} \langle\alpha|\Pi|\beta_i\rangle \langle\beta_i\,|\beta_j\rangle \langle\beta_j|\Pi^+|\alpha\rangle \times \\ &\quad \times \sum_{\nu_i\varrho_j} \langle\mu_\alpha|\nu_i\rangle \langle\nu_i|\mathcal{H}_L - E_{\mu_\alpha}|\varrho_j\rangle \langle\varrho_j|\mu_\alpha\rangle + \\ &\quad + \mathrm{av}_{\mu_\alpha} \sum_{ij} \langle\alpha|\Pi|\beta_i\rangle \langle\beta_i|\,\mathcal{H}_J\,|\beta_j\rangle \langle\beta_j|\Pi^+|\alpha\rangle \times \\ &\quad \times \sum_{\nu_i\varrho_j} \langle\mu_\alpha|\nu_i\rangle \langle\nu_i|\varrho_j\rangle \langle\varrho_j|\mu_\alpha\rangle \,. \end{aligned}$$

We can simplify the first term using $\langle\beta_i|\beta_j\rangle = \delta_{ij}$, which also leads to $\nu_i = \varrho_j$. The vibrational sums are unity because $\sum_{\nu_i} |\nu_i\rangle \langle\nu_i| = 1$ etc. Therefore,

$$\langle E^1\rangle = \bar{E} + \frac{1}{\langle E^0\rangle} \sum_{ij} \langle\alpha|\Pi|\beta_i\rangle \langle\beta_i|\mathcal{H}_J|\beta_j\rangle \langle\beta_j|\Pi^+|\alpha\rangle ,$$

where

$$\bar{E} = \mathrm{av}_{\mu_\alpha} \sum_{\nu_i} |\langle\mu_\alpha|\nu_i\rangle|^2 (E_{\nu_i} - E_{\mu_\alpha}) .$$

Assuming $E_{\mathrm{JT}} \gg \hbar\omega_0$, $\bar{E} = 2E_{\mathrm{JT}}$, since it is the first moment of a Gaussian peaked at that energy, independent of temperature [14].

In applying these expressions to the optical absorption of the V^- centres, we note that $\boldsymbol{e}$ can be analysed into its components parallel and perpendicular to the centre axis. In Table 1 we list $\langle E^0\rangle$ and $\langle E^1\rangle$ for these cases.

Table 1

The lower moments of the lineshapes

	$\boldsymbol{e} \parallel$ centre axis	$\boldsymbol{e} \perp$ centre axis
$\langle E^0\rangle$	$4|\langle\alpha|\Pi_\parallel|\beta_1\rangle|$	$2|\langle\alpha|\Pi_\perp|\beta_1\rangle|$
$\langle E^1\rangle$	$2E_{\mathrm{JT}} + 2J$	$2E_{\mathrm{JT}}$

We assume that $\boldsymbol{l}$ (equation (3)) exactly lies along [110] type directions. This means $\langle\alpha|\Pi_\parallel|\beta_1\rangle = \langle\alpha|\Pi_\perp|\beta_1\rangle$. Judging from $\langle E^0\rangle_\parallel$ and $\langle E^0\rangle_\perp$, it seems then that the perpendicular shapefunction has half the area of the parallel one. However, it is twice as probable that $\boldsymbol{e}$ cuts the centre axes perpendicularly than parallel, since they are randomly distributed over the six [100] type directions. So both curves have the same area. All higher moments do not depend on J. This can be seen by noting that the n-th central moment contains n factors of the type $\langle\beta_i\nu_i|\ \mathcal{H} - E_{\mu_\alpha} - \langle E^1\rangle\ |\beta_j\nu_j\rangle$. On carrying out the appropriate summations and noting that, e.g., $\langle\alpha|\Pi_\perp|\beta_1\rangle = -\langle\alpha|\Pi_\perp|\beta_3\rangle$ one sees that for the perpendicular case $\mathcal{H}$ does not lead to J contributions; for the other case, the J contributions of $\mathcal{H}$ and $\langle E^1\rangle_\parallel$ cancel.

Because of this independence of J, the parallel and perpendicular lineshapes can be inferred from the special case $J = 0$. Here the shape is a mere superposition of those corresponding to transitions from well α to any of the β_i. Each of these leads to a Gaussian, as demonstrated above. The *exact* lineshapes therefore are

$$I_\parallel = 4m^2 \sqrt{W/\pi}\ \exp\left(-W(\hbar\omega - 2E_{\mathrm{JT}} - 2J)^2\right) \tag{5}$$

and

$$I_\perp = 4m^2 \sqrt{W/\pi}\ \exp\left(-W(\hbar\omega - 2E_{\mathrm{JT}})^2\right)$$

with $m = |\langle\alpha|\Pi_\parallel|\beta_1\rangle|$ and $W^{-1} = 4E_{\mathrm{JT}}\hbar\omega_0$.

The absorption curve of a crystal with a random distribution of centre axes along [100] type directions is proportional to $(I_\parallel + I_\perp)\ \hbar\omega$ [13].

For untreated MgO crystals, it has been shown [7] that the Mg vacancies generally are associated with Al^{3+} ions, assumed to lie in next nearest neighbour positions, forming V_{Al}^- centres. In order to describe the observed [3] V_{Al}^- absorption by the above model, it has to be considered that the hole states $|\beta_i\rangle$ are

Table 2

Ranges within which the parameters fitting the absorption curve can be chosen (in eV)

$\hbar\omega_0$	0.06 to 0.08
ε	0.0 to 0.2
E_{JT}	0.84 to 1.05
J	0.12 to 0.23

higher than $|\alpha\rangle$ by an energy ε, caused by the association of Al^{3+} next to and screened by the O^{2-} ion γ. In (5), $\hbar\omega$ thus has to be replaced by $\hbar\omega - \varepsilon$.

2. Comparison with Experiment

The low energy part of the optical absorption of V_{Al}^- can be fitted well with parameters lying in the ranges given in Table 2. Neither E_{JT} nor $\hbar\omega_0$ depend critically on the assumed values of J and ε. Fig. 2a shows a fit with a representative set of parameters.

From a correlation of the oscillator strength of the band, 0.1 [3], with (3), one finds $J = 0.19$ eV, in agreement with the range of values given in Table 2. The size of $\hbar\omega_0$, 0.06 to 0.08 eV, corresponds to an energy lying in the LO phonon band of MgO [15].

Qualitative verification of the predicted shape of $(I_\perp - I_\parallel)$ was obtained with MgO crystals, enriched with V^- centres by heating to 1000 °C and quenching rapidly [6], at 6 K under an unspecified uniaxial stress, Fig. 2b. It can be assumed that only holes bound to isolated vacancies reorient under these conditions. It is known that the V^- axes tend to be aligned along the stress direction [16]. In that way it was possible to conclude from this experiment that $(I_\perp - I_\parallel)$ varies in a way predicted by the theory with increasing energy. Probably because of uncontrollable stress birefringence, the zero of $(I_\perp - I_\parallel)$ differed from crystal to crystal and thus is not shown in Fig. 2b.

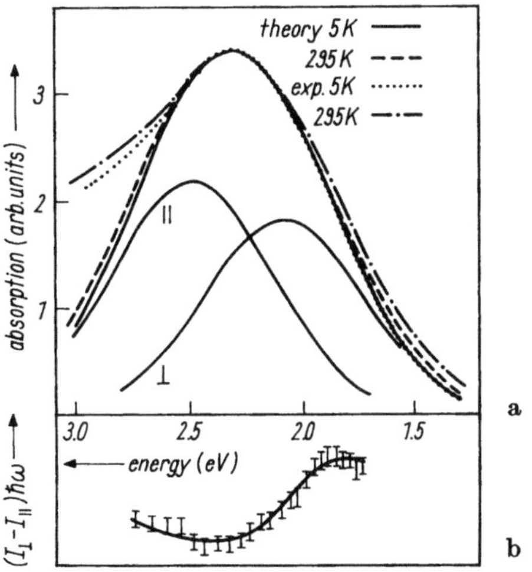

Fig. 2. a) Fit of the low energy onset of the 5 K optical absorption band of the V_{Al}^- centre. Parameter set used: $\varepsilon = 0.2$, $\hbar\omega_0 = 0.07$, $E_{JT} = 0.91$, $J = 0.208$ eV. The theoretical 295 K ordinates were multiplied by 1.04 to give coincidence with the experimental maximum. b) Variation of $(I_\perp - I_\parallel)$ with energy. The bars denote the uncertainties of the measurement

So far, a harmonic approximation for the vibrational properties of the V^- type centres was used. This seems not to be justified completely, since the intensity as well as the width of the 295 K V_{Al}^- absorption are slightly larger than those predicted by the 5 K parameters (Fig. 2a).

For the temperature dependence of the width the following expression was used [14]:

$$W^{-1}(T) = W^{-1}(0) \operatorname{Coth} (\hbar\omega_0/2kT) .$$

Here it should be realized that the oscillators in reality are somewhat anharmonic by the — so far neglected — action of $\mathcal{H}_J$ on the ground state. This will flatten the potential towards the origin and along the [110] type directions in Q space where higher lying levels are approaching the ground state. The effect of anharmonicity on the intensity can be considered by taking account of the vibrational dependence of the hole function in $\langle E^0 \rangle$:

$$\langle E^0 \rangle_{\text{anh}} = \text{av}_{\mu_\alpha} \sum_{\nu_i \beta_i} |\langle \alpha(\mu_\alpha) \mu_\alpha | \Pi | \beta_i(\nu_i) \nu_i \rangle|^2 .$$

A separation of hole and vibrational functions now is not possible any more. However, qualitatively it can be seen that a lowering of the potential towards the origin will shorten the distances between the hole functions $|\alpha\rangle$ and $|\beta_i\rangle$, if the excited states are occupied thermally. This will increase the exchange and thus the intensity. The increased width of the absorption can be understood by noting that the flattening of the potential leads to a smaller spacing of the excited vibrational levels than $\hbar\omega_0$, an energy consistent with the curvature of the potential near its minimum.

3. Conclusion

With the present theory we give a consistent model for the V_{Al}^- absorption band, pointing out the reasons for the position, high oscillator strength, and large width of the band. The analysis can be extended to the absorption of other V^- type centres in MgO, CaO, and SrO, which show rather similar features.[4])

It should be noted, however, that these bands generally show short wavelength tails not explained in the present simple version of the model. They could be attributed to the fact that not only transitions to the $O^{2-}p_z$ orbitals considered so far contribute to the polaronic absorption but also to all other orbitals having finite exchange with $|\alpha\rangle$. All conceivable such Franck-Condon transitions have higher energy than those to the $O^{2-}p_z$ functions, because now electronic excitations have to be added to the lattice distortion energy. These higher transition energies also tend to decrease the transition moments to such orbitals, because their admixture to $|\alpha\rangle$, analogous to equations (1) and (2), now contain larger energy denominators. The present approach thus does not exclude bands at higher energy, which can have lower intensity. In the case of MgO, overlap with another band of unknown origin, centered at 3.1 eV [1] and independent of V^- and V_{Al}^-, also has to be considered. This can explain the discrepancy between theory and experiment at short wavelengths in Fig. 2a.

---

[4]) It is observed that centres having the Mg vacancy filled with Li^+ or Na^+ have an absorption maximum at lower energy than the corresponding V^- centres. E_{JT} of the filled centres thus is smaller than of the empty ones. This is in accord with the much higher reorientation time observed [16] for V^- (MgO) (100 ms at 4.2 K) as compared to that of V_{Li} (MgO (2 ms)).

Acknowledgements

We thank A. Hervé and G. Rius for communicating unpublished results and supplying some crystals.

References

[1] B. Henderson and J. E. Wertz, Adv. Phys. **17**, 749 (1968).
[2] J. E. Wertz, G. S. Saville, P. Auzins, and J. W. Orton, J. Phys. Soc. Japan **18**, Suppl. II, 305 (1963).
[3] Y. Chen and W. A. Sibley, Phys. Rev. **154**, 842 (1967).
[4] H. T. Tohver, B. Henderson, Y. Chen, and M. M. Abraham, Phys. Rev. B **5**, 3276 (1972).
[5] M. M. Abraham, Y. Chen, J. L. Kolopus, and H. T. Tohver, Phys. Rev. B **5**, 4945 (1972).
[6] L. E. Halliburton, L. A. Kappers, D. L. Cowan, F. C. Dravnieks, and J. E. Wertz, Phys. Rev. Letters **30**, 607 (1973).
[7] W. P. Unruh, Y. Chen, and M. M. Abraham, Phys. Rev. Letters **30**, 446 (1973).
[8] P. Sauer, O. Schirmer, and J. Schneider, phys. stat. sol. **16**, 79 (1966).
[9] F. S. Ham, Phys. Rev. Letters **28**, 1048 (1972).
[10] F. S. Ham, in: Paramagnetic Resonance, Ed. S. Geschwind, Plenum Press, New York 1972 (p. 106).
[11] G. F. Koster, J. O. Dimmock, R. G. Wheeler, and H. Statz, Properties of the Thirty Two Point Groups, MIT Press, Cambridge (Mass.) 1963.
[12] See e.g. J. Appel, in: Solid State Phys. **21**, 193 (1968).
[13] C. H. Henry and C. P. Slichter, in: Physics of Color Centers, Ed. W. P. Fowler, Academic Press, 1968.
[14] M. Lax, J. chem. Phys. **20**, 1752 (1952).
[15] M. J. L. Sangster, G. Peckham, and D. H. Saunderson, J. Phys. C **3**, 1026 (1970).
[16] A. Hervé and G. Rius, to be published.

(Received November 30, 1973)

phys. stat. sol. (b) 62, 393 (1974)
Subject classification: 6 and 8; 22.5.2

Institut Max von Laue — Paul Langevin, Grenoble

Dispersion Relations of Mode Grüneisen Parameters in Rubidium Halides

By
H. JEX[1])

Dispersion relations of microscopic mode Grüneisen parameters have been calculated from an evaluation of a shell model to third order. Leading terms such as the anharmonic shell–shell interaction between first and second neighbours and Coulomb contributions are taken into account. In the phonon dispersion curves breathing terms are included. The thermal expansion and mean Grüneisen constants of RbF, RbCl, RbBr, and RbI are analysed with this model.

Die Dispersionsrelationen der mikroskopischen Grüneisen-Parameter von RbF, RbCl, RbBr und RbJ wurden aus einem anharmonisch erweiterten Schalenmodell berechnet. Als Hauptbeiträge werden die anharmonische Schale–Schale-Wechselwirkung zwischen ersten und zweiten Gitternachbarn und die Coulomb-Beiträge berücksichtigt. Das Modell für die Gitterdynamik enthält außerdem die „breathing"-Anteile. Die mikroskopischen Grüneisen-Parameter der Rubidiumhalogenide werden mit experimentellen Daten aus optischen, Ultraschall- und Neutronenmessungen verglichen. Der lineare thermische Ausdehnungskoeffizient und der mittlere Grüneisen-Parameter werden ebenfalls analysiert.

1. Introduction

The phonon dispersion of rubidium halides has been investigated with coherent inelastic neutron scattering by Rolandson and Raunio [1]. Most of their data have been obtained at a sample temperature of 80 K; but also room temperature values have been reported including the temperature-induced shifts of the phonon energies. Some interesting aspects have been worked out, which are not completely understood up to now. There is a violance of the Lyddane-Sachs-Teller relation in RbCl at 80 K and there are significant discrepancies between the elastic constants determined from neutron scattering and ultrasonic experiments. These deviations are about 30% for c_{12} in RbBr and RbI and also 30% for c_{44} in RbI. Part of this discrepancy may be attributed to the transition from the zero-sound regime in the neutron experiments to the first-sound regime of the ultrasonic experiments. Loidl [2] has estimated this effect from a calculation of the mass operators and anharmonic renormalization of the phonon energies in the long-wavelength limit. The RbBr dispersion seems to be very sensitively influenced by the polarizabilities.

Rolandson and Raunio [1] have also pointed out that the specific heat and Debye temperatures determined from integrations over their one-phonon density of states should be remeasured with more precision. Some integrated quantities such as mean square vibrational amplitudes and thermal expansion at high temperatures from 300 to 700 K have been investigated recently by Stamm [3] for RbCl, RbBr, and RbI. However, anharmonic properties con-

---

[1]) Permanent address: Institut für Kernphysik der Universität Frankfurt am Main, FRG.

nected with the phonon dispersion are rather incompletely known from experiment. Thermal expansion and Grüneisen constants are available in the low-temperature range for the following materials. Schuele and Smith [4] and White [5] have reported mean Grüneisen constants and linear thermal expansion coefficients of RbI and other alkali halides. In particular White determined negative values for RbI and RbBr below 10 K. These negative values were attributed to the negative values of the individual mode Grüneisen parameters in the X-direction of the Brillouin zone. James and Yates [6] and Cooper and Yates [7] have determined mean Grüneisen constants of RbBr and RbCl, respectively, by experiment. Additional data for RbCl, RbBr, and RbI are available from measurements of the pressure dependence of the elastic constants by Chang and Barsch [8].

Pressure experiments have also been applied to the rubidium halides in order to test their stability against phase transformations. Vukcevich [9] has explained these results in more detail. It was found that RbCl, RbBr, and RbI perform a polymorphic phase transition from the NaCl structure to the CsCl structure. The connection between the throughout negative mode Grüneisen parameters of the TA branches in the [100] direction of the Brillouin zone has been pointed out. No phase transition could be observed in RbF [9].

Phonon energy shifts under hydrostatic pressure leading to the individual mode Grüneisen parameters have been investigated in most cases at the Γ-point from ultrasonic experiments in RbCl, RbBr, and RbI by Chang and Barsch [8] and for RbCl by Voronov et al. [10]. Optical values at the Γ-point are reported for RbI in [13] and by Lowndes (cited in [12]). No data are available for RbF. There is also a neutron scattering experiment from the TA phonon at the X-point of the Brillouin zone in RbI by Saunderson [11]. A large negative value of $\gamma_{TA}(X) = -3.32$ was deduced from the scattering data. The aim of these experiments was to look for relations between the phase transition and the phonon energy renormalization under hydrostatic pressure. No soft mode was observed at the X-point.

A theoretical interpretation of the individual mode Grüneisen parameters and of the averaged Grüneisen constants and the coefficients of linear thermal expansion has been given within simple shell models with pressure-dependent input parameters. Barsch and Achar [12] have investigated selected phonons in the Brillouin zone in their very first paper, while Namjoshi et al. [13] and again Barsch and Achar [12] afterwards have calculated the complete dispersion of the mode Grüneisen parameters. Namjoshi et al. have restricted their calculation to RbI and some other alkali halides. Although the negative mode Grüneisen parameters of the TA branches in the [100] direction were shown to be responsible for the negative thermal expansion of RbI below 10 K, this treatment seems to be not sufficient, because the shell model used is too simple to predict the measured phonon energies satisfactorily. This has been achieved by Kress [14] with a breathing shell model which is in excellent agreement with the phonon data. This model then was taken for a calculation of the dispersion of the mode Grüneisen parameters and the integrated quantities of RbI. Macroscopic input parameters and their pressure derivatives were used equivalently to derive the interior model parameters.

In this article the breathing shell model was also used for all rubidium halides as a starting model to represent the measured phonon dispersion with sufficient agreement. However, the dispersion of the mode Grüneisen parameters was evaluated by restricting to the leading terms in a generalization of the breathing shell model to anharmonic terms. The aim was to minimize the number of introduced new parameters and by the way having an applicable method when the set of pressure derivatives is not known from experiment. Leading contributions are the anharmonic shell–shell interaction between first and second neighbours and the long-ranged Coulomb interaction. These contributions are determined from two new parameters that may be adjusted to represent the mean Grüneisen constants or the coefficient of linear thermal expansion over the total temperature range. This method was applied to LiH and ^{7}LiD by Jex [15] where the pressure dependence of the macroscopic input parameters was missing. In most cases the second-neighbour contributions may be neglected. The calculation is then restricted to the leading terms and avoids the six pressure derivatives of the macroscopic input parameters involved in Namjoshi's et al. [13] calculation or the nine parameters introduced by Kress [14].

Besides the anharmonic shell–shell interaction involved in this paper recently also anharmonic shell–core force constants have been proposed for the explanation of Raman scattering experiments. However, it has been pointed out by Haberkorn et al. [16], that this intra-ionic anharmonicity may be less dominant for alkali halides rather than, e.g., for MgO. These parameters and the anharmonic inter-ionic ones obtained from mode Grüneisen parameters may be important for further calculations of the phonon lifetimes and energy renormalization with realistic well determined input data. They may also be involved in Loidl's [2] calculations at long phonon wavelengths where e.g. the third-order ladder diagrams have been summed up to infinite order when looking for the difference between the first- and zero-sound regime.

2. Theory

In a microscopic picture the coefficient of linear thermal expansion and the mean Grüneisen constants are determined by an average over the isothermal logarithmic volume change of the lattice modes $\omega(\lambda)$ characterized by quantum numbers $\lambda = \begin{pmatrix} q \\ j \end{pmatrix}$. This volume derivative may equivalently be defined by the pressure derivative and the isothermal bulk modulus B^T:

$$\gamma(\lambda) = -\left[\frac{\partial \ln \omega(\lambda)}{\partial \ln V}\right]_T = B^T \left[\frac{\partial \ln \omega(\lambda)}{\partial p}\right]_T .$$

This quantity may be obtained from the equation of motion after applying linear perturbation theory. In linear perturbation the eigenvectors $\boldsymbol{e}(k|\lambda)$ remain unchanged,

$$\omega^2(\lambda) = \sum_{\substack{\alpha\beta \\ kk'}} e_\alpha^*(k|\lambda)\, D_{\alpha\beta}\begin{pmatrix} \boldsymbol{q} \\ kk' \end{pmatrix} e_\beta(k'|\lambda)\,,$$

$$\gamma(\lambda) = -\frac{1}{2\omega^2(\lambda)} \sum_{\substack{\alpha\beta \\ kk'}} e_\alpha^*(k|\lambda) \left[V \frac{\mathrm{d}}{\mathrm{d}V} D_{\alpha\beta}\begin{pmatrix} \boldsymbol{q} \\ kk' \end{pmatrix}\right] e_\beta(k'|\lambda)\,.$$

If all volume or pressure dependences of the macroscopic parameters involved in the dynamical matrix are known, the Grüneisen parameters may be determined from the above equation with the maximum knowledge available. However, in many cases it is necessary to treat only the leading terms because of the reduced information available. As a guide for their selection we choose the importance of the equivalent terms in the description of the lattice dynamics. The leading terms will then be the short-range shell–shell interaction between first neighbours, the Coulomb term, the second-neighbour shell–shell interaction, and dipole and breathing terms. From an analysis of Raman scattering experiments [16] it was concluded that the anharmonic shell–core interaction seems to be less important for alkali halides than, e.g., for MgO.

The short-range part may be evaluated in terms of the Fourier transformed third potential derivatives [17, 18]

$$\gamma_{\mathrm{R}}(\lambda) = -\frac{2}{3\omega(\lambda)} \sum_{\alpha\beta} V_{\alpha\beta}(\lambda, -\lambda, -)\,\delta_{\alpha\beta}\,,$$

where

$$V_{\gamma\gamma}(\lambda, -\lambda, -) = \\ = \frac{1}{4\omega(\lambda)}\,\frac{1}{2} \sum_{\substack{\alpha\beta\\ kk'}} \{\varphi^{kk'}_{\alpha\beta\gamma,\gamma}(0)\,[m_k^{-1} e_\alpha(k|\lambda)\, e^*_\beta(k|\lambda) + m_{k'}^{-1} e_\alpha(k'|\lambda)\, e^*_\beta(k'|\lambda)] - \\ - \varphi^{kk'}_{\alpha\beta\gamma,\gamma}(\boldsymbol{q})\,(m_k m_{k'})^{-1/2}\,[e_\alpha(k|\lambda)\, e^*_\beta(k'|\lambda) + e_\alpha(k'|\lambda)\, e^*_\beta(k|\lambda)]\}$$

and

$$\varphi^{kk'}_{\alpha\beta\gamma,\gamma}(\boldsymbol{q}) = \sum_{l-l'} \varphi_{\alpha\beta\gamma}\begin{pmatrix} ll' \\ kk' \end{pmatrix} x_\gamma \begin{pmatrix} ll' \\ kk' \end{pmatrix} \mathrm{e}^{i\boldsymbol{q}\boldsymbol{x}\binom{ll'}{kk'}}\,,$$

m_k are the ion masses. The third derivative of the potential function $\varphi = \varphi(r)$ is given by

$$\varphi_{\alpha\beta\gamma} = \frac{x_\alpha x_\beta x_\gamma}{r^3}\left(\varphi''' - \frac{3}{r}\varphi'' + \frac{3}{r^2}\varphi'\right) + \frac{1}{r^2}\left(\varphi'' - \frac{1}{r}\varphi'\right)(x_\alpha\delta_{\beta\gamma} + x_\beta\delta_{\gamma\alpha} + x_\gamma\delta_{\alpha\beta})\,.$$

We restrict to the NaCl type lattices. The quantity $\varphi^{kk'}_{\alpha\beta\gamma,\gamma}(\boldsymbol{q})$ may then be evaluated from the following components by permutation of the indices:

$$\varphi^{kk'}_{\alpha\alpha\alpha,\alpha}(\boldsymbol{q}) = (1 - \delta_{kk'})\, a\varphi''' \cos\frac{1}{2} q_\alpha a\,,$$

$$\varphi^{kk'}_{\alpha\beta\beta,\beta}(\boldsymbol{q}) = 0\,,$$

$$\varphi^{kk'}_{\alpha\alpha\beta,\beta}(\boldsymbol{q}) = (1 - \delta_{kk'})\, 2\left(\varphi'' - \frac{1}{r}\varphi'\right) \cos\frac{1}{2} q_\beta a\,,$$

$$\varphi^{kk'}_{\alpha\beta\gamma,\gamma}(\boldsymbol{q}) = 0\,.$$

This is the contribution from the first neighbours; a is the lattice constant. The second-neighbour short-range contribution may be evaluated from the following components:

$$\varphi^{kk'}_{\alpha\alpha\alpha,\alpha}(\boldsymbol{q}) = \delta_{kk'}\left[\frac{a}{\sqrt{2}}\varphi''' + 3\left(\varphi'' - \frac{1}{r}\varphi'\right)\right] \times \\ \times \left[\cos\frac{1}{2} q_\alpha a \cos\frac{1}{2} q_\beta a + \cos\frac{1}{2} q_\alpha a \cos\frac{1}{2} q_\gamma a\right],$$

$$\varphi^{kk'}_{\alpha\alpha\beta,\beta}(\boldsymbol{q}) = \delta_{kk'}\left[\frac{a}{\sqrt{2}}\varphi''' - \left(\varphi'' - \frac{1}{r}\varphi'\right)\right] \cos\frac{1}{2} q_\alpha a \cos\frac{1}{2} q_\beta a\,,$$

$$\varphi^{kk'}_{\alpha\beta\beta,\beta}(\boldsymbol{q}) = \delta_{kk'}\left[\frac{a}{\sqrt{2}}\varphi''' - \left(\varphi'' - \frac{1}{r}\varphi'\right)\right]\sin\frac{1}{2}q_\alpha a \sin\frac{1}{2}q_\beta a\,,$$

$$\varphi^{kk'}_{\alpha\beta\gamma,\gamma}(\boldsymbol{q}) = 0\,.$$

The remaining elements are obtained after cyclic permutation of the indices. By the way three new parameters φ'''_{12}, φ'''_{11}, and φ'''_{22} have been introduced, while the second and first derivatives may be determined from A_{12}, B_{12}, A_{11}, B_{11}, A_{22}, and B_{22} introduced in the usual shell model analysis. In the above equations the indices k, k' referring the sublattices have been omitted from φ''', φ'', φ' for the reason of simplicity.

The Coulomb part of the dynamical matrix can be ruled out explicitely; the result may also be obtained much easier when differentiating the e^2/V_a dependence and keeping the charges Z_k fixed. This contribution to the Grüneisen parameter reads as follows:

$$\gamma_C(\lambda) = \frac{1}{2\omega^2(\lambda)}\frac{e^2}{V_a}\sum_{\substack{\alpha\beta\\ kk'}} e^*_\alpha(k|\lambda)\, Z_k m_k^{-1/2} C_{\alpha\beta}\begin{pmatrix}\boldsymbol{q}\\ kk'\end{pmatrix} m_{k'}^{-1/2} Z_{k'} e_\beta(k'|\lambda)\,,$$

where $C_{\alpha\beta}\begin{pmatrix}\boldsymbol{q}\\ kk'\end{pmatrix}$ is the Coulomb matrix introduced. This may also be a guide, when the volume dependence of the ionic charge or shell contributions are to be taken into account.

$$\tilde{\gamma}_C(\lambda) = -\frac{1}{2\omega^2(\lambda)}\frac{e^2}{V_a}\sum_{\substack{\alpha\beta\\ kk'}} e^*_\alpha(k|\lambda)\, m_k^{-1/2}\left[V\frac{\mathrm{d}Z_k}{\mathrm{d}V} C_{\alpha\beta}\begin{pmatrix}\boldsymbol{q}\\ kk'\end{pmatrix} Z_{k'} + \right.$$

$$\left. + Z_k C_{\alpha\beta}\begin{pmatrix}\boldsymbol{q}\\ kk'\end{pmatrix} V\frac{\mathrm{d}Z_{k'}}{\mathrm{d}V}\right] m_{k'}^{-1/2}\, e_\beta(k'|\lambda)\,.$$

In this article we have neglected these terms.

3. Numerical Calculations

The numerical calculations have been performed with the breathing shell model for the lattice dynamics of the rubidium halides. Its application is discussed in detail by Kress [14]. The parameters involved in the evaluation of the third-order coefficient $V_{\alpha\beta}(\lambda, -\lambda, -)$ are listed in Table 1. First and second derivatives were taken from the breathing shell model fits. From this model, then, the dispersion relations of the mode Grüneisen parameters may be

Table 1

Transition pressures in kbar and third potential derivatives in $e^2/2V_a r$ for rubidium halides

	RbF	RbCl	RbBr	RbI
transition pressure [9]	—	4.9	4.5	4.0
NaCl–CsCl [10]	—	6.5 to 7	—	—
third potential derivatives				
φ'''_{12}	−94.92	−143.1	−147.2	−158.9
φ'''_{11}	0	0	0	0
φ'''_{22}	0	0	− 4.86	− 5.12
r (Å)	2.815	3.2971	3.445	3.629

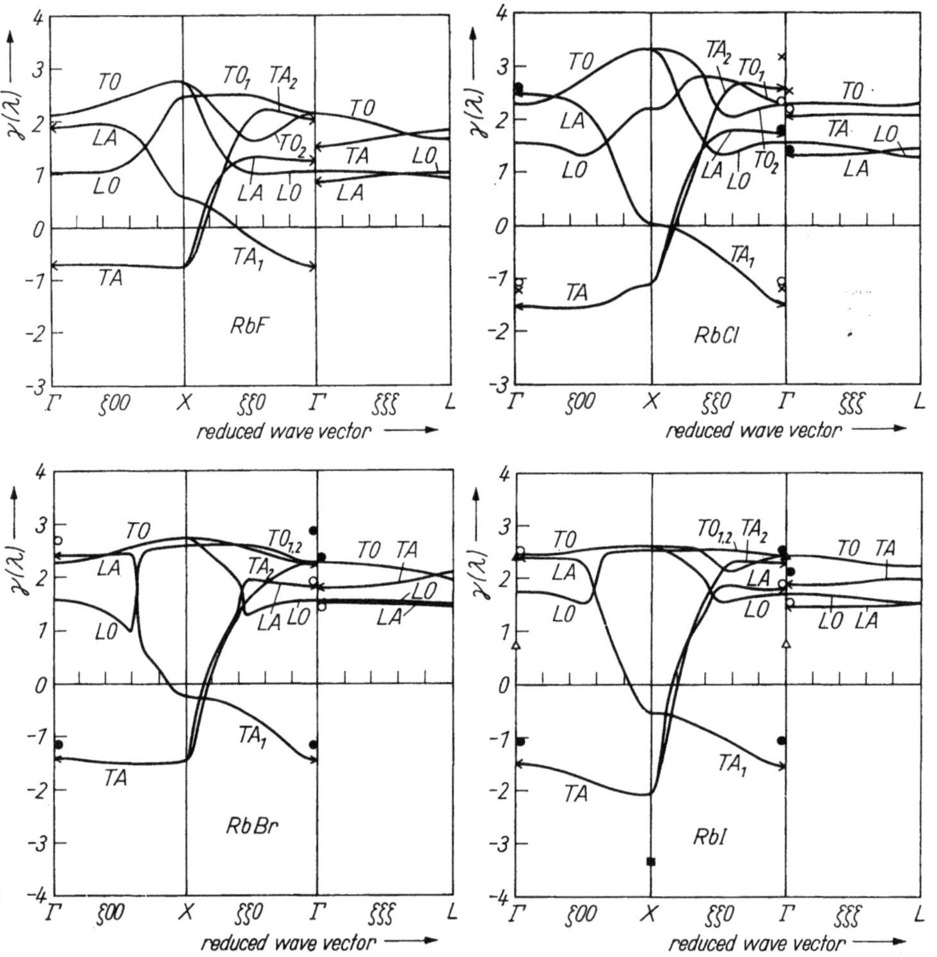

Fig. 1. Dispersion of mode Grüneisen parameters $\gamma(\lambda)$ for rubidium halides. Ultrasonic data by Chang and Barsch [8] (● transverse and ○ longitudinal modes), Voronov et al. [10] (× transverse modes), ▲ transverse optical data cited in Namjoshi et al. [13], △ longitudinal optical data by Lowndes cited in [12], and ■ neutron experiment by Saunderson [11]

obtained (Table 2 and Fig. 1). The discrepancy between the theoretical calculation and experiments is in most cases less than 10% for RbCl, RbBr, and RbI. No data are available for RbF; therefore we expect these Grüneisen parameters to be more uncertain. Also, averaged quantities are not available to compare with. The discrepancy seems to be larger for the TA branch in RbI. Specially the dispersion is not sufficient to fit the value at the Γ-point and Saunderson's [11] neutron experiment at the X-point. The same difficulties are pointed out by Barsch and Achar [12] with the six-parameter shell model and they arise as well with the nine-parameter breathing shell model. It seems that the measured value at the X-point is too negative; when one tries to move the TA branch towards more negative values, then the averaged behaviour gives too large negative mean Grüneisen constants or thermal expansion coefficients when compared with White's [5] experiments. The γ(TA) values at the X-point are listed in Table 3; there is satisfactory agreement with Barsch and Achar's [12] calculation also in the general shape of the mode Grüneisen dispersion curves. Since the transition pressures listed in Table 3 are rather low one might think of non-linear pressure effects being important in the neutron experiments. This may be estimated from Barsch and Achar's data by assuming the pressure derivative of the individual $\gamma_{TA}(X)$ being the same as for the averaged values. Taking the non-linear data from Table 3 one obtains

Table 2

Theoretical and experimental [8] mode Grüneisen parameters at the Γ-point

		acoustic modes							optical modes	
		100		110			111			
		TA	LA	TA1	TA2	LA	TA	LA	TO	LO
RbF	exp.	—	—	—	—	—	—	—	—	—
	theor.	−0.72	1.93	−0.72	2.07	1.24	1.55	0.92	2,17	1.06
RbCl	exp.	−1.20 a)	—	−1.20 a)	3.14 a)	—	2.55 a)	—	—	—
		−1.05	2.62	−1.05	2.37	1.86	2.23	1,44	—	—
	theor.	−1.64	2.48	−1.64	2.57	1.74	2.06	1.33	2.27	1.56
RbBr	exp.	−1.13	2.70	−1.13	2.89	1.92	2.39	1.48	—	—
	theor.	−1.42	2.42	−1.42	2.27	1.85	1.82	1.52	2.30	1.57
RbI	exp.	−1.15 b)								
		−1.06	2.53	−1.06	2.56	1.91	2.15	1.54	2.37 c)	0.76 d)
	theor.	−1.52	2.37	−1.52	2.29	1.80	1.88	1.47	2.45	1.71

a) Voronov et al. [10].
b) Saunderson [11].
c) Namjoshi et al. [13].
d) Cited by Achar and Barsch [12] (phys. stat. sol.).

Table 3

Characteristic data of rubidium halides: thermal expansion $\alpha(T)$ in 10^{-6} deg^{-1}, bulk modulus B^T in 10^{11} dyn cm^{-2}, and transition pressure in kbar

	$\gamma_{TA}(x)$		$\bar{\gamma}$ (T = 300K)		p_{trans}	α (T = 300K)		B^T	$B^T\left[\frac{d\bar{\gamma}}{dp}\right]_T$
	exp.	theor.	exp.	theor.	exp.	exp.	theor.	exp.	exp.
RbF	—	−0.76	—	1.37	—	—	26.4	—	—
RbCl	—	−1.36	1.45	1.56	6.5 to 7*)	35.4	37.7	1.5582	−2.8
		−1.79**)			4.9				
RbBr	—	−1.48	1.47	1.52	4.5	37.6	39.0	1.3016	−2.6
		−1.45**)							
RbI	−3.32	−2.06	1.51	1.51	4.0	39.5	41.7	1.0532	−2.4
		−2.32**)							
ref.	[11]		[8]		[9]	[8]		[8]	[8]

*) Cited in [10].
**) Cited in [12].

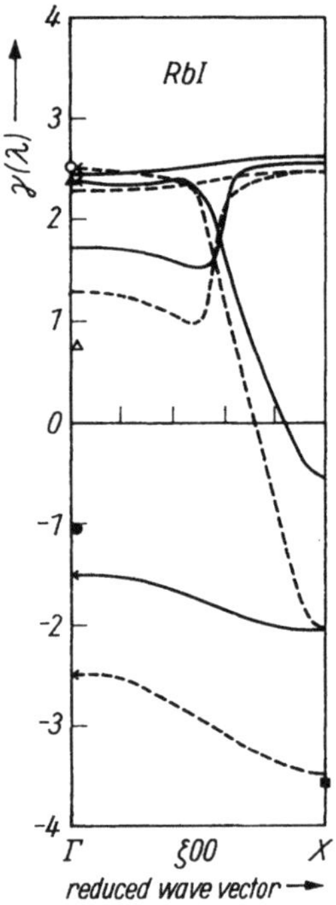

Fig. 2. Dispersion of the mode Grüneisen parameter $\gamma(\lambda)$ of RbI including (———) and excluding (— — —) Coulomb interaction

a correction $\Delta\gamma_{TA}(X) = -0.07$ for a pressure of 4 atm which cannot explain the full discrepancy.

In the RbI calculation two free anharmonic parameters have been used, one for the first neighbours and the other one for the second-neighbour interaction, which was restricted to the I^-–I^- ions. The contributions arising from the Coulomb anharmonicity may be demonstrated in Fig. 2. The shifts of the TA and LA branch look similar to the results discussed by Cowley and Cowley [17] for KBr. The order of magnitude of the three effects taken into account in this calculation are the first-neighbour contribution scaled by 100, the Coulomb part giving 30% correction, and the second neighbours contributing of the order of 10%. Since the input parameters where also chosen to achieve reasonable agreement with the mean Grüneisen constants we have a check for the applicability of a model to represent averaged quantities and differential behaviour as well. The averaged Grüneisen constants are presented in Fig. 3a to d for RbF, RbCl, RbBr, and RbI, respectively. The experimental data are rather complete for RbI only. From the scattering of these experimental results, however, one may get a feeling for the errors. The agreement seems to be

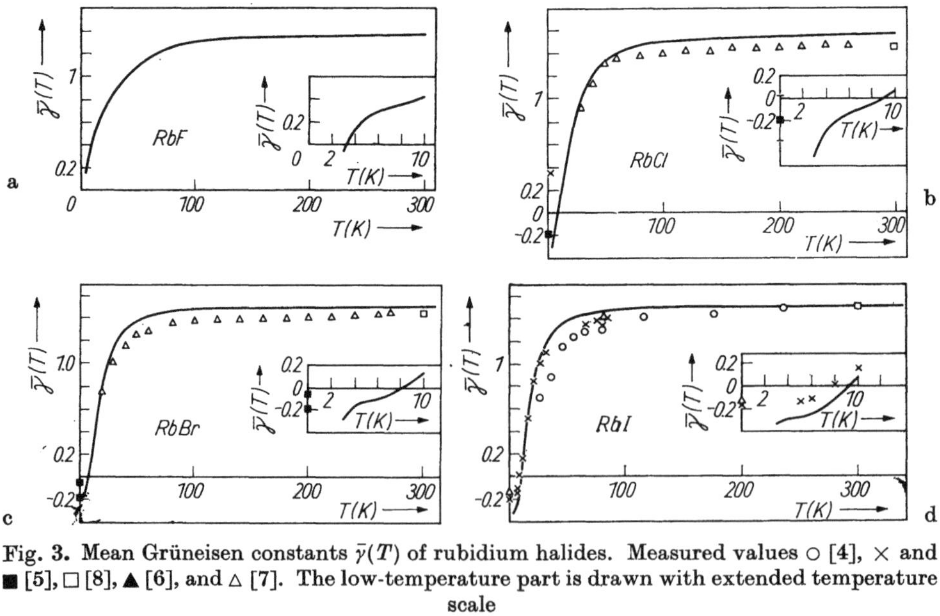

Fig. 3. Mean Grüneisen constants $\bar{\gamma}(T)$ of rubidium halides. Measured values ○ [4], × and ■ [5], □ [8], ▲ [6], and △ [7]. The low-temperature part is drawn with extended temperature scale

satisfactory. In the low-temperature region below 10 K ($\approx \theta/10$) the Grüneisen constant becomes negative. However, it is too negative indicating that the TA branches are too negative. The averaged data have been obtained from a fine mesh of 916 equally spaced q-points in the irreducible part of the Brillouin zone. This may be sufficient for all calculations down to 4 K while below that temperature the theoretical curve slopes off steeply as may be observed in the low-temperature parts of Fig. 3 drawn separately. At these temperatures the statistics is no longer sufficient and one needs a continuum theory. Negative $\bar{\gamma}(T)$ may also be expected for RbCl and RbBr in agreement with the measurements and estimates made by White [5] for $\bar{\gamma}(T = 0)$. The experimental data of RbCl by Cooper and Yates [7] and of RbBr by James and Yates [6] deviate of the order of 7% from the theoretical predictions. However, more refined results of RbBr obtained by White (1968) [5] indicate that the discrepancy is reduced for this crystal. If one tries to fit these data the agreement with the mode Grüneisen parameters becomes worse. The room temperature value of the Grüneisen constants $\bar{\gamma}(T = 300\text{ K})$ is listed in Table 3. The deviation from Chang and Barsch [8] is less than 7%. There is one value for RbF of $\bar{\gamma} = 1.28$ cited in the book of Born and Huang [20] which seems to be a little low, we have obtained $\bar{\gamma} = 1.37$. However, these data are rather uncertain up to now.

Thermal expansion data have been evaluated from

$$\alpha(T) = \frac{1}{3B^T V} \sum_{\lambda} C(\lambda, T)\, \gamma(\lambda)$$

by the way avoiding the specific heat data which are not known with sufficient accuracy as has been pointed out by Rolandson and Raunio [1] from their analysis of the frequency spectra obtained from their neutron scattering data. In this article we compare only the room temperature data of $\alpha(T)$ listed in Table 3. The agreement with Chang and Barsch [8] is of the order of 6%. $C(\lambda, T)$ is the specific heat function.

4. Discussion

The phonon dispersion of the rubidium halides has been studied with inelastic neutron scattering by Rolandson and Raunio [1]. A careful analysis of these data has elucidated some results not yet understood which may be connected with anharmonicity. There is a discrepancy between the ultrasonic- and the neutron-determined elastic constants, the breakdown of the Lyddane-Sachs-Teller relation at 80 K in RbCl, the strong dependence of the phonon dispersion fit on the polarizabilities in RbBr. In addition energy renormalization between 80 and 300 K scattering data are available.

In this article we have studied those anharmonic interactions connected with thermal strain. From the individual mode Grüneisen parameters, from the mean Grüneisen constants, and from the coefficients of thermal expansion anharmonic force constants have been deduced which may be of interest for further calculations. The aim was to restrict to the leading terms and to minimize the number of parameters introduced. We have discussed the first- and second-neighbour shell–shell short-range interaction and the long-ranged Coulomb part. Recently [16] the importance of the intra-ionic shell–core anharmonicity has been pointed out for an explanation of the Raman spectra in MgO. Its importance for the mode Grüneisen parameters is an open question. Some aspects of the cation–anion influences on the mode Grüneisen parameters and Grüneisen constants are elucidated in the article of Barsch and Achar (1971) [12]. The long-ranged Coulomb contributions seem to be rather important for the TA and LA branches. From this result it can be concluded that Coulomb anharmonicity is not negligible in phonon lifetime calculations and in energy renormalization due to temperature change. Some results of Knohl [19] emphasize this for optical phonons at the Γ-point in LiF. It was found by Knohl that the Coulomb contribution is destructive. It is desirable to include these contributions also for phonons with wave vector $\boldsymbol{q} \neq 0$.

When comparing with the shell model [12, 13] and the breathing shell model calculations [14], the treatment outlined in this article gives satisfactory results when the number of parameters is regarded. A critical discussion with reference [14] will be published in [21]. Within the frame of all models proposed so far, it was not possible to reproduce a sufficiently large dispersion on the γ(TA) in the X-direction of RbI in order to fit the values measured at the Γ-point and at the X-point.

From this present analysis it seems desirable to have more data from this branch and perhaps further neutron measurements of mode Grüneisen parameters are stimulated.

Acknowledgements

I would like to thank the director of the ILL-Grenoble, Prof. Dr. R. Mössbauer, for the hospitality of the institute. Stimulating discussions with Dr. W. Kress are kindly acknowledged.

References

[1] S. Rolandson and G. Raunio, J. Phys. C 4, 958 (1971); J. Phys. C 3, 1013 (1970); phys. stat. sol. **40**, 749 (1970).
[2] A. Loidl, private communication.
[3] H. Stamm, Diplomarbeit, Universität Frankfurt, 1974.
[4] D. E. Schuele and C. S. Smith, J. Phys. Chem. Solids **25**, 801 (1964).
[5] G. K. White, Proc. Roy. Soc. **A286**, 204 (1965); Proc. 11th Internat. Conf. Low-Temp. Phys., St. Andrews 1968, Vol. 1, Ed. J. F. Allen, D. M. Finlayson, and D. M. McCall (p. 541).
[6] B. W. James and B. Yates, Phil. Mag. **12**, 253 (1965).
[7] R. F. Cooper and B. Yates, J. Phys. C 4, L113 (1971).
[8] Z. P. Chang and G. R. Barsch, J. Phys. Chem. Solids **32**, 27 (1971).
[9] M. R. Vukcevich, phys. stat. sol. (b) **54**, 435 (1972).
[10] F. F. Voronov, V. A. Goncharova, and T. A. Agapova, Soviet Phys. — Solid State 8, 2726 (1966).
[11] D. H. Saunderson, Phys. Rev. Letters **17**, 530 (1966).
[12] G. R. Barsch and B. N. N. Achar, phys. stat. sol. **35**, 881 (1969).
B. N. N. Achar and G. R. Barsch, Phys. Rev. B 3, 4352 (1971); phys. stat. sol. (a) **6**, 247 (1971).
[13] K. V. Namjoshi, S. S. Mitra, and J. F. Vetelino, Phys. Rev. B 3, 4398 (1971).
[14] W. Kress, phys. stat. sol. (b) **62**, 403 (1974).
[15] H. Jex, J. Phys. Chem. Solids, to be published.
[16] R. Haberkorn, M. Buchanan, and H. Bilz, Solid State Commun. **12**, 681 (1973).
[17] E. R. Cowley and R. A. Cowley, Proc. Roy. Soc. **A287**, 259 (1965).
[18] H. Jex, phys. stat. sol. (b) **45**, 343 (1971).
[19] U. Knohl, phys. stat. sol. (b) **53**, 295 (1972).
[20] M. Born and K. Huang, Dynamical Theory of Crystal Lattices, Clarendon Press, Oxford 1966 (p. 52).
[21] W. Kress and H. Jex, phys. stat. sol. (b), to be published.

(Received November 21, 1973)

phys. stat. sol. (b) 62, 403 (1974)

Subject classification: 6 and 8; 12.2; 22.5.2

Institut Max von Laue — Paul Langevin, Grenoble, and Physik-Department E13 der Technischen Universität München

Calculation of Phonon Dispersion Curves and Mode Grüneisen Parameters in RbI

By

W. KRESS

The phonon dispersion curves of RbI are calculated using the breathing shell model, allowing for the force constants of the isotropic deformation to be unequal to the shell-core force constants. The model is extended to include the proper third order anharmonic contributions. The mode Grüneisen parameters are calculated within this model using first order perturbation theory. The parameters entering the calculation of the mode Grüneisen parameters are the isothermal pressure derivatives of the harmonic breathing shell model parameters and are obtained from the pressure derivatives of the long wavelength measurements in a similar way as are the harmonic parameters. Using this model the specific heat and the linear thermal expansion coefficient as functions of temperature are calculated.

Les courbes de dispersion de phonons sont calculées pour le RbI en utilisant un „breathing shell model" où les constantes de force de la déformation isotropique peuvent prendre des valeurs différentes des forces coquille-coeur. Le modèle est généralisé de façon à inclure toutes les contributions anharmoniques du troisième ordre. Les paramètres de Grüneisen individuels sont calculés à partir de ce modèle en utilisant une perturbation du premier ordre. Les paramètres intervenant dans le calcul des paramètres de Grüneisen individuels sont les dérivées isothermiques par rapport à la pression des paramètres harmoniques du „breathing shell model". Ils sont obtenus d'une manière similaire à celle des paramètres harmoniques, à partir des dérivées par rapport à la pression des quantités mesurées aux grandes longueurs d'onde. Utilisant ce modèle, la chaleur spécifique et le coefficient linéaire de dilatation thermique ont été calculés en fonction de la température.

1. Introduction

During recent years the lattice dynamics of the alkali halides have been studied both experimentally and theoretically by many authors. Rubidium iodide seems to be one of the most interesting substances of this set and was chosen as an example for the present investigation for the following reasons.

Of the alkali halides showing NaCl structure at atmospheric pressure, RbI has the largest electronic polarizabilities for both the Rb^+ and the I^- ions [1]. The treatment of these polarizabilities by an adequate model for the calculation of the phonon dispersion relations therefore seems to be important.

The mode Grüneisen parameter for the long wavelength transverse acoustical (TA) phonon in the Δ-direction of the Brillouin zone (BZ), determined from the measurement of the pressure dependence of the elastic constant c_{44}, has a large negative value compared with other alkali halides [2 to 4].

Below 8 °K the linear thermal expansion coefficient is negative [5]. Therefore the mode Grüneisen parameter for low-energy phonons is expected to be negative in an extended region of the BZ.

The phonon dispersion curves of RbI have been measured and calculated by Raunio and Rolandson [12]. The calculation is based on a shell model [13, 14] where short-range forces up to second-nearest neighbours are taken into account. These forces are assumed to be axially symmetric and to act only through the shells. In general this leads to eleven parameters which were determined by a least-squares fit in the main symmetry directions of the BZ. This procedure leads to nice agreement with the measurements [12], except for the elastic constants measured by ultrasonic techniques [15], where the difference arises because of anharmonic effects leading to different sound velocities in the first- and zero-sound regime [16].

Within the quasiharmonic approximation the mode Grüneisen parameters can be obtained from the phonon dispersion curves calculated at different pressures. Since a least-squares fit is necessary to obtain the eleven parameters of the shell model mentioned above, such a model cannot be employed for calculations of the mode Grüneisen parameters as there are insufficient data to make a fit. Barsch and Achar [3] therefore used a simplified six-parameter shell model for the calculation of the mode Grüneisen parameters. Namjoshi et al. [17] did similar calculations. Unfortunately this model leads only to qualitative agreement with the measurements [12] if it is used to calculate the phonon dispersion curves [17].

2. The Breathing Shell Model

The breating shell model (BSM) was first proposed by Schröder [20] and used to calculate the phonon dispersion curves of many ionic crystals [20 to 23]. The advantage of this model is that the parameters are determined from the elastic constants, the dielectric constants, and the transverse optical frequency $\omega_{TO}(\Gamma)$. Therefore no least-squares fit to all measured phonons is necessary in order to achieve good agreement with the measurements.

The BSM was chosen for the present calculations. This model allows for isotropic deformations of the ions, adding to the displacement coordinates of the centres of mass ("cores") and the centres of electronic charge distribution ("shells") a scalar electronic degree of freedom ("shell radius"). It is assumed that the short-range non-Coulomb forces act only via the shells. Within the quasi-harmonic approximation using the translation symmetry of the lattice and eliminating the electronic degrees of freedom using the adiabatic approximation the equation of motion is

$$M\omega^2(\boldsymbol{q})\, U(\boldsymbol{q}) = [\tilde{R} + ZCZ - (\tilde{R} + ZCY)\,(\tilde{R} + K + YCY)^{-1}\,(\tilde{R}^{+} + YCZ)]\, U(\boldsymbol{q})\,. \tag{2.1}$$

The matrix $\tilde{R}$ is given by

$$\tilde{R} = R - PG^{-1}P^{+}\,. \tag{2.2}$$

The short-range part R of the dynamical matrix is given by

$$R\begin{pmatrix}\boldsymbol{q}\\ \varkappa\varkappa'\end{pmatrix} = \sum_{N=0}^{NN} \sum_{i=1}^{I} D_{iN} \hat{R}\begin{pmatrix}N\\ \varkappa\varkappa'\end{pmatrix} D_{iN}^{+} e^{-i\boldsymbol{q}D_{iN}r\begin{pmatrix}N\\ \varkappa\varkappa'\end{pmatrix}}\,, \tag{2.3}$$

where N labels the neighbours shells ($N = 0$ contribution of a ion due to its own displacement, $N = 1$ nearest neighbours, $N = 2$ second-nearest neighbours) and $\varkappa$ counts the ions in the elementary cell ($\varkappa = 1$ alkali ion, $\varkappa = 2$ halide ion).

D_{iN} are the transformation operators creating the star of I neighbour positions of the shell of neighbours N if applied to $\mathsf{r}\begin{pmatrix} N \\ \varkappa\varkappa' \end{pmatrix}$. Assuming axially symmetric forces and using the notation

$$\left(\frac{\partial^2 V_{\varkappa\varkappa'}}{\partial r^2}\right)_{\parallel}\Bigg|_{r=r_0(\varkappa\varkappa')} = \frac{e^2}{v_a}\frac{A_{\varkappa\varkappa'}}{2}, \qquad \left(\frac{\partial^2 V_{\varkappa\varkappa'}}{\partial r^2}\right)_{\perp}\Bigg|_{r=r_0(\varkappa\varkappa')} = \frac{e^2}{v_a}\frac{B'_{\varkappa\varkappa'}}{2},$$

$$\frac{1}{r_0(\varkappa\varkappa')}\left(\frac{\partial V_{\varkappa\varkappa'}}{\partial r}\right)_{\parallel}\Bigg|_{r=r_0(\varkappa\varkappa')} = \frac{e^2}{v_a}\frac{B_{\varkappa\varkappa'}}{2}, \qquad \text{where} \quad A_{\varkappa\varkappa'},\ B_{\varkappa\varkappa'};\ B'_{\varkappa\varkappa'},$$

and $r_0(\varkappa\varkappa')$ are symmetric in $\varkappa\varkappa'$, for

$N = 1$:
$$\mathsf{r}\begin{pmatrix} 1 \\ \varkappa\varkappa' \end{pmatrix} = r_0(1, 0, 0)^T$$

$$\hat{\mathsf{R}}\begin{pmatrix} 1 \\ \varkappa\varkappa' \end{pmatrix} = -\frac{1}{2}\frac{e^2}{v_a}\begin{pmatrix} A_{\varkappa\varkappa'} & 0 & 0 \\ 0 & B'_{\varkappa\varkappa'} & 0 \\ 0 & 0 & B'_{\varkappa\varkappa'} \end{pmatrix},$$

for $N = 2$:
$$\mathsf{r}\begin{pmatrix} 2 \\ \varkappa\varkappa \end{pmatrix} = r_0(1, 1, 0)^T,$$

$$\hat{\mathsf{R}}\begin{pmatrix} 2 \\ \varkappa\varkappa \end{pmatrix} = -\frac{1}{4}\frac{e^2}{v_a}\begin{pmatrix} A_{\varkappa\varkappa} + B_{\varkappa\varkappa} & A_{\varkappa\varkappa} - B_{\varkappa\varkappa} & 0 \\ A_{\varkappa\varkappa} - B_{\varkappa\varkappa} & A_{\varkappa\varkappa} + B_{\varkappa\varkappa} & 0 \\ 0 & 0 & 2B_{\varkappa\varkappa} \end{pmatrix},$$

and for $N = 0$: $r\begin{pmatrix} 0 \\ \varkappa\varkappa \end{pmatrix} = (0, 0, 0)$ using the translation invariance of the lattice

$$R_{\alpha\beta}\begin{pmatrix} 0 \\ \varkappa\varkappa \end{pmatrix} = \frac{e^2}{v_a}\left(A_{\varkappa\varkappa'} + 2B'_{\varkappa\varkappa'} + 2A_{\varkappa\varkappa} + 4B'_{\varkappa\varkappa}\right)\delta_{\alpha\beta}$$

is obtained, where e is the electronic charge, $a = 2r_0$ the lattice constant, and v_a the volume of the elementary cell.

Assuming further that the shell–shell coupling and the deformation–shell coupling is due to the same force constants,

$$\mathsf{P}\begin{pmatrix} \boldsymbol{q} \\ \varkappa\varkappa' \end{pmatrix} = \mathsf{P}_0\begin{pmatrix} \boldsymbol{q} \\ \varkappa\varkappa \end{pmatrix} + \sum_{N=1}^{NN}\sum_{i=1}^{I} \mathsf{D}_{iN}\hat{\mathsf{R}}\begin{pmatrix} N \\ \varkappa\varkappa' \end{pmatrix}\frac{\mathsf{r}\begin{pmatrix} N \\ \varkappa\varkappa' \end{pmatrix}}{\left|\mathsf{r}\begin{pmatrix} N \\ \varkappa\varkappa' \end{pmatrix}\right|}\,\mathrm{e}^{-i\boldsymbol{q}\mathsf{D}_{iN}\mathsf{r}\begin{pmatrix} N \\ \varkappa\varkappa' \end{pmatrix}} \tag{2.4}$$

and

$$\mathsf{G}\begin{pmatrix} \boldsymbol{q} \\ \varkappa\varkappa' \end{pmatrix} = \mathsf{G}_0\begin{pmatrix} \boldsymbol{q} \\ \varkappa\varkappa \end{pmatrix} + \sum_{N=1}^{NN}\frac{\mathsf{r}^T\begin{pmatrix} N \\ \varkappa\varkappa' \end{pmatrix}\hat{\mathsf{R}}\begin{pmatrix} N \\ \varkappa\varkappa' \end{pmatrix}\mathsf{r}\begin{pmatrix} N \\ \varkappa\varkappa' \end{pmatrix}}{\left|\mathsf{r}\begin{pmatrix} N \\ \varkappa\varkappa' \end{pmatrix}\right|^2}\sum_{i=1}^{I}\mathrm{e}^{-i\boldsymbol{q}\mathsf{D}_{iN}\mathsf{r}\begin{pmatrix} N \\ \varkappa\varkappa' \end{pmatrix}} \tag{2.5}$$

are obtained. In the system already transformed to relative displacements

$$\mathsf{P}_0\begin{pmatrix} \boldsymbol{q} \\ \varkappa\varkappa \end{pmatrix} = 0$$

and

$$G_0\begin{pmatrix} \boldsymbol{q} \\ \varkappa\varkappa \end{pmatrix} = \frac{e^2}{v_a}(G_\varkappa + 3A_{\varkappa\varkappa'} + 6A_{\varkappa\varkappa}) , \tag{2.6}$$

where $G_\varkappa$ is the force constant for the isotropic deformation of the ion $\varkappa$. The Coulomb matrix is denoted by C and Z, Y are the diagonal matrices

$$Z_{\alpha\beta}(\varkappa\varkappa') = Z_\varkappa \delta_{\alpha\beta}\delta_{\varkappa\varkappa'} , \qquad Y_{\alpha\beta}(\varkappa\varkappa') = Y_\varkappa \delta_{\alpha\beta}\delta_{\varkappa\varkappa'}$$

of the ionic charges and the shell charges, respectively. The shell charges are assumed to be equal [24]. M and K are the diagonal matrices of the atomic masses and the force constants $k_\varkappa$ for the coupling of the shells to its own core, respectively.

The parameters of the model were determined by the following relations:

$$c_{11} = \frac{1}{2r_0}\frac{e^2}{v_a}[-5.112\,Z^2 + A_{12} + A_{22} + B_{22} - E] , \tag{2.7}$$

$$c_{12} = \frac{1}{2r_0}\frac{e^2}{v_a}\left[1.392\,Z^2 + \frac{1}{2}(A_{22} - B_{22}) - B_{12}'' - E\right], \tag{2.8}$$

$$c_{44} = \frac{1}{2r_0}\frac{e^2}{v_a}\left[1.392\,Z^2 + \frac{1}{2}(A_{22} - B_{22}) + B_{12}''\right], \tag{2.9}$$

where $B_{12}'' = B_{12}' - B_{12}$ is the non-central part of the nearest-neighbour force constant. The central part B_{12} has been eliminated using the stability condition

$$B_{12} + 2B_{22} = -\frac{2}{3}\alpha_M Z^2 , \tag{2.10}$$

where α_M is the Madelung constant and

$$E = \frac{A_{12}^2 G_1 + 4\sqrt{2}A_{12}A_{22}G_1 + A_{12}^2(G_2 + 12A_{22}) + 8A_{22}^2(G_1 + 3A_{12})}{G_1G_2 + 3A_{12}(G_1 + G_2) + 12A_{22}(G_1 + 3A_{12})} \tag{2.11}$$

is the contribution of the isotropic deformation to the elastic constants.

In the approximation $k_\varkappa \gg A_{12} + 2B_{12}'$, the electronic polarizabilities $\alpha_\varkappa$ are given by

$$\alpha_\varkappa = \frac{Y_\varkappa^2 v_a}{k_\varkappa + A_{12} + 2B_{12}'} . \tag{2.12}$$

From the reststrahlen frequency ω_R and the dielectric constants ε_0 and ε_∞ two additional conditions

$$Z'^2 = \frac{9}{4\pi}\frac{v_a}{e^2}\mu\omega_R^2\frac{\varepsilon_0 - \varepsilon_\infty}{(\varepsilon_\infty + 2)^2} \tag{2.13}$$

and

$$R_0' = \frac{v_a}{e^2}\mu\omega_R^2\frac{\varepsilon_0 + 2}{\varepsilon_\infty + 2} \tag{2.14}$$

are obtained, where $\mu = m_1m_2/(m_1 + m_2)$ is the reduced mass,

$$Z' = Z + \frac{R_0(Y_2k_1 - Y_1k_2)}{k_1k_2 + R_0(k_1 + k_2)} , \tag{2.15}$$

$$R_0' = R_0 - \frac{R_0^2(k_1 + k_2)}{k_1 k_2 + R_0(k_1 + k_2)}, \tag{2.16}$$

and R_0 is given by $R_0 = A_{12} + 2B_{12}'$.

Neglecting the second-nearest neighbour interactions between the alkali ions, the phonon dispersion curves of RbI were calculated and are shown along the symmetry directions in Fig. 1.

The agreement between the measurements [12] and the calculations is good. The differences are a few per cent which is comparable with the experimental errors.

Compared with the calculations of Namjoshi et al. [17] the agreement of the LA and the TA branch in the Δ-direction is considerably improved. Improvements were obtained for the T_1A branch in the Σ-direction and the LO and

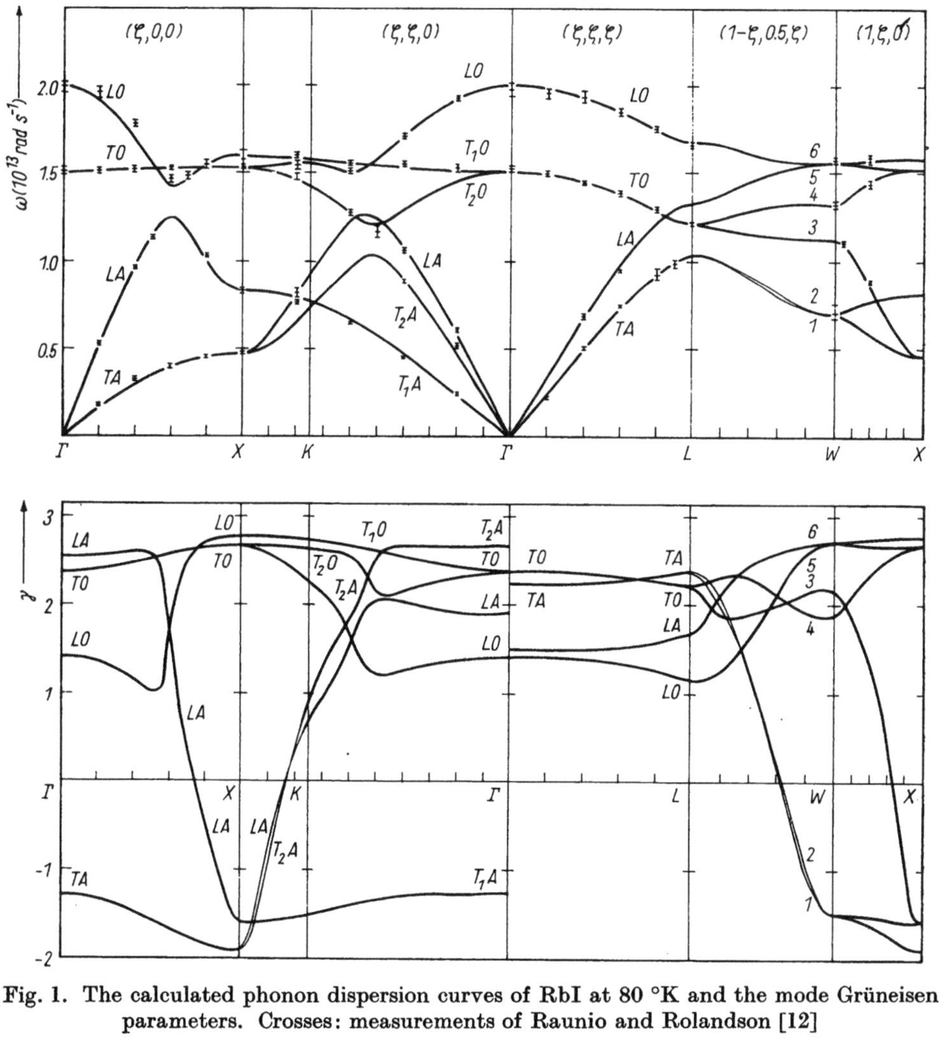

Fig. 1. The calculated phonon dispersion curves of RbI at 80 °K and the mode Grüneisen parameters. Crosses: measurements of Raunio and Rolandson [12]

LA branches in the Λ-direction. The dispersion curves presented here are similar to those obtained by Raunio and Rolandson [12] using an eleven-parameter shell model, fitted by a least-squares method to their measurements. The deviations are of the same size and sign.

Table 1

Input parameters used for the calculation of the phonon dispersion curves

m_1 (at. units)	m_2 (at. units)	r_0 (Å)	c_{11} (10^{11} dyn/cm^2)	c_{12} (10^{11} dyn/cm^2)	c_{44} (10^{11} dyn/cm^2)
85.467	126.904	3.642	3.430	0.47	0.362

ε_0	ε_∞	ω (10^{13} rad s^{-1})	α_1 (Å^3)	α_2 (Å^3)	Z (e)	F_2
4.57	2.6018	1.51	1.797	6.199	0.87953	3

The input parameters used for the calculation are given in Table 1, where α_1, α_2, and ε_∞ are related by the Clausius-Mosotti formula

$$\alpha_1 + \alpha_2 = \frac{3}{4\pi} v_a \frac{\varepsilon_\infty - 1}{\varepsilon_\infty + 2} \tag{2.17}$$

and the splitting into the polarizabilities of the two ions in the elementary cell was taken from the paper of Tessman et al. [1]. Because of the big difference of the sound velocities in the first- and zero-sound regime, the elastic constants c_{ij} have been fitted to the neutron data [12]. In Table 2 the values c_{ij} used in the calculations are compared with the ultrasonic measurements [15] and with the values used by Raunio and Rolandson [12] in different models.

Table 2

Comparison of the elastic constants in units of 10^{11} dyn/cm^2 used in the present calculation with those given by Raunio and Rolandson [12] and the ultrasonic results of Lewis et al. [15]

	present	model 1 [12]	model 2 [12]	model 3 [12]	ultrasonic [15]
c_{11}	3.430	3.374	3.422	3.424	3.061
c_{12}	0.470	0.406	0.446	0.489	0.35
c_{44}	0.362	0.381	0.382	0.371	0.289

In contrast to Schröder's original assumption that the breathing force constant $G_\varkappa$ and the core–shell force constant $k_\varkappa$ are equal, an improvement for the longitudinal phonons at the L-point was achieved with $G_2 = 3k_2$. A factor of this order of magnitude is also indicated by calculations of the second-order Raman spectra in some alkali halides [26]. The isotropic deformation of the Rb^+ ions is small and its contribution to the dispersion curves is negligible. For convenience $G_1 = k_1$ was assumed since k_1 is large.

3. Calculation of the Mode Grüneisen Parameters

Applying hydrostatic pressure p to a perfect crystal results in an isotropic homogeneous deformation of the lattice. Since the RbI lattice has inversion symmetry no sublattice displacements $u(\varkappa)$ are induced applying hydrostatic pressure, and the relative change of the lattice constant $\Delta r_0/r_0$ may be used to specify the internal strain. The dependence of the phonon frequencies $\omega\begin{pmatrix} q \\ j \end{pmatrix}$ on this external parameter is usually defined by the mode Grüneisen parameter

$$\gamma\begin{pmatrix} q \\ j \end{pmatrix} = -\frac{\mathrm{d} \ln \omega\begin{pmatrix} q \\ j \end{pmatrix}}{\mathrm{d} \ln v_a}. \tag{3.1}$$

In order to calculate the mode Grüneisen parameters the dynamical matrix is expanded with respect to the relative change in volume $\mathrm{d} \ln v_a$ at constant temperature T:

$$D = D_0 + D_1 \mathrm{d} \ln v_a\,. \tag{3.2}$$

Here D_0 is the dynamical matrix in the quasi-harmonic approximation at the temperature T. Expanding the eigenfrequencies ω^2 in the same way,

$$\omega^2 = \omega_0^2 + \omega_1^2 \,\mathrm{d} \ln v_a \tag{3.3}$$

and using first-order perturbation theory the mode Grüneisen parameter

$$\gamma\begin{pmatrix} q \\ j \end{pmatrix} = -\frac{1}{2} e_0^+\begin{pmatrix} q \\ j \end{pmatrix} D_0^{-1}\begin{pmatrix} q \\ \varkappa\varkappa' \end{pmatrix} D_1 \begin{pmatrix} q \\ \varkappa\varkappa' \end{pmatrix} e_0\begin{pmatrix} q \\ j \end{pmatrix} \tag{3.4}$$

is obtained. The matrix D_1 is given by

$$D_1 = \tilde{R}_1 - \tilde{T}_1\tilde{S}_0^{-1}\tilde{T}_0^+ - \tilde{T}_0\tilde{S}_0^{-1}\tilde{T}_1^+ + \tilde{T}_0\tilde{S}_0^{-1}\tilde{S}_1\tilde{S}_0^{-1}\tilde{T}_0^+\,, \tag{3.5}$$

where

$$\tilde{T} = R + ZCY - PG^{-1}P^+\,, \qquad \tilde{S} = R + YCY + K - PG^{-1}P^+\,,$$

$$\tilde{R}_1 = R_1 + Z_1C_0Z_0 + Z_0C_0Z_1 + Z_0C_1Z_0 - P_1G_0^{-1}P_0 - P_0G_0^{-1}P_1 + P_0G_0^{-1}G_1G_0^{-1}P_0 \tag{3.6}$$

and $\tilde{T}_1$ and $\tilde{S}_1$ are given by analogous expressions. Since $r\begin{pmatrix} 1 \\ \varkappa \end{pmatrix}$ and q are co- and contravariant vector components the product $r\begin{pmatrix} 1 \\ \varkappa \end{pmatrix}q$ is invariant. The Coulomb matrices C are, apart from the factor v_a^{-1}, volume-independent, and the short-range force constant matrices $\hat{R}_1\begin{pmatrix} N \\ \varkappa\varkappa' \end{pmatrix}$ contain the derivatives of the force constants with respect to $\mathrm{d} \ln v_a$ at constant temperature T taken at the equilibrium position $r = r_0(\varkappa\varkappa')$ and are apart from the factor $\frac{1}{3}r_0$ the third-order anharmonicities of the potential.

In the long wavelength acoustical limit the mode Grüneisen parameter is constant and given by

$$\gamma_{ij} = -\frac{1}{6} - \frac{1}{2}\frac{\mathrm{d} \ln c_{ij}}{\mathrm{d} \ln v_a}\bigg|_T\,. \tag{3.7}$$

The parameters of the perturbation matrix D_1 are determined by the derivatives of (2.7) to (2.14).

$$\gamma_{11} = \frac{1}{2}\left[1 - \frac{e^2}{2r_0v_a}\frac{1}{c_{11}}(-10.224\, Z\dot{Z} + \dot{A}_{12} + \dot{A}_{22} + \dot{B}_{22} - \dot{E})\right], \tag{3.8}$$

$$\gamma_{12} = \frac{1}{2}\left[1 - \frac{e^2}{2r_0v_a}\frac{1}{c_{12}}(2.784\, Z\dot{Z} + \frac{1}{2}(\dot{A}_{22} - \dot{B}_{22}) - \dot{B}_{12}'' - \dot{E})\right], \tag{3.9}$$

$$\gamma_{44} = \frac{1}{2}\left[1 - \frac{e^2}{2r_0v_a}\frac{1}{c_{44}}\left(2.784\, Z\dot{Z} + \frac{1}{2}(\dot{A}_{22} - \dot{B}_{22}) + \dot{B}_{12}^{||}\right)\right], \tag{3.10}$$

$$\dot{\alpha}_\varkappa = \alpha_\varkappa\left[1 + 2\frac{\mathrm{d}\ln Y_\varkappa}{\mathrm{d}\ln v_a}\bigg|_T - \frac{\mathrm{d}\ln}{\mathrm{d}\ln v_a}(k_\varkappa + A_{12} + 2B_{12}')\bigg|_T\right], \tag{3.11}$$

$$\dot{Z}' = Z'\left[\frac{1}{2} + \frac{\mathrm{d}\ln\omega_R}{\mathrm{d}\ln v_a}\bigg|_T + \frac{1}{2}\frac{\mathrm{d}\ln}{\mathrm{d}\ln v_a}(\varepsilon_0 - \varepsilon_\infty)|_T - \frac{\mathrm{d}\ln}{\mathrm{d}\ln v_a}(\varepsilon_\infty + 2)|_T\right], \tag{3.12}$$

$$\dot{R}' = R_0\left[1 + 2\frac{\mathrm{d}\ln\omega_R}{\mathrm{d}\ln v_a}\bigg|_T + \frac{\mathrm{d}\ln}{\mathrm{d}\ln v_a}(\varepsilon_0 - 2)|_T - \frac{\mathrm{d}\ln}{\mathrm{d}\ln v_a}(\varepsilon_\infty + 2)|_T\right] \tag{3.13}$$

is obtained where the dot denotes the operator $\mathrm{d}/\mathrm{d}\ln v_a|_T$. The input parameters used for the calculation of the mode Grüneisen parameters of RbI are given in Table 3. The values γ_{ij} were taken from [4], [27], and [28].

Table 3

Input parameters used for the calculation of the mode Grüneisen parameters

$\gamma_{11} = 2.55$	$\gamma_{44} = -1.27$	$\frac{\mathrm{d}\ln\varepsilon_0}{\mathrm{d}\ln v}\Big\|_T = 1.41$	$\frac{\mathrm{d}\alpha_{Rb}}{\mathrm{d}\ln v}\Big\|_T = 0.286$ Å^3
$\gamma_{12} = 1.86$	$\gamma_{TO}(\Gamma) = 2.37$	$\frac{\mathrm{d}n_\infty}{\mathrm{d}\ln v}\Big\|_T = -0.407$	$\frac{\mathrm{d}\alpha_I}{\mathrm{d}\ln v}\Big\|_T = 3.438$ Å^2

Since in the present model the polarizabilities of both the rubidium and the iodine ions are included it is necessary to split the change of the polarizability due to isothermal relative volume changes $\mathrm{d}\alpha/\mathrm{d}\ln v_a|_T$ into anion and cation contributions:

$$\frac{\mathrm{d}\alpha}{\mathrm{d}\ln v_a}\bigg|_T = \frac{\mathrm{d}\alpha_1}{\mathrm{d}\ln v_a}\bigg|_T + \frac{\mathrm{d}\alpha_2}{\mathrm{d}\ln v_a}\bigg|_T. \tag{3.14}$$

Using the Clausius-Mosotti equation (2.17)

$$\frac{\mathrm{d}\alpha}{\mathrm{d}\ln v_a}\bigg|_T = \alpha_1 + \alpha_2 + \alpha\frac{6\sqrt{\varepsilon_\infty}}{(\varepsilon_\infty - 1)(\varepsilon_\infty + 2)}\frac{\mathrm{d}n}{\mathrm{d}\ln v_a}\bigg|_T \tag{3.15}$$

is obtained.

The pressure dependence of the refractive index n at the wavelength $\lambda_D = 5893$ Å was measured for many alkali halides [29 to 35]. Table 4b shows the change of electronic polarizability due to the relative change in volume calculated from the measurements using the dielectric constants ε_D given by Tessman et al. [1]. In the third column the values of the electronic polarizability obtained as the sum of the ionic polarizabilities fitted by Tessman et al. [1] are given.

Table 4a

Derivatives of the electronic polarizabilities with respect to volume changes of alkali ions and halide ions for $\lambda = \lambda_D$

Ion	F^-	Cl^-	Br^-	I^-	Li^+	Na^+	K^+	Rb^+
$\left.\frac{d\bar{\alpha}_D}{d \ln v_a}\right\|_T$ (Å^3)	0.454	1.655	2.278	3.438	0.021	0.275	0.205	0.286

Table 4b

Measured dielectric constants, averaged electronic polarizabilities, and their derivatives with respect to volume changes for $\lambda = \lambda_D$

compound	ε_D [1]	$\bar{\alpha}_D$ (Å^3) [1]	$\left.\frac{d \ln \alpha_D}{d \ln v_a}\right\|_T$
LiF	1.936	0.673	0.706 [33],
NaF	1.758	1.052	0.693 [35],
NaCl	2.385	3.368	0.573 [35], 0.586 [31]
KCl	2.221	4.294	0.428 [35], 0.419 [31]
KBr	2.432	5.492	0.454 [35], 0.470 [31]
KI	2.779	7.765	0.471 [35], 0.520 [31]
RbCl	2.231	4.939	0.398 [34], 0.395 [32]
RbBr	2.411	6.137	0.416 [35], 0.409 [32]
RbI	2.714	8.410	0.441 [32]

These polarizabilities have been used to calculate the third term of (3.15). Splitting this term into anion and cation contributions the values of Table 4a have been obtained.

For LiF, NaF, and NaCl agreement is obtained since no cross check was possible with the data given in Table 4b. For the Rb^+ and K^+ ions a check was possible and the data given in [35] and [34] together with the RbI value of [32] are found to show more consistency than the values of [31]. To determine the absolute values of $[d\alpha/d \ln v_a]_T$ it was assumed that

$$\frac{\alpha_{Li}}{\alpha_F} = \left.\frac{d\alpha_{Li}}{d \ln v_a}\right|_T \left[\left.\frac{d\alpha_F}{d \ln v_a}\right|_T\right]^{-1}.$$

Since the polarizabilities and their derivatives are small for LiF compared with the values for RbI the deviations from this assumption will lead only to small corrections in the case of RbI. Since for the first three compounds no cross check was possible the values for Li^+, Na^+, and F^- may have considerable errors.

It is further supposed that the values obtained for λ_D are an adequate approximation for the λ_∞-values of the polarizability changes which enter the theory. For consistency reasons the value $(dn_\infty/d \ln v_a)|_T$ given in Table 3 has been calculated from the split values of $(d\bar{\alpha}/d \ln v_a)|_T$.

The value $(d \ln \varepsilon_0/d \ln v_a)|_T$ was taken from Jones [36] and Lowndes [37] and the value $\gamma_{TO}(\Gamma)$ as cited by Namjoshi et al. [17]. The calculated mode Grüneisen parameters in the symmetry directions of the BZ are given in Fig. 1.

4. Calculation of the Linear Thermal Expansion Coefficient

Within the quasiharmonic approximation [7] the mode Grüneisen parameter $\gamma\begin{pmatrix} \boldsymbol{q} \\ j \end{pmatrix}$ is related to the Helmholtz free energy F by

$$-\left.\frac{\partial F}{\partial V}\right|_T = \frac{\hbar}{2V} \sum_{\boldsymbol{q}j} \gamma\begin{pmatrix} \boldsymbol{q} \\ j \end{pmatrix} \omega\begin{pmatrix} \boldsymbol{q} \\ j \end{pmatrix} \coth \frac{1}{2} \beta\hbar\omega\begin{pmatrix} \boldsymbol{q} \\ j \end{pmatrix} \tag{4.1}$$

and the Grüneisen parameter $\bar{\gamma}$ is obtained by averaging the mode Grüneisen parameters with the Einstein specific heat function

$$\bar{\gamma}(T) = \frac{\sum_{\boldsymbol{q}j} \gamma\begin{pmatrix} \boldsymbol{q} \\ j \end{pmatrix} c\left(T, \begin{matrix} \boldsymbol{q} \\ j \end{matrix}\right)}{\sum_{\boldsymbol{q}j} c\left(T, \begin{matrix} \boldsymbol{q} \\ j \end{matrix}\right)}, \tag{4.2}$$

where $c(\lambda) = \left(\frac{1}{2}\lambda\right)^2 \left(\sinh \frac{1}{2}\lambda\right)^{-2}$ with $\lambda = \beta\hbar\omega\begin{pmatrix} \boldsymbol{q} \\ j \end{pmatrix}$.

Using the thermodynamic relation

$$-\frac{\partial^2 F}{\partial V \partial T} = \frac{\partial p}{\partial T} = -\frac{\partial V/\partial T}{\partial V/\partial p} = 3\alpha(T)\, B_T \tag{4.3}$$

the linear thermal expansion coefficient $\alpha(T)$ is obtained:

$$\alpha(T) = \frac{k}{3B_T V} \sum_{\boldsymbol{q}j} \gamma\begin{pmatrix} \boldsymbol{q} \\ j \end{pmatrix} c\left(T, \begin{matrix} \boldsymbol{q} \\ j \end{matrix}\right). \tag{4.4}$$

Using the definition of the Grüneisen parameter $\bar{\gamma}$ equation (4.2) may be written in the form

$$\alpha(T) = \frac{\bar{\gamma}(T)\, c_V(T)}{3B_T(T)\, V(T)}, \tag{4.5}$$

where c_V is the specific heat at constant volume and B_T is the adiabatic bulk modulus.

The linear thermal expansion coefficient $\alpha(T)$ and the specific heat at constant volume were calculated using a mesh of q-points in the irreducible part of the BZ which corresponds to 32000 points in the BZ. The temperature dependence of the volume was taken into account by extrapolating the volume from one tem-

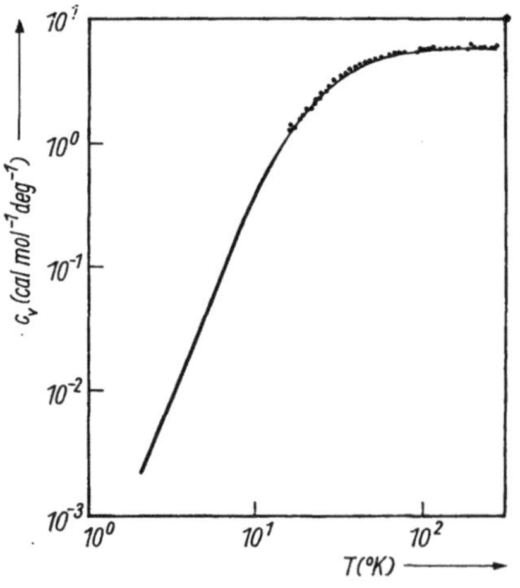

Fig. 2. The calculated specific heat of RbI. ● Measurements of Clusius et al. [9]

perature point to the next using the calculated expansion coefficient $\alpha(T)$. The temperature intervals where 5 deg between 300 and 20 °K and 1 deg below. Using the temperature dependence of the elastic constants measured by Haussühl [25] the adiabatic bulk modulus

$$B_S = \tfrac{1}{3}\,(c_{11}^S + 2c_{12}^S) \tag{4.6}$$

was obtained. The isothermal bulk modulus was calculated using the approximation

$$\frac{B_T}{B_S} = 1 - \frac{9\alpha^2 v_a B_T}{c_p}\,T \approx 1 - A c_V T\,. \tag{4.7}$$

The constant A was determined from the measurements at $T = 300$ °K.

The specific heat $c_V(T)$ was calculated and is compared in Fig. 2 with the measurements of Clusius et al. [9]. A double-logarithmic plot was favoured instead of plotting the temperature dependence of the Debye temperature $\theta_D(T)$ since the relative errors of the Debye temperature and the relative errors of the specific heat are related by a temperature-dependent factor $f(T)$:

$$\frac{\Delta\theta_D}{\theta_D} = -f(T)\,\frac{\Delta c_V}{c_V}\,. \tag{4.8}$$

For temperatures small compared to the Debye temperature $f(T)$ is $\tfrac{1}{3}$ where as in the high-temperature limit

$$f(T \gg \theta_D) \to 10\left(\frac{T}{\theta_D}\right)^2$$

is obtained. Since RbI has low Debye temperatures compared to other alkali halides the comparison with experimental results in a $\theta_D(T)$ plot may be misleading.

The Debye temperatures calculated from the present model are in agreement with the results of Raunio and Rolandson. This shows that both models give a quite similar description of the phonon spectrum.

In Fig. 3 the calculated linear thermal expansion coefficient $\alpha(T)$ is given together with the measurements of White [5] and those of Schuele and Smith [6]. Quite good agreement with the measurements of White was obtained over the

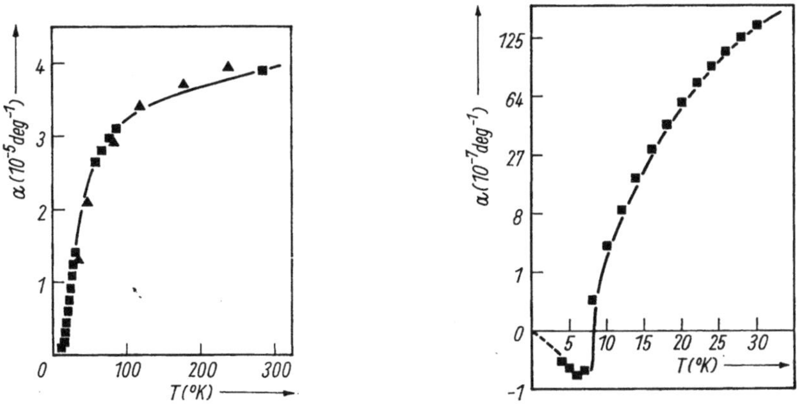

Fig. 3. The calculated linear thermal expansion coefficient of RbI. ■ Measurements of White [5], ▲ measurements of Schuele and Smith [6]

whole temperature region. In the low-temperature region a small shift in temperature of about 0.5 °K between the calculations and the measurements remains. This may be related to the use of the phonon spectrum at 80 °K for the calculation. Nevertheless the use of the phonon spectrum calculated with the breathing shell model seems to be a more realistic approach than the calculations with simplified shell model versions showing not negligible discrepancies with the measurements especially for the TA and LA branches in the Δ-direction of the BZ. In the low-temperature region these branches enter quite sensitively into the calculation of the linear thermal expansion coefficient. The negative values of $\alpha(T)$ are related to the negative mode Grüneisen parameters of the acoustic branches. For low temperatures the chosen mesh for the calculations should be fine enough to take into account the anisotropy of the elastic constants and the mode Grüneisen parameters in the region $\boldsymbol{q} \to 0$. In the present calculation this numerically critical region is estimated to be below 4 °K.

5. Discussion and Conclusion

Using the breathing shell model good agreement between the measured and calculated phonon dispersion curves was obtained. It is shown that the splitting of the isothermal pressure dependence of the electronic polarizability into contributions of individual ions is possible. This fact together with the knowledge of the isothermal isotropic pressure dependence of the adiabatic elastic constants (c_{11}^S, c_{12}^S, c_{44}^S), the dielectric constants (ε_0, ε_∞), and the reststrahlen frequency ω_R allowed the calculation of the mode Grüneisen parameters in the whole BZ. The results for the main symmetry directions are given in Fig. 1. From the experimental data and the Lyddane-Sachs-Teller [10] relation the mode Grüneisen parameters for $q = 0$ are completely determined. At the present time only one mode Grüneisen parameter for $\boldsymbol{q} \neq 0$ has been measured [11]. It was not possible to obtain agreement with this measured value of -3.32 for the TA phonon at the X-point of the BZ without changing γ_{44} which is taken from the ultrasonic measurements. The change of γ_{44} results mainly a shift of the whole branch of the TA mode Grüneisen parameters of the Δ-direction. The chosen value of γ_{44} in the present calculation is the lower limit of different experimental results for $(\mathrm{d}c_{44}^S/\mathrm{d}p)_T$.

It was found that the mode Grüneisen parameter for the LA branch at the X-point is strongly dependent on the change of ionic change $(\partial Z/\partial r)_T$. Smaller values of this parameter lead to a larger Grüneisen parameter. Consequently the mode Grüneisen parameter of the LA branch in the Δ-directions is falling more slowly and reaches the value zero at bigger q-values. The value $(\partial Z/\partial r)_T$ was adjusted to fit the linear thermal expansion coefficient at 283 °K.

Quite good agreement with the measured thermal expansion coefficient was found over the whole temperature region. To a certain extent this agreement proves that the calculated mode Grüneisen parameters are realistic. Nevertheless direct measurements of the mode Grüneisen parameters in the whole BZ would be desirable.

The parameters deduced from the experimental data are given in Table 5. The main contributions arise from the short-range force constants and the ionic charge. If a Born-Mayer repulsive potential $u(r) = \mathrm{B} \exp(-r/\varrho)$ to nearest neighbours is assumed, $\mathrm{d} \ln A_{12}/\mathrm{d} \ln r = -r_0/\varrho$. The value $-\mathrm{d} \ln A_{12}/\mathrm{d} \ln r = 10.36$ given in Table 5 in fact gives good agreement with $r_0/\varrho = 10.40$ obtained by Born and Huang [8] using the compressibility and the lattice constant

Table 5
Shell model parameters and their derivatives for RbI

	X	$\frac{\mathrm{d}X}{\mathrm{d}\ln r}$	$\frac{\mathrm{d}\ln X}{\mathrm{d}\ln r}$
A_{12}	$14.40\ e^2/v_a$	$-149.2\ e^2/v_a$	-10.36
B'_{12}	$-0.9981\ e^2/v_a$	$+15.84\ e^2/v_a$	-15.87
A_{22}	$0.7991\ e^2/v_a$	$-9.240\ e^2/v_a$	-11.56
B_{22}	$-0.1834\ e^2/v_a$	$-1.797\ e^2/v_a$	9.798
Z	$0.8795\ e$	$1.296\ e$	1.473
$Y_{(Y_1=Y_2)}$	$-3.888\ e$	$40.03\ e$	-10.30
k_1	$800.3\ e^2/v_a$	$-17002\ e^2/v_a$	-21.25
k_2	$223.2\ e^2/v_a$	$-5123\ e^2/v_a$	-22.96
E_1	1.000	0.000	0.000
E_2	3.000	0.000	0.000
d_1	$0.05934\ e$	$0.07690\ e$	1.296
d_2	$0.2047\ e$	$0.5085\ e$	2.484
α_1	$0.01860\ v_a$	$0.008880\ v_a$	0.4774
α_2	$0.06416\ v_a$	$0.1069\ v_a$	1.666

only. The change in ionic charge cancels partly the strengthening of the Coulomb interaction due to the change in the volume of the elementary cell ($\mathrm{d}\ln v_a^{-1}/\mathrm{d}\ln r = -3$). The core shell anharmonic contributions $\mathrm{d}\ln k_\varkappa/\mathrm{d}\ln r$ are of the same order of magnitude as the short range anharmonic contributions, but the influence on the dispersion of the mode Grüneisen parameters is weak since these contributions cancel partly with the change of the shell charge. This can be seen from the short-range polarizability changes $\mathrm{d}\ln d_\varkappa/\mathrm{d}\ln r$ and the polarizability changes $\mathrm{d}\ln \alpha_\varkappa/\mathrm{d}\ln r$, keeping in mind that the rigid part already gives the main contribution to the phonon dispersion and that these terms give rise to the change of the shell terms. Excluding these anharmonic terms related to the shells, however, it is impossible to describe the measurements for $\boldsymbol{q} = 0$ within the experimental errors. Restricting the anharmonic parameters to two leading terms Jex [38] calculated the dispersion relations of the mode Grüneisen parameters for the rubidium halides. A critical comparison of the methods employed here and in his paper as well as of the results obtained in both approximations will be given in a forthcoming paper [39].

Some anharmonic parameters determined in the present calculation could be used together with the fourth-order contributions in the calculation of the second order Raman effect as well as the phonon damping function $\Gamma(\omega)$ and the frequency-dependent phonon shift $\Delta(\omega)$.

Acknowledgements

I am gratefully indebted to Prof. H. Bilz for his interest and for discussions and to Prof. E. Lüscher for his support. The work was initiated by proposals of Drs. G. Quittner, R. Lechner, and J. Kalus and has benefitted from discussions with Dr. H. Jex. I wish to thank the Bundesministerium für Wissenschaft und Forschung, Bonn, for financial support. The critical reading of the manuscript by Dr. M. Buchanan is gratefully acknowledged.

References

[1] J. R. Tessman, A. H. Kahn, and W. Shockley, Phys. Rev. **92**, 890 (1953).
[2] W. B. Daniels and C. S. Smith, Papers Symp. Phys. Chem. High Pressures, London 1963 (No. 50).
[3] G. R. Barsch and B. N. N. Achar, phys. stat. sol. **35**, 881 (1969).
B. N. N. Achar and G. R. Barsch, phys. stat. sol. (a) **6**, 247 (1971).
[4] Z. P. Chang and G. R. Barsch, J. Phys. Chem. Solids **32**, 27 (1971).
[5] G. K. White, Proc. Roy. Soc. **A286**, 204 (1965).
[6] D. E. Schuele and C. S. Smith, J. Phys. Chem. Solids **25**, 801 (1964).
[7] G. Leibfried and W. Ludwig, Solid State Phys. **12**, 275 (1961).
[8] M. Born and K. Huang, Dynamical Theory of Crystal Lattices, Oxford University Press, Oxford 1954 (p. 26).
[9] K. Clusius, J. Goldmann, and A. Perlick, Z. Naturf. **4a**, 424 (1949).
[10] R. H. Lyddane, R. G. Sachs, and E. Teller, Phys. Rev. **59**, 673 (1941).
[11] D. H. Saunderson, Phys. Rev. Letters **17**, 530 (1966).
[12] G. Raunio and S. Rolandson, phys. stat. sol. **40**, 749 (1970).
[13] A. D. B. Woods, W. Cochran, and B. N. Brockhouse, Phys. Rev. **119**, 980 (1960).
[14] R. A. Cowley, W. Cochran, B. N. Brockhouse, and A. D. B. Woods, Phys. Rev. **131**, 1030 (1963).
For further references see: W. Cochran, CRC Critical Reviews in Solid State Sciences **2**, 1 (1971).
[15] J. T. Lewis, A. Lehoczky, and C. V. Briscoe, Phys. Rev. **161**, 877 (1967).
[16] R. A. Cowley, Proc. Phys. Soc. **90**, 1127 (1967).
[17] K. V. Namjoshi, S. S. Mitra, and J. V. Vetelino, Phys. Rev. B **3**, 4398 (1971).
[18] B. W. Jones, Phil. Mag. **16**, 1085 (1967).
[19] G. O. Jones, D. H. Martin, P. A. Mawer, and C. H. Perry, Proc. Roy. Soc. **A261**, 10 (1961).
[20] U. Schröder, Solid State Commun. **4**, 347 (1966).
[21] V. Nüsslein and U. Schröder, phys. stat. sol. **21**, 309 (1967).
[22] U. Schröder, Habilitationsschrift, München 1969.
[23] M. J. L. Sangster, G. Peckham, and D. H. Saunderson, J. Phys. C **3**, 1026 (1970).
[24] J. E. Hanlon and A. W. Lawson, Phys. Rev. **113**, 472 (1959).
[25] S. Haussühl, Z. Phys. **159**, 223 (1960).
[26] R. Haberkorn, private communication.
[27] M. Ghafelehbashi, D. P. Dandekar, and A. L. Ruoff, J. appl. Phys. **41**, 652 (1970).
[28] J. J. Fontanella and D. E. Schuele, J. Phys. Chem. Solids **31**, 647 (1970).
[29] E. Burstein and P. L. Smith, Phys. Rev. **74**, 229 (1948).
[30] T. S. Narasimhamarty, J. Opt. Soc. Amer. **51**, 914 (1961).
[31] K. G. Bansigir and K. J. Iyengar, Acta cryst. **14**, 727 (1961).
[32] K. G. Bansigir, Acta cryst. **23**, 505 (1967).
[33] K. Vedam and E. D. D. Schmidt, Solid State Commun. **3**, 373 (1965).
[34] K. Vedam and E. D. D. Schmidt, J. Mater. Sci. **1**, 310 (1966).
[35] K. Vedam, E. D. D. Schmidt, and J. L. Kirk (to be published in Phys. Rev. as cited in [3]).
[36] B. W. Jones, Phil. Mag. **16**, 1085 (1967).
[37] R. P. Lowndes, Proc. Roy. Soc. **A316**, 351 (1970).
[38] H. Jex, phys. stat. sol. (b) **62**, 393 (1974).
[39] W. Kress and H. Jex, phys. stat. sol. (b) (to be published).

(Received January 15, 1974)

phys. stat. sol. (b) 62, 417 (1974)

Subject classification: 14

Sektion Physik, Bergakademie Freiberg

On the Definition of the Current Operator Application to the Hubbard Model

By

J. MONECKE

A disagreement existing in the literature between various calculations of the electrical conductivity in the single-band Hubbard model is resolved. It is replied to an article by Cabib and Kaplan who tried to disprove that there are no absorption and emission lines near $\omega = U$ due to the translational invariance. The physical reason for this result and some of its consequences are discussed.

Es werden Widersprüche, die in der Literatur zwischen unterschiedlichen Berechnungen der elektrischen Leitfähigkeit im Ein-Band-Hubbardmodell auftreten, gelöst. Damit wird auf eine Arbeit von Cabib und Kaplan erwidert, in der zu widerlegen versucht wurde, daß auf Grund der Translationsinvarianz keine Absorptions- und Emissionslinien in der Nähe von $\omega = U$ auftreten können. Die physikalische Grundlage dieses Resultates und einige seiner Konsequenzen werden diskutiert.

1. Introduction

In order to explain the metal–insulator transitions in some transition metal compounds [1] there have been several attempts [2 to 8] to calculate the frequency-dependent conductivity $\sigma(\omega)$ in the one band Hubbard model [9] given by the Hamiltonian

$$H = \sum_{\sigma, i, j = \text{n.n. of } i} t_{i,j} c^+_{i,\sigma} c^-_{j,\sigma} + U \sum_i n_{i,\uparrow} n_{i,\downarrow} \,. \tag{1}$$

Beside that the results differ due to the different approximations used and due to their different ranges of applicability there remains one serious discrepancy. Whereas some authors (e.g. [2, 6]) found that the complex conductivity shows an absorption and emission spectrum with lines at $\omega = 0$ and $\omega = U$ (in the atomic limit) we proved [7] that the line at $\omega = U$ corresponding to an excitation across the Hubbard gap [9] occurs only due to approximations which do not take into account properly the phases of the interacting electrons and, therefore, violate the intrinsic translational invariance of the problem. We found a line at $\omega = 0$ only, the strength of which depends of course on the interaction. A metal-insulator transition, then, can be explained only by a jump in the optical mass [7]

$$\frac{1}{m_{xx}} = \frac{1}{N} \sum_{k,\sigma} \langle n_{k,\sigma} \rangle \frac{\partial^2 \varepsilon(\boldsymbol{k})}{\partial k_x^2} \tag{2}$$

as indicated also by some approximate calculations [8, 10, 11]. A similar result (with another expression for m_{xx}) was found by Brinkmann and Rice [12].

In a recent paper Cabib and Kaplan [13] tried to disprove the result of [7] pretending that we did use an incorrect (not translational invariant) expression for the current operator. Further they tried to prove [14] that the zero temperature dc conductivity vanishes. With this article we reply to their arguments. We discuss the physical reasons for the absence of excitations across the Hubbard gap and point out some consequences of this result.

2. How to Define the Current Operator?

In the case of one single band the expression for the total current operator is

$$\boldsymbol{j} = \frac{e}{i} [\boldsymbol{R}, H] = - ie \sum_{\sigma, i, j = \text{n.n. of } i} t_{i,j}(\boldsymbol{R}_i - \boldsymbol{R}_j)\, c^+_{i,\sigma} c^-_{j,\sigma} , \tag{3}$$

where the position operator $\boldsymbol{R}$ has been replaced by

$$\boldsymbol{R} = \sum_{i,\sigma} \boldsymbol{R}_i n_{i,\sigma} . \tag{4}$$

Below we prove that for potentials with an inversion centre $\boldsymbol{R}$ is diagonal in the Wannier representation and, therefore, is given by (4).

In the case of an infinite crystal this current obviously is translational invariant. Cabib and Kaplan tried to prove that this is not the case for a finite chain with periodic boundary conditions as well as for a ring.

Let us first consider the case of a chain with N atoms (denoted by the index $i = 1, \ldots, N$) with periodic boundary conditions. Supposing periodic boundary conditions means that we actually perform calculations for an infinite chain of $n \cdot N$ atoms ($n \to \infty$) which is divided into n periodicity volumes and that we consider one of these periodicity volumes as a representative of the finite chain. Because the summation over j in the expression for the current (3) involves terms with the indices $j = 0$ and $j = N + 1$ we have to answer the question how the matrix elements $X_i = \langle i| x |i\rangle$ have to be defined for $i = 0$ and $i = N + 1$ (due to the periodicity of the Wannier functions $\psi(x - x_{i+N}) = \psi(x - x_i)$ we have $c^\pm_{0,\sigma} = c^\pm_{N,\sigma}$ and $c^\pm_{N+1,\sigma} = c^\pm_{1,\sigma}$ for the operators). Because the Wannier functions are normalized with respect to one periodicity volume the X_i for the infinite chain of nN atoms are given by

$$X_i = \lim_{n\to\infty} \frac{1}{n} \int\limits_{(n \text{ period. vol.})} \psi^*(x - x_i)\, x \psi(x - x_i)\, \mathrm{d}x . \tag{5}$$

Using $\psi(x - x_{i+N}) = \psi\big((x - L) - x_i\big)$ we get

$$X_{i+N} = X_i + L , \tag{6}$$

where L is the length of one periodicity volume. With these X_i the current, given by (3), is translationally invariant and by Fourier transformation we get the usual expression

$$\boldsymbol{j} = \sum_{k,\sigma} \frac{\partial}{\partial \boldsymbol{k}} \varepsilon(\boldsymbol{k})\, n_{k,\sigma} . \tag{7}$$

Instead of $\psi(x - x_{i+N}) = \psi\big((x - L) - x_i\big)$ we could, however, use the periodicity of the Wannier functions

$$\psi(x - x_{i+N}) = \psi(x - x_i)$$

and would get from (5)

$$X_{i+N} = X_i .$$

Using these values of X_i the current (3) would not be translationally invariant.

There is no unique continuation for the matrix elements X_i outside the periodicity volume. The correct choice of the X_i for $i = 0$ and $i = N + 1$ has to be determined rather by physical than by mathematical arguments: the operator $\boldsymbol{R}$ occurs in the interaction term

$$H' = -e\boldsymbol{E}\boldsymbol{R}$$

of the Hamiltonian. Choosing the X_i according to $X_0 = X_N$, $X_{N+1} = X_1$ would mean to apply an inhomogeneous field to the infinite chain. Hence the right continuation is given by (6).

The situation in the case of a ring is essentially the same. Here we have

$$|\boldsymbol{j}| = -ie \sum_{\sigma, i, j\, =\, \text{n.n. of } i} t_{i,j}\, |\boldsymbol{R}_i - \boldsymbol{R}_j|\, c^+_{i,\sigma} c^-_{j,\sigma}\,, \tag{8}$$

where $\boldsymbol{R}_i$ is the vector from the centre to the i-th atom (a "direction" of the total current as in [13] has no meaning in the case of a ring, we have to do with an essentially linear problem). If we would apply a field to the ring which after one turn around the ring would jump back to its initial value we would get, of course, some strange current. In the case of a ring a homogeneous electric field can be produced by a linearly increasing or decreasing magnetic field perpendicular to the ring. In this case we get again a translationally (better: rotational) invariant current.

Whether or not the expression (3) for the total current is the correct one, hence, is only the question how the $\boldsymbol{R}_i$ are continued outside the periodicity volume. In both our papers [7, 8] we explicitly have used the correct choice.

3. Can There Be Absorption and Emission Lines near $\omega = U$?

The physical reason for the difference between the various results mentioned in the introduction is the following: A line near $\omega = U$ corresponds to an excitation across the Hubbard gap and has to be interpreted as being due to a dipole transition of an electron from an otherwise empty site to a site already occupied by another electron, this would just correspond to an increase in energy of U in the atomic limit. Our calculation [7], however, results in the conclusion that dipole transitions of this kind are forbidden.

The situation is essentially the same as in the case of noninteracting Bloch electrons. Due to the translational invariance we get for the matrix elements

$$\langle i|\, \boldsymbol{r}\, |j\rangle = \boldsymbol{R}(i,j) = \lim_{n\to\infty} \frac{1}{n} \int\limits_{(n \text{ period. vol.})} \psi^*(\boldsymbol{r} - \boldsymbol{r}_i)\, \boldsymbol{r}\psi\,(\boldsymbol{r} - \boldsymbol{r}_j)\, \mathrm{d}^3r \tag{9}$$

the relation

$$\boldsymbol{R}(i + 1, j + 1) = \boldsymbol{R}(i, j) + \boldsymbol{a}\delta_{ij}\,, \tag{10}$$

where a is the lattice spacing. For $i \neq j$ we have

$$\boldsymbol{R}(i, j) = R(i - j)\,. \tag{11}$$

For potentials with central symmetry the Wannier functions obey the relation [6, 15]

$$\psi(\boldsymbol{r} - \boldsymbol{r}_i) = \pm\psi^*(\boldsymbol{r}_i - \boldsymbol{r}) \tag{12}$$

from which we get

$$\boldsymbol{R}(i-j) = -\boldsymbol{R}(-j+i) = 0 \quad \text{if} \quad i \neq j\,. \tag{13}$$

The matrix $\langle i|\,\boldsymbol{r}\,|j\rangle$ is diagonal. Bloch electrons do not move simply due to dipole transitions. The electric field, coupled to the Hamiltonian by the term $H' = -e\boldsymbol{Er}$ does not change a Wannier function centered at the site i to another one centred at a neighbouring site. It alters, however, the weights of the basic Wannier states in the Bloch function,

$$\psi_{\boldsymbol{k}}(\boldsymbol{r}) = \frac{1}{\sqrt{N}} \sum_{i=1}^{N} e^{i\boldsymbol{kr}_i}\, \psi(\boldsymbol{r}-\boldsymbol{r}_i) = \frac{1}{n\sqrt{N}} \sum_{\text{all } i} e^{i\boldsymbol{kr}_i}\, \psi(\boldsymbol{r}-\boldsymbol{r}_i)\,,$$

in such a way that the probability of finding an electron is increased in the direction of lower potential energy:

$$-\,e\boldsymbol{E}\boldsymbol{r}\psi_{\boldsymbol{k}}(\boldsymbol{r}) = -e\boldsymbol{E}\,\frac{1}{n\sqrt{N}} \sum_{\text{all } i} \boldsymbol{R}_i\, e^{i\boldsymbol{kr}_i}\, \psi(\boldsymbol{r}-\boldsymbol{r}_i)\,, \quad \boldsymbol{R}_i = \langle i|\,\boldsymbol{r}\,|i\rangle\,. \tag{14}$$

The same is true in the case of interacting electrons (if we restrict ourselves to one single band). Any state $|n\rangle$ is a linear combination built up from the set

$$\{|i,0\rangle, |i,1\rangle, |i,2\rangle\} \tag{15}$$

of states with zero, one or two electrons at the i-th atom. Due to (6) and (13) the position operator is given by the expression

$$\boldsymbol{R} = \frac{1}{n} \sum_{\sigma, \text{all } i} \boldsymbol{R}_i n_{i,\sigma}\,, \tag{16}$$

where the factor $1/n$ provides the correct normalization to N atoms. It changes the weights of the basis states (15) but cannot perform a transition of the kind $|i,1\rangle \to |i+1,1\rangle$ or produce a state with one empty and one doubly occupied atom from two single occupied ones. The action of the perturbation $H' = -e\boldsymbol{ER}$, therefore, cannot cause an increase in energy of U in the atomic limit.

If we adopt the intuitive picture [11, 17, 18] that in the case of a half-filled s-band the charge carriers are the polar states $|i,0\rangle$ and $|i,2\rangle$ we see that these charge carriers can be moved, but not produced, by the external field. There is no excitation across the Hubbard gap and we have to expect an absorption line only at $\omega = 0$.

In the case of more than one band and/or degenerate bands the situation is much more complicate and a dipole transition of an electron from one site to another can be allowed by hybridization effects: The Hubbard subbands may be built up from states with different band index, the transition between them being allowed. There are indeed some experimental hints that such an "inter-site hop" occurs in α-$TiCl_3$ [1]. The situation may be similar as in the case of a normal exciton. An exciton state is caused by Coulomb interaction and built up by a hole state and an electron state from different bands. A dipole transition into an exciton state is allowed in general.

In the same way the difference in the low temperature dc conductivity between simple and transition metals may be qualitatively explained: The Coulomb interaction between s- and d-electrons in transition metals gives rise to the T^2-term in the resistivity. Due to the results of [7] the s–s interaction in simple metals cannot cause a similar term in the resistivity, and a T^2-behaviour which should be dominant at low temperatures never has been observed in simple metals, indeed [18].

The absence of dipole transitions between different eigenstates of the Hubbard Hamiltonian explained qualitatively above may be proved rigorously by the same type of arguments as in [7]:

Let us denote by P one of the lattice translation operators:

$$P^{-1}c^{\pm}_{i,\sigma}P = c^{\pm}_{i+1,\sigma}\,.$$

For the position operator R in the same direction we get

$$P^{-1}RT = \frac{1}{n}\sum_{\sigma,\,\mathrm{all}\,i} R_i\, n_{i+1,\sigma} = \frac{1}{n}\sum_{\sigma,\,\mathrm{all}\,i} (R_{i+1} - a)\, n_{i+1,\sigma} = R - aN\,,$$

where N is the number of electrons in one periodicity volume. If $|n\rangle$ are the common eigenstates of H and P we get

$$\langle n|\, R\, |m\rangle = \langle n|\, P^{-1}RP\, |m\rangle = \alpha^*_n\alpha_m\langle n|\, R\, |m\rangle$$

due to the conservation of the total number of electrons. We have, therefore, either $\alpha^*_n\alpha_m = 1$ or $\langle n|\, R|\, m\rangle = 0$. In the first case we write

$$\langle n|\, R|\, m\rangle = \frac{1}{n}\sum_{\sigma,\,\mathrm{all}\,i} R_i\,\langle n\,|n_{i,\sigma}|\, m\rangle\,.$$

Due to the conservation of the total number of electrons we get again

$$\langle n|\, n_{i,\sigma}\, |m\rangle = \langle n|\, n_{i+1,\sigma}\, |m\rangle = \frac{1}{N}\langle n|\, N\, |m\rangle = 0 \qquad \text{if } n \neq m$$

and therefore

$$\langle n|\, \boldsymbol{R}\, |m\rangle = 0 \qquad \text{if} \qquad n \neq m\,. \tag{17}$$

Dipole transitions between different eigenstates are forbidden as well for non-interacting as for interacting electrons in the case of one single band.

We want to point out that from $\langle n|\, \boldsymbol{R}\, |m\rangle = 0$ if $n \neq m$ it does not follow that $[\boldsymbol{R}, H] = 0$ which is clearly not the case because of $\boldsymbol{j} = \frac{e}{i}[\boldsymbol{R}, H]$. Such a conclusion could be drawn if the set of states $|n\rangle$ would be a complete one and if $\boldsymbol{R}$ would be a bounded operator. Both suppositions are not fulfilled. The $|n\rangle$ as the eigenfunctions of the periodic operator H have the same periodicity as the Wannier functions: $\psi(\boldsymbol{r} - \boldsymbol{r}_{i+N}) = \psi(\boldsymbol{r} - \boldsymbol{r}_i)$. $\boldsymbol{R}$, however, not being a periodic operator is defined not only on the set of periodic basis functions. Further, from (16) it is easy to see that $\langle n|\, \boldsymbol{R}\, |n\rangle = \infty$ for any periodic function with $\langle n|\, N\, |n\rangle \neq 0$ which means $\boldsymbol{R}$ to be unbounded. The situation essentially is the same as in ordinary quantum mechanics. Here we have

$$\int_{-\infty}^{+\infty} \mathrm{e}^{-ikx}\, x\, \mathrm{e}^{ipx}\, \mathrm{d}x = 0 \qquad \text{if} \quad p \neq k$$

but, of course, $[x, p] \neq 0$.

From $\boldsymbol{j} = \frac{e}{i}\,[\boldsymbol{R}, H]$ we get

$$\langle n|\, \boldsymbol{j}\, |m\rangle = \frac{e}{i}\,(E_m - E_n) \lim_{n\to\infty} \frac{1}{n} \sum_{\sigma,\,\mathrm{all}\,i} \boldsymbol{R}_i\langle n|\, n_{i,\sigma}\, |m\rangle\,.$$

Due to the unboundness of $\boldsymbol{R}$ we cannot conclude from this equation $\langle n| \boldsymbol{j} |n\rangle = 0$ as stated by Cabib and Kaplan [13], we would get $0 \cdot \infty$. In general there will be difficulties if one calculates diagonal matrix elements of an operator which is defined as the difference of two unbounded ones.

In [14] Cabib and Kaplan stated: "For an even number of particles the time reversal operator θ and H can be simultaneously diagonalized and since $\boldsymbol{j}$ changes sign under time reversal, $\boldsymbol{j}_{nn} = 0$ in this basis." We do not see how this result could be proved and, therefore, merely give a simple counterexample: For two free electrons the mean value of the total current operator

$$\boldsymbol{j} = \frac{e}{m} \frac{\hbar}{i} \left(\frac{\partial}{\partial \boldsymbol{x}_1} + \frac{\partial}{\partial \boldsymbol{x}_2} \right)$$

in a state with the eigenenergy $E_{k_1,k_2} = \hbar^2/2m\ (k_1^2 + k_2^2)$ is

$$\frac{\hbar e}{m} (\boldsymbol{k}_1 + \boldsymbol{k}_2)$$

which, in general, is not zero.

4. Conclusions

We disproved the arguments of Cabib and Kaplan against the result that due to the translational invariance there is no absorption or emission near $\omega = U$. The physical reason behind is the impossibility of dipole transitions across the Hubbard gap due to the external field. This behaviour is quite different from that of a normal semiconductor. We, therefore, have to expect for a Mott insulator a different dielectric behaviour in comparison with a semiconductor, too. Using a simplified version of the approach of Brinkman and Rice [19] we found indeed [20] that the screening on the metallic site of the Mott transition is reduced in comparison with the case of noninteracting electrons and that the intraband part of the plasma frequency vanishes at the metal–insulator transition. On the insulating site we got simply $\varepsilon(q = 0, \omega = 0) = 1$ for the intraband part instead of Penn's two band result [21] $\varepsilon(q = 0, \omega = 0) = 1 + (\omega_{pl}^2/E_g^2)$ with $E_g \approx U$ in the atomic limit. This behaviour is well explained by the absence of transitions across the Hubbard gap.

The influence of electron–electron collisions on the conductivity is reduced to changes in the optical mass m_{xx} (2) in the expression $\sigma_{xx} = ie^2n/(\omega + i\varepsilon)\, m_{xx}$, there is no real part in σ_{xx}. If other scattering processes are included we get of course for the real dc conductivity $\sigma_{xx} = e^2n\tau/m_{xx}$ where τ is determined by these scattering processes. This result is similar to that of Brinkman and Rice [12]. But because in our expression there occurs the optical mass instead of a mass averaged over the Fermi surface the cancellation between the lower velocity and the larger relaxation time in σ_{xx} as discussed by Brinkman and Rice does not take place. As a consequence a metal–insulator transition may be well described by a sudden change in m_{xx}.

Acknowledgements

We would like to express our gratitude to Dr. W. Haubenreisser and to all members of the research group "Theory of Magnetism" of the Technical University of Dresden for their stimulating interest. We are particularly obliged to Dr. W. Gasser and to Mr. K.-H. Heinig for many fruitful discussions.

References

[1] J. A. Wilson, Adv. Phys. **21**, 143 (1972).
[2] K. Kubo, J. Phys. Soc. Japan **31**, 30 (1971).
[3] R. Bari, D. Adler, and R. Lange, Phys. Rev. B **2**, 2898 (1970).
[4] R. Kubo and N. Ohata, J. Phys. Soc. Japan **28**, 1402 (1970).
[5] W. F. Brinkman and T. M. Rice, Phys. Rev. B **2**, 1324 (1970).
[6] R. Bari and T. A. Kaplan, Phys. Rev. B **6**, 4623 (1972).
[7] J. Monecke, phys stat. sol. (b) **51**, 369 (1972).
[8] K.-H. Heinig and J. Monecke, phys. stat. sol. (b) **50**, K117 (1972).
[9] J. Hubbard, Proc. Roy. Soc. A **276**, 238 (1963).
[10] J. Monecke, phys. stat. sol. (b) **51**, K81 (1972).
[11] J. Monecke, phys. stat. sol. (b) **57**, K21 (1972).
[12] T. M. Brinkman and W. F. Rice, Phys. Rev. B **5**, 4350 (1972).
[13] D. Cabib and T. A. Kaplan, phys. stat. sol. (b) **58**, 85 (1973).
[14] D. Cabib and T. A. Kaplan, AIP Conf. Proc. Magnetism and Magnetitc Materials, Denver 1972, American Institute of Physics, 1973 (p. 1504).
[15] W. Brauer, Einführung in die Elektronentheorie der Metalle, Akademische Verlagsgesellschaft Geest & Portig, Leipzig 1972.
[16] N. F. Mott and Z. Zinamon, Rep. Progr. Phys. **33**, 881 (1970).
[17] N. N. Bogolyubov, Lektsii po kvantovoi statistike, Izbrannye trudy, Vol. 2, Naukova dumka, Kiev 1970.
[18] F. J. Blatt, Physics of Electronic Conduction in Solids, McGraw-Hill Publ. Co., New York 1968.
[19] W. F. Brinkman and T. M. Rice, Phys. Rev. B **2**, 4302 (1970).
[20] J. Monecke, phys. stat. sol. (b) **53**, K109 (1972).
[21] D. R. Penn, Phys. Rev. **128**, 2093 (1962).

(Received December 17, 1973)

phys. stat. sol. (b) 62, 425 (1974)

Subject classification: 2 and 13; 21

Zentralinstitut für Elektronenphysik der Akademie der Wissenschaften der DDR, Abteilung Theoretische Physik, Berlin

A General Cluster Expansion and the Definition of Self-Consistent n-Site Approximations in the Theory of Disordered Systems

By

H. J. FISCHBECK

A general cluster expansion of the averaged transition operator of a disordered system in terms of the distribution functions of its constituents is given. Self-consistent n-site approximations are defined by requiring that a truncated version of this series should be zero in an effective medium. In the alloy case these are shown to be equivalent to previously derived approximations. For amorphous systems a new type of n-site approximations is obtained.

Eine allgemeine Cluster-Entwicklung des gemittelten Streuoperators eines ungeordneten Systems nach den Verteilungsfunktionen seiner Konstituenten wird angegeben. Selbstkonsistente n-Zentren-Approximationen werden definiert durch die Forderung, daß eine abgebrochene Version dieser Reihe in einem effektiven Medium gleich Null sein soll. Im Fall von Legierungen wird gezeigt, daß diese zu früher abgeleiteten Approximationen äquivalent sind. Für amorphe Systeme wird eine neue Art von n-Zentren-Approximationen erhalten.

1. Introduction

The problem of generalizing CPA has attracted much attention in the last years. At present there is a rather confusing variety of different theories which brings about the need of comparing and selecting according to suitable criteria [1, 2]. The question of what is the best n-site approximation [2] is open and can perhaps be answered only by extensive numerical studies. A rough classification of the different approaches may be given without pretension to completeness as follows:

a) Diagrammatic expansions of the self-energy in terms of corrected cumulants [3] of the scatterer density [2, 4 to 7].

b) Cluster extensions of CPA in the sense that an effective medium is determined by the requirement that a specified cluster causes, on the average, no scattering in this medium [1, 8 to 13].

c) Pair and higher-order decouplings of the averaged multiple-scattering series [14 to 17] that are exact to second and higher powers of the concentration, respectively.

A necessary condition on any self-consistent approximation to the self-energy $\Sigma(z)$ is the analyticity for $z = E \pm i0$ and the correct asymptotic behaviour for large z. Recently this requirement has been shown to be violated by several approximations of type (a) and (b) [1, 18] whereas no such breakdown has been reported on the approximations of type (c). The molecular CPA of Tsukada has been tested in a wide range of parameters by Mertsching [1] and Nickel and Butler [18]. It proved to be always analytic. This is plausible because it is the proper CPA of a cluster of neighbouring atoms (molecule) as scattering entity with a cluster-diagonal self-energy. Recently, a general proof of analyticity has been found by Ducastelle [19]. Unfortunately this approximation breaks the full lattice symmetry of the averaged alloy [1, 20].

In this paper we propose a very simple and physically reasonable definition of self-consistent cluster approximations. It is based on a general expansion of the averaged transition operator $\mathcal{T}$ of a disordered system in terms of the distribution functions of its constituents. In the special case of a disordered alloy it turns out to be equivalent to the cluster approximation of Morita and Chen [16], which is in the pair case identical with the approximations of Schwartz and Ehrenreich [15] and Cyrot-Lackmann and Ducastelle [14]. The present consideration provides a simplified and generalized version of the approaches of type (c) [14 to 17]. The most interesting novel feature of our $\mathcal{T}$ operator expansion is its validity for any type of disorder. With it we have a common frame for defining self-consistent n-site approximations not only for alloys but also for amorphous systems. In the alloy case it is already known [14 to 17], in the other case, however, to the author's knowledge, not.

2. Cluster Expansion

Let $n^i(\boldsymbol{r}) = \sum_{\boldsymbol{R}^i} \delta(\boldsymbol{r} - \boldsymbol{R}^i)$ be the density of atoms of the species i at the positions $\boldsymbol{R}^i$ in a special configuration of the disordered system in question. Then, we can consider the sequence

$$g^{i_1}(\boldsymbol{r}_1) \equiv n^{i_1}(\boldsymbol{r}_1) ,$$

$$g^{i_1 i_2}(\boldsymbol{r}_1, \boldsymbol{r}_2) = g^{i_1}(\boldsymbol{r}_1) \left(n^{i_2}(\boldsymbol{r}_2) - \delta_{i_1,i_2}\delta(\boldsymbol{r}_1 - \boldsymbol{r}_2)\right) ,$$

$$g^{i_1 i_2 i_3}(\boldsymbol{r}_1, \boldsymbol{r}_2, \boldsymbol{r}_3) = g^{i_1 i_2}(\boldsymbol{r}_1, \boldsymbol{r}_2) \left(n^{i_3}(\boldsymbol{r}_3) - \delta_{i_1,i_3}\delta(\boldsymbol{r}_1 - \boldsymbol{r}_3) - \delta_{i_2,i_3}\delta(\boldsymbol{r}_2 - \boldsymbol{r}_3)\right) ,$$

of the single, pair, triple, ... densities [21], which is recursively defined by

$$g^{[k+1]} = g^{[k]}(n_{(k+1)}) - \sum_{j=1}^{k} \delta_{(j),(k+1)} , \qquad k \geqq 0 , \; g^{[0]} = 1 . \tag{2.1}$$

This short-hand notation should be read as follows:

$$g^{[k]} = g^{i_1 \dots i_k}(\boldsymbol{r}_1 \cdots \boldsymbol{r}_k) , \quad n_{(k)} = n^{i_k}(\boldsymbol{r}_k) ; \quad \delta_{(j),(k)} = \delta_{i_j,i_k}(\boldsymbol{r}_j - \boldsymbol{r}_k) .$$

That means, a superscript $[k]$ in square brackets refers to a set of k variables $i_1\boldsymbol{r}_1, \dots, i_k\boldsymbol{r}_k$ characterizing a cluster of k atoms, whereas a subscript (k) in round brackets refers to a single variable $i_k\boldsymbol{r}_k$ characterizing only one atom. Accordingly, summations over a set of k variables and over a single variable are abbreviated by $\sum_{i_1 \dots i_k} \int \mathrm{d}\boldsymbol{r}_1 \cdots \mathrm{d}\boldsymbol{r}_k = \int \mathrm{d}[k]$ and $\sum_{i_k} \int \mathrm{d}\boldsymbol{r}_k = \int \mathrm{d}(k)$ respectively.

Let $H = H_0 + V$ be the Hamiltonian of a special configuration with

$$V = \sum_{i_1} \int \mathrm{d}\boldsymbol{r}_1 n^{i_1}(\boldsymbol{r}_1)\, v^{i_1}(\boldsymbol{r}_1) \equiv \int \mathrm{d}(1)\, n_{(1)} v_{(1)} \tag{2.2}$$

being its statistical part, where $v^{i_1}(\boldsymbol{r}_1) \equiv v_{(1)}$ is the potential of an atom of the species i_1 at the position $\boldsymbol{r}_1$. Then, the Green's function $G(z) = (z - H)^{-1}$ is given in terms of the non-statistical propagator $G_0(z) = (z - H_0)^{-1}$ by

$$G_0^{-1} G - \int \mathrm{d}(1)\, n_{(1)} v_{(1)} G = 1 . \tag{2.3}$$

Furthermore we introduce the products

$$\Gamma^{[k]} = g^{[k]} G = G g^{[k]} . \tag{2.4}$$

After renaming the summation variable of (2.3) according to $(1) \to (k+1)$, we multiply (2.3) by $g^{[k]}$ and apply (2.1) and (2.4). In this way we obtain

$$\left(G_0^{-1} - \sum_{j=1}^{k} v_{(j)}\right) \Gamma^{[k]} - \int \mathrm{d}(k+1)\, v_{(k+1)} \Gamma^{[k+1]} = g^{[k]}.$$

Introducing the cluster Green's function

$$G^{[k]} = \left(G_0^{-1} - \sum_{j=1}^{k} v_{(j)}\right)^{-1} \tag{2.5}$$

we can rewrite this equation as a recursive definition

$$\Gamma^{[k]} = g^{[k]} G^{[k]} + \int \mathrm{d}(k+1)\, G^{[k]} v_{(k+1)} \Gamma^{[k+1]}, \qquad k \geqq 0 \tag{2.6}$$

of the quantities $\Gamma^{[k]}$. For $k = 0$ this equation is equivalent to (2.3). This recursion defines an artificial hierarchy. The only use of it consists in the fact that it produces an expansion of $\Gamma^{[0]} = G$ in terms of the Green's functions $G^{[k]}$ and the multiple densities $g^{[k]}$ of clusters of $1, 2, \ldots, k, \ldots$ atoms. Thus, starting with $\Gamma^{[0]} = G$, repeated application of (2.6) yields

$$G = G_0 \left(1 + \sum_{k=1}^{\infty} \int \mathrm{d}[k]\, g^{[k]} \prod_{j=1}^{k} v_{(j)} G^{[j]}\right).$$

In this way the statistical dependences of G are completely expressed by the multiple densities $g^{[k]}$. Taking now the average of $G_0^{-1} G - 1 = VG$, we obtain the following cluster expansion:

$$\Sigma \mathcal{G} = \langle VG \rangle = \sum_{k=1}^{\infty} \int \mathrm{d}[k]\, \langle g^{[k]} \rangle \prod_{j=1}^{k} v_{(j)} G^{[j]}. \tag{2.7}$$

It can be rewritten as a cluster expansion of the averaged transition operator

$$\mathcal{T}_0 = \langle VG \rangle\, G_0^{-1} = \sum_{k=1}^{\infty} \int \mathrm{d}[k]\, \langle g^{[k]} \rangle\, \tau_0^{[k]}, \tag{2.8}$$

where $\tau_0^{[k]}$ is something like a "reduced" scattering operator

$$\left.\begin{aligned} \tau_0^{[k]} &= t_0^{[1]} \prod_{j=2}^{k} G_0 v_{(j)} (1 - G_0 V^{[j]})^{-1}, \qquad k \geqq 2, \\ \tau_0^{[1]} &= t_0^{[1]} = v_{(1)} (1 - G_0 v_{(1)})^{-1}, \end{aligned}\right\} \tag{2.9}$$

of a cluster of k atoms with the cluster potential $V^{[k]} = \sum_{l=1}^{k} v_{(l)}$. Regarding the invariance of $\langle g^{[k]} \rangle$ with respect to permutations of its k arguments, $\tau_0^{[k]}$ may under the integral (2.8) also be written as

$$\tau_0^{[k]} = \sum_{j=1}^{k} \frac{(-1)^{k-j}}{j!(k-j)!}\, t_0^{[j]}, \tag{2.10}$$

where $t_0^{[j]} = V^{[j]} (1 - G_0 V^{[j]})^{-1}$ is the transition operator of a cluster of j atoms.

All the statistical information about the system under consideration is contained in the distribution functions $\langle g^{[k]} \rangle$ which are considered to be known from experiment. This is the most appealing aspect of the expansions (2.7) and (2.8) from the physical point of view. By normalization, $\langle g^{[k]} \rangle$ can be shown to be proportional to the k-fold product $c^{i_1} c^{i_2} \cdots c^{i_k}$ of concentrations c^i of the components of the system. Thus, the expansions (2.7) and (2.8) proceed in terms of increasing powers of the concentrations.

3. Self-Consistent n-Site Approximation

So far the statistical part V of the Hamiltonian $H = H_0 + V$ has not been specified except that it must be expressible in terms of the density as in (2.2). We are free to write $H = H_0 + \Sigma + (V - \Sigma)$ and to take $W = V - \Sigma$ as the statistical part. The corresponding non-statistical propagator is just the averaged Green's function $\mathcal{G}(z) = (z - H_0 - \Sigma)^{-1}$ we are looking for. Then, the self-consistency requirement is $\langle WG \rangle = \langle (V - \Sigma) G \rangle = 0$ or $\mathcal{T} = \langle T \rangle = 0$.

Supposing now $W = V - \Sigma$ to be written as

$$W = \int \mathrm{d}(1)\, n_{(1)} w_{(1)}\,, \tag{3.1}$$

we obtain a self-consistent n-site *approximation* by imposing an *approximate* self-consistency condition in the shape of a truncated version of (2.7)

$$\langle WG \rangle \approx \sum_{k=1}^{n} \int \mathrm{d}[k]\, \langle g^{[k]} \rangle \prod_{j=1}^{k} w_{(j)} \mathcal{G}^{[j]} = 0\,, \tag{3.2}$$

where $\mathcal{G}^{[j]} = (\mathcal{G}^{-1} - W^{[j]})^{-1}$ is the Green's function of a cluster $W^{[j]} = \sum_{l=1}^{j} w_{(l)}$ of j atoms in the effective medium represented by the averaged propagator $\mathcal{G}$. An equivalent formulation of this n-site approximation provides (2.8)

$$\mathcal{T} = \langle WG \rangle\, \mathcal{G}^{-1} \approx \sum_{k=1}^{n} \int \mathrm{d}[k]\, \langle g^{[k]} \rangle\, \tau^{[k]} = 0\,. \tag{3.3}$$

The definition of the "reduced" scattering operator $\tau^{[k]}$ in the effective medium may be inferred from (2.9) by the replacements $v_{(k)} \to w_{(k)}$, $G_0 \to \mathcal{G}$, $V^{[k]} \to W^{[k]}$.

3.1 Substitutional disorder

In the alloy case only the occupation of the lattice sites $\boldsymbol{R}$ is random, i.e. we have $n^i(\boldsymbol{r}) = \sum_{\boldsymbol{R}} \eta^i_{\boldsymbol{R}} \delta(\boldsymbol{r} - \boldsymbol{R})$, where the occupation number $\eta^i_{\boldsymbol{R}}$ is 1 if an atom of the species i is at $\boldsymbol{R}$ and 0 otherwise. Because of $\sum_i \eta^i_{\boldsymbol{R}} = 1$ the decomposition of the self-energy into local self-energies $\sigma^i(\boldsymbol{r})$ [21] can be rewritten in the way we need in (3.1):

$$\Sigma = \sum_i \int \mathrm{d}\boldsymbol{r} \langle n^i(\boldsymbol{r}) \rangle\, \sigma^i(\boldsymbol{r}) = \sum_{\boldsymbol{R}} \sigma(\boldsymbol{R}) = \sum_i \int \mathrm{d}\boldsymbol{r}\, n^i(\boldsymbol{r}) \sigma(\boldsymbol{r})\,,$$

where $\sigma(\boldsymbol{r}) = \sum_i c^i \sigma^i(\boldsymbol{r})$ is the mean of the local self-energies and $c^i = \langle \eta^i_{\boldsymbol{R}} \rangle$ the concentration of the atoms of the species i. In case of an alloy we thus have

$$w_{(1)} \equiv w^{i_1}(\boldsymbol{r}_1) = v^{i_1}(\boldsymbol{r}_1) - \sigma(\boldsymbol{r}_1)\,. \tag{3.4}$$

The single-site approximation ($n = 1$) given by (3.2) or (3.3) is just CPA. The n-site approximation (3.3) has recently been proposed by Morita and Chen [16], which can be seen from the form (2.10) of the "reduced" transition operator. The corresponding pair approximation has already been considered by Cyrot-Lackmann and Ducastelle [14] and by Schwartz and Ehrenreich [15]. This may be realized by noting that

$$\tau^{[2]} \equiv t_{(1)} \mathcal{G} w_{(2)} (1 - \mathcal{G} W^{[2]})^{-1} = t_{(1)} \mathcal{G} t_{(2)} (1 - \mathcal{G} t_{(1)} \mathcal{G} t_{(2)})^{-1}\, (1 + \mathcal{G} t_{(1)})\,.$$

3.2 Positional disorder

In amorphous systems there is no other possibility to write the self-energy in the appropriate form than the trivial identity

$$\Sigma = \int \mathrm{d}(1)\, n_{(1)} \frac{1}{N}\, \Sigma\,,$$

where $N = \int d(1)\, n_{(1)}$ is the total number of atoms of the system. Therefore,

$$w_{(1)} = v_{(1)} - \frac{1}{N} \Sigma . \quad (3.5)$$

The n-site approximation of an infinite amorphous system is to be taken from (3.2) as

$$\lim_{N\to\infty} \sum_{k=1}^{n} \int d[k] \langle g^{[k]} \rangle \prod_{j=1}^{k} \left(v_{(j)} - \frac{1}{N} \Sigma \right) \left(\mathcal{G}^{-1} - V^{[j]} + \frac{j}{N} \Sigma \right)^{-1} = 0 . \quad (3.6)$$

The compensation of inverse powers of N comes from the fact that Σ is independent of the integration variables. In the k-fold product in (3.6) there are $\binom{k}{l}$ terms being proportional to $N^{-l}\Sigma^{l}$. Each of these terms contains a contribution that is independent of l integration variables, so that the integration

$$\int d[l]\, g^{[k]} = (N - k + 1)\,(N - k + 2) \ldots . (N - k + l)\, g^{[k-l]} \approx N^{l} g^{[k-l]}$$

compensates the factor N^{-l}. Other compensations are not possible. Therefore, the term $(j/N)\Sigma$ in the denominators of (3.6) always disappears in the limit $N \to \infty$. The single-site approximation ($n = 1$) deriving from (3.6) is the well-known self-consistent t matrix approximation

$$\int d(1) \langle g^{[1]} \rangle\, t^{[1]} - \Sigma = 0 , \quad (3.7)$$

where $t^{[1]} = t_{(1)} = v_{(1)}(1 - \mathcal{G} v_{(1)})^{-1}$. After having performed the limit $N \to \infty$, we conveniently define the cluster t operators in the effective medium generally as

$$t^{[k]} = V^{[k]}(1 - \mathcal{G} V^{[k]})^{-1} \quad (3.8)$$

contrary to the alloy case, where the "corrected" potentials $W^{[k]}$ are maintained, because the potential correction in (3.4) can be localized at the lattice points.

The result (3.7) has been obtained by Faulkner [22] by a similar $N \to \infty$ argument. On the other hand it can be derived from the Schwartz-Ehrenreich equation [23] in the special case of complete disorder. Unlike the single-site decoupling of Schwartz and Ehrenreich [23] our prescription offers no possibility to include short-range order information. Therefore, the single-site approximation (3.7) makes sense only for gas-like arrangements of impurities. In dense systems like liquids or glasses short-range order is constitutive of the system as a whole. Thus, at least a pair approximation is required in order to keep the atoms in the correct distances by means of the pair distribution function $\langle g^{[2]} \rangle$. The pair approximation given by (3.6) reads

$$\int d[2] \langle g^{[2]} \rangle t^{[1]} \mathcal{G} v_{(2)} (1 + \mathcal{G} t^{[2]}) + \int d[1] \langle g^{[1]} \rangle \{ t^{[1]} - \Sigma \mathcal{G} t^{[1]} - t^{[1]} \mathcal{G} \Sigma - t^{[1]} \mathcal{G} \Sigma \mathcal{G} t^{[1]} \} + \Sigma \mathcal{G} \Sigma - \Sigma = 0 . \quad (3.9)$$

The pair term of (3.6) gives rise to several single-site corrections. This is also a consequence of the complete smearing out of the potential correction in (3.5) due to the randomness of the atomic positions. Formally it would be possible to describe also an alloy in the way of (3.5). But this would certainly be worse than the usual way (3.4) because all the single-site corrections arising from any higher-order term of (3.6) are already included in the single-site term with (3.4) being used. However, it should be stressed once more that in the context of this work the use of (3.5) is inevitable in amorphous systems, which simply reflects the higher degree of positional disorder.

4. Concluding Remarks

The occurrence of non-analyticity in self-consistent cluster approximations seems to be the rule rather than the exception [1, 18, 20]. Therefore, extensive tests or even a proof of analyticity [19] is a necessary and perhaps the most crucial step in establishing a self-consistent n-site approximation. The hitherto existing numerical applications of a two-site version of (3.3) to tight-binding alloy models [15, 17] do not suffice in this respect. Only further work can decide on the reliability of the approximation scheme discussed here. For the time being we have no other argument in favour of this scheme than its conceptual simplicity and consequence and the natural way in which the statistical information about a disordered system enters. On the other hand, practical calculations are rather difficult because the sums or integrals of (3.3) or (3.6) are not restricted to nearest-neighbour clusters. Further approximations such as restriction to specific clusters should be made with caution in order not to destroy the possibly existing analyticity.

Acknowledgements

The author is grateful to Prof. Dr. W. Brauer, Dr. J. Mertsching, Dr. H. Puff, and Dr. H. W. Streitwolf for many helpful discussions. He is especially indebted to Dr. H. W. Streitwolf for suggesting a considerable simplification of the argument.

References

[1] J. Mertsching, phys. stat. sol. (b) **59**, 227 (1973).
[2] F. Ducastelle, J. Phys. F **2**, 468 (1972).
[3] F. Yonezawa, Progr. theor. Phys. (Kyoto) **40**, 734 (1968).
[4] R. N. Aiyer, R. J. Elliott, J. A. Krumhansl, and P. L. Leath, Phys. Rev. **181**, 1006 (1969).
[5] B. Nickel and J. A. Krumhansl, Phys. Rev. B **4**, 4354 (1971).
[6] R. Mills, Phys. Rev. B **8**, 3650 (1973).
[7] J. P. Gaspard and F. Cyrot-Lackmann, J. Phys. C **5**, 3047 (1972).
[8] E-Ni Foo, H. Amar, and M. Ausloos, Phys. Rev. B **4**, 3350 (1971).
[9] V. Čápek, phys. stat. sol. (b) **43**, 61 (1971).
[10] W. H. Butler, Phys. Letters A **39**, 203 (1972).
[11] E-Ni Foo, S. M. Bose, and M. Ausloos, Phys. Rev. B **7**, 3454 (1973).
[12] M. Tsukada, J. Phys. Soc. Japan **32**, 1475 (1972).
[13] J. Zittartz, Solid State Commun. **14**, 51 (1974)
[14] F. Cyrot-Lackmann and F. Ducastelle, Phys. Rev. Letters **27**, 429 (1971).
[15] L. Schwartz and H. Ehrenreich, Phys. Rev. B **6**, 2923 (1972).
[16] T. Morita and C. C. Chen, J. Phys. Soc. Japan **34**, 1136 (1973).
[17] T. Horiguchi, C. C. Chen, and T. Morita, Solid State Commun. **12**, 1243 (1973); **13**, 957 (1973).
[18] B. G. Nickel and W. H. Butler, Phys. Rev. Letters **30**, 373 (1973).
[19] F. Ducastelle, preprint.
[20] P. L. Leath, J. Phys. C **6**, 1559 (1973).
[21] H. J. Fischbeck, phys. stat. sol. (b) **53**, 527 (1972).
[22] J. S. Faulkner, Phys. Rev. B **1**, 934 (1970).
[23] L. Schwartz and H. Ehrenreich, Ann. Phys. (Germany) **64**, 100 (1971).

(Received January 17, 1974)

phys. stat. sol. (b) 62, 431 (1974)

Subject classification: 13.4 and 20.1; 22.5.3

Zentralinstitut für Elektronenphysik der Akademie der Wissenschaften der DDR, Berlin

Jahn-Teller Effect on Excited Orbital Triplet States of $3d^n$ Ions in Eightfold Cubic Coordination

Co^{2+} and Fe^{2+} in CdF_2

By

W. ULRICI

The ions Co^{2+} and Fe^{2+} can be substitutionally incorporated on Cd^{2+} sites in CdF_2. The optical absorption bands of the transitions $^4A_{2g} \rightarrow {}^4T_{1g}(F)$, $\rightarrow {}^4T_{1g}(P)$ and $^5E_g \rightarrow {}^5T_{2g}$ are no single bands but exhibit a pronounced triplet structure measured up to 500 K. The splittings increase linearly with $\sqrt{T}$ at high temperatures. These results are explained as due to a strong dynamical Jahn-Teller effect (interaction with τ_{2g} modes) on the excited orbital triplet states based on the theory by Toyozawa and Inoue [3]. The frequencies of the τ_{2g} modes and the coupling constants (Jahn-Teller energies) are determined. The doublet structure of the $^4A_{2g} \rightarrow {}^4T_{2g}$ absorption band and additional fine structures of the $\rightarrow {}^4T_{1g}(F)$ band at low temperatures are discussed as probably being caused by the simultaneous action of first order spin–orbit coupling and strong $T \otimes \tau_{2g}$ coupling.

Co^{2+}- and Fe^{2+}-Ionen können substitutionell auf Cd^{2+}-Plätzen im CdF_2 eingebaut werden. Die Absorptionsbanden der Übergänge $^4A_{2g} \rightarrow {}^4T_{1g}(F)$, $\rightarrow {}^4T_{1g}(P)$ und $^5E_g \rightarrow {}^5T_{2g}$ sind keine einzelnen Banden, sondern zeigen eine ausgeprägte Triplettstruktur, die bis 500 K gemessen wurde. Die Aufspaltungen nehmen bei hohen Temperaturen linear mit $\sqrt{T}$ zu. Diese Ergebnisse werden mit Hilfe der Theorie von Toyozawa und Inoue [3] erklärt durch das Wirken eines starken dynamischen Jahn-Teller-Effekts (Wechselwirkung mit τ_{2g}-Schwingungen) an den angeregten Bahntriplettzuständen. Die Frequenzen der τ_{2g}-Schwingungen und die Kopplungskonstanten (Jahn-Teller-Energien) werden bestimmt. Die Dublettstruktur der $^4A_{2g} \rightarrow {}^4T_{2g}$ Bande sowie zusätzliche Feinstrukturen der $\rightarrow {}^4T_{1g}(F)$ Bande bei tiefen Temperaturen werden diskutiert und sind wahrscheinlich durch die gleichzeitige Wirkung von Spin–Bahn-Kopplung und $T \otimes \tau_{2g}$ Kopplung bedingt.

1. Introduction

The interpretation of optical spectra of defect centres in crystals requires in many cases to take account the electron–lattice coupling where the coupling with non totally symmetrical modes called Jahn-Teller effect (JTE) and Ham-effect are most important (see the reviews by Sturge [1] and Englman [2]). Of particular interest is the absorption band shape associated with a transition from a singlet to an orbital triplet state. It was intensively investigated on F-centres in halides, on the F^+-centre in CaO and especially on the transitions $^1A_{1g} \rightarrow {}^1T_{1u}$ and $\rightarrow {}^3T_{1u}$ (C- and A-band) of heavy-metal ions in alkali halides. The latter problem was treated theoretically in a semiclassical manner by Toyozawa and Inoue [3] and by Cho [4]. They calculated the absorption line shape with the Franck-Condon approximation and found that strong linear interaction of an electronic triplet state with a triply degenerate τ_2 mode (symbolized by $T \otimes \tau_2$) results in a symmetrical triple band whereas the additional linear interaction with α_1 and ε modes broadens this structure. Within the framework of this theory the structure and the temperature dependence of the splitting of the C-band could be explained. By refinements including quadratic coupling the authors were able to explain the doublet structure of the A-band [3] and the asymmetry of the C-band [5].

All of the examples concerning the strong $T \otimes \tau_2$ coupling considered hitherto (both experimentally and theoretically) refer to triplet states including p-electrons. In order to investigate to what extent such a strong JTE acts also on orbital triplet states including d-electrons, it seems appropriate to investigate the optical absorption spectra of $3d^n$ transition-metal ions substituted on Me^{2+} sites in CaF_2-type lattices. In this case the triply degenerate electronic states form σ-bonds with the ligands and a strong coupling with the radial τ_{2g} mode is expected [1].

In a preliminary note [6] it has been shown for the first time that absorption bands due to transitions to excited orbital triplet states of $3d^n$ ions exhibit a line shape as predicted by the semiclassical theory. The transitions in consideration are ${}^4A_{2g} \rightarrow {}^4T_{1g}(F)$ and $\rightarrow {}^4T_{1g}(P)$ of Co^{2+} and ${}^5E_g \rightarrow {}^5T_{2g}$ of Fe^{2+} where the ions are substituted for Cd^{2+} in CdF_2. In the present paper a detailed investigation of these absorptions is made and an interpretation is given in terms of the strong electron–lattice coupling on the base of the theory of Toyozawa and Inoue [3]. This interpretation is considerably supported by a detailed investigation of the electron spin resonance of CdF_2:Co^{2+} [7] and the measurement of the magnetic circular dichroism of the ${}^4A_{2g} \rightarrow {}^4T_{1g}(P)$ band of CdF_2:Co^{2+} [8] carried out simultaneously, and by the Mössbauer investigations on CdF_2:Fe^{2+} published recently by Steger and Kostiner [9].

2. Experimental

Single crystals of CdF_2 were grown by the Bridgman technique using high purity CdF_2 powder precipitated from Cd acetate solution and HF as starting material [10] to which were added metallic cobalt or iron. The boules obtained in this way were homogeneously coloured. The crystals for the optical measurements (12 mm in diameter, length 2 to 25 mm) were cut from the central part of the boules, the parallel faces were mechanically polished.

The concentration values of the dopants mentioned in this paper are taken from the quantities added to the melt. In the range used (100 to 5000 ppm) the absorption coefficient in the maximum of the Co^{2+} and Fe^{2+} absorption bands was proportional to the concentration.

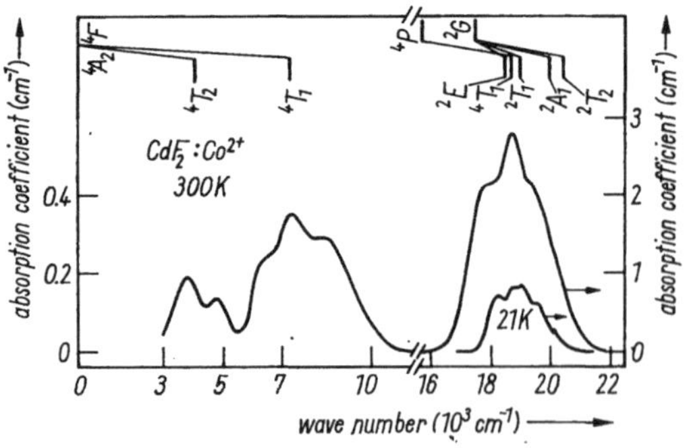

Fig. 1. Absorption spectrum of a CdF_2:3100 ppm Co^{2+} crystal

The optical spectra were measured with a Beckman DK-2A spectrophotometer, where the crystals are mounted in metal cryostats applicable in the temperature ranges 20 to 500 K and 5 to 300 K.

3. Experimental Results

3.1 $CdF_2:Co^{2+}$

The absorption spectrum of $CdF_2:Co^{2+}$ measured at 300 K is shown in Fig. 1. It consists of three structured band systems[1]) centered at 4200, 7250, and 18700 cm^{-1}. The EPR investigations on these crystals [7] have proved that the Co^{2+} ions are situated on Cd^{2+} sites in an undisturbed cubic environment of the eight fluorine ions. It follows that the band systems can be assigned to the spin-allowed transitions $^4A_{2g} \rightarrow {}^4T_{2g}(F)$ (4200 cm^{-1}), $\rightarrow {}^4T_{1g}(F)$ (7250 cm^{-1}), and $\rightarrow {}^4T_{1g}(P)$ (18700 cm^{-1}). The calculation of the parameters of the crystal field theory using the transitions $\rightarrow {}^4T_{1g}(F)$ and $\rightarrow {}^4T_{1g}(P)$ yields the values $Dq = 400$ cm^{-1} and $B = 930$ cm^{-1}.

As can be seen from the energy level diagram in Fig. 1 (calculated with the parameters mentioned above and $C/B = 3.9$) there are four states arising from the ^2G term, all of which lie in the same energy region as the $^4T_{1g}(P)$ term. The weak sharp absorption band at 20100 cm^{-1} and the weak broad band at 19800 cm^{-1} (at 21 K) are probably due to the transitions $\rightarrow {}^2T_2$ and $\rightarrow {}^2A_1$,

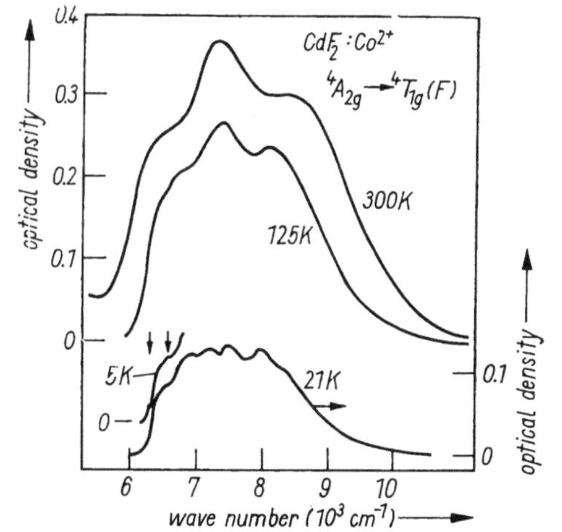

Fig. 2

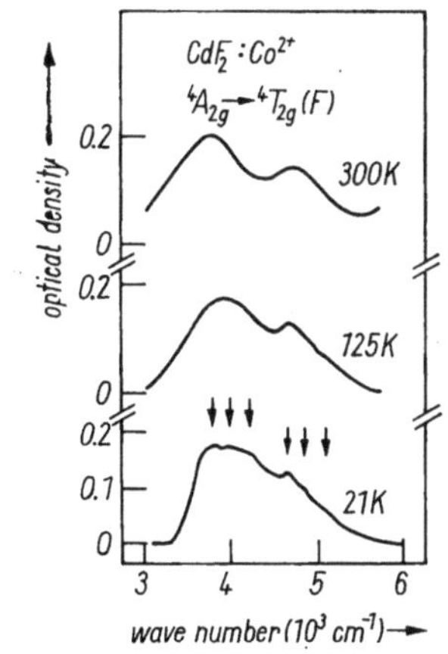

Fig. 3

Fig. 2. Absorption spectra of a CdF_2: 3100 ppm Co^{2+} crystal (thickness $d = 22.0$ mm) in the region of the $^4A_{2g} \rightarrow {}^4T_{1g}(F)$ transition at various temperatures. The curve at 5 K is shown only in that region in which two new weak absorptions appear (indicated by arrows). In the remaining part the spectrum at 5 K is identical with that at 21 K

Fig. 3. Absorption spectra of a CdF_2: 3100 ppm Co^{2+} crystal ($d = 22.0$ mm) in the region of the $^4A_{2g} \rightarrow {}^4T_{2g}(F)$ transition. The arrows indicate the additional fine structure

1) The spectrum of $CdF_2:Co^{2+}$ published by Simo et al. [11] is identical with that reported here, but no interpretation was given by the authors. The absorption spectrum of $CdF_2:Co^{2+}$ was published partly also already by Stolov and Yakovleva [12].

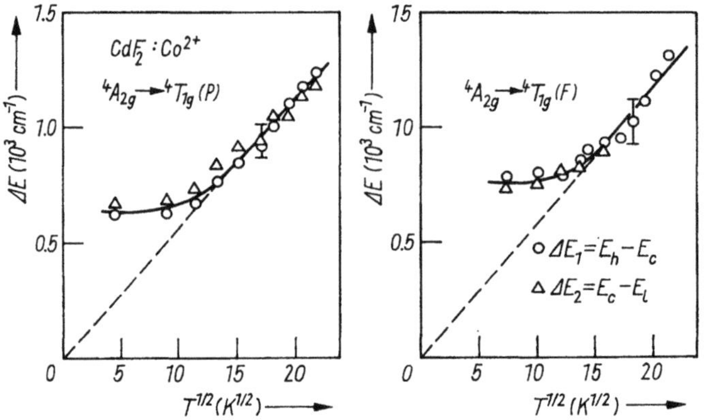

Fig. 4. Temperature dependence of the separations ΔE between the central peak (position E_c) and the low energy (E_l) and high energy side peak (E_h), respectively, of the $\rightarrow {}^4T_{1g}(P)$ and $\rightarrow {}^4T_{1g}(F)$ triplet bands. The dashed lines are the extrapolation to the origin of the linear variation of ΔE with $\sqrt{T}$ at high temperatures

respectively (cf. the MCD spectrum [8], whereas the weak sharp peak on the central peak of the triplet band (19000 cm^{-1}) is due to $\rightarrow {}^2E$ or $\rightarrow {}^2T_1$.

The essential feature of the Co^{2+} spectrum measured here is the appearence of a triplet structure of the $\rightarrow {}^4T_{1g}(P)$ and $\rightarrow {}^4T_{1g}(F)$ band and a doublet structure of the $\rightarrow {}^4T_{2g}(F)$ band instead of a single Gaussian shaped band as expected for these transitions. Additionally at low temperatures both the $\rightarrow {}^4T_{1g}(F)$ band and the $\rightarrow {}^4T_{2g}(F)$ band are split into at least six components as shown in Fig. 2 and 3. Above about 100 K these fine structures vanish and only the triplet and the doublet structure, respectively, are measured up to temperatures of 500 K. That the fine structures are phonon structures can be excluded by their temperature dependence. They show no change during cooling down from 21 to 5 K. Exceptions are the two weak sharp absorptions on the low-energy side of the $\rightarrow {}^4T_{1g}(F)$ absorption (see Fig. 2) which appear at temperatures below 10 K (6270 and 6540 cm^{-1}).

In Fig. 4 the temperature dependence of the separations between the central peak and the side peaks of the $\rightarrow {}^4T_{1g}(P)$ and $\rightarrow {}^4T_{1g}(F)$ triplet bands are plotted versus $\sqrt{T}$. It can be seen that the separations are symmetrical to the position of the central peak and increase at high temperatures linearly with $\sqrt{T}$. The peak positions of the subbands of the $\rightarrow {}^4T_{1g}(P)$ band were obtained by computer decomposition into three Gaussian bands. The asymmetry of this triplet band is small and independent of the temperature ($A_l/A_h = 1.1 \pm 0.1$; A_l and A_h are the heights of the low-energy and the high-energy subband, respectively). Because of the complicated fine structure of the $\rightarrow {}^4T_{1g}(F)$ band at low temperatures such a decomposition was not carried out. The peak positions of the triplet subbands at high temperatures were taken from the spectrum. At high temperatures only $\Delta E_1 = E_h - E_c$ was determined as the low-energy peak is less pronounced and overlaps with the $\rightarrow {}^4T_{2g}(F)$ band.

Fig. 5 shows the temperature dependence of the separation ΔE between the peaks of the $\rightarrow {}^4T_{2g}(F)$ doublet band. The peak positions were taken from the spectrum. ΔE increases with increasing temperature and varies also linearly

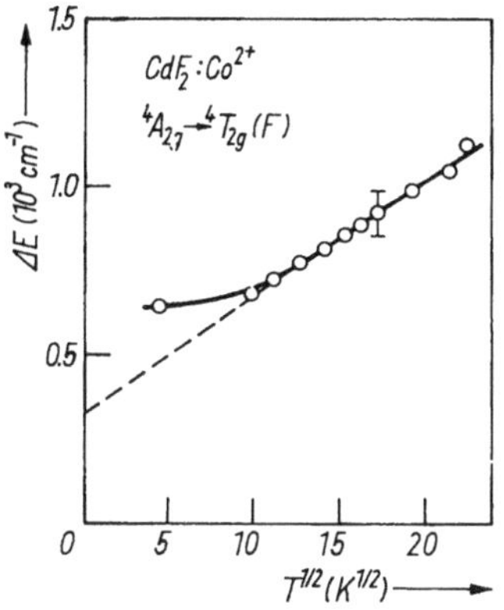

Fig. 5

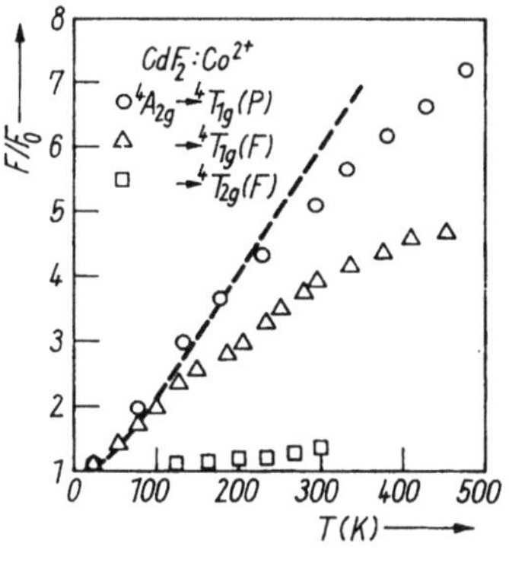

Fig. 6

Fig. 5. Temperature dependence of the separation ΔE between the components of the $\rightarrow {}^4T_{2g}(F)$ doublet band. The dashed line represents the extrapolation to $T = 0$ K of the linear variation of ΔE with $\sqrt{T}$ at high temperatures

Fig. 6. Temperature dependence of the oscillator strength $f/f_0 = F/F_0$ (F area under the absorption band) of the three absorption band systems of Co^{2+} in CdF_2. The dashed line is calculated with (8) and $\hbar\omega_u = 70$ cm^{-1}

with $\sqrt{T}$ at high temperatures, but this straight line does not go through the origin. A reduction of the error of ΔE by using a computer decomposition was not possible because of the overlapping with the $\rightarrow {}^4T_{1g}(F)$ band and the low-energy limit (3000 cm^{-1}) of the spectrometer.

In Fig. 6 the temperature dependence of the area under the three band systems is shown (the area F is proportional to the oscillator strength f, that means $F/F_0 = f/f_0$; F_0, f_0 are extrapolated values at 0 K). Two facts are remarkable:

(i) The oscillator strength of the $\rightarrow {}^4T_{1g}(P)$ and $\rightarrow {}^4T_{1g}(F)$ bands increase strongly with increasing temperature as it is expected for these vibration-allowed transitions. However, the measured values cannot be fitted by the well-known law $f/f_0 = \coth(\hbar\omega_u/2kT)$. With increasing temperature the deviations from this law become stronger and stronger.

(ii) The oscillator strength of the $\rightarrow {}^4T_{2g}(F)$ band is nearly independent of the temperature.

As the EPR results have shown [7] that the environment of the Co^{2+} ions has exactly the symmetry O_h, i.e. no static distortion is present, merely the spin–orbit coupling would be able to produce a splitting of the orbital triplet states within the framework of the crystal field theory. However, this interpretation of the measured triplet and doublet structure, respectively, must be excluded because

(i) the measured overall splittings are larger by a factor of 2 to 3 than the calculated ones with the highest possible spin–orbit coupling constant $|\lambda| = 178$ cm^{-1} (free ion value of the ^{4}F state),

(ii) the measured splittings increase with increasing temperature, unreasonable for a spin–orbit splitting.

Therefore only the strong interaction of the orbital triplet states with lattice vibrations can be responsible for the measured splittings.

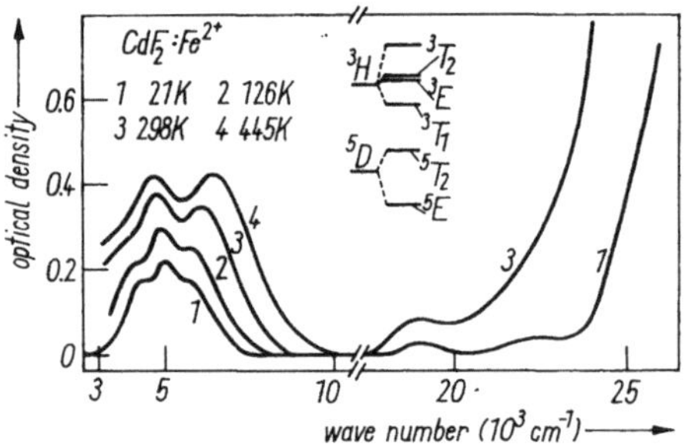

Fig. 7

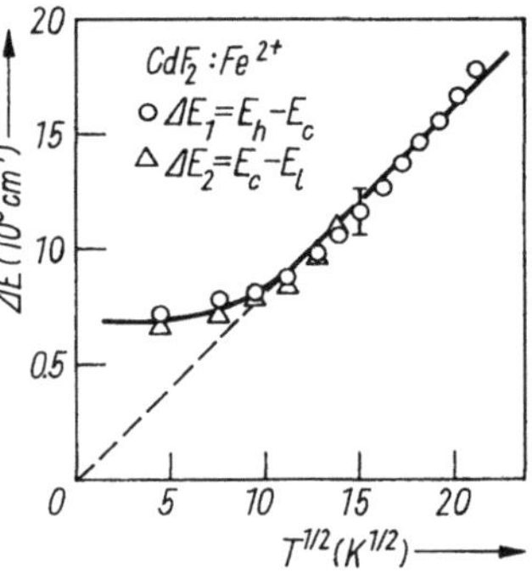

Fig. 8

Fig. 7. Absorption spectra of a CdF_2:4700 ppm Fe^{2+} crystal ($d = 18.5$ mm) at various temperatures. The energy level scheme shows the lowest free ion levels of Fe^{2+} (^5D ground state and ^3H) and their crystal field splittings in eightfold cubic coordination

Fig. 8. Temperature dependence of the separation between the central peak (E_c) and the high energy (E_h) and low energy (E_l) side peak, respectively, of the $^5E_g \to {}^5T_{2g}$ triplet band of Fe^{2+} in CdF_2

3.2 CdF_2:Fe^{2+}

Fe^{2+} has been investigated hitherto neither in CdF_2 nor in other fluorite-type host lattices by optical absorption or EPR. The Mössbauer investigations on iron-doped CdF_2 crystals show [9] that the Fe^{2+} ions are situated on Cd^{2+} sites and homogeneously distributed (according to the lattice statistics) and that no static distortion around the isolated Fe^{2+} ions is present.

Fig. 7 shows the optical absorption spectrum of an iron-doped CdF_2 crystal at 21 K. It consists of a triplet band centered at 4950 cm^{-1} and two weak bands at 19000 and 22200 cm^{-1}. This spectrum can be interpreted as due to Fe^{2+} in eightfold cubic environment (see the energy level scheme in Fig. 7), where the triplet band in the near IR is assigned to the $^5E_g \to {}^5T_{2g}$ transition yielding $Dq = 495$ cm^{-1} and the weak absorptions are assumed being due to the spin-forbidden transitions $^5E_g \to {}^3T_{1g}$ and $\to {}^3E_g$, $^3T_{2g}$ (a fit is possible with the reasonable values $B = 910$ cm^{-1} and $C = 3700$ cm^{-1}). As can be seen from Fig. 7 the oscillator strength of the $^5E_g \to {}^5T_{2g}$ band increases with increasing temperature.

Fig. 8 shows the temperature dependence of the separations between the side peaks and the central peak of the $^5E_g \to {}^5T_{2g}$ triplet band. At high temperatures ΔE_1 increases linearly with $\sqrt{T}$ (ΔE_2 could be determined only at low temperatures). Whereas at low temperatures the triplet band is symmetrical with respect to the central peak, the band becomes more and more asymmetric with increasing temperature (see Fig. 7). No fine structure of the triplet band was measured even at the lowest temperatures.

The arguments suggesting the interpretation of the triplet structure of the $^5E_g \to {}^5T_{2g}$ band of Fe^{2+} as due to strong electron–lattice interaction are the same as in the case of Co^{2+}. The Mössbauer investigations provide evidence for the undisturbed eightfold cubic environment of the Fe^{2+} ions. Therefore only spin–orbit coupling would be able to produce a splitting of the $^5T_{2g}$ state yielding an overall splitting of $5|\lambda| = 520$ cm^{-1}. However, the measured overall splittings are 1400 cm^{-1} at 21 K and 2750 cm^{-1} at 300 K.

4. Discussion

4.1 Theory

In order to treat the electron–lattice interaction of an impurity centre with point symmetry O_h usually the following Hamiltonian is used

$$H = H_e + H_p + H_{ep}. \quad (1)$$

H_e is the Hamiltonian of the impurity electrons the solution of which yields also the triply degenerate levels considered here (crystal field theory without spin–orbit coupling). H_p describes the lattice vibrations and H_{ep} the interaction of the 3d electrons with the lattice vibrations. The even lattice vibrations are treated in the harmonic approximation and will be described by the set of coordinates $\{Q\} = Q_1, \ldots, Q_6$ which transform according to the irreducible representations $\alpha_{1g}(Q_1)$, $\varepsilon_g(Q_2, Q_3)$, and $\tau_{2g}(Q_4, Q_5, Q_6)$, respectively, of the point group O_h. The Hamiltonian $H_p + H_{ep}$ describing the linear interaction with only triply degenerate modes (τ_{2g}) has the form

$$H_p + H_{ep} = T_\tau + \frac{1}{2}\omega_\tau^2 \sum_{i=4}^{6} Q_i^2 + c \begin{pmatrix} 0 & Q_6 & Q_5 \\ Q_6 & 0 & Q_4 \\ Q_5 & Q_4 & 0 \end{pmatrix}, \quad (2)$$

where T_τ is the nuclear kinetic energy, ω_τ is the frequency of the τ_{2g} modes, and c is the coupling constant of the linear interaction with τ_{2g} modes. Equation (2) acts on the state vector of the wave function $\Psi = \sum_{i=1}^{3} a_i \psi_i$ of a triply orbitally degenerate electron state T (ψ_i are the wave functions of the substates of T). The solution of the static problem (T_τ is regarded as a parameter) yields the potential hyperface $V = V(Q_4, Q_5, Q_6)$.

In the case of the eightfold cubic coordination of the impurity ion there are two kinds of τ_{2g} modes: $\tau_{2g}^a(Q_4, Q_5, Q_6)$ and $\tau_{2g}^b(Q_7, Q_8, Q_9)$. However, we neglect τ_{2g}^b as the coupling of the electronic T state with these tangential modes will be considerably smaller than the coupling with the radial τ_{2g}^a modes. The formal difficulties can be overcome by the concept of interaction mode coordinates introduced by Toyozawa and Inoue [3] leading to the same Hamiltonian (2) where Q_4, Q_5, Q_6 represent all vibrations of symmetry τ_{2g}.

The optical absorption line shape $F(E)$ of a transition from a singlet ground state (Ψ_g) to the degenerate excited state (Ψ_e) is calculated by using the Franck-Condon approximation, meaning

$$F(E) = |\langle\Psi_g| \boldsymbol{P} |\Psi_e\rangle|^2 \, I(E), \quad (3)$$

where $\boldsymbol{P}$ is the electronic transition operator. The line shape function $I(E)$ is given by

$$I(E) = \int \prod_n \mathrm{d}Q_n W_g\{Q\} \, \delta(E - V_e\{Q\} + V_g\{Q\}), \quad (4)$$

where $\prod_n \mathrm{d}Q_n \, W_g\{Q\}$ represents the probability that when the system is in the electronic ground state g it will be found in the volume element $\prod_n \mathrm{d}Q_n$ near the point $\{Q\}$ of the vibrational coordinate space. n runs over all vibrational coordinates of the set $\{Q\}$ to be taken into account. $V_g\{Q\}$ and $V_e\{Q\}$ are the potential faces of the ground and excited electronic state, respectively, where $V_e\{Q\}$ is the solution of the static Jahn-Teller problem in the degenerate excited state.

In this way Toyozawa and Inoue [3] have calculated the line shape $I(E)$ of the transition from the $^1A_{1g}$ ground state to the $^1T_{1u}$ excited state interacting only with τ_{2g} modes. They obtained a symmetrical three-peaked absorption line shape with a logarithmic singularity at $E = E_0$, where E_0 is the electronic transition energy at $\{Q\} = 0$ (see Fig. 6 in [3]). The distance of both side peaks to the central peak was found to be

$$\Delta E(T) = \Delta E(0) \sqrt{\coth \frac{\hbar\omega_\tau}{2kT}} \tag{5}$$

with

$$\Delta E(0) = 0.91 \sqrt{\frac{3}{2} E_{JT} \hbar\omega_\tau} \tag{6}$$

the distance at zero temperature. E_{JT} is the Jahn-Teller energy of the $T \otimes \tau_{2g}$ interaction and is given by $E_{JT} = c^2/3$. From (5) and (6) follows in the limit of high temperatures

$$\Delta E(T) = 0.91 \sqrt{c^2kT}\,. \tag{7}$$

If the linear interaction with τ_{2g} and α_{1g} modes is taken into account, a smoothed symmetrical three-peaked absorption line shape is obtained if $a/c \ll 1$ holds (a is the coupling constant with the α_{1g} mode) as it is shown in Fig. 7 of [3]. Coupling of the T-state with α_{1g} and/or ε_g modes alone yields an unsplitted Gaussian-shaped absorption band.

Cho [4] has carried out numerical calculations of this problem including all possible linear interactions (α_{1g}, ε_g, and τ_{2g} modes). In his paper the line shapes for a great variety of coupling constants a, b, and c are given. However, often very similar line shapes were obtained from different sets of coupling constants i.e. the determination of the constants a and b from the comparison of measured and calculated line shapes is not unique.

As we deal with transitions between electron states arising from 3d states in centres with inversion symmetry, the fact must be considered that these transitions are parity-forbidden and partly allowed by coupling with odd vibrations resulting in a small ($\approx 10^{-4}$) and temperature-dependent oscillator strength f [13] according to

$$f = f_0 \coth \frac{\hbar\omega_u}{2kT}\,. \tag{8}$$

f_0 is the oscillator strength at zero temperature and ω_u the effective frequency of the odd vibrations. After Cho [4] (3) remains valid also for these transitions supposed that only linear coupling with even and odd vibrations is taken into account. That means the electronic transition matrix element $|\langle\Psi_g|\,\boldsymbol{P}\,|\Psi_e\rangle|^2$ in (3) (≈ 1 in the case of the $^1A_{1g} \rightarrow {}^1T_{1u}$ transition of Tl^+ type centres in alkali halides) is replaced by (8).

4.2 Triplet structures

In Sections 3.1 and 3.2 arguments are given that the measured triplet structures can be explained only by electron–lattice interaction. As mentioned in the foregoing section only the strong linear interaction of the electronic triplet states with τ_{2g} modes provides a triplet structure of the absorption band where the splittings increase linearly with $\sqrt{T}$ at high temperatures. Therefore the triplet structure of the transitions $^4A_{2g} \rightarrow {}^4T_{1g}(P)$ and $\rightarrow {}^4T_{1g}(F)$ of Co^{2+} and $^5E_g \rightarrow {}^5T_{2g}$ of Fe^{2+} must be interpreted as due to such a strong interaction with τ_{2g} modes. The analysis of the temperature dependence of the splittings enables us to determine the coupling constants c^2 (from the slope of the $\Delta E = f(\sqrt{T})$ plot at high temperatures according (7)) and the frequencies ω_τ (from $\Delta E(0)$ with help of c^2 or by fitting the measured temperature dependence in the whole range with (5)). Both procedures of determining ω_τ yield the same values listed in Table 1 but with large errors because of the errors of the ΔE values. The

Table 1

Parameters of the strong Jahn-Teller coupling ($T \otimes \tau_{2g}$) obtained from the analysis of the temperature dependence of the splittings with (5) to (7). $\Delta E(0)$ is the separation between the central peak and the side peaks of the triplet band extrapolated to $T = 0$ K and $S = E_{JT}/\hbar\omega_\tau$ the Huang-Rhys factor

	transition	$\Delta E(0)$ (cm^{-1})	c^2 (eV)	$\hbar\omega_\tau$ (cm^{-1})	E_{JT} (cm^{-1})	S_τ
Co^{2+}	$^4A_{2g} \rightarrow {}^4T_{1g}(P)$	630	0.71	170 ± 40	1900	11
	$^4A_{2g} \rightarrow {}^4T_{1g}(F)$	750	0.79	210 ± 40	2100	10
Fe^{2+}	$^5E_g \rightarrow {}^5T_{2g}$	700	1.39	110 ± 30	3750	35

Huang-Rhys factors $S_\tau = E_{JT}/\hbar\omega_\tau$ listed also in Table 1 show that in all cases a strong coupling is present, i.e. the use of the Franck-Condon approximation is justified [3]. As the triplet structure is fairly pronounced, the coupling with the modes α_{1g} and ε_g must be relatively small, i.e. $a/c \ll 1$ and $b/c < 1$ (see [4]) so that the determination of c^2 with (7) is permissible.

The frequencies ω_τ lie within the frequency range of vibrations of the undisturbed CdF_2 lattice [14] i.e. the interacting modes are no local modes as would be expected from the pure isotopic effect ($M'/M \approx 0.5$, M' mass of the transition metal ion, M mass of the Cd ion). However, the ionic radii of the $3d^n$ ions (r') are smaller than the radius r of the Cd^{2+} ion ($r'/r \approx 0.75$) and no excess charge is present, so that the force constants f' can become considerably smaller than f, the force constant in the undisturbed lattice. Thus the occurrence of a resonant band mode seems to be possible.

Considering only the linear interaction with τ_{2g} modes a symmetrical line shape is predicted. Honma [5] has shown that by quadratic interaction with τ_{2g} modes, which results from the possiblity that the curvatures of potentials of the excited T-state may be different from those of the ground state, an asymmetric triplet structure will be obtained. If the coupling constant of this quadratic interaction (γ) is positive (the curvatures of the potentials of the excited state are less than those of the ground state) the high-energy side peak will be more intensive and more pronounced than the low-energy one and this asymmetry increases with increasing temperature. The triplet band of the $^5E_g \rightarrow {}^5T_{2g}$ transition of Fe^{2+} exhibit these features as can be seen from Fig. 7. Therefore we can conclude that this asymmetry can be ascribed to the presence of a difference in the force constant between the $^5T_{2g}$ and ^5E_g states ($\gamma > 0$). The triplet structures of the $^4A_{2g} \rightarrow {}^4T_{1g}$ absorptions of Co^{2+} show only a small and nearly temperature-independent asymmetry, i.e. this kind of quadratic interaction can be neglected for these Co^{2+} transitions.

4.3 *Fine structures at low temperatures and doublet structure of the $^4A_{2g} \rightarrow {}^4T_{2g}$ transition*

A speciality of the orbital triplet states arising from 3d electron states consists in their first-order spin–orbit splitting. That means that possibly also in the case of strong Jahn-Teller coupling (in the sense $S \gtrapprox 5$, i.e. related to phonon energies) the spin–orbit coupling cannot be neglected. The occurence of the fine structures at low temperatures for the $\rightarrow {}^4T_{1g}(F)$ transition seems to be a strong suggestion that the validity of the simple theory used here becomes restricted when first-order spin–orbit coupling within the T-state takes place.

Table 2

Ratio of spin–orbit coupling constant λ and Jahn-Teller energy E_{JT} for the three $T \otimes \tau_{2g}$ systems. The value $|\lambda| = 70\ cm^{-1}$ for the $^4T_{1g}(P)$ state is taken from the MCD measurements [8], the other ones are the free ion values

	transition	$\lvert\lambda\rvert$ (cm^{-1})	$10\,\lvert\lambda\rvert/E_{JT}$
Fe^{2+}	$^5E_g \rightarrow {}^5T_{2g}$	104	0.28
Co^{2+}	$^4A_{2g} \rightarrow {}^4T_{1g}(P)$	70	0.37
	$^4A_{2g} \rightarrow {}^4T_{1g}(F)$	178	0.84

In Table 2 the values $10|\lambda|/E_{JT}$ are listed for the three $T \otimes \tau_{2g}$ systems investigated here, the behaviour of which can be described by the simple theory at temperatures above 100 K. It can be seen that for the transitions showing no fine structures $10|\lambda|/E_{JT}$ is considerably smaller than for the $\rightarrow {}^4T_{1g}(F)$ transition of Co^{2+}.

The $^4A_{2g} \rightarrow {}^4T_{2g}(F)$ absorption band shows an unexpected behaviour which cannot be explained on the base of the linear $T \otimes \tau_{2g}$ interaction in the whole temperature range. Spin–orbit coupling alone is not able to produce the splitting (the maximum overall splitting should be 360 cm^{-1} but $\Delta E(0) = 650\ cm^{-1}$ is measured). Moreover ΔE increases with $\sqrt{T}$ at high temperatures. From this follows that also the doublet structure must be caused by electron–lattice interaction. However, the straight line $\Delta E \sim \sqrt{T}$ does not go through the origin indicating that an additional interaction must be effective.

For a rough estimation we assume that also for the splitting of the doublet band $\Delta E \sim \sqrt{c^2kT}$ holds (as only τ_{2g} modes are able to produce a measurable splitting of a T-state), and obtain from the slope (see Fig. 5) that c of the $^4T_{2g}$ state is smaller by a factor 2 than c of the $^4T_{1g}$ states. According $E_{JT} \sim c^2$ the Jahn-Teller energy of the $T_{2g} \otimes \tau_{2g}$ interaction will be smaller by a factor 4, i.e. $E_{JT} \approx 500\ cm^{-1}$. With $|\lambda| = 178\ cm^{-1}$ we get $10|\lambda|/E_{JT} \approx 3.5$ for the $^4T_{2g}$ state, a value which is larger by an order of magnitude than the values for other $T \otimes \tau_{2g}$ systems (cf. Table 2). This estimation shows that probably in the case of the $^4T_{2g}$ state in the whole temperature range spin–orbit coupling and Jahn-Teller coupling are of comparable magnitude, i.e. neither of the coupling mechanism dominates the other. At low temperatures similar fine structures are resolved in the $^4T_{2g}$ band arising likewise from spin–orbit coupling as in the $^4T_{1g}(F)$ band.

The $^4T_{1g}(F)$ transition represents obviously a change-over case between the cases for which spin–orbit coupling is negligible ($10|\lambda|/E_{JT} \lesssim 0.5$) and those cases for which the absorption line shape in the whole temperature range is affected by spin–orbit coupling ($10|\lambda|/E_{JT} \gtrsim 2$).

The explanation of the fine structures and the doublet structure as due to the simultaneous action of Jahn-Teller coupling and first-order spin–orbit coupling of the orbital triplet states is very speculative but no other possibility seems available. A proof must be carried out by detailed theoretical investigation of the problem of the strong $T \otimes \tau_{2g}$ interaction including spin–orbit coupling for 3d T-states.

Acknowledgements

The author wishes to thank Dr. E. Brink for growing the crystals, Dipl. Phys. W. Gehlhoff for permission to quote his EPR results before publication and for discussions and Mrs. D. Nicolaus for technical assistance.

References

[1] M. D. Sturge, Solid State Phys. **20**, 91 (1967).
[2] R. Englman, The Jahn-Teller Effect in Molecules and Crystals, Wiley/Interscience Publishers, London 1972.
[3] Y. Toyozawa and M. Inoue, J. Phys. Soc. Japan **21**, 1663 (1966).
[4] K. Cho, J. Phys. Soc. Japan **25**, 1372 (1968).
[5] A. Honma, Sci. Light **18**, 33 (1969).
[6] W. Ulrici, phys. stat. sol. (b) **44**, K29 (1971).
[7] W. Gehlhoff, to be published.
[8] J. Donecker and J. Kluge, phys. stat. sol. (b) **59**, 163 (1973).
[9] J. Steger and E. Kostiner, J. chem. Phys. **58**, 3389 (1973).
[10] E. Brink, D. Becker, J. Teltow, and K. Tempelhoff, Kristall und Technik **7**, 1101 (1972).
[11] C. Simo, E. Banks, and S. Holt, Inorg. Chem. **8**, 1446 (1969).
[12] A. L. Stolov and Z. S. Yakovleva, Fiz. tverd. Tela **10**, 1513 (1968).
[13] A. D. Liehr and C. J. Ballhausen, Phys. Rev. **106**, 1161 (1957).
[14] P. Denham, G. R. Field, P. L. R. Morse, and G. R. Wilkinson, Proc. Roy. Soc. **A317**, 55 (1970).

(Received January 21, 1974)

phys. stat. sol. (b) 62, 443 (1974)

Subject calssification: 13.4; 20.3; 22.1.1; 22.1.2

Physikalisches Institut der Universität Stuttgart

Effects of Doping on the Condensed Electron–Hole State in Germanium and Silicon

By

R. W. Martin and R. Sauer

Radiative recombination in electron–hole drops has been investigated for doping levels up to $\approx 10^{18}$ cm^{-3} in Ge and Si. It was found that the energy of the condensed state and the density of electrons and holes become dependent on excitation and on impurity concentration above 10^{16} cm^{-3}. Characteristic differences of the lineshapes for the case of Ge and Si are discussed.

Die strahlende Rekombination in Elektron–Loch-Tröpfchen ist in Ge und Si für Dotierungskonzentrationen bis zu etwa 10^{18} cm^{-3} untersucht worden. Es stellt sich heraus, daß bei über 10^{16} cm^{-3} die Dichte und die Energie des kondensierten Elektron–Loch-Zustands abhängig werden von der Anregung und der Höhe der Dotierung. Charakteristische Unterschiede in den Linienformen für Ge und Si werden diskutiert.

1. Introduction

Since 1968, when Keldysh proposed for the first time the idea of electron–hole drop (EHD) formation, nearly all photoluminescence experiments on the condensed electron–hole state in Ge and Si have been performed with "pure" or unintentionally doped material with very low impurity concentrations. Most of these early experiments which established the concept of a liquid phase of electrons and holes are summed up in the review article by Pokrovskii [1]. The presence of shallow donors or acceptors up to concentrations of $n_d \approx 10^{15}$ cm^{-3} rather causes sharp lines in the photoluminescence spectra in addition to the well known bound exciton (BE) line instead of having profound influence on the EHD state. The new features of these multi-structured spectra were described by Kaminskii and Pokrovskii [2] and by Sauer [3] in the case of Si. Similar additional line structure in photoluminescence spectra due to impurities exists in Ge [4]. Essentially new results in more highly doped samples have been reported by Göbel et al. [5] and by Benoît à la Guillaume and Voos [6] in Si and Ge. These results are concerned with the appearance of EHD emission without phonon participation (NP emission) in the doping range of 10^{15} to 10^{18} cm^{-3} in both materials. Recently, Novikov et al. [7] studied the influence of impurities on energies and shapes of the EHD emission from Ge crystals with impurity concentrations up to 10^{16} cm^{-3}. Energy shifts of the EHD bands in phosphorus-doped silicon were reported for very high doping levels ($\approx 10^{18}$ cm^{-3}) by Halliwell and Parsons [8].

In this work, we give a detailed study of lineshapes and energies of the condensed e–h state as a function of excitation and doping level and report on a number of new effects which have not yet been described. We limit ourselves to a maximum impurity concentration of $\approx 10^{18}$ cm^{-3}. A partial interpretation of the results is given including the influence of surface energy on the EHD and possible sample filling with e–h liquid at high excitation and high impurity concentration.

2. Germanium

Phonon-free (NP) emission is observed in Ge already at an impurity concentration of about 4×10^{13} cm^{-3} [5]. The intensity of the NP line was found to grow linearly with doping level relative to the LA-phonon-assisted line in the case of Ge:As [6] and Ge:P. The intensity ratio I_{NP}/I_{LA} is approximately 1 for $n_{As} \approx 2\times10^{16}$ cm^{-3} and $n_P \approx 3\times10^{16}$ cm^{-3}, respectively. It is generally one or two orders of magnitude smaller for Sb, In, and Ga and shows a sublinear dependence on doping level.

The liquid–gas phase diagram [9 to 11] predicts that the density n_c within the condensed phase is constant at a certain temperature. A variation of the laser intensity only changes the sum of condensed carriers (size or number of droplets) but not their density n_c. As the lineshapes and energetic positions of the luminescence lines depend only on n_c [12, 13] they are expected to be independent of laser power. This is indeed observed for intrinsic samples and doped samples up to a doping level near 10^{16} cm^{-3}.[1]) Remarkable changes are observed at doping levels between 10^{16} and 10^{17} cm^{-3} where both the linewidths and energetic positions show a strong dependence on excitation intensity. This is illustrated in Fig. 1 where the EHD^{LA} band of Ge:In (6×10^{16} cm^{-3}) is shown for three different excitation levels. In Fig. 1, the linewidths diminish strongly with decreasing laser power whereby the peak positions shift slightly to higher energies. From the variation of the high energy edge (chemical potential μ) it follows that the binding energy per particle E_B relative to the free exciton [14] increases with decreasing excitation (this is correct under the assumption that the formalism

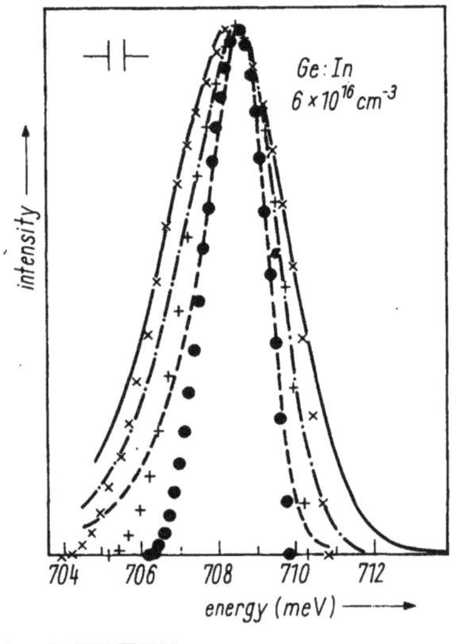

Fig. 1. EHD^{LA} bands of Ge:In (6×10^{16} cm^{-3}) at 1.6 K and at different excitation levels (normalized). —— experiment, ××× theory, 440 W/cm^2; —·—·— experiment, +++ theory, 58 W/cm^2; — — — experiment, ••• theory, 1.9 W/cm^2. Values of the laser power are corrected for transmission losses of windows and lenses, and for reflection losses at sample and mirrors

[1]) Novikov et al. [7] reported changes of the EHD lines already for $n_d < 10^{16}$ cm^{-3}. This is probably due to different excitation sources. Novikov et al. used a He–Ne laser with $\lambda = 1.52$ µm (volume excitation) whereas we used a Kr laser with $\lambda = 647$ nm (surface excitation), maximum power 650 mW.

Table 1

EHD parameters of Fig. 1: Density within electron–hole drops (n_c), Fermi energy of electrons and holes (E_F), chemical potential (μ), and binding energy (E_B). First three lines: Ge:In, 6×10^{16} cm^{-3}, thickness 3.5 mm; fourth line: intrinsic Ge; fifth line: Ge:In, 6×10^{16} cm^{-3}, thickness 0.1 mm

exc. intensity (W/cm²)	n_c (cm^{-3})	E_F (meV)	μ (meV)	$h\nu_{max}$ (meV)	E_B (meV)
440	2.59×10^{17}	6.9	710.8	708.4	2.4
58	1.65×10^{17}	5.1	710.3	708.6	2.9
1.9	9.65×10^{16}	3.6	709.8	708.7	3.4
	2.2×10^{17}	6.1	710.6	708.5	2.6
440	3.54×10^{17}	8.5	711.4	708.3	1.8

developed for pure samples [12, 13] is still valid for samples with impurity concentrations discussed in this section). Fitting the theoretical lineshapes according to Pokrovskii et al. [15] to the experimental ones yields the Fermi energy of electrons and holes, E_F, the density, n_c, and the chemical potential, μ [16]. Values for the sample of Fig. 1 are compiled in the first three lines of Table 1. The calculation was carried out using the masses quoted by Brinkman and Rice [17].

The halfwidths for samples doped with Sb, Ga, and In are plotted versus laser power in Fig. 2. They are independent of excitation intensity for doping levels n_d below 10^{16} cm^{-3} (curves 1). The linewidths grow with excitation for $n_d > 10^{16}$ cm^{-3}. For Ge:In and Ge:Ga (Fig. 2b, c) they even exceed those of intrinsic material if the laser power is beyond a critical value. This critical excitation intensity was found to be higher for Ge:Sb (Fig. 2a) than for Ge:In and Ge:Ga. Because of limited

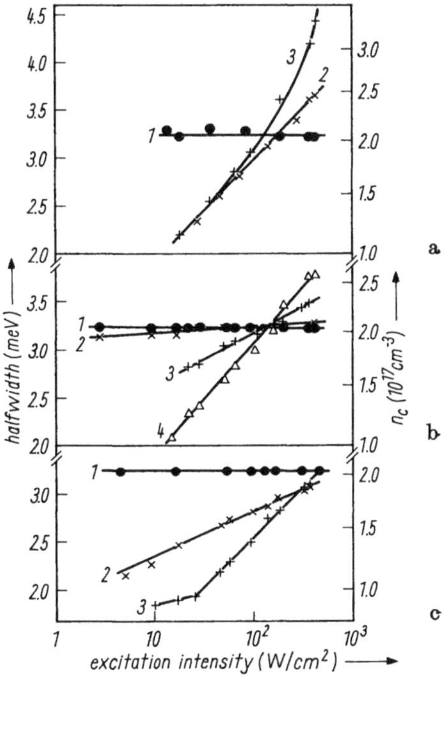

Fig. 2. Halfwidths of EHD^{LA} bands of differently doped crystals as a function of laser power. a) Ge:Sb; (1) 3×10^{15} cm^{-3}, (2) 2×10^{16} cm^{-3}, (3) 7×10^{16} cm^{-3}. b) Ge:Ga; (1) 3×10^{14} cm^{-3}, (2) 1.5×10^{16} cm^{-3}, (3) 3×10^{16} cm^{-3}. (4) 5×10^{16} cm^{-3}. c) Ge:In; (1) 3.5×10^{15} cm^{-3}, (2) 6×10^{16} cm^{-3} (thickness 3.5 mm), (3) 6×10^{16} cm^{-3} (thickness 0.1 mm). The density n_c (right-hand side scale) was obtained using the proportionality between halfwidth and $n_c^{2/3}$

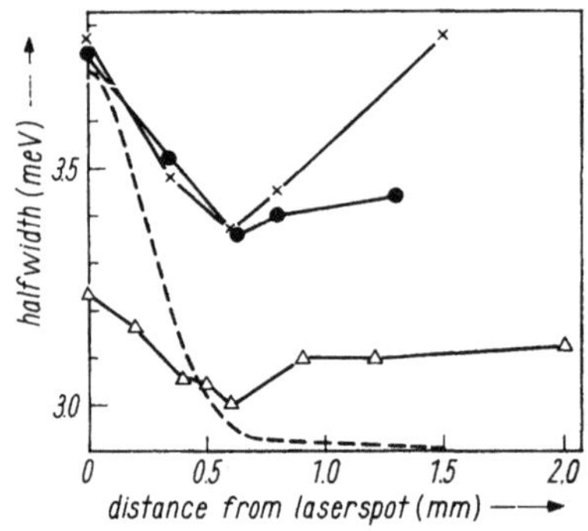

Fig. 3. Halfwidths of EHD^{LA} bands of differently doped crystals as a function of the distance from the laser spot. × Ge:In; 6×10^{16} cm^{-3}; ● Ge:Ga; 5×10^{16} cm^{-3}; △ Ge:Sb; 2×10^{16} cm^{-3} ———. Intensity profile of Ge:In; 6×10^{16} cm^{-3} (intensity in arb. units). Laser power ≈ 410 W/cm^2, $T = 1.6$ K

laser power we were therefore not able to observe this broadening in Ge:Sb. Note that the halfwidths decrease with increasing n_d at "low" laser power whereas they increase at "high" laser power. Benoît à la Guillaume and Voos [6] only observed a reduction of n_c with increasing doping level. Their laser power obviously was below the critical value mentioned above.

The effects observed at "low" excitation intensity can be explained by assuming that in highly doped samples many small droplets are created instead of only a few big ones in pure samples [18, 7]. Indeed, the role of impurity atoms as nucleation centres suggests that more droplets are formed with enhanced doping. They must be small as the volume of droplets is roughly proportional to the reciprocal value of their number. If surface effects like surface energy and curvature energy [19 to 21] play a role in the energy balance of the system, their influence will be the stronger the smaller the droplets are. Therefore one must expect deviations from the theoretical calculations [12, 13] if e.g. the surface energy becomes comparable to the binding energy. For a droplet of radius r, the surface energy per electron–hole pair, E_S/N, is given by $E_S/N = 3\gamma/(n_c r)$, where γ is the surface energy per cm^2.

An estimation using $\gamma = 10^{-4}$ erg/cm^2 [20, 21] yields $E_S/N = 0.2$ meV for a droplet with radius 500 Å and $n_c = 2 \times 10^{17}$ cm^{-3}. This value can amount to 0.5 meV if densities smaller than 2×10^{17} cm^{-3} (compare Table 1) are used. Experimentally determined values of the droplet radii for relatively pure samples are typically 5 μm [1]. The influence of the surface term E_S/N for such large drops is too small to be detected in experiment.

The hypothesis of the influence of the surface on EHD properties is supported by the following experiment: When the drops diffuse away from the place of their creation, they evidently will shrink due to recombination and evaporation. Thus, luminescence light coming from droplets which have moved away a certain distance from the laser spot should show the typical features of emission from small droplets. This was actually found and is shown in Fig. 3 where the halfwidths of the EHD^{LA} bands are plotted versus the distance from the laser spot (experimental details are described in [22]). The halfwidths and therefore n_c diminish with growing distance from the laser spot. The subsequent increase for $x \gtrsim 0.6$ mm has a different reason: From the intensity profile (dashed line in Fig. 3) it can be seen that the minimum coincides approximately with the point where the EHD have practically disappeared. The extremely weak signals detected for $x > 0.6$ mm are due to luminescence light scattered within the sample [22, 23]. As this scattered light bears mainly the spectroscopic features of the emission at maximum intensity near $x = 0$, the linewidths increase again for $x > 0.6$ mm.

From the model described it is expected that the halfwidths for doped samples asymptotically approach those of pure ones with increasing laser power (larger droplet size). Instead, a line broadening is observed for very high excitation intensity (Fig. 2). This broadening at high laser power can be understood assuming that a certain volume of the crystal is entirely filled with electron–hole liquid. In this situation, according to the phase diagram the density can be larger than n_c depending on laser power. A crude estimation shows that with our maximum laser power we can maintain a stationary density of 5×10^{17} cm^{-3} in a hemisphere of 0.3 mm radius which is a realistic value for the diffusion length in the samples used here [22]. The steep rise of the halfwidth of the thin crystal (0.1 mm) in Fig. 2c supports this idea. In this crystal, the diffusion perpendicular to the sample surface is prevented and the critical density n_c is reached at lower laser power than necessary for thick samples (the EHD parameters of the thin sample are listed in the last line of Table 1). The two mechanisms proposed here qualitatively explain the available data. The observed smooth variation of the lineshapes (e.g. curves 2 to 4 in Fig. 2b) remains a problem and might be pure accidental.

If the doping level is raised above $\approx 10^{17}$ cm^{-3}, the peaks show a redshift at constant laser power, finally leading to donor–valence band (acceptor–conduction band) transitions. We conclude this from energetic arguments, lineshape, and temperature dependence. The energy shifts with doping (Table 2) are much

Table 2

Change from EHD emission to donor–valence band transitions at high doping levels

doping (cm^{-3})	luminescence	$h\nu_{max}^{LA}$ (meV)	$h\nu_{max}^{NP}$ (meV)
$< 10^{16}$	EHD	708.5	736
Sb; 7×10^{16}	EHD	707.8	735
Sb; 3×10^{17}	donor–valence band	706	733.2
P; 1×10^{17}	EHD	707.3	734.7
P; 5×10^{18}	donor–valence band	702	730

larger than those which we observed as a function of excitation level (Table 1). Table 2 shows that the energy of the donor–valence band transition is approached at high doping levels. The lineshapes deviate strongly from the usual EHD case. They are much broader (depending on doping) and show an extended low-energy wing typical for donor–valence band transitions involving tail states [24]. Finally, at increasing temperature these lines could be observed with only slightly reduced intensity up to 30 K whereas the EHD bands very rapidly vanish for $T > 4.2$ K.

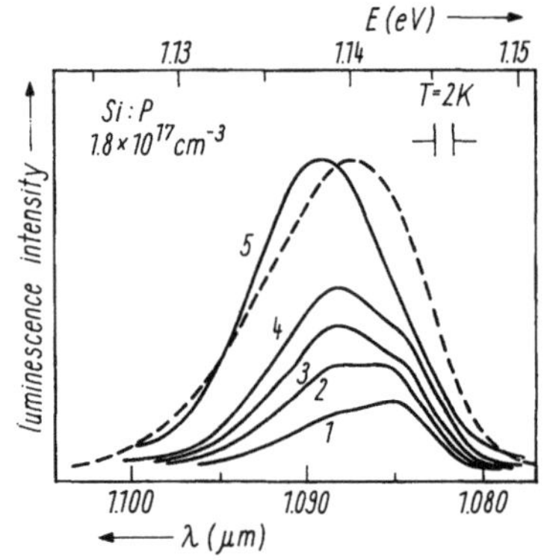

a

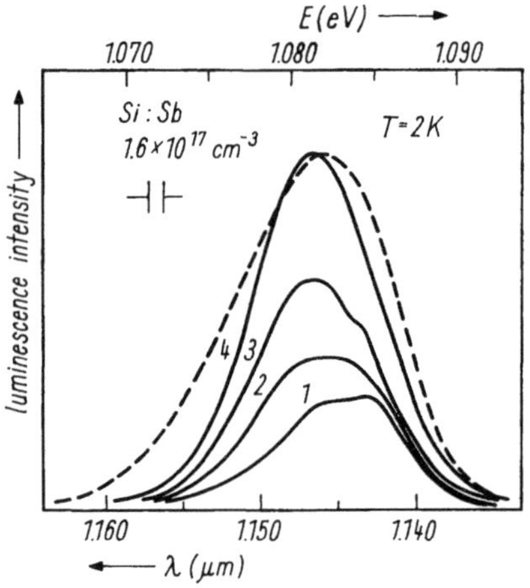

b

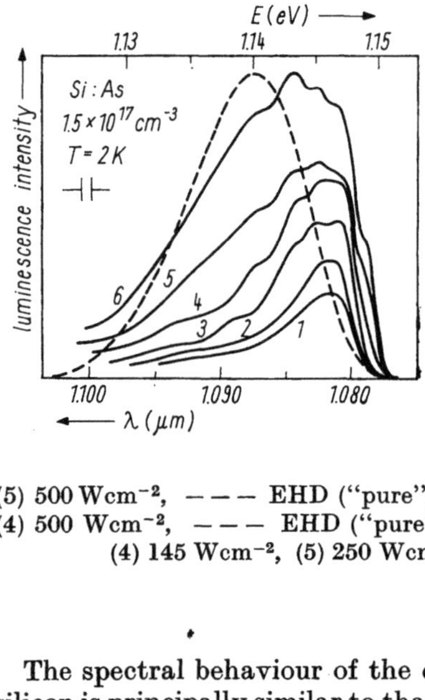

c

Fig. 4. EHD spectra of highly doped Si:P, Si:Sb, and Si:As at $T = 2$ K as a function of excitation. The spectra are shown with b) TO phonon participation or a) and c) phonon-free (NP spectra), depending upon which of the spectra is better detected by the S-1 type cathode of the photomultiplier [5]. Intensity units are arbitrary, intensities measured at a) $\lambda = 1.089$ μm, b) $\lambda = 1.147$ μm, and c) $\lambda = 1.082$ μm increase nearly linearly with excitation. The dashed-line EHD spectrum of a pure sample is always given for comparison. In a) and c), these spectra were measured as TO replicas and transferred to the NP region by adding the TO phonon energy of 58.2 meV. a) (1) 15 Wcm^{-2}, (2) 38 Wcm^{-2}, (3) 100 Wcm^{-2}, (4) 215 Wcm^{-2}, (5) 500 Wcm^{-2}, – – – EHD ("pure"); b) (1) 25 Wcm^{-2}, (2) 50 Wcm^{-2}, (3) 200 Wcm^{-2}, (4) 500 Wcm^{-2}, – – – EHD ("pure"); c) (1) 25 Wcm^{-2}, (2) 50 Wcm^{-2}, (3) 70 Wcm^{-2}, (4) 145 Wcm^{-2}, (5) 250 Wcm^{-2}, (6) 500 Wcm^{-2}, – – – EHD ("pure")

3. Silicon

The spectral behaviour of the condensed electron–hole state in highly doped silicon is principally similar to that of germanium. Important differences appear: The lineshapes deviate from the characteristic shapes which are typical for EHD emission of pure Ge and Si samples. Moreover, the bands are in some cases modified by additional structure. We consider two striking features of the high-concentration spectra:

a) The lineshapes and energies of the EHD bands are no longer independent of the excitation level as in "pure" samples.

b) The lineshapes of the EHD bands at high excitation vary with the doping level in a characteristic manner whereas they do not depend on the doping level for $n_d \lessapprox 10^{15}$ cm^{-3}.

These properties are now described in detail.

a) Spectra of the EHD state as a function of the excitation level are shown in Fig. 4, a to c for Si:P, Si:Sb, and Si:As. These spectra consist of emission bands which at low excitation level exhibit a maximum at an energy definitely larger than the maximum energy of EHD luminescence of pure crystals. In Si:P and Si:Sb (Fig. 4a, b), a weaker second maximum appears ≈3 meV below the first one. At increasing excitation levels, both maxima change their role. The lower one becomes dominant until a broad and structureless emission band at even lower energy develops at the highest excitation levels. Deviating from

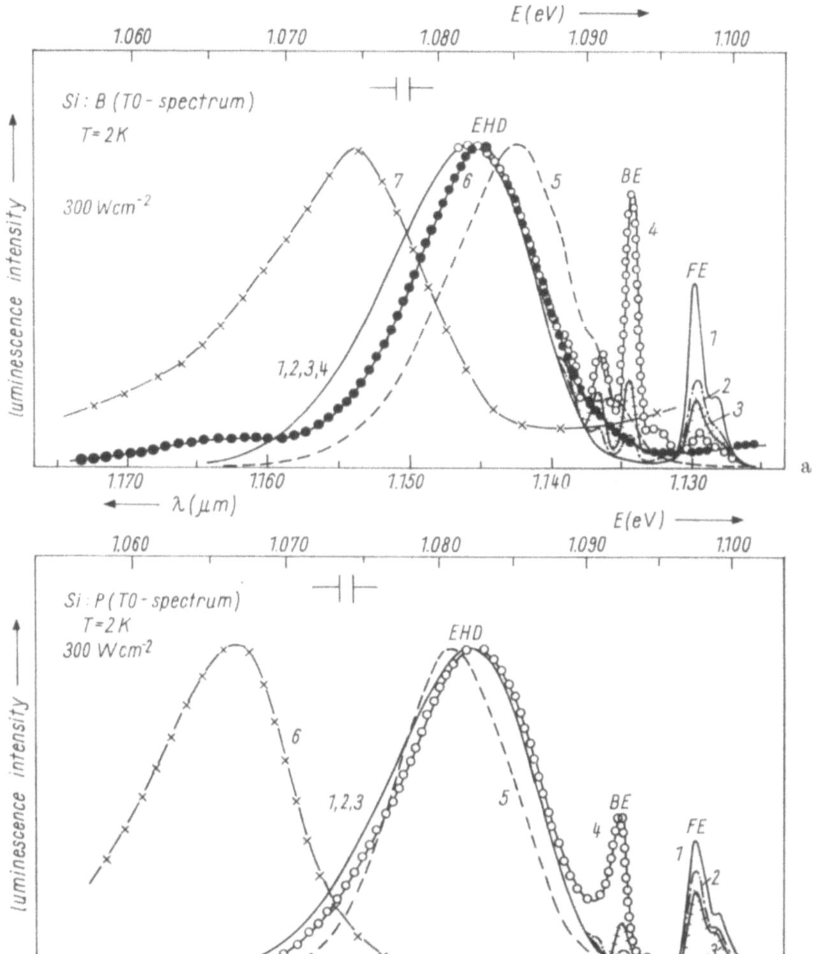

Fig. 5. EHD^{TO} spectra of a) Si:B and b) Si:P at high excitation as a function of doping level. The spectra are normalized, the measured intensities differ by less than 20%. a) (1) 3.3×10^{12} cm^{-3}, (2) 2.2×10^{13} cm^{-3}, (3) 1.3×10^{14} cm^{-3}, (4) 2.6×10^{15} cm^{-3}, (5) 7×10^{16} cm^{-3}, (6) 3.6×10^{17} cm^{-3}, (7) 1.2×10^{18} cm^{-3}; b) (1) 1.2×10^{12} cm^{-3}, (2) 8×10^{12} cm^{-3}, (3) 4.8×10^{13} cm^{-3}, (4) 6×10^{16} cm^{-3}, (5) 1.8×10^{17} cm^{-3}, (6) 6.4×10^{17} cm^{-3}

the case of Ge, these spectra show lineshapes which are markedly different from those of pure samples. They can no longer be fitted satisfactorily by the simple convolution integral method of Pokrovskii et al. [15] which was still used above for the spectra of Ge. An analogous behaviour was observed in the case of Si:Al and Si:B[2]) at doping levels of $\approx 10^{17}\,cm^{-3}$. The halfwidths of all spectra in Fig. 4a, b are smaller than those of pure samples. The spectra of Si:Sb and Si:P resemble each other. In contrast, Si:As in Fig. 4c exhibits spectra where at increasing excitation level the high-energy components of the emission do not vanish and thus produce halfwidths which highly exceed the EHD halfwidths. Besides, the totally modified lineshapes of the curves at higher excitation level indicate that the character of the electronic states must have been altered profoundly. The peaks superimposed on the bands probably originate from bound exciton states as was described in a previous paper for Si:B and Si:P at low impurity level [3] and for Si:B at high impurity concentration [25], Thus, compared to Fig. 4a, b and Fig. 5 Si:As seems to be an exception. A clear interpretation of these spectra is therefore not possible.

Returning to the case of Si:P and Si:Sb, we may interprete the general features of these spectra (neglecting their structure) similarly as it was done for germanium. The impurities serve as condensation centres. At low excitation level, they determine the number of droplets which do not fuse to drops of large radii. Thus, surface terms have considerable influence on the energy and simultaneously on the halfwidth of the emission. In Si we could not observe halfwidths larger than those of the pure EHD bands (omitting the pathological Si:As) as we did in Ge. This is probably only due to experimental limitations: The lifetime of electrons and holes in the condensed phase are $\approx 40\,\mu s$ [9] and 69 ns [26] for Ge and Si, respectively. Therefore at the same laser power the stationary density of carriers is smaller in Si than in Ge by a factor of $\approx 1.8 \times 10^{-3}$. On the other hand, the density in the liquid phase in Si ($\approx 3.4 \times 10^{18}\,cm^{-3}$) is much higher than in Ge ($2.2 \times 10^{17}\,cm^{-3}$). Consequently, we do not achieve densities high enough to fill a part of the sample with an electron–hole liquid.

b) To obtain the influence of the doping concentration on the lineshapes and energetic positions of the EHD bands, we investigated the spectra of samples doped with boron or phosphorus at different concentrations and at identical experimental conditions (constant temperature and constant excitation level, Fig. 5). The situation is quite analogous for both Si:B and Si:P even in details. Up to doping levels of $10^{15}\,cm^{-3}$ the samples exhibit the usual EHD emission bands whose lineshapes coincide exactly. On the high-energy part of the spectrum, the lines of the free exciton (FE) and the bound exciton (BE) are observed. Energetically below the BE emission, additional lines arise which for lower excitation levels were described more detailed in [3, 25]. The intensity of the FE line is reduced with increasing doping level until it vanishes completely at $n_d \approx 10^{17}\,cm^{-3}$ (Fig. 5). Contrarily, the BE line first grows and is most intense at $n_d \approx 10^{16}\,cm^{-3}$. At higher doping it decreases and at $n_d \approx 10^{17}\,cm^{-3}$ it has practically disappeared.

---

[2]) The spectra of Si:B at $n_d = 7 \times 10^{16}\,cm^{-3}$ excellently show the mentioned additional structure. They are subject of a seperate paper in the context of bound multiple-exciton complexes [25]. The sample of Si:Al contains a considerable amount of boron and is therefore omitted here.

An essential influence of the impurities on the EHD state is apparent at $n_d > 10^{16}$ cm^{-3}. Beginning above 10^{16} cm^{-3}, the EHD bands contract and shift to higher energies. With slightly increased n_d they shift back to lower energies maintaining their narrow halfwidths. Eventually, upon further increase in doping level up to $\approx 10^{18}$ cm^{-3}, they shift to even lower energies. This latter rapid shift was also observed by Halliwell and Parsons [8] in the case of Si:P connected with a large line broadening. These authors claim that even at impurity concentrations higher than 10^{18} cm^{-3} the emission bands are due to radiative recombination of electron–hole drops. We rather assume that these bands originate from donor band–valence band (acceptor–conduction band) transitions involving tail states. This idea is supported by the considerable and rapid shift of the bands at $n_d \gtrsim 10^{18}$ cm^{-3} and by the extremely large halfwidths reported by Halliwell and Parsons [8]. A decision based on temperature-dependent measurements is not possible as we only investigated the low-temperature spectra whereas the critical temperature of the condensed phase is at least 20 K [27] for Si.

4. Summary

We have shown that for both, Ge and Si, the EHD luminescence spectra change significantly in the presence of impurities. For not too high excitation levels, the halfwidths of the EHD bands diminish with increasing doping level whereby the peak positions show first a blueshift and then a redshift. For very high doping concentrations ($\gtrsim 10^{17}$ cm^{-3} for Ge, $\gtrsim 10^{18}$ cm^{-3} for Si) a change from EHD emission to donor–valence band (acceptor–conduction band) transitions is observed. Unlike pure material, the halfwidths become dependent upon excitation intensity and may even exceed those of intrinsic samples (Ge:In, Ge:Ga). The effects are explained assuming the creation of very small droplets in doped crystals: In small droplets a considerable influence of surface effects on the EHD properties is expected.

There are remarkable differences in the lineshapes of Si and Ge. In Ge, the EHD bands are always smooth (at 1.6 K) and fit satisfactorily with the theoretical lineshape. This is not the case for Si where the EHD bands show additional structure. The structure is presumably due to bound exciton complexes and complicates the analysis of the spectra.

Acknowledgements

The authors would like to thank M. H. Pilkuhn and R. Conradt for helpful discussions. The financial support of the Deutsche Forschungsgemeinschaft in the framework of the SFB 67 is gratefully acknowledged.

References

[1] Ya. E. Pokrovskii, phys. stat. sol. (a) **11**, 385 (1972).
[2] A. S. Kaminskii and Ya. E. Pokrovskii, Zh. eksper. teor. Fiz., Pisma **11**, 381 (1970); Soviet Phys. — J. exper. theor. Phys., Letters **11**, 255 (1970).
A. S. Kaminskii, Ya. E. Pokrovskii, and N. V. Alkeev, Zh. eksper. teor. Fiz. **59**, 1937 (1971); Soviet Phys. — J. exper. theor. Phys. **32**, 1048 (1971).
[3] R. Sauer, Phys. Rev. Letters **31**, 376 (1973).
[4] R. W. Martin, State Solid Commun., to be published.
[5] E. Göbel, R. W. Martin, M. H. Pilkuhn, and R. Sauer, Proc. 11th Internat. Conf. Phys. Semicond., Warsaw 1972 (p. 691).

[6] C. Benoît à la Guillaume and M. Voos, Solid State Commun. **11**, 1585 (1972).
[7] B. V. Novikov, R. L. Korchazhkina, and N. S. Sokolov, Fiz. tverd. Tela **15**, 459 (1973); Soviet Phys. — Solid State **15**, 326 (1973).
[8] R. E. Halliwell and R. R. Parsons, Solid State Commun. **13**, 1245 (1973).
[9] C. Benoît à la Guillaume, M. Voos, and F. Salvan, Phys. Rev. B **5**, 3079 (1972).
[10] R. W. Martin and M. H. Pilkuhn, Solid State Commun. **11**, 571 (1972).
[11] R. N. Silver, Phys. Rev. B **8**, 2403 (1973).
[12] M. Combescot and P. Nozières, J. Phys. C **5**, 2369 (1972).
[13] W. F. Brinkman, T. M. Rice, P. W. Anderson, and S. T. Chui, Phys. Rev. Letters **28**, 961 (1972).
[14] C. Benoît à la Guillaume and M. Voos, Solid State Commun. **12**, 1257 (1973).
[15] Ya. E. Pokrovskii, A. Kaminskii, and K. I. Svistunova, Proc. 10th Internat. Conf. Phys. Semicond. Cambridge Mass. 1970 (p. 504).
[16] C. Benoît à la Guillaume and M. Voos, Phys. Rev. B **7**, 1723 (1973).
[17] W. F. Brinkman and T. M. Rice, Phys. Rev. B **7**, 1508 (1973).
[18] R. W. Martin, Verh. DPG(VI) **8**, 296 (1973).
[19] H. Büttner, Festkörperprobleme **13**, 145 (1973).
[20] L. M. Sander, H. B. Shore, and L. J. Sham, Phys. Rev. Letters **31**, 533 (1973).
[21] T. M. Rice, to be published.
[22] R. W. Martin, phys. stat. sol. (b) **61**, 223 (1974).
[23] C. Benoît à la Guillaume, M. Voos, and F. Salvan, Phys. Rev. Letters **27**, 1214 (1971).
[24] C. Benoît à la Guillaume and J. Cernogora, phys. stat. sol. **35**, 599 (1969).
[25] R. Sauer, Solid State Commun., to be published.
[26] J. D. Cuthbert, Phys. Rev. B **1**, 1552 (1970).
[27] B. M. Ashkinadze, I. P. Kretsu, S. M. Ryvkin, and I. D. Yaroshetskii, Zh. eksper. teor. Fiz. **58**, 507 (1970); Soviet Phys. — J. exper. theor. Phys. **31**, 271 (1970).

(Received December 28, 1973)

Note added in proof:

In agreement with our results at "low" laser power, C. Benoît à la Guillaume and M. Voos have observed similar shifts of the peak and high energy edge of the EHD^{LA} band of Ge doped with As and Sb (3×10^{16} cm^{-3}). (Private communication by M. Voos.)

phys. stat. sol. (b) **62**, 453 (1974)

Subject classification: 13 and 20; 22.1.2

Sektion Physik der Martin-Luther-Universität Halle

Reconstructed Three-Dimensional Valence Electron Momentum Density of Si from Compton Profile Measurements

By

W. Schülke

Recently measured anisotropies of Si valence electron Compton profiles with the scattering vector in the [100], [110], and [111] direction are analyzed together with calculated anisotropies of the corresponding Compton profiles from a modified free electron model. From both sets of profiles the momentum density $\varrho(\boldsymbol{p})$ is reconstructed. In comparison with the modified free electron model the experimental momentum density is strongly reduced in the vicinity of $p = 0.6$ to 1.0 at. units along the [100] direction and enhanced mainly in the vicinity of $p = 1.1$ at. units along the [110] direction. These main features of the reconstructed experimental momentum density are well reproduced by a pseudopotential band calculation of valence electron momentum densities along the [100], [110], and [111] directions. Also the influence of the orthogonalization term on the momentum density is estimated.

Die kürzlich gemessenen Anisotropien der Valenzelektronen-Comptonprofile von Si mit einer [100]-, [110]- und [111]- Streuvektorrichtung werden zusammen mit den berechneten Anisotropien der zugehörigen Comptonprofile eines modifizierten freien Elektronenmodels analysiert. Aus diesen beiden Sätzen von Profilen wurde die Impulsdichte $\varrho(\boldsymbol{p})$ rekonstruiert. Im Vergleich zum modifizierten freien Elektronenmodell ist die experimentelle Impulsdichte im Bereich von $p = 0{,}6$ bis 1,0 at. Einheiten längs der [100]-Richtung stark reduziert, jedoch angehoben hauptsächlich im Bereich von $p = 1{,}1$ at. Einheiten längs der [110]-Richtung. Diese Haupteigenschaften der rekonstruierten experimentellen Impulsdichte werden auch gut durch eine Pseudopotential-Bandrechnung der Valenzelektronenimpulsdichten längs der [100]-, [110]- und [111]-Richtungen wiedergegeben. Der Einfluß der Orthogonalisierungsterme wird ebenfalls abgeschätzt.

1. Introduction

Recently the importance of the spectral analysis of inelastically scattered X-rays (Compton profile (CP) measurements) has been recognized again [1 to 4]. The main reason is to get information about the momentum density especially of valence electrons in solids.

Recently CP measurements on Si single crystals are done by Schülke et al. [3] and Schülke and Lucht [5] independently of measurements by Weiss [6] and Reed and Eisenberger [7]. The valence electron parts of the profiles given in [5] with the scattering vector in the [100], [110], and [111] direction exhibit anisotropies above the statistical error of the results and are analyzed in the following sections. The CP's of [3] show nearly the same anisotropy but additionally some strong oscillations in the region of small p_z, that are mainly due to the overestimation of small oscillations by the deconvolution procedure. The measured anisotropies of [5] are in a rather good agreement with Reed's results [7], whereas they disagree with those of Weiss [6]. It could be shown in [5] that the averaged experimental valence CP is in good agreement with a theoretical profile calculated on the basis of hybridized atomic wave functions.

But this hybridization model does not yield any anisotropy of this profiles. Thus it seems necessary to analyse this anisotropy in more detail and to compare the results of this analysis with calculations using better wave functions.

2. Basic Formulas for CP Calculation

Assuming the validity of the impulse approximation [8] the differential Compton scattering cross-section for an electron system described by one-electron Bloch wave functions $\psi_{\boldsymbol{k},\nu}$ is given by

$$\frac{\mathrm{d}\sigma}{\mathrm{d}\omega\,\mathrm{d}\Omega} = \left(\frac{\mathrm{d}\sigma}{\mathrm{d}\Omega}\right)_{\text{Thomson}} \left(\frac{\omega_1}{\omega_2}\right) \int \mathrm{d}^3p\varrho(\boldsymbol{p})\,\delta\left(\omega - \frac{K^2}{2m} - \frac{\boldsymbol{K}\boldsymbol{p}}{m}\right), \tag{1}$$

where, according Lundqvist and Lydén [9], the electron momentum density $\varrho(\boldsymbol{p})$ is approximated by

$$\varrho(\boldsymbol{p}) = (2\pi)^{-3} \sum_{\boldsymbol{k},\nu} n(\boldsymbol{k}, \nu)\, |\textstyle\int \mathrm{d}^3r\, \psi_{\boldsymbol{k},\nu} \exp(-i\boldsymbol{p}\boldsymbol{r})|^2 \tag{2}$$

neglecting all nondiagonal elements of the averaged occupation number. $n(\boldsymbol{k}, \nu)$ are the diagonal elements of the averaged occupation number of the state $|\boldsymbol{k}, \nu\rangle$. $\omega = \omega_1 - \omega_2$ is the energy transferred to the scattering system (ω_1 is the photon energy before the scattering process, ω_2 is the photon energy after scattering); $\boldsymbol{K}$ is the momentum transfered to the scattering system.

Thus the measurement of the Compton profile $I(l)$, where l is the wavelength shift from the centre of the CP, yields

$$I(l) \sim I(p_z) \sim \textstyle\int\int \varrho(\boldsymbol{p})\, \mathrm{d}p_x\, \mathrm{d}p_y \tag{3}$$

assuming that the scattering vector $\boldsymbol{K}$ is the z-axis of Cartesian coordinates and p_z is well approximated by

$$q = \frac{mcl}{2\lambda_0 \sin(\theta/2)}, \tag{4}$$

where λ_0 is the primary wavelength and θ is the scattering angle; (since $\omega \ll \omega_1$, ω_2 under experimental conditions used here, (ω_1/ω_2) is approximated by 1).

The substitution of a plane wave expansion of $\psi_{\boldsymbol{k},\nu}(\boldsymbol{r})$

$$\psi_{\boldsymbol{k},\nu}(\boldsymbol{r}) = V^{-1/2} \sum_{\boldsymbol{G}} a_{\boldsymbol{G},\nu}(\boldsymbol{k}) \exp[i(\boldsymbol{k} + \boldsymbol{G}), \boldsymbol{r}] \tag{5}$$

into (2) yields

$$\varrho(\boldsymbol{p}) = \frac{V}{(2\pi)^3} \sum_{\nu} n(\boldsymbol{p} - \boldsymbol{G}', \nu)\, |a_{\boldsymbol{G}',\nu}(\boldsymbol{p} - \boldsymbol{G}')|^2, \tag{6}$$

where $\boldsymbol{G}'$ is a reciprocal lattice vector that translates $\boldsymbol{p}$ into the first Brillouin zone.

Equation (6) may also be reexpressed in terms of the extended zone scheme by

$$\varrho(\boldsymbol{p}) = \frac{V}{(2\pi)^3} \sum_{\boldsymbol{G}''} n(\boldsymbol{p} - \boldsymbol{G}'')\, |a_{\boldsymbol{G}''}(\boldsymbol{p} - \boldsymbol{G}'')|^2, \tag{7}$$

where $\boldsymbol{G}''$ are those reciprocal lattice vectors, for which $\boldsymbol{p} - \boldsymbol{G}''$ lies within the occupied region of the extended zone scheme.

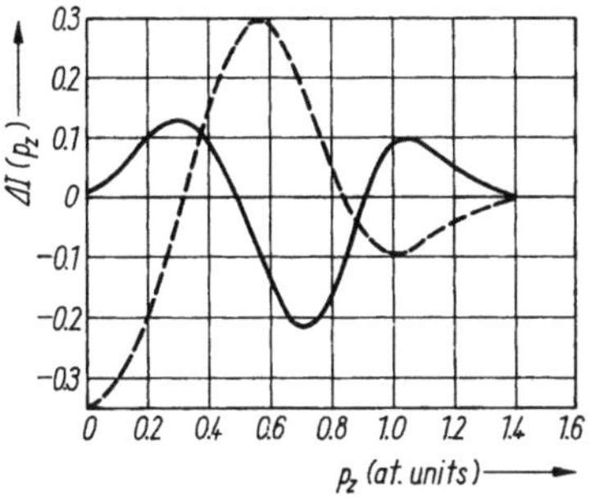

Fig. 1. Calculated anisotropy of CP's from the modified free-electron model given as $I(p_z)_{[100]} - I(p_z)_{[110]}$ (solid curve) and $I(p_z)_{[111]} - I(p_z)_{[100]}$ (dashed curve)

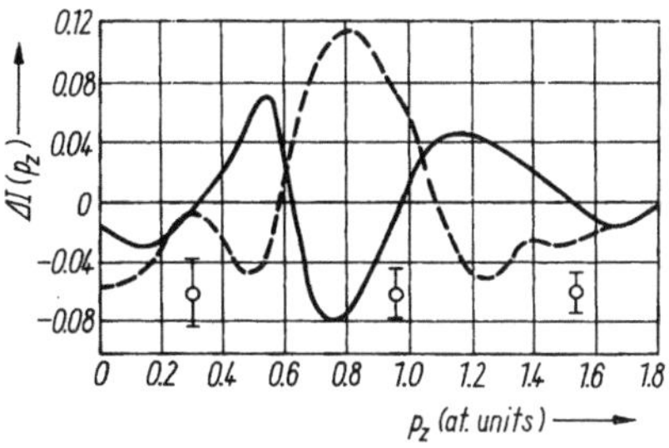

Fig. 2. Anisotropy of measured CP's given as $I(p_z)_{[100]} - I(p_z)_{[110]}$ (solid curve) and $I(p_z)_{[111]} - I(p_z)_{[100]}$ (dashed curve)

3. Modified Free Electron Model

A first approach to explain the anisotropy of the CP's of Si is suggested by (7) neglecting all $a_{G''}(\boldsymbol{k})$ against $a_0(\boldsymbol{k})$ in a first approximation. Thus the CP (see (3)) is only determined by the shape of the occupied region in the extended zone scheme, provided that $n(\boldsymbol{k}) = 1$ within this region and $n(\boldsymbol{k}) = 0$ outside (modified free electron model). In the case of Si and other semiconductors with diamond structure this occupied region is the so-called Jones zone, formed by the set of $\{220\}$ planes [10], [11].

According to this modified free electron model the CP's of Si with the scattering vector in the [100], [110], and [111] direction were calculated and normalized in the same manner as the experimental ones (see [5]). For the purpose of a comparison with experimental CP's, they were convoluted with the resolution function of the experimental profiles. These modified free electron CP's exhibit a considerable anisotropy which is shown in Fig. 1. One sees that the experimental anisotropy given in Fig. 2 is only 1/3 the calculated one and that there is agreement of both figures only with respect to the rough tendency.

Thus indeed one part of the experimentally obtained anisotropy of the Si CP's is influenced by the shape of the occupied Jones zone, but the neglect of all higher momentum components $|a_{G''}(\boldsymbol{p} - \boldsymbol{G}'')|^2$ in (8) is a too rough approximation.

4. Reconstruction of the Three Dimensional Momentum Density from the Experimental CP's

To get more insight into the details of the experimentally obtained deviations from the modified free-electron model the three dimensional momentum density was reconstructed according to Mijnarends [12] by the inversion of the CP's measured with different directions of the scattering vector.

Since the momentum density $\varrho(\boldsymbol{p})$ has to remain invariant under all operations of the crystal point group, $\varrho(\boldsymbol{p})$ may be expanded in lattice harmonics $F_l(\vartheta, \varphi)$

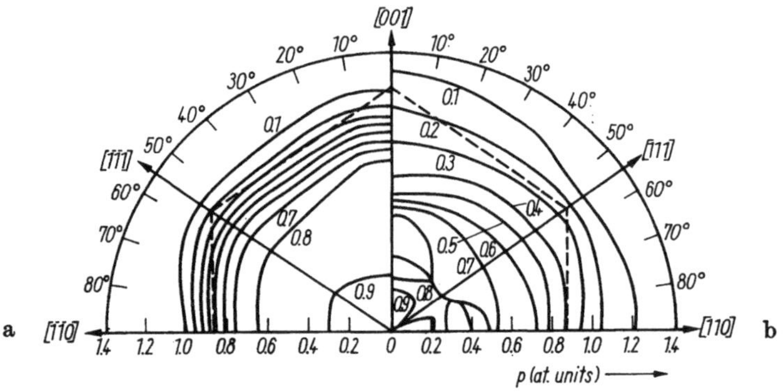

Fig. 3. a) Experimental valence electron momentum density contour diagram in the (1$\bar{1}$0) plane. b) Valence electron momentum density contour diagram from the modified free electron model in the (1$\bar{1}$0)-plane

of cubic symmetry with coefficients $\varrho_l(p)$. These $\varrho_l(p)$ can be calculated from

$$\varrho_l(p) = -\frac{1}{p}\left[\frac{\mathrm{d}g_l(p)}{\mathrm{d}p} - \frac{l(l+1)}{2p}\,g_l(p) + \frac{1}{p^2}\int\limits_0^p g_l(p_z)\,P_l''\left(\frac{p_z}{p}\right)\mathrm{d}p_z\right], \qquad (8)$$

where the $g_l(p)$ are determined by the experimental values $I_{\vartheta,\varphi}(p_z)$ of the CP with scattering vector $\boldsymbol{K}$ in a direction described by the azimuthal ϑ and polar angle φ according to

$$I_{\vartheta,\varphi}(p_z) \sim 2\pi \sum_l F_l(\vartheta, \varphi)\, g_l(p_z) \qquad (9)$$

and $P''(p_z/p)$ is the second derivative of the l order Legendre polynominal with respect to its argument.

Using cubic harmonics of α-type for $l = 0, 4, 6$ [13] the $\varrho_l(p)$ and then $\varrho(\boldsymbol{p})$ were determined from the $I(p_z)$ in the [100], [110], and [111] direction and plotted in a contour diagram. $\int \varrho(\boldsymbol{p})\,\mathrm{d}^3p$ was normalized to four times the volume of the Si Brillouin zone.

The Fig. 3a shows the experimental momentum density in the (1$\bar{1}$0) plane. In the contour diagram of the Fig. 3b, which is derived from the modified free-electron CP's for the purpose of comparison, one can easily see both the smearing of the angular variations in the momentum density due to the truncation at $l = 6$ and the smearing from the resolution function. The dashed line on Fig. 3a and b gives the intersection of the Jones zone with the (1$\bar{1}$0) plane.

Comparing Fig. 3a with Fig. 3b the difference between the modified free-electron momentum density and the experimental one is evident. The main feature consists in the transfer of momentum density from the vicinity of $p = 0.6$ to 1.0 at.units along the [100] direction to the vicinity of $p = 1.1$ at. units along the [110] direction. Besides this the experimental momentum density is more smeared over the whole momentum space.

Thus it must be concluded that a satisfactory explanation of the experimental CP anisotropy can only be found by a band computation of wave functions, which is given in the next section.

5. Band Computation of Si Momentum Density by a Pseudopotential Approach

It is well known that the band structure of Si can be calculated with sufficient accuracy using a local pseudopotential with only few nonvanishing Fourier coefficients [14, 15]. Thus it is assumed that also the Si valence electron wave functions can be obtained as eigenfunctions of the pseudopotential band structure. By doing so, the contribution of the valence electrons in the core region is misrepresented, since the terms that orthogonalize the valence and core wave functions are omitted. In Section 6 the contribution of the orthogonalization terms to the momentum density is estimated. The pseudopotential coefficients used here are those of the Brust scheme [15]: $V_{111} = -0.105$, $V_{220} = +0.02$, $V_{311} = +0.04$ at. units. Depending on $\boldsymbol{k}$ 15 to 18 plane waves are included directly and more are taken into account by perturbation theory. The coefficients $a_{G,\nu}(\boldsymbol{k})$ thus calculated were used to get the momentum density $\varrho(\boldsymbol{p})$ according to (6) with $\boldsymbol{p}$ in the [100], [110], and [111] direction. In Fig. 4a and b the calculated momentum density for the [100] and [110] direction, respectively, is plotted together with the corresponding valence bands, which contribute to the momentum density. The contribution of the different valence bands can be read directly from these figures by the small letters belonging to one another. It can be seen that some bands do not contribute to the momentum density for $\boldsymbol{p}$ in a given direction although other bands with the same $\boldsymbol{k}$ do. This fact can be understood without any difficulty by group theoretical arguments as given by Mijnarends [16].

In Fig. 5a and b the calculated momentum density (b) with $\boldsymbol{p}$ in the [100] and [110] direction, respectively, is plotted together with the momentum density of the modified free-electron model (a) in the upper part of the figures. Both are convoluted with the resolution function of the experimental CP to make possible a comparison with the experimental momentum density (d) as it was obtained by a Mijnarends analysis, which is shown together with the momentum

a 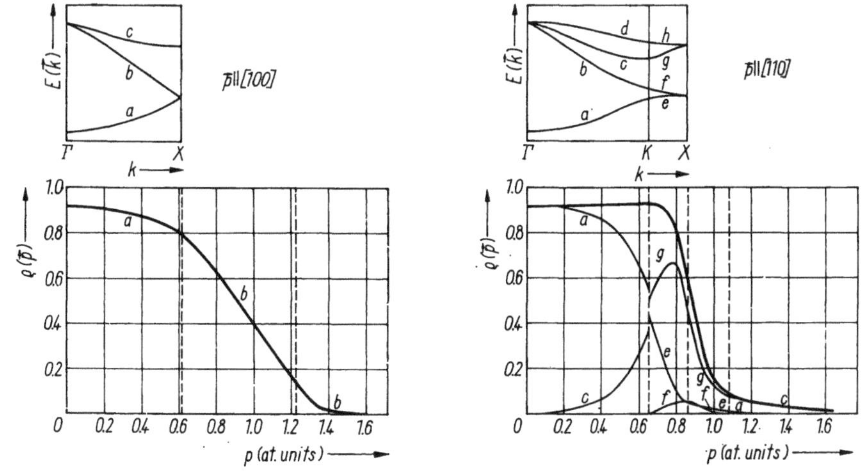 b

Fig. 4. Valence electron momentum density as obtained by pseudopotential band calculation. The contribution of the various bands to the momentum density can be seen by the small letters. a) $\boldsymbol{p} \parallel [100]$; b) $\boldsymbol{p} \parallel [110]$

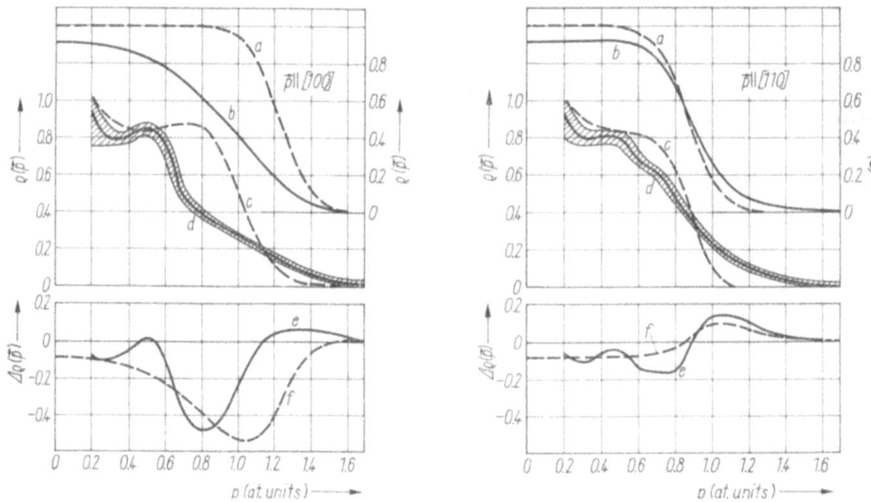

a b

Fig. 5. Upper part: momentum density of the modified free-electron model (a) and momentum density calculated by the pseudopotential approach (b); both curves are convoluted with the experimental resolution function. Middle part: momentum density of the modified free-electron model as obtained by a Mijnarends analysis (c) and experimental momentum density as obtained by a Mijnarends analysis (d). The uncertainty in determining $\varrho(\boldsymbol{p})$ from experimental CP's is denoted by the cross-hatched region. Lower part: (e) difference of the curves (c) and (d), (f) difference of the curves (a) and (b) to see the experimental determined (e) and the theoretical calculated (f) reduction and enhancement of momentum density with respect to the modified free electron momentum density. a) $\boldsymbol{p} \parallel [100]$; b) $\boldsymbol{p} \parallel [110]$

density of the modified free-electron model (c) also obtained by a Mijnarends analysis in the middle of these figures. In the lowest part the difference (a)—(b) and (c)—(d) is given in the curves (f) and (e), respectively.

Comparing these curves for the different orientation of $\boldsymbol{p}$ one sees that the main features of the experimental difference curve (e) agree with the corresponding theoretical difference curve (f). Especially the large negative difference in momentum density with $\boldsymbol{p} \parallel [100]$ in the vicinity of 0.8 at.units and the considerable large positive difference with $\boldsymbol{p} \parallel [110]$ in the vicinity of 1.1 at.units already mentioned as the main feature of the contour diagrams are well reproduced by the theoretical curves.

The fact that especially the [100] Mijnarends curves (c) and (d) are compressed so that the minimum of the corresponding difference curve (e) is shifted to smaller momenta is due to the truncation at $l = 6$ of the expansion in lattice harmonics.

6. Influence of the Orthogonalization Terms on the Momentum Density

In the band calculations of the momentum density of Si, given in Section 5, all orthogonalization terms are omitted. To estimate their influence on the momentum density a proper orthogonalization term was added to the pseudo-

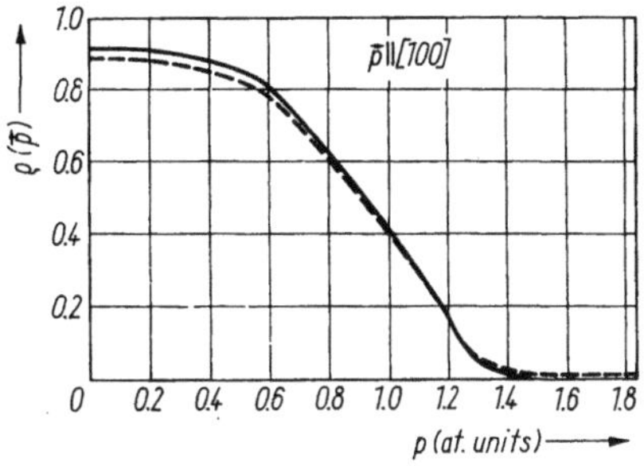

Fig. 6. Valence electron momentum density as calculated neglecting (solid curve) and including (dashed curve) the orthogonalization terms

wave functions, calculated in Section 5, thus obtaining

$$\varrho(\boldsymbol{p}) = \frac{V}{(2\pi)^3} \sum_{\boldsymbol{k},\nu} \delta(\boldsymbol{k} + \boldsymbol{G}' - \boldsymbol{p})\, n(\boldsymbol{k},\nu)\, c_{\boldsymbol{k},\nu}^2 \Big| \sum_{\boldsymbol{G}} a_{\boldsymbol{G},\nu}(\boldsymbol{k})\, \delta(\boldsymbol{G} - \boldsymbol{G}') -$$

$$- 4\pi V_0^{-1} F(\boldsymbol{G} - \boldsymbol{G}')\, a_{\boldsymbol{G},\nu}(\boldsymbol{k}) \sum_{n,l} (2l+1)\, A_{nl}(\boldsymbol{k} + \boldsymbol{G})\, A_{nl}(\boldsymbol{p})\, P_l(\cos \vartheta_p) \Big|^2 , \qquad (10)$$

where V_0 is the unit cell volume, $\boldsymbol{G}'$ has the same meaning as in (6), $A_{nl}(\boldsymbol{q})$ is given by

$$A_{nl}(\boldsymbol{q}) = \int_0^\infty r R_{nl}(r)\, j_l(qr)\, \mathrm{d}r , \qquad (11)$$

where $R_{nl}(r)$ is the radial part of the core wave functions $u_{nlm}(\boldsymbol{r})$, designated in the usual way by

$$u_{nlm}(\boldsymbol{r}) = \frac{Y_{lm}(\vartheta, \varphi)\, R_{nl}(r)}{r} . \qquad (12)$$

ϑ_p is the azimuthal angle of the momentum $\boldsymbol{p}$ with respect to a coordinate system, having its z-axis in the direction of $\boldsymbol{k} + \boldsymbol{G}$. $F(\boldsymbol{q})$ is the structure factor.

Thus $\varrho(\boldsymbol{p})$ was calculated with $\boldsymbol{p}$ in the [100] direction using the following approximation for the normalization constant $c_{\boldsymbol{k},\nu}$:

$$c_{\boldsymbol{k},\nu} \approx \left[1 - \frac{4\pi}{V_0} \sum_{n,l} F(0)\, (2l+1)\, A_{nl}^2(\boldsymbol{p})\right]^{-1/2} \qquad (13)$$

(this would be correct only for the case of *one* orthogonal plane wave) and radial parts of orthogonalized Slater wave functions given by Woodruff [17]. From Fig. 6 one sees that the correction of the pseudowave function momentum density of Section 5 is only small, thus justifying the omission of the orthogonalization terms throughout the whole calculation of Section 5.

7. Concluding Remarks

Summarizing the results of this analysis of the Si CP's one can conclude that the main features of the Compton profile anisotropy can be understood as due to both the anisotropy of the occupied region in the extended zone scheme and special properties of the momentum density, which are well reproduced by a pseudopotential band calculation of the momentum density. The remarkable differences between the CP anisotropies of Si and the anisotropy of the angular correclation curves of positron annihilation in Si single crystals as given by Erskine and McGervey [10] and analyzed by Stroud and Ehrenreich [18] could arise from the influence of the positron wave function.

References

[1] W. C. Phillips and R. J. Weis, Phys. Rev. **171**, 790 (1968).
[2] M. Cooper, Adv. Phys. **20**, 453 (1971).
[3] W. Schülke, U. Berg, and O. Brümmer, in: Reinststoffe in Wissenschaft u. Technik, Akademie-Verlag, Berlin 1972 (p. 83).
[4] W. Schülke and O. Brümmer, in: Elektronenstruktur, VEB Deutscher Verlag für Grundstoffindustrie, Leipzig 1972 (p. 95).
[5] W. Schülke and W. Lucht, phys. stat. sol. (b) **59**, K107 (1973).
[6] R. J. Weiss, Phil. Mag. **26**, 153 (1972).
[7] W. A. Reed and P. Eisenberger, Phys. Rev. B **6**, 4596 (1972).
[8] P. Eisenberger and P. M. Platzmann, Phys. Rev. A **2**, 415 (1970).
[9] B. I. Lundqvist and C. Lydén, Phys. Rev. B **4**, 3360 (1971).
[10] J. C. Erskine and J. D. McGervey, Phys. Rev. **151**, 615 (1966).
[11] P. E. Mijnarends, Phys. Rev. B **4**, 2820 (1971).
[12] P. E. Mijnarends, Phys. Rev. **160**, 512 (1967).
[13] F. C. von der Lage and H. A. Bethe, Phys. Rev. **71**, 612 (1947).
[14] L. Kleinman and J. C. Phillips, Phys. Rev. **112**, 685 (1958).
[15] D. Brust, Phys. Rev. **134**, A1337 (1964).
[16] P. E. Mijnarends, Physica (Utrecht) **63**, 235 (1973).
[17] T. O. Woodruff, Solid State Phys. **4**, 367 (1957).
[18] D. Stroud and H. Ehrenreich, Phys. Rev. **171**, 399 (1968).

(Received January 25, 1974)

phys. stat. sol. (b) 62, 461 (1974)

Subject classification: 13.5.2

Michelson Laboratory, China Lake, California (a), Health Physics Division, Oak Ridge National Laboratory, Oak Ridge, Tennessee and Department of Physics, University of Tennessee, Knoxville (b)

Diffuse Scattering and Surface-Plasmon Generation by Photons at a Rough Dielectric Surface[1])

By

J. M. ELSON (a) and R. H. RITCHIE (b)

First-order perturbation theory has been employed to compute the scattering and conversion to surface plasma oscillations of photons incident normally on a nearly planar vacuum–dielectric surface. The calculation proceeds through a classical solution of Maxwell's equations. Transformation to a non-orthogonal system of coordinates is effected. The Green function tensor for the Hertz vector potential appropriate to the plane surface geometry is found and applied to a determination of the scattered fields. The results are valid for arbitrarily large damping of fields due to electronic transitions in the dielectric.

Mit Störungstheorie erster Ordnung wurde die Streuung und Konversion in Oberflächenplasmaschwingungen von senkrecht zu einer nahezu ebenen Grenzschicht Vakuum–Dielektrikum einfallenden Photonen berechnet. Die Berechnung erfolgte durch eine klassische Lösung der Maxwellschen Gleichungen. Eine Transformation in ein nichtorthogonales Koordinatensystem wurde durchgeführt. Der Tensor der Greenschen Funktion für das Hertzsche Vektorpotential, der für die ebene Oberflächengeometrie anwendbar ist, wurde gefunden und zur Bestimmung der gestreuten Felder angewendet. Die Ergebnisse sind gültig für beliebig große Dämpfung der Felder durch elektronische Übergänge in dem Dielektrikum.

1. Introduction

Study of scattering and absorption of photons incident on a dielectric having a surface which displays irregular deviations from planarity is quite old [1]. The processes of interest in this paper include (i) diffuse scattering of photons and (ii) photon-surface plasmon coupling. Reference to the recent literature on this subject have been given in a paper by the authors [2]. We reported there the results of calculations which were implemented with the aid of a non-orthogonal coordinate transformation and quantum mechanical first-order perturbation theory. The approach taken involved approximating all absorptive processes as small perturbations upon the time evolution of undamped field oscillators representing the photon field and the surface plasmon field. To extend our treatment to include the possibility of appreciable damping due to electronic transitions in the dielectric, we use here classical perturbation theory based upon a similar transformation to a non-orthogonal coordinate system.

The system is taken to consist of two semi-infinite media: vacuum in the upper half-space $(z > \zeta(x, y))$ and a medium of dielectric permittivity ε in the lower

[1]) Research sponsored in part by the U.S. Atomic Energy Commission under contract with Union Carbide Corporation. Also funded in part by the Naval Air Systems Command Airtask No. A310310B/008A/3R02103002.

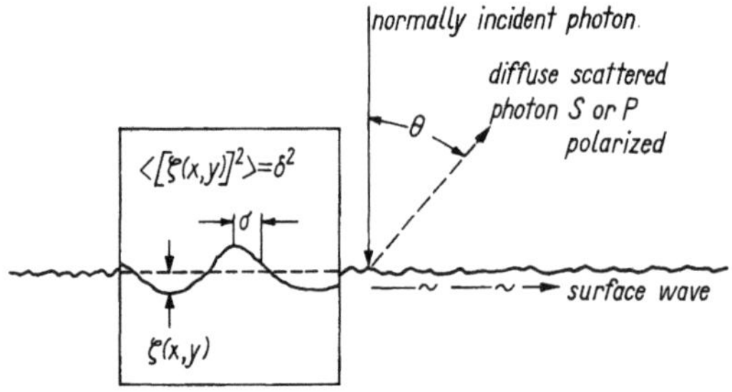

Fig. 1. Schematic representation of geometry involved in the diffuse scattering and surface plasmon conversion of photons at a rough dielectric surface

half-space $(z < \zeta(x, y))$. The interface between the two media is defined by $z - \zeta(x, y) = 0$, where $\zeta(x, y)$ is a roughness function specifying the shape of the interface (see the inset of Fig. 1). The dotted line represents the location of the $z = 0$ plane and may be chosen arbitrarily. The interface is taken to be between $-L/2 \leqq x \leqq L/2$ and $-L/2 \leqq y \leqq L/2$, where in the final result the limit $L \to \infty$ is taken. Fig. 1 also depicts a photon beam striking the surface. Due to the presence of surface irregularity, some of these photons may be scattered into non-specular directions, or may be converted into surface plasmon modes propagating along the surface.

2. Theory

Maxwell's equations in terms of the Hertz vector $\boldsymbol{\pi}(\boldsymbol{r}, t)$ may be written in Cartesian coordinates as

$$\nabla\times(\nabla\times\boldsymbol{\pi}) - \nabla(\nabla\cdot\boldsymbol{\pi}) + \frac{1}{c^2}\,\varepsilon\big(z-\zeta(\boldsymbol{\varrho})\big)\,\frac{\partial^2}{\partial t^2}\,\boldsymbol{\pi} = 0\,, \tag{2.1}$$

where $\varrho = |\boldsymbol{\varrho}|$ and $\boldsymbol{\varrho}$ is a two-dimensional position vector parallel with the average planar surface. The dielectric permittivity $\varepsilon(\alpha) = \varepsilon$ if $\alpha < 0$ and is unity otherwise. As usual, $\boldsymbol{B} = \frac{\varepsilon}{c}\nabla\times\boldsymbol{\pi}$ and $\boldsymbol{E} = \nabla(\nabla\cdot\boldsymbol{\pi}) - \frac{1}{c^2}\,\varepsilon\,\ddot{\boldsymbol{\pi}}$. Making a change of variables in (2.1) defined by

$$u_1 = x\,,\qquad u_2 = y\,,\qquad u_3 = z - \zeta(\boldsymbol{\varrho}) \tag{2.2}$$

yields an equation of the form

$$\left[\nabla_u^2 - \frac{\varepsilon(u_3)}{c^2}\,\frac{\partial^2}{\partial t^2}\right]\boldsymbol{\pi} = \boldsymbol{\sigma}(\boldsymbol{\pi})\,. \tag{2.3}$$

∇_u^2 has the form of a Cartesian Laplacian operator, viz.,

$$\nabla_u^2 = \frac{\partial^2}{\partial u_1^2} + \frac{\partial^2}{\partial u_2^2} + \frac{\partial^2}{\partial u_3^2}\,. \tag{2.4}$$

All vectors are written in the form $\boldsymbol{V} = \sum_{i=1}^{3} \boldsymbol{e}_1 V_i$, where the unit vectors $(\boldsymbol{e}_1, \boldsymbol{e}_2, \boldsymbol{e}_3)$ are tangential to the coordinate curves in the (u_1, u_2, u_3) system. The inhomogeneous term $\boldsymbol{\sigma}(\boldsymbol{\pi})$ of (2.3) is introduced by the coordinate transformation. It may be written in the form

$$\boldsymbol{\sigma}(\boldsymbol{\pi}) = \sum_{n=1}^{\infty} \boldsymbol{\sigma}^{(n)}(\boldsymbol{\pi}) , \tag{2.5}$$

where the term of index n in the sum is proportional to the nth power of $\zeta(x, y)$. At this point we make the assumption that the surface irregularity constitutes a weak perturbation and that (2.5) may be approximated by

$$\left[\nabla^2 - \frac{\varepsilon(u_3)}{c^2} \frac{\partial^2}{\partial t^2}\right] \boldsymbol{\pi}^{(1)} = \boldsymbol{\sigma}^{(1)}(\boldsymbol{\pi}^{(0)}) , \tag{2.6}$$

where $\boldsymbol{\pi}^{(1)}$ is a small correction to a zero-order solution $\boldsymbol{\pi}^{(0)}$. The latter is obtained by setting $\zeta = 0$. In this study the zero-order field is taken to correspond to a photon normally incident with respect to the mean surface. Assuming that $\boldsymbol{\pi}^{(0)}$ has only an $\boldsymbol{e}_1$ component and depends only on u_3 and t, the general first-order source term may be simplified to the following form:

$$\boldsymbol{\sigma}^{(1)}(\boldsymbol{\pi}^{(0)}) = \boldsymbol{e}_1 \left(\frac{\partial^2 \zeta}{\partial u_1^2} + \frac{\partial^2 \zeta}{\partial u_2^2}\right) \frac{\partial \pi_1^{(0)}}{\partial u_3} - \boldsymbol{e}_3 \left(\frac{\partial^3 \zeta}{\partial u_1^3} + \frac{\partial^3 \zeta}{\partial u_1 \, \partial u_2^2}\right) \pi_1^{(0)} . \tag{2.7}$$

If the interface were perfectly smooth, the induced polarization currents would all have the same polarization as the incident beam (for an isotropic dielectric) and would give rise only to specular reflection. Equation (2.7) shows that surface irregularity causes polarization currents in directions other than that of the incident polarization. This differs from previous theories [3, 4] in which the source currents, for normally incident light, have directions which parallel that of the $\boldsymbol{E}$ vector of the incident light.

We must require that the tangential components of the $\boldsymbol{E}$ and $\boldsymbol{H}$ fields be continuous at each point on the surface $u_3 = 0$. That is, if $\boldsymbol{n}$ is a unit vector at the point $(u_1, u_2, 0)$, then

$$\boldsymbol{n} \times [\boldsymbol{E}_+ - \boldsymbol{E}_-] = \boldsymbol{n} \times [\boldsymbol{H}_+ - \boldsymbol{H}_-] = 0 . \tag{2.8}$$

The subscript $(\pm)$ on the field quantities indicates that each is to be evaluated at the point $(u_1, u_2, 0 \pm)$. If we express $\boldsymbol{n}$ as a power series in ζ and its derivatives and retain only zeroth and first powers we find $\boldsymbol{n} \sim \boldsymbol{e}_3 - \left(\boldsymbol{e}_1 \frac{\partial \zeta}{\partial u_1} + \boldsymbol{e}_1 \frac{\partial \zeta}{\partial u_2}\right)$. Since $\boldsymbol{E}$ and $\boldsymbol{H}$ are also implicitly developed in similar power series in ζ we may collect terms of the same order in ζ and equate the coefficients to zero independently.[2]) When this is done it is found that for the case where the zero order Hertz vector has only a component in the $\boldsymbol{e}_1$ direction, it is necessary only to require continuity of the tangential components of the first-order fields at $u_3 = 0$ and similarly for the zero order fields. This is done below.

Equation (2.6) may be solved by Green function techniques. The Green function tensor $\mathfrak{G}$ may be taken as a solution of

$$\left[\nabla^2 - \frac{\varepsilon(u_3)}{c^2} \frac{\partial^2}{\partial t^2}\right] \mathfrak{G} \boldsymbol{e}_i = \boldsymbol{e}_i \delta(\boldsymbol{u} - \boldsymbol{u}') \, \delta(t - t') , \tag{2.9}$$

where in general $i = 1, 2, 3$. The vectors $\boldsymbol{u}$ and $\boldsymbol{u}'$ denote locations of the observation point and the source, respectively. The various terms of the tensor

[2]) Similar but more complicated conditions result if one works in a Cartesian coordinate system, see [5].

may be obtained by incorporating the proper discontinuity at $u_3' = 0$ and by matching boundary conditions at the interface $u_3 = 0$.

The solution to (2.6) is given by

$$\boldsymbol{\pi}^{(1)}(\boldsymbol{u}, t) = \int \mathrm{d}^3u' \, \mathrm{d}t' \, \mathbf{G}(\boldsymbol{u}, t; \boldsymbol{u}', t') \, \boldsymbol{S}(\boldsymbol{u}', t') , \tag{2.10}$$

where it is convenient to take $\boldsymbol{S}(\boldsymbol{u}, t)$ to be a real source function defined by

$$\boldsymbol{S}(\boldsymbol{u}, t) = \mathrm{Re} \, \{\boldsymbol{\xi}(\boldsymbol{u}) \, \mathrm{e}^{-i\omega_0 t}\} \tag{2.11}$$

corresponding to the assumption that the zero-order field is monochromatic with frequency ω_0. Further, the Green tensor may be Fourier analyzed as

$$\mathbf{G}(\boldsymbol{u}, t; \boldsymbol{u}', t') = (2\pi)^{-3} \int \mathrm{d}^2k \, \mathrm{d}\omega \, \mathbf{g}_{\boldsymbol{k},\omega}(u_3, u_3') \exp{[i\boldsymbol{k}(\boldsymbol{\varrho} - \boldsymbol{\varrho}') - i\omega(t - t')]} , \tag{2.12}$$

where u_1 and u_2 are the components of the vector $\boldsymbol{\varrho}$. Using (2.11) and (2.12) in (2.10) yields, after integration over t' and ω,

$$\boldsymbol{\pi}^{(1)}(\boldsymbol{u}, t) = \mathrm{Re} \, \{\boldsymbol{\pi}^{(1)}(\boldsymbol{u}) \, \mathrm{e}^{-i\omega_0 t}\} ,$$

where

$$\boldsymbol{\pi}^{(1)}(\boldsymbol{u}) = (2\pi)^{-2} \int \mathrm{d}^2k \int \mathrm{d}^3u' \, \mathbf{g}_{\boldsymbol{k},\omega_0}(u_3, u_3') \, \boldsymbol{\xi}(\boldsymbol{u}') \exp{[i\boldsymbol{k}(\boldsymbol{\varrho} - \boldsymbol{\varrho}')]} . \tag{2.13}$$

In obtaining (2.13), use of the relation $\mathbf{g}_{\boldsymbol{k},\omega_0} = \mathbf{g}_{-\boldsymbol{k},-\omega_0}$ has been made.

A quantity of interest is

$$W = \int \mathrm{d}^3u \, \boldsymbol{EJ} , \tag{2.14}$$

where $\boldsymbol{J} = [(\varepsilon - 1)/4\pi] \, \partial\boldsymbol{E}/\partial t$ and the integration is over all space of the dielectric ε. Equation (2.14) represents the rate of absorption of energy from an incident electromagnetic disturbance due to currents induced by the fields. If we assume $\boldsymbol{E} = \mathrm{Re}\,\{\boldsymbol{E}^{(1)}\mathrm{e}^{-i\omega_0 t}\}$, average over time and take the real part, one contribution to (2.14) becomes

$$\overline{W}^{(1)} = \frac{\varepsilon_2\omega_0}{8\pi} \int \mathrm{d}^3u \, |\boldsymbol{E}^{(1)}(\boldsymbol{u})|^2 , \tag{2.15}$$

where ε_2 is the imaginary part of $\varepsilon = \varepsilon_1 + i\varepsilon_2$. Equation (2.15) represents the average rate of energy deposition in the dielectric by currents associated with the first-order field, which may be associated with the existence of surface plasma oscillations. Energy residing in the latter goes finally into thermal motion in the solid or it may be converted into electromagnetic field energy. As $\varepsilon_2 \to 0$, it will be seen below that although electronic absorption vanishes in this limit, surface plasmon oscillations may still be excited.

3. The Green Function Tensor

If we substitute (2.12) into (2.9) and use the fact that

$$\delta(u_1 - u_1') \, \delta(u_2 - u_2') \, \delta(t - t') = (2\pi)^{-3} \int \mathrm{d}^2k \, \mathrm{d}\omega \mathrm{e}^{i[\boldsymbol{k}(\boldsymbol{\varrho} - \boldsymbol{\varrho}') - \omega(t - t')]}$$

we get

$$\left\{\frac{\mathrm{d}^2}{\mathrm{d}u_3^2} - \alpha^2(u_3)\right\} \mathbf{g}_{\boldsymbol{k},\omega} \cdot \boldsymbol{e}_i = \boldsymbol{e}_i \delta(u_3 - u_3') , \tag{3.1}$$

where

$$\alpha^2(u_3) = \begin{cases} \nu^2 = k^2 - \varepsilon\omega^2/c^2 , & (u_3 < 0) , \\ \eta^2 = k^2 - \omega^2/c^2 , & (u_3 > 0) . \end{cases} \tag{3.2}$$

A typical element of g may be written in the form g_{31}^{mv} where the subscripts $\boldsymbol{k}$, ω are dropped henceforth. The m and v denote dielectric medium ($u_3 < 0$) and vacuum ($u_3 > 0$), respectively. The first superscript denotes the location of the source and the second the observation point. The subscript pair denotes a particular element of the tensor where the row and column numbers also specify the direction of the Hertz vector produced by the source and the direction of the source vector, respectively. Thus, the example shown above denotes the Green tensor element for sources in the dielectric and observation points in the vacuum. Only the 11, 31, 22, and 33 elements need be considered here, while the superscript pair may take on any of the four possible permutations of the letters m and v.

For sources in the vacuum, (3.1) may be solved applying the usual conditions of continuity of tangential fields at the surface $u_3 = 0$. It is found that for directions of propagation in the ($\boldsymbol{e}_1$, $\boldsymbol{e}_3$) plane

$$g_{11}^{\mathrm{vv}} = g_{22}^{\mathrm{vv}} = \frac{1}{2\eta}\left[-\mathrm{e}^{-\eta|u_3-u_3'|} + \frac{\nu-\eta}{\nu+\eta}\,\mathrm{e}^{-\eta(u_3+u_3')}\right], \tag{3.3}$$

$$g_{31}^{\mathrm{vv}} = \frac{ik(1-\varepsilon)}{(\eta+\nu)\,(\eta\varepsilon+\nu)}\,\mathrm{e}^{-\eta(u_3+u_3')}, \tag{3.4}$$

$$g_{11}^{\mathrm{vm}} = g_{22}^{\mathrm{vm}} = \frac{-1}{\varepsilon(\eta+\nu)}\,\mathrm{e}^{-\eta u_3'+\nu u_3}, \qquad g_{31}^{\mathrm{vm}} = \frac{ik(1-\varepsilon)}{\varepsilon(\eta+\nu)\,(\nu+\eta\varepsilon)}\,\mathrm{e}^{-\eta u_3'+\nu u_3}, \tag{3.5}$$

and

$$g_{33}^{\mathrm{vv}} = \frac{1}{2\eta}\left[-\mathrm{e}^{-\eta|u_3-u_3'|} + \frac{\nu-\varepsilon\eta}{\nu+\varepsilon\eta}\,\mathrm{e}^{-\eta(u_3+u_3')}\right], \qquad g_{33}^{\mathrm{vm}} = \frac{-1}{\nu+\varepsilon\eta}\,\mathrm{e}^{-\eta u_3+\nu u_3}. \tag{3.6}$$

When the sources are located in the ε-medium, the tensor elements become

$$\left.\begin{aligned}
g_{11}^{\mathrm{mm}} &= g_{22}^{\mathrm{mm}} = \frac{1}{2\nu}\left[-\mathrm{e}^{-\nu|u_3-u_3'|} + \frac{\eta-\nu}{\eta+\nu}\,\mathrm{e}^{\nu(u_3+u_3')}\right],\\
g_{31}^{\mathrm{mm}} &= \frac{ik(1-\varepsilon)}{(\eta+\nu)\,(\eta\varepsilon+\nu)}\,\mathrm{e}^{\nu(u_3+u_3')},\\
g_{11}^{\mathrm{mv}} &= g_{22}^{\mathrm{mv}} = \frac{-\varepsilon}{\eta+\nu}\,\mathrm{e}^{\nu u_3'-\eta u_3},\\
g_{31}^{\mathrm{mv}} &= \frac{ik\varepsilon(1-\varepsilon)}{(\eta+\nu)\,(\nu+\eta\varepsilon)}\,\mathrm{e}^{\nu u_3'-\eta u_3},\\
g_{33}^{\mathrm{mm}} &= \frac{1}{2\nu}\left[-\mathrm{e}^{-\nu|u_3-u_3'|} + \frac{\nu-\varepsilon\eta}{\nu+\varepsilon\eta}\,\mathrm{e}^{\nu(u_3+u_3')}\right],\\
g_{33}^{\mathrm{mv}} &= \frac{-\varepsilon}{\nu+\varepsilon\eta}\,\mathrm{e}^{\nu u_3'-\eta u_3}.
\end{aligned}\right\} \tag{3.7}$$

4. Photon–Surface Plasmon Coupling

The solution corresponding to the zero-order Hertz vector may be written in complex form as

$$\pi_1^{(0)} = I e^{-i\omega_0 t} \begin{cases} e^{-\frac{i\omega_0}{c} u_3} + \dfrac{1-\sqrt{\varepsilon}}{1+\sqrt{\varepsilon}} e^{i\frac{\omega_0}{c} u_3}, & (u_3 > 0), \\ \dfrac{2}{\varepsilon(1+\sqrt{\varepsilon})} e^{\frac{\omega_0}{c}\sqrt{-\varepsilon}\, u_3}, & (u_3 < 0), \end{cases} \tag{4.1}$$

where I is the amplitude of the incident wave and where ε is evaluated for argument ω_0 everywhere it appears. Substitution of (4.1) into (2.7), (2.10), and (2.13) yields, for points of observation lying in the ε-medium,

$$\pi_1^{(1)} = -\frac{2I\omega_0^2(1-\varepsilon)}{(2\pi c)^2\, \varepsilon(1+i\sqrt{-\varepsilon})} \int d^2k\, \zeta_{\boldsymbol{k}}\, e^{i\boldsymbol{k}\boldsymbol{\varrho}} \frac{e^{\nu u_3}}{(\eta+\nu)} \tag{4.2}$$

and

$$\pi_3^{(1)} = \frac{2iI\omega_0^2(1-\varepsilon)^2}{(2\pi c)^2\, \varepsilon(1+i\sqrt{-\varepsilon})} \int d^2k\, \zeta_{\boldsymbol{k}}\, e^{i\boldsymbol{k}\boldsymbol{\varrho}} \frac{k_1\, e^{\nu u_3}}{(\eta+\nu)\,(\eta\varepsilon+\nu)}, \tag{4.3}$$

where k_1 is the component of $\boldsymbol{k}$ (the wave vector parallel to the surface) parallel to the $\boldsymbol{e}_1$ axis, $\zeta_{\boldsymbol{k}} = \int d^2\varrho\, \zeta(\varrho)\, e^{-i\boldsymbol{k}\boldsymbol{\varrho}}$, and terms corresponding to the zero-order field have been dropped. Computing the electric field corresponding to this expression and substituting into (2.15) yields

$$\overline{W}^{(1)} = \frac{\omega_0^5 \varepsilon_2 I_0^2}{2^4\pi^3 c^4} \left|\frac{1-\varepsilon}{1+\sqrt{\varepsilon}}\right|^2 \int d^2k\, |\zeta_{\boldsymbol{k}}|^2\, B(\boldsymbol{k}),$$

where $\boldsymbol{k} = k(\boldsymbol{e}_1 \cos\varphi + \boldsymbol{e}_2 \sin\varphi)$ and

$$B(\boldsymbol{k}) = \frac{1}{\mathrm{Re}(\nu)} \left\{ \frac{(|\nu|^2 + k^2)\,\eta^2}{|\nu+\eta\varepsilon|^2} \cos^2\varphi + \frac{(\omega_0/c)^4}{|\eta+\nu|^2} \sin^2\varphi \right\}. \tag{4.4}$$

To consider the limiting case as $\varepsilon_2 \to 0$, it is straightforward to show that $\lim\limits_{\varepsilon_2\to 0} \varepsilon_2\, |\eta\varepsilon+\nu|^{-2} = \dfrac{-2\pi\, \varepsilon_1 k_0 \delta(k-k_0)}{(\omega_0/c)^2\,(1-\varepsilon_1)^2}$, where $k_0 = (\omega_0/c)\left(\dfrac{\varepsilon_1}{\varepsilon_1+1}\right)^{1/2}$. Recognizing that k_0 is the surface plasmon wave vector reveals that $\overline{W}^{(1)}$ represents absorption of energy from the incident beam into surface plasma oscillations. If $\overline{W}^{(1)}$ is divided by the average incident energy, $\dfrac{c}{8\pi}(\omega_0/c)^4 L^2 I^2$, where L^2 is the total area of illumination, then the result is equal to the probability P_{SP} of photon–surface plasmon conversion. If the surface irregularity is random in nature, then the ensemble average $\langle P_{SP}\rangle$ is needed to describe the average behavior from a large number of surface elements [2]. The probability of photon–surface plasmon conversion may be written as the decrease in reflectance of the specular beam, viz.

$$\Delta R_{SP} = \left(\frac{\omega_0}{c}\right)^4 \frac{\varepsilon_1^2}{[-(1+\varepsilon_1)]^{5/2}} \frac{\langle|\zeta_{\boldsymbol{k}}|^2\rangle}{L^2}. \tag{4.5}$$

The ensemble average of $|\zeta|^2$ may be expressed as $L^2\delta^2 g(k)$, where δ^2 is the r.m.s. variation of the surface height about its mean position and $g(k)$ is the surface scattering factor [2]. In these terms (4.5) becomes

$$\Delta R_{\mathrm{SP}} = \delta^2 \left(\frac{\omega_0}{c}\right)^4 \frac{\varepsilon_1^2}{[-(1+\varepsilon_1)]^{5/2}}\, g\left(\frac{\omega_0}{c}\left[\frac{\varepsilon_1}{1+\varepsilon_1}\right]^{1/2}\right). \tag{4.6}$$

It is interesting that if ε_1 is taken to be $\varepsilon_1 = 1 - \omega_{\mathrm{p}}^2/\omega_0^2$ appropriate to a classical free-electron gas, the expression for ΔR_{SP} reduces exactly to that obtained in equation (24) of reference [6] using a different form of perturbation theory.[3]) However, (4.6) does not reduce to that obtained by the authors [2] for the same electron gas model, but using quantized fields for the photon and SP modes. The reason for this discrepancy will be treated in a later publication.

5. Diffuse Scattering

Considering once again an incident beam as given by equation (4.1) and considering points of observation in the vacuum yields expressions similar to (4.2) and (4.3) except that the factor $\mathrm{e}^{\nu u_s}$ is replaced by $\varepsilon\,\mathrm{e}^{-\eta u_s}$. From these equations the electric and magnetic fields may be determined. Then the radial component of the Poynting flux in the far zone may be calculated, and in a standard way one finds the energy radiated into the direction specified by (θ, φ). Finally, dividing by the total incident energy as before yields

$$\frac{\mathrm{d}P^{(1)}}{\mathrm{d}\Omega} = \frac{\omega_0^4 |\zeta_{\boldsymbol{k}}|^2\, |1-\varepsilon|^2 \cos^2\theta}{\pi^2 c^4 L^2 \left|1 + i\sqrt{-\varepsilon}\right|^2} \left\{\frac{|\nu|^2 \cos^2\varphi}{|\nu - iq\varepsilon|^2} + \frac{\omega_0^2}{c^2}\,\frac{\sin^2\varphi}{|\nu - iq|^2}\right\}, \tag{5.1}$$

where $q = (\omega_0/c)\cos\theta$. The first term in (5.1) represents the probability that p-polarized light is observed per unit solid angle around the direction (θ, φ), whereas the second term corresponds to a s-polarized light. In the limit $\varepsilon_2 \to 0$, (5.1) takes the form

$$\frac{\mathrm{d}P^{(1)}}{\mathrm{d}\Omega} = \left(\frac{\omega_0}{c}\right)^4 \frac{|\zeta_{\boldsymbol{k}}|^2}{\pi^2 L^2} \cos^2\theta \left\{\frac{\sin^2\theta - \varepsilon_1}{\sin^2\theta - \varepsilon_1\cos^2\theta}\cos^2\varphi + \sin^2\varphi\right\}. \tag{5.2}$$

Expressing $\langle|\zeta_{\boldsymbol{k}}|^2\rangle$ as above, we find that for the s-polarized component

$$\frac{\mathrm{d}P_{\mathrm{s}}^{(1)}}{\mathrm{d}\Omega} = \frac{\delta^2}{\pi^2}\left(\frac{\omega_0}{c}\right)^4 \cos^2\theta \sin^2\varphi\, g\left(\frac{\omega_0}{c}\sin\theta\right) \tag{5.3}$$

which agrees exactly with equation (27) of reference [2]. For the p-polarized component

$$\frac{\mathrm{d}P_{\mathrm{p}}^{(1)}}{\mathrm{d}\Omega} = \frac{\delta^2}{\pi^2}\left(\frac{\omega_0}{c}\right)^4 \cos^2\theta\cos^2\varphi \left[\frac{\sin^2\theta - \varepsilon_1}{\sin^2\theta - \varepsilon_1\cos^2\theta}\right] g\left(\frac{\omega_0}{c}\sin\theta\right). \tag{5.4}$$

The discrepancy between this result and equation (28) of reference [2] will be discussed in a later publication. However, we note that as $\theta \to 0$, (5.3) and (5.4) approach the same limit, as they should.

[3]) Note that A. Daude et al. [7] find that their data on ΔR_{SP} may be fitted using equation (24) of reference [6] choosing quite reasonable values of the parameters δ and σ. Endriz and Spicer [8] have fitted their data on diffuse scattering and absorption at rough Al surfaces using equation (27), (28), and (24) of reference [2] by the present authors. We note that since there are two adjustable parameters, viz., δ and σ, in rough surface theories, and since $g(k)$ is usually not known a priori, the results are not particularly sensitive to the form of the "dielectric factor" appearing in the various theories. (The "dielectric factor" of equation (4.5) above is $\varepsilon_1^2[-(1+\varepsilon_1)]^{-5/2}$.)

If ν and q are expressed in terms of ω_0 and θ, one finds that (5.1) is equivalent to the more complete expressions

$$\frac{\mathrm{d}P_\mathrm{p}^{(1)}}{\mathrm{d}\Omega} = \frac{\delta^2}{\pi^2}\left(\frac{\omega_0}{c}\right)^4 \cos^2\theta \cos^2\varphi \left|\frac{1-\varepsilon}{1+i\sqrt{-\varepsilon}}\right| \frac{|\varepsilon - \sin^2\theta|^2}{\left|\sqrt{\varepsilon - \sin^2\theta} + \varepsilon\cos\theta\right|^2}\, g\left(\frac{\omega_0}{c}\sin\theta\right) \tag{5.5}$$

and

$$\frac{\mathrm{d}P_\mathrm{s}^{(1)}}{\mathrm{d}\Omega} = \frac{\delta^2}{\pi^2}\left(\frac{\omega_0}{c}\right)^4 \cos^2\theta \sin^2\varphi \left|\frac{1-\varepsilon}{1+i\sqrt{-\varepsilon}}\right|^2 \frac{1}{\left|\sqrt{\varepsilon - \sin^2\theta} + \cos\theta\right|^2}\, g\left(\frac{\omega_0}{c}\sin\theta\right) \tag{5.6}$$

which are valid for complex ε.

6. Conclusion

The probability of photon–surface plasmon conversion at a weakly rough dielectric surface characterized by arbitrary complex dielectric permittivity has been computed using classical perturbation theory. In addition, the probability of diffuse scattering on a surface of the same nature has been evaluated. The results appear to be more comprehensive than those heretofore available in the literature. In particular, it is straightforward to apply the ensemble averaged form of equation (4.4) to compute the probability of photon absorption at a rough dielectric surface for arbitrarily large damping of the fields. One needs only to divide (4.4) by $\frac{c}{8\pi}(\omega_0/c)^4 L^2 I^2$ for normalization. Such a calculation requires an integration over $|\boldsymbol{k}|$. Since the integrand of (4.4) contains $g(k)$ as a factor, and since the latter is not usually well-known but is often fitted to the data on ΔR_SP, we have not employed this equation here, but will consider such applications in a later publication.

Acknowledgements

The authors are grateful for many helpful conversations with Dr. J. L. Stanford and for his continued interest in the work reported here. Thanks are also due E. Kretschmann and A. Otto for helpful communications concerning the matters discussed above.

References

[1] P. Beckmann and A. Spizzichino, The Scattering of Electromagnetic Waves from Rough Surfaces, Pergamon Press, London/Oxford 1973.
[2] J. M. Elson and R. H. Ritchie, Phys. Rev. B 4, 4129 (1971).
[3] E. Kretschmann, Z. Phys. **227**, 412 (1969).
E. Kröger and E. Kretschmann, Z. Phys. **237**, 1 (1970).
[4] D. Beaglehole and O. Hunderi, Phys. Rev. B 2, 309 (1970).
[5] S. O. Rice, Commun. pure appl. Math. 4, 351 (1951).
H. J. Juranek, Z. Phys. **233**, 324 (1970).
E. Kröger and E. Kretschmann, Z. Phys. **237**, 1 (1970).
[6] J. Crowell and R. H. Ritchie, J. Opt. Soc. Amer. **60**, 794 (1970).
[7] A. Daude, A. Savary, and S. Robin, J. Opt. Soc. Amer. **62**, 1 (1972).
[8] J. G. Endriz and R. Spicer, Phys. Rev. B 4, 4144 (1971).

(Received November 19, 1973)

phys. stat. sol. (b) 62, 469 (1974)

Subject classification: 19; 13.4; 22.5.3

Solid State Physics Laboratory, University of Groningen

EPR of $BaCl_2$:Gd^{3+}

By

T. OOSTERHOFF and H. W. DEN HARTOG

The g-value and the most important crystal field parameters of Gd^{3+} in orthorhombic (D_{2h}^{16}) $BaCl_2$ are determined at room temperature: $g = 1.994$ (isotropic), $b_2^0 = \pm 90 \times 10^{-4}\ cm^{-1}$, and $b_2^2 = \mp 444 \times 10^{-4}\ cm^{-1}$. The corresponding theoretical values are $g = 1.993$, $b_2^0 = 85 \times 10^{-4}\ cm^{-1}$, and $b_2^2 = -324 \times 10^{-4}\ cm^{-1}$. A comparison of the results with values for Eu^{2+} and Gd^{3+} ions at Pb^{2+} sites in $PbCl_2$ and $PbBr_2$ is made.

Der g-Faktor und die wichtigsten Kristallfeldparameter des Gd^{3+} in orthorhombischem (D_{2h}^{16}) $BaCl_2$ wurden bei Zimmertemperatur bestimmt: $g = 1{,}994$ (isotrop), $b_2^0 = \pm 90 \times 10^{-4}\ cm^{-1}$ und $b_2^2 = \mp 444 \times 10^{-4}\ cm^{-1}$. Die entsprechenden berechneten Werte sind $g = 1{,}993$, $b_2^0 = 85 \times 10^{-4}\ cm^{-1}$ und $b_2^2 = -324 \times 10^{-4}\ cm^{-1}$. Die Ergebnisse werden mit Werten für Eu^{2+}- und Gd^{3+}-Ionen auf Pb^{2+}-Gitterplätzen in $PbCl_2$ und $PbBr_2$ verglichen.

1. Introduction

Gd^{3+} is like Eu^{2+} a lanthanide ion with a $4f^7$ configuration, which implies that one is dealing with an $^8S_{7/2}$ ground state. In a magnetic field this level is split into eight equally spaced sublevels when only the Zeeman interaction is taken into account. The presence of the crystalline field, however, displaces each of the sublevels m_J in a characteristic way, giving rise to a seven-line EPR spectrum.

$BaCl_2$ has an orthorhombic lattice structure (space group D_{2h}^{16} [1]). The point group symmetry of the Ba^{2+}-sites is monoclinic (C_s). The study of the crystal field splitting with EPR provides information about the surrounding of the Gd^{3+}-ion. The present experimental results indicate that the Gd^{3+}-ion is built in substitutionally for Ba^{2+}-ions. A schematic two-dimensional representation of the Gd^{3+} ion at site I and the surrounding Ba^{2+} and Cl^- ions is given in Fig. 1. The experimental and theoretical crystal field parameters are compared with each other and it turns out that on the basis of the simple point-ion-lattice approximation reasonable agreement is found. The best fit between theory and experiment is obtained for Eu^{2+} (especially, when the lattice is allowed to relax) [2], probably because there is no difference in charge between the ions of the host crystal and the impurity ion. For Gd^{3+}, however, which has an effective charge $+1$, it seems that besides distortions of the environment also effects due to lattice and electronic polarization should be considered.

2. Experimental Procedure

Single crystals of $BaCl_2$ were prepared by a Bridgman technique. The raw material $BaCl_2 \cdot 6H_2O$ was first heated in vacuum to remove the crystal water. Subsequently, the $BaCl_2$ powder was treated with dry Cl_2 gas at 1050 °C in order to prevent impurities like O^{2-} and OH^- from entering the crystal. Before the

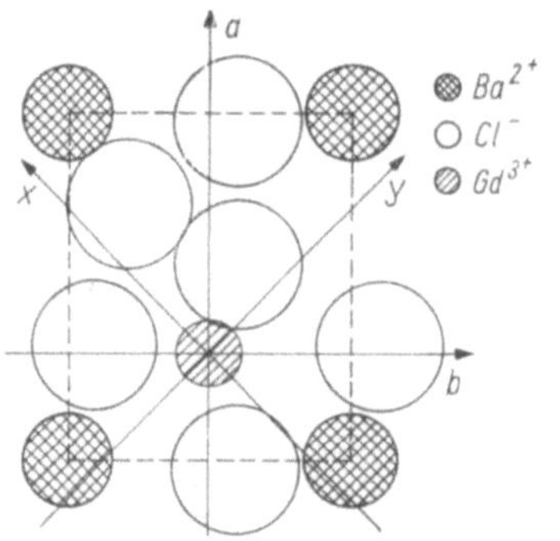

Fig. 1. Two-dimensional schematic representation of a Gd^{3+} ion at a Ba^{2+} (I) site. The crystallographic directions a and b and the principal directions x and y of the crystal field are indicated

treatment with Cl_2 approximately 0.1 mol% $GdCl_3$ was added. Because $BaCl_2$ is hygroscopic the material was treated in a properly dried atmosphere or in vacuum.

$BaCl_2$ has reasonable cleavage properties perpendicular to the crystallographic a-axis ($a_0 = 9.333$ Å). The other directions b and c could be determined accurately by means of a polarization microscope. The EPR-signals were measured as a function of the angle of rotation about the a-, b-, and c-axes at intervals of 6°. The EPR-experiments were carried out at room temperature with a Varian V4500 EPR spectrometer. The modulation frequency was 100 kHz. The resonance conditions were determined by means of reference samples DPPH and Mn^{2+} in ZnS and with the help of a Sievers cavity frequency meter.

3. Experimental Results

Fig. 2 shows the EPR signal of $BaCl_2$:Gd^{3+} for H along the z-axis of the centre. It turns out that this direction coincides with the crystallographic c-axis of

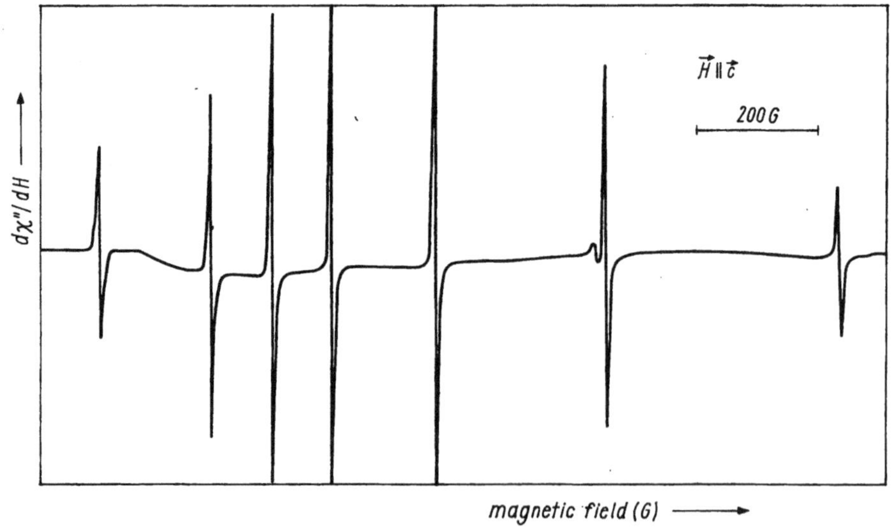

Fig. 2. The EPR-signal of $BaCl_2$:Gd^{3+} for $\boldsymbol{H} \parallel \boldsymbol{c}$

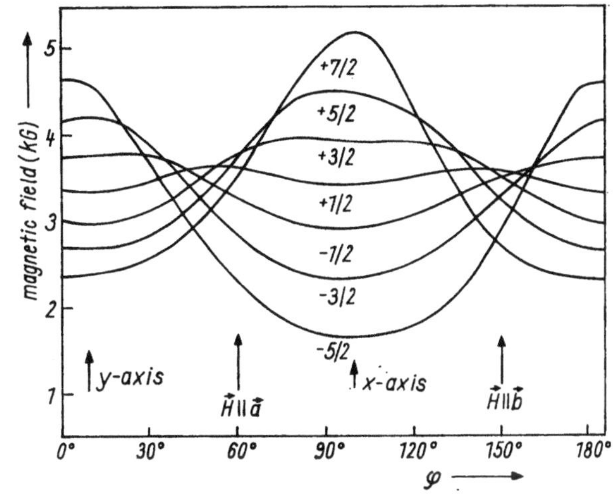

Fig. 3. Fine structure rotational diagram of the EPR-spectrum of $BaCl_2:Gd^{3+}$ for rotation around the *c*-axis. Here *H* is in the (*a*, *b*)-plane; the resonance lines of Gd (I) are only considered. A curve labeled *M* corresponds with a transition $M \leftrightarrow M-1$ (b_0^2 is assumed to be positive)

$BaCl_2$. The spectrum consists of seven lines; in this direction the Gd-ions at sites I and II give the same EPR spectrum.

In the (x, y)-plane in general fourteen lines are observed as it is expected for two non equivalent Gd-positions. The (x, y)-plane of the centres is parallel to the crystallographic (a, b)-plane. In Fig. 3 we show the angular dependence of the EPR signal for rotation in the (a, b)-plane. Here, only the spectrum corresponding with Gd (I) is given; the curves associated with Gd (II) can be obtained by reflection of the drawing about the *a*- or *b*-axis.

The *x*- and *y*-directions were taken from the curves given in Fig. 2. It was shown by Vrehen [3] that in this way these directions can be determined rather accurately. In our case the angle φ_0 between the *x*- and the crystallographic *a*-axis is 39°. The magnetic parameters g_x, g_y, g_z, b_2^0, and b_2^2 were calculated with a least squares fit procedure using the resonance field in the principle directions (*x*, *y*, and *z*). The analysis was carried out until second order in b_2^0 and b_2^2. The results are compiled in Table 1.

Table 1
Fine structure parameters and *g*-values for $BaCl_2:Gd^{3+}$

$g_x = 1.997$	$b_2^0 = \pm 90 \times 10^{-4}$ cm^{-1}
$g_y = 1.995$	$b_2^2 = \mp 444 \times 10^{-4}$ cm^{-1}
$g_z = 1.990$	
$g_{iso} = \bar{g} = 1.994$	$\varphi_0 = 39°$

4. Discussion

4.1 Theory

The spin Hamiltonian describing the $^8S_{7/2}$ state of Gd^{3+} in $BaCl_2$ can be written as

$$\mathcal{H} = g\beta \boldsymbol{H} \cdot \boldsymbol{S} + \sum_{\substack{l=2 \\ \text{even}}}^{6} \sum_{\substack{m=0 \\ \text{even}}}^{l} [B_l^m O_l^m + B_l^{-m} O_l^{-m}] \,. \tag{1}$$

Here, the first term is the Zeeman interaction and the second one is due to the interaction with the crystal field of the lattice surrounding the Gd-ion. The nature of the equivalent operators O_l^m is elucidated by Abragam and Bleaney [4]. In (1) we have taken g to be isotropic which is consistent with our observation that within the experimental accuracy $g_x = g_y = g_z$. Equation (1) can be reduced to the form

$$\mathcal{H} = g\beta \boldsymbol{H} \cdot \boldsymbol{S} + B_2^0 O_2^0 + B_2^2 O_2^2 , \tag{2}$$

where the x-, y- and z-axes are now the principle axes of the crystal field. This implies that $B_2^{-2} \equiv 0$; it is assumed that higher degree terms of the crystal field are small. As usual in the literature we shall express our results in terms of b_l^m instead of B_l^m. b_2^0 and b_2^2 are connected with the parameters occurring in (2) by the following relations $b_2^0 = 3B_2^0$ and $b_2^2 = 3B_2^2$. The crystal field at the Gd^{3+} site can be expressed in a sum of Coulomb potentials due to the surrounding point charges (when the point-ion-lattice approximation is applied).

$$V(\boldsymbol{r}) = \sum_j \frac{q_j}{|\boldsymbol{R}_j' - \boldsymbol{r}|} . \tag{3}$$

Here, q_j and $\boldsymbol{R}_j'$ denote the charge and position vector of the j-th ion, respectively. The potential can be expanded in a series of spherical harmonics $Y_l^m(\theta, \varphi)$ of which we only take into account the predominant second degree terms

$$V_2(\boldsymbol{r}') = c_2^{0\prime}(3z'^2 - r'^2) + c_2^{2\prime}(x'^2 - y'^2) + c_2^{-2\prime} 2x'y' . \tag{4}$$

The reference frame of axes $\hat{x}'$, $\hat{y}'$, and $\hat{z}'$ are along the crystallographic a-, b-, and c-axes. After a proper rotation about the z'-axis over an angle of φ_0 (4) can be transformed into the simplified form

$$V_2(\boldsymbol{r}) = c_2^0(3z^2 - r^2) + c_2^2(x^2 - y^2) , \tag{5}$$

where $\hat{x}$, $\hat{y}$ and $\hat{z}$ are the principle axes of the second degree crystal field and

$$\left. \begin{aligned} c_2^0 &= \frac{1}{4} \sum_j \frac{q_j}{R_j^5} [3Z_j^2 - R_j^2] , \\ c_2^2 &= \frac{3}{4} \sum_j \frac{q_j}{R_j^5} [X_j^2 - Y_j^2] , \end{aligned} \right\} \tag{6a}$$

and

$$\varphi_0 = \frac{1}{2} \operatorname{arctg} \left(\frac{c_2^{-2\prime}}{c_2^{2\prime}} \right). \tag{6b}$$

The crystal structure of $BaCl_2$ is the same as that of $PbCl_2$ (orthorhombic; space group D_{2h}^{16}). The positions of the ions in the unit cell have not yet been determined for $BaCl_2$ and therefore we shall assume that the normalized coordinates within the unit cell are equal to those of $PbCl_2$ [5]. The parameters a_0, b_0, and c_0 for $BaCl_2$ are 9.333, 7.823, and 4.705 Å, respectively. The corresponding values for $PbCl_2$ are 9.030, 7.608, and 4.525 Å. By an inspection of (6) it is easily seen that the values of c_2^0 and c_2^2 are reduced by a factor of

$$p = \frac{a_0(\mathrm{PbCl_2})}{a_0(\mathrm{BaCl_2})} \frac{b_0(\mathrm{PbCl_2})}{b_0(\mathrm{BaCl_2})} \frac{c_0(\mathrm{PbCl_2})}{c_0(\mathrm{BaCl_2})} = 0.91 . \tag{7}$$

Recently, Willemsen and Hommels [2] calculated numerically c_2^0 and c_2^2 for $PbCl_2$. By applying the factor p to these results we find $c_2^0(BaCl_2) = 3.3\times \times 10^{18}$ V/m² and for $c_2^2 = 10.3\times 10^{18}$ V/m². Furthermore it can be seen in (6b) that in this approximation $\varphi_0(BaCl_2) = \varphi_0(PbCl_2) = 18°$.

It is now of interest to connect the values c_2^0 and c_2^2 with the experimentally observed crystal field parameters b_2^0 and b_2^2. Unfortunately, no theoretical treatment for situations with low symmetry (such as Gd^{3+} in $BaCl_2$) have been published. Therefore, we shall in accordance with Vrehen [3], Willemsen and Hommels [2] and Beyerinck and Willemsen [6] assume that it is allowed to use the quantitative relation between b_2^0 and A_2^0 which was derived by Hutchison et al. [7].

$$b_2^0 = -\frac{12}{5}\frac{\zeta^3}{W_P^2 W_D}\gamma_2 A_2^0\langle r^2\rangle\,, \tag{8}$$

where ζ is the one-electron spin–orbit coupling parameter, W_P and W_D are the energy separations between the ground state $^8S_{7/2}$ and the excited states denoted by $^6P_{7/2}$ and $^6D_{7/2}$, respectively. γ_2 is the screening factor and $\langle r^2\rangle$ is the average of the operator r^2 for a 4f-electron of Gd^{3+}. A_2^0 is related with c_2^0 through the equation $A_2^0\langle r^2\rangle = -e\, c_2^0\langle r^2\rangle$.

In addition to the relation given by (8) there is a significant contribution to b_2^0 from a mechanism proposed by Wybourne [8] who introduced relativistic Hartree-Fock wave functions in the calculations of the ground state splitting. The effect of this mechanism can be written as

$$b_2^0 = \frac{16}{245}\frac{\gamma_2 A_2^0\zeta\,(-5R_{++}^2 + 3R_{+-}^2 + 2R_{--}^2)}{W_P}\,, \tag{9}$$

where R_{++}^2, R_{+-}^2, and R_{--}^2 are relativistic integrals. R_{++}^2 and R_{--}^2 are calculated by Waber and Cromer [9] for Gd^{3+}, while R_{+-}^2 is obtained by interpolation [8].

Using the following numerical values for the parameters appearing in (8) and (9) $\zeta = 1534\ \mathrm{cm^{-1}}$ [10], $W_P = 32200\ \mathrm{cm^{-1}}$, $W_D = 40000\ \mathrm{cm^{-1}}$, $\gamma_2 = 0.33$ [2], $\langle r^2\rangle = 0.785\ a_0^2$ [11], $R_{++}^2 = 0.824\,a_0^2$, $R_{--}^2 = 0.804\,a_0^2$, and $R_{+-}^2 = 0.814\ a_0^2$ we obtain for $b_2^0 = 85\times 10^{-4}\ \mathrm{cm^{-1}}$. From the ratio $c_2^2/c_2^0 = b_2^2/b_2^0$ we find $b_2^2 = = -324\times 10^{-4}\ \mathrm{cm^{-1}}$.

The g-value of the ground state of the $4f^7$ system in Gd^{3+} is expected to be isotropic and rather close to the free electron value $g_e = 2.0023$ because we are dealing with almost a pure $^8S_{7/2}$ state. Deviations from the free electron g-value are indicative for a coupling between the ground state and excited $^6P_{7/2}$, $^6D_{7/2}$, etc. states. It turns out that the g-shift for Eu^{2+} and Gd^{3+} is approximately independent of the host crystal [12 to 14, 2, 3, 6, 7]. In the case of Eu the g-values of the $4f^7$ configuration in the various host crystals even closely agree with the g-value obtained from an atomic beam experiment with neutral Eu-atoms [4]. Therefore, it is highly probable that also for Gd^{3+} in $BaCl_2$ the g-value is isotropic and approximately equal to the g-value of Gd^{3+} in other substances.

Theoretically, the admixture of the $^6P_{7/2}$ state to the ground state can be written as $(1-\alpha^2)^{1/2}\ {}^8S_{7/2} + \alpha\,{}^6P_{7/2}$, where α is a measure of the amount of admixture. The g-value of this state is then

$$g = (1-\alpha^2)\,g(^8S_{7/2}) + \alpha^2 g(^6P_{7/2})\,, \tag{10}$$

where $g(^8S_{7/2}) = 2.0023$ and $g(^6P_{7/2}) = \frac{12}{7}$. In accordance with Baker and Williams [12] we take

$$\alpha^2 = 14 \frac{\zeta^2}{W_P^2} \approx 0.0318 .$$

Inserting this value into (10) we obtain $g = 1.993$, which agrees with the average value of g within experimental accuracy.

4.2 Discussion of the experimental results

In this section we shall compare the experimental results with the theoretically obtained parameters and we shall particularly concentrate on the crystal field parameters. In addition to this we shall also compare our results with those of Eu^{2+} and Gd^{3+} in $PbCl_2$ and $PbBr_2$ [2, 3, 6]. In Table 2 we give a survey

Table 2

Experimental and theoretical parameters of $^8S_{7/2}$ ions in orthorhombic crystals

system	lattice parameters (Å)	g-value*)	crystal field-parameters (10^{-4} cm^{-1})**) $b_2^{-2} = 0$			ref.
$PbCl_2:Eu^{2+}$	$a_0 = 9.030$ $b_0 = 7.608$ $c_0 = 4.525$	1.993	$b_2^0 = 104$ $b_2^2 = -514$	(115) (−358)	(115) (−512)	[2]
$PbCl_2:Gd^{3+}$	$a_0 = 9.030$ $b_0 = 7.608$ $c_0 = 4.525$	1.989 (1.993)	$b_2^0 = 99$ $b_2^2 = -437$	(94) (−292)	(93) (−307)	[2, 6]
$PbBr_2:Eu^{2+}$	$a_0 = 9.5255$ $b_0 = 8.0461$ $c_0 = 4.7249$	1.990	$b_2^0 = 177$ $b_2^2 = -527$	(68) (−453)	(178) (−521)	[2]
$BaCl_2:Gd^{3+}$	$a_0 = 9.333$ $b_0 = 7.823$ $c_0 = 4.705$	1.994 (1.993)	$b_2^0 = \pm 90$ $b_2^2 = \mp 444$	(85) (−324)	— —	this work

*) Values in parenthesis are calculated ones.
**) Values in parenthesis are calculated ones; the first column corresponds with the point-ion-lattice approximation, while in the second column are values corrected for lattice distortions.

of the magnetic parameters obtained for S-state lanthanides in the orthorhombic host lattices.

From Table 2 it is clear that the agreement between the experimental and theoretical values for the crystal field parameters is reasonable especially when the crudeness of the model is taken into consideration. Improvements on the calculated values of b_2^0 and b_2^2 for Eu^{2+} and Gd^{3+} in $PbCl_2$ and $PbBr_2$ were obtained by Willemsen and Hommels [2] by the introduction of small distortions of the immediate environment of the impurity. The ultimate results for Eu^{2+} in $PbCl_2$ and $PbBr_2$ are in very good agreement with experiment. For the system $PbCl_2$: Gd^{3+}, however, the corrected crystal field parameter b_2^2 is still unsatisfactory. Although our theoretical crystal field parameters for $BaCl_2:Gd^{3+}$ are certainly not very accurate because of the lack of knowledge about the exact positions of the ions in the unit cell we have the impression that the present system shows even more clearly the same features as $PbCl_2:Gd^{3+}$. The agreement between theory and experiment is acceptable for b_2^0, but there is a large discrepancy for b_2^2. This discrepancy can not be removed by taking into account lattice distortions. It is expected that a simple relaxation of the nearest neighbour Cl^- ions with respect to the Gd^{3+} ion will only affect the value of b_2^0. This relaxation will probably take place because there is an appreciable difference in ion size between Ba^{2+} and Gd^{3+} ($r_{Ba^{2+}} = 1.34$ Å, $r_{Gd^{3+}} = 0.94$ Å). Furthermore, the charge of the two types of ions is different. This may lead to an extra relaxation of the surrounding ions. In addition to this effect one should also consider contributions to the crystal field produced by electronic polarization. The excess positive charge of the Gd-ion will polarize the surrounding ions. The dipoles produced in this way will contribute to the crystal field at the Gd^{3+} site. Without a detailed calculation of these effects it is not clear whether they also contribute to b_2^2 or not. This will be the subject of further investigations. In order to obtain more evidence for the correctness of the above interpretation the study of the systems $BaCl_2:Eu^{2+}$, $SrBr_2:Gd^{3+}$ and $SrBr_2:Eu^{2+}$ is planned for the near future.

Acknowledgement

The authors are very grateful to Mr. P. Wesseling for growing the crystals.

References

[1] R. W. G. Wyckoff, Crystal Structures, Interscience Publishers, 1948.
[2] B. Willemsen and W. C. Hommels, phys. stat. sol. (a) **10**, 183 (1972).
[3] Q. H. F. Vrehen, Thesis, Utrecht 1963.
[4] A. Abragam and B. Bleaney, EPR of Transition Ions, Clarendon Press, Oxford 1970.
[5] R. L. Sass, E. B. Brackett, and T. E. Brackett, J. phys. Chem. **67**, 2863 (1963).
[6] H. C. W. Beyerinck and B. Willemsen, Physica (Utrecht) **47**, 515 (1970).
[7] C. A. Hutchison, B. R. Judd, and D. F. D. Pope, Proc. Phys. Soc. B **70**, 514 (1957).
[8] B. G. Wybourne, Phys. Rev. **148**, 317 (1966).
[9] J. I. Waber and D. T. Cromer, J. chem. Phys. **42**, 4116 (1965).
[10] B. R. Judd and I. Lindgren, Phys. Rev. **122**, 1802 (1961).
[11] A. J. Freeman and R. E. Watson, Phys. Rev. **127**, 2058 (1962).
[12] J. M. Baker and F. I. B. Williams, Proc. Roy. Soc. **A 267**, 283 (1962).
[13] W. Low, Phys. Rev. **109**, 265 (1958).
[14] S. Geschwind and J. P. Remeika, Phys. Rev. **122**, 757 (1961).

(Received November 27, 1973)

phys. stat. sol. (b) 62, 477 (1974)

Subject classification: 13; 6; 21

Department of Physics, Waseda University, Tokyo

Quantum-Mechanical Study of Rosette-Motion Channeling

By

A. TAMURA and Y. H. OHTSUKI

Quantum mechanical treatment of the "Rosette-Motion" channeling of energetic electron beams is developed. Energy levels and scattering angles corresponding to channeling are determined. Detailed calculations are performed for W⟨111⟩ for various energies. The lifetime of such quantum states is discussed making use of the mean free path due to lattice vibrations.

Es wird eine quantenmechanische Behandlung der „Rosettenbewegungs"-Kanalisierung energiereicher Elektronen entwickelt. Energieniveaus und Streuwinkel, die der Kanalisierung entsprechen, werden bestimmt. Detaillierte Berechnungen werden für W⟨111⟩ für verschiedene Energien durchgeführt. Die Lebensdauer solcher Quantenzustände wird diskutiert, wobei die mittlere freie Weglänge infolge von Gitterschwingungen herangezogen wird.

1. Introduction

In connection with the experiment of a new type of electron-channeling (here we call it rosette-motion channeling) by Sizmann and coworkers [1], some theoretical considerations have been performed by several authors [1 to 3] from the classical mechanical point of view. However, it is worthwhile to note that quantum mechanics is important for the rosette-motion channeling, since the potential energy acting on the electron beam is comparable with the transverse energy of the channeling electrons.

Here we develop a quantum mechanical theory of rosette-motion channeling assuming a string potential, and determine the quantum mechanical levels (or quantized channeling directions). In Section 2 and 3, we solve the Schrödinger equation by the power series expansion, and determine the energy levels (or the quantized channeling directions). In Section 4 detailed numerical calculations are performed for various energies between 1 to 3.5 MeV into W⟨111⟩. In Section 5, we discuss the stability of the channeling levels considering the position-dependent mean free path.

2. Solution of the Schrödinger Equation

For simplicity, we consider only one string which is represented by the string potential [1],

$$\bar{V}(r) = -\frac{Ze^2}{d} \ln\left[\left(\frac{ca}{r}\right)^2 + 1\right], \tag{1}$$

where a is the Thomas-Fermi screening length, d the distance between atoms in the string, and $c \approx \sqrt{3}$. Then we obtain the Schrödinger equation for the radial component, $R(\varrho)$ with changing the variable as $\varrho \equiv r/a$,

$$\frac{\mathrm{d}^2R}{\mathrm{d}\varrho^2} + \frac{1}{\varrho}\frac{\mathrm{d}R}{\mathrm{d}\varrho} + \left(\frac{2\mu a^2}{\hbar^2} E_\perp - \frac{2\mu a^2}{\hbar^2}\bar{V}(\varrho) - \frac{\lambda}{\varrho^2}\right) R = 0\,, \tag{2}$$

where $-|E_\perp| = E - (\hbar^2 k_z^2/2\mu)$ is the transverse energy of channeling electrons with the mass μ to the string, E being the incident energy, and $\lambda = m^2$, $m = 0, \pm 1, \pm 2, \ldots$ Changing the variable as $R = \varrho^{-1/2} Y$, we obtain

$$\frac{d^2 Y}{d\varrho^2} + \left\{\beta \ln\left[\left(\frac{c}{\varrho}\right)^2 + 1\right] - \frac{\lambda - (1/4)}{\varrho^2}\right\} Y = \alpha Y\,, \tag{3}$$

where $\alpha = 2\mu a^2 |E_\perp|/\hbar^2$ and $\beta = 2\mu a^2 Ze^2/d$.

For solving (3), we use the power series expansion at the minimum point ϱ_0 of the effective potential

$$W(\varrho) = -\beta \ln\left[\left(\frac{c}{\varrho}\right)^2 + 1\right] + \frac{\lambda - (1/4)}{\varrho^2}\,. \tag{4}$$

When we stop the expansion at the power $(\varrho - \varrho_0)^2$ and noting that $(dW/d\varrho)|_{\varrho=\varrho_0} = 0$ in our case, we easily obtain the usual harmonic oscillator type solution,

$$Y(\varrho) = \frac{1}{D}\, e^{-(1/2)\sqrt{B}(\varrho-\varrho_0)^2}\, H_n\big(B^{1/4}(\varrho - \varrho_0)\big) \tag{5}$$

and

$$E_\perp = -|E_\perp| = \frac{\hbar^2}{2\mu a^2}\big[A + (2n+1)\sqrt{B}\big]\,, \tag{6}$$

where

$$A = -\beta \ln\left(\frac{\beta c^2}{(\lambda - (1/4))\,\epsilon}\right) - \frac{\lambda - (1/4)}{c^2} < 0 \quad \text{and} \quad B = \frac{2(\beta c^2 - \lambda + (1/4))^3}{\beta c^6(\lambda - (1/4))}\,.$$

In (5), $H_n(x)$ is the Hermite polynomial of n-th order. D is a normalization factor.

3. Quantized Directions

Now we calculate the expectation value of the kinetic energy $E_\perp(\text{K.E}) = \hat{p}_\varrho^2/(2\mu a^2)$, where $\hat{p}_\varrho$ is the momentum operator conjugate to ϱ, $[\hat{p}_\varrho, \varrho] = -i\hbar$:

$$\langle E_\perp(\text{K.E})\rangle = \frac{\hbar^2}{2\mu a^2}\left[\frac{\sqrt{B}}{2} + B\varrho_0\delta\varrho_0 + B\delta\varrho_0^2\right], \tag{7}$$

where

$$\delta\varrho_0 = \frac{e^{-\sqrt{B}\varrho_0^2}}{2\sqrt{B}\int\limits_0^\infty e^{-\sqrt{B}(\varrho-\varrho_0)^2}\, d\varrho}\,. \tag{8}$$

As shown in the next section, $\delta\varrho_0$ in (7) is very small.

By the transverse energy conservation, the emerging angle θ of channeled beams from the crystal surface is obtained by (7),

$$\theta = \sqrt{\frac{\langle E_\perp(\text{K.E})\rangle}{E}}\,. \tag{9}$$

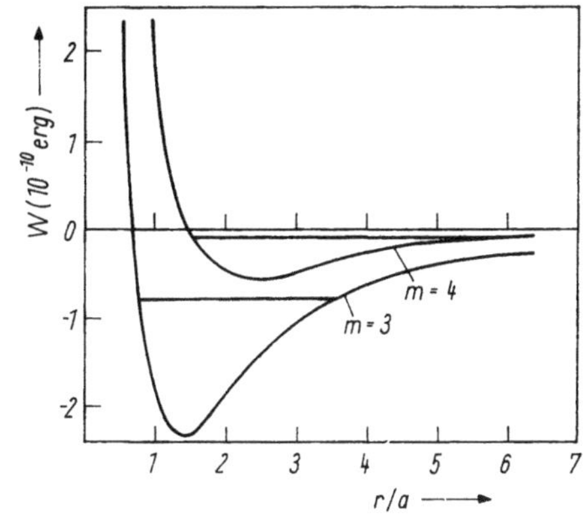

Fig. 1. Two quantized states of the rosette-motion channeling of 3.0 MeV electrons in W⟨111⟩ axis

4. Numerical Estimates

Now we consider in detail the W⟨111⟩ case for various energies, which is summarized in Table 1. The bound states are selected by the condition

$$E_n < 0 \,, \qquad 1 < \varrho_0 < \frac{d^*}{2a} \tag{10}$$

(d^* being the distance between strings). However, the bound states satisfying the above condition have a mean free path for decaying which is discussed in the next section. We note that allowed bound states are $\pm m = 3$ and 4 for 3.0 MeV and 3.5 MeV, but $\pm m = 3$ for 2.0 MeV and $\pm m = 2$ for 1 MeV. The quantized scattering angles are also shown in Table 1.

Table 1

Results of calculations

incident energy (MeV)	$\pm m$	ϱ_0	$\delta\varrho_0$	$\langle E_\perp(\text{K.E.})\rangle$ (10^{-11} erg)	$\theta \times 10^{-3}$ (rad)	l (Å)
1.0	2	1.30	0.03	12.8	8.92	
2.0	3	1.76	0.02	5.17	4.02	
3.0	3	1.31	0.01	7.86	4.05	0.9×10^4
	4	2.38	0.01	2.41	2.24	4.5×10^4
3.5	3	1.18	0.01	8.69	3.94	
	4	2.00	0.01	3.21	2.39	

In Fig. 1, we show the two energy levels corresponding to $\pm m = 3$ and 4 for 3.0 MeV electrons.

5. Stability of Quantum States

The quantum states determined in the previous section have mean lifetime τ due to phonon and electron excitations. We define the mean free path by $l = \tau \cdot v_0$, v_0 being the incident-electron velocity. l is calculated making use of the diffusion coefficient function, $D(r)$ (or local diffusion coefficient) proposed by Ohtsuki [4]

$$l^{-1} = \frac{2}{d^{*2}} \int_0^{d^*} D(r)\, |R(r)|^2\, r \, \mathrm{d}r \,, \tag{11}$$

where

$$D(r) = -\frac{\mu}{2\pi^3\hbar^3 v_0} \operatorname{Im} \int \mathrm{d}K' \frac{(1 - K_z'/K')}{K'^2 - k_n^2} \mathrm{e}^{-i(K_x'x + K_y'y)} \times$$
$$\times H_{0n}'(K_x', K_y', K_z' - K) \, H_{n0}'(x, y, K - K_z') \,. \tag{12}$$

In the above expression, $H_{0n}'(x, y, z)$ is the matrix element between the 0- and n-th states of the crystal. The lower limit of the integration in (12) should be taken as $K_c = \sqrt{2\mu E_\perp/\hbar^2}$. The detailed calculation of $D(r)$ has been performed by Kitagawa and Ohtsuki [5] and Waho and Ohtsuki [6]. Here we give a rough estimate of the mean free path l in Table 1. We note that in the cases of high energy the quantum states have long mean free paths.

6. Concluding Remarks

Recently, a ring pattern has been discovered by Fujimoto et al. [7], and interpreted as quantized rosette-motion channeling by themselves. The angular separation of two quantized directions of the channeled beam in the 3.0 MeV case determined here (see Table 1) is of the same order in separation between the two experimental rings.

If we consider higher levels, the power series method for solving the Schrödinger equation is not good enough. However, the quantum mechanical states considered here are treated well by the method, since the allowed range of ϱ in the rosette-motion is at most between 1 and 3.

The mean free paths calculated in Table 1 are rough estimates, because we considered only phonon excitations. However, we expect that electronic excitations do not change the order of magnitude of the mean free path.

Acknowledgement

The authors thank Prof. Fujimoto for his private communication on his experimental work of the quantized ring before publication.

References

[1] H. J. KREINER, F. BELL, R. SIZMANN, D. HARDER, and W. HÜTTL, Phys. Letters A **33**, 135 (1970).
F. BELL, H. J. KREINER, and R. SIZMANN, Phys. Letters A **38**, 373 (1972).

[2] H. C. H. NIP, M. J. HOLLIS, and J. C. KELLY, Phys. Letters A **28**, 324 (1968).
H. C. H. NIP and J. C. KELLY, Atomic Collision Phenomena in Solids, North-Holland Publ. Co., Amsterdam 1970.
H. C. H. NIP, R. L. DALGLISH, A. P. M. CHANG, and J. C. KELLY, Phys. Letters A **34**, 257 (1971).

[3] U. SCHIEBEL, A. NEUFERT, and G. CLAUSNITZER, Phys. Letters A **42**, 45 (1972).

[4] Y. H. OHTSUKI, J. Phys. Soc. Japan **34**, 673 (1973).

[5] M. KITAGAWA and Y. H. OHTSUKI, Phys. Rev. B **8**, 3117 (1973).

[6] T. WAHO and Y. H. OHTSUKI, Radiation Effects, in the press.

[7] F. FUJIMOTO, K. KOMAKI, H. FUJITA, N. SUMITA, Y. UCHIDA, K. KAMBE, and G. LEHMPFUHL, private communications.

(Received November 19, 1973)

phys. stat. sol. (b) 62, 481 (1974)

Subject classification: 11

Department of Physics and Astrophysics, University of Delhi

Ionized-Beam Energy Loss to Electron Gas

By

S. SHAH, M. P. SRIVASTAVA, and L. S. KOTHARI

The stopping power and other transport coefficients have been calculated for ionized beams channeling through a crystal using a statistical method developed by Chandrasekhar. Only the interaction of beam particles with conduction electrons, treated as a free gas, is considered. Unlike earlier investigations, also the effect on the test particles of encounters with faster field particles is considered. Two types of potentials have been used, a Debye and a Coulomb one. Both potentials give nearly the same results. It is found that the values of energy loss and transport coefficients increase with increasing energy of the incident beam. The stopping power is nearly the same as that obtained by earlier workers.

Die Bremsenergie und andere Transportkoeffizienten werden für ionisierte Strahlen, die durch einen Kristall kanalisieren, berechnet, wobei eine von Chandrasekhar entwickelte statistische Methode benutzt wird. Es wird nur die Wechselwirkung von Strahlteilchen mit den als freies Elektronengas behandelten Leitungselektronen berücksichtigt. Im Gegensatz zu früheren Untersuchungen wurde auch der Einfluß von Stößen der Testpartikel mit schnelleren Feldteilchen berücksichtigt. Zwei Potentialtypen, Debye- und Coulombpotential, wurden benutzt und geben nahezu dieselben Ergebnisse. Es wird gefunden, daß die Energieverlustwerte und Transportkoeffizienten mit wachsender Energie des einfallenden Strahls ansteigen. Die Bremsenergie ist nahezu dieselbe wie die von früheren Autoren gefundene.

1. Introduction

There is at present considerable interest in the problem of channeling of particles through crystals. Ionized atoms in a channeled beam can loose energy either to phonons or to electrons. The electrons can be either the conduction electrons or the core electrons. Erginsoy [1] has shown that, for a high energy ion beam channeling through a crystal, the energy loss to phonons is small and that the energy loss to conduction electrons is the only important process. Energy loss to core electrons is also small since the beam particles do not come close enough to nuclei of the crystal [1, 2].

Pathak and Yussouff [2] and Lidiard [3] have studied the scattering of energetic charged beams by conduction electrons, using the dynamic dielectric function $\varepsilon(q, \omega)$ for the electron gas. This problem can also be studied in a more direct manner using techniques developed in plasma physics and astrophysics by various workers [4 to 8]. However, in these studies contributions from collisions with large impact parameter — the so-called nondominant terms, have been neglected. Moreover, it is assumed that the velocity of the beam particles — henceforth called test particles, is greater than the maximum speed of the field particles i.e. scatterer particles. Srivastava and Devanathan [9] have evaluated the transport properties without separating dominant and nondominant terms. However, their work is for temperatures and densities commonly found in plasmas.

Following the method of Srivastava and Devanathan we have calculated the stopping power and transport coefficients of singly ionized beams due to conduction electrons in matter treating them as free particles. The great advantage of the present method lies in the fact that one can calculate various transport coefficients apart from mean energy loss, which alone has been calculated by other workers.

For the interaction between ions and electrons we have used the screened Coulomb potential. The limit of screening length going to infinity gives a pure Coulomb potential and this limit too has been considered for the sake of comparison.

In Section 2 we have derived the expressions for the rate of energy loss and the Fokker-Planck coefficients (which relate to transport properties) by averaging over all possible encounters with the field particles. It may also be mentioned that, unlike earlier investigators, we have not put any restriction on the speed of the test particles. The effect on the test particles of encounters with faster field particles has also been considered.

In Section 3 numerical results are discussed for the energy loss and diffusion coefficients in various cases. We have considered Cu^+ in Cu, Kr^+ in Cu, Cu^+ in Ge and Xe^+ in W and compared our results with those of other workers. We find that the mean energy loss is proportional to E^n where n varies from about 0.5 to approximately 1 as the energy of the incident beam increases from 0.1 to 500 keV. Mean square energy loss and other transport coefficients also show an increase with increasing energy of the incident beam.

2. Calculation of Energy Loss and Transport Coefficients

We consider a test particle of mass m_1 and charge Z_1e of a beam travelling with velocity v_1 and encountering field particles of mass m_2, charge Z_2e and velocity distribution given by

$$f(v_2) = N\left(\frac{m_2}{2\pi k_B T}\right)^{3/2} e^{-\frac{mv_2^2}{2k_B T}}, \tag{1}$$

where T is the temperature, N the density of field particles, and k_B the Boltzmann constant.

The angle χ through which the test particle will be scattered is given by [9]

$$\chi = \pi - 2\psi = \pi - 2\int_0^{v_{00}} \left(1 - v^2 - F(v, v_0)\right)^{-1/2} dv\ , \tag{2}$$

where $F(v, v_0)$ is given by

$$F(v, v_0) = \frac{2}{\nu - 1}\left(\frac{v}{v_0}\right)^{\nu - 1} \exp\left(-\frac{\alpha b}{v}\right),$$

$v = b/r$ where b is the impact parameter and r denotes the distance between interacting particles; $v_0 = b(\mu V^2/K)^{1/(\nu-1)}$, μ being the reduced mass and $\boldsymbol{V} = \boldsymbol{v}_1 - \boldsymbol{v}_2$, the relative velocity between test and field particles (these velocities are expressed in units of $\sqrt{k_B T/m_2}$). K denotes the force constant for central potentials $-\left(K/(\nu - 1)\right) r^{-\nu+1} \exp(-\alpha r)$ (the exponent ν is taken as 2 in the present case). α is the inverse of the Debye length (λ_D) in the case of screened potentials and is zero for the Coulomb case. The upper limit of integration

v_{00} in (2) is given by the positive root of the equation

$$1 - v^2 - F(v, v_0) = 0\,. \tag{3}$$

Expressions for energy loss, momentum transfer etc. in a single collision are well known [10]. In order to study the passage of a test particle through a system of field particles one must multiply the relevant expression for a single collision by the number of encounters in a time interval $\mathrm{d}t$ with impact parameter lying between b and $b + \mathrm{d}b$, orbital plane range ε, $\varepsilon + \mathrm{d}\varepsilon$, velocity range v_2, $v_2 + \mathrm{d}v_2$ and direction confined to solid angle $\sin\theta\,\mathrm{d}\theta\,\mathrm{d}\varphi$. This number is given by

$$f(v_2)\, v_2^2\, \mathrm{d}v_2 \sin\theta\, \mathrm{d}\theta\, \mathrm{d}\varphi\; V\mathrm{d}t\; b\, \mathrm{d}b\, \mathrm{d}\varepsilon\,.$$

The expressions for energy loss and mean square energy loss are

$$-\left\langle\frac{\mathrm{d}E}{\mathrm{d}t}\right\rangle = \frac{8\pi^2 m_1 m_2}{(m_1+m_2)^2} \iiint \frac{v_2}{v_1} f(v_2) \cos^2\psi\; b\, \mathrm{d}b\, \mathrm{d}v_2 \times$$
$$\times \left[m_1 v_1^2 - m_2 v_2^2 - \left(\frac{m_2 - m_1}{2}\right)(V^2 - v_2^2 - v_1^2)\right] V^2\, \mathrm{d}V\,,$$

$$\left\langle\frac{\mathrm{d}E^2}{\mathrm{d}t}\right\rangle = \frac{8\pi^2 m_1^2 m_2^2}{(m_1+m_2)^2} \left[\iiint 2f(v_2) \cos^4\psi\; b\, \mathrm{d}b\, \frac{v_2}{v_1}\, V^2\, \mathrm{d}V\, \mathrm{d}v_2 \times\right.$$
$$\times \frac{(m_1 v_1^2 - m_2 v_2^2 + v_1 v_2 (m_2 - m_1)\cos\theta)^2}{(m_1+m_2)^2} +$$
$$\left. + \iiint f(v_2) \frac{v_2}{v_1} \cos^2\psi \sin^2\psi\, V^2\, \mathrm{d}V\, b\, \mathrm{d}b\, \mathrm{d}v_2 \left(v_1^2 v_2^2 - \frac{(v_1^2 + v_2^2 - V^2)^2}{4}\right)\right].$$

Integration over φ and ε from zero to 2π has already been carried out and integration over θ has been transformed to variable V, using the relation

$$V^2 = v_1^2 + v_2^2 - 2v_1 v_2 \cos\theta\,.$$

It is necessary to split the integral over V into two ranges (i) $v_2 < v_1$, (ii) $v_2 \geqq v_1$, since the integrands in the two cases are different. The upper limit for integration over b is chosen according to the nature of the problem itself. Since we are discussing the channeling of particles the maximum impact parameter can be of the order of interatomic spacing. We have taken the upper cut-off for integration over the impact parameter to be $(\lambda/2)\,a$, where $\lambda = 2\pi/k_\mathrm{F} = 2\pi\,(1/3\pi^2 N)^{1/3}$ [11] is the Fermi length and a is the parameter which is chosen to be of the order of unity.

The final expressions for energy loss, stopping power, and mean square energy loss are

$$-\left\langle\frac{\mathrm{d}E}{\mathrm{d}t}\right\rangle = \alpha\left[2Av_1^3 I(2, 2, 0) + \left(\frac{4}{3}A - \frac{8}{3}\right) v_1 I(4, 2, 0) -\right.$$
$$- \left(\frac{2}{15}A + \frac{8}{15}\right) v_1^{-1} I(6, 2, 0) + \left(\frac{8}{15}A + \frac{2}{15}\right) v_1^4 J(1, 2, 0) +$$
$$\left. + \left(\frac{8}{3}A - \frac{4}{3}\right) v_1^2 J(3, 2, 0) - 2J(5, 2, 0)\right], \tag{4}$$

$$-\left\langle\frac{\mathrm{d}E}{\mathrm{d}x}\right\rangle = -\frac{1}{v_1}\left\langle\frac{\mathrm{d}E}{\mathrm{d}t}\right\rangle, \tag{5}$$

$$\left\langle\frac{\mathrm{d}E^2}{\mathrm{d}t}\right\rangle = \beta\Bigg[\left(1+\left(\frac{A-1}{A+1}\right)^2+2\left(\frac{A-1}{A+1}\right)\right)v_1^5 I(2,4,0)+$$
$$+\left(-\frac{5}{3}+5\left(\frac{A-1}{A+1}\right)^2+2\left(\frac{A-1}{A+1}\right)\right)v_1^3 I(4,4,0)+$$
$$+\left(\frac{1}{3}+3\left(\frac{A-1}{A+1}\right)^2-\frac{18}{5}\left(\frac{A-1}{A+1}\right)\right)v_1 I(6,4,0)+$$
$$+\left(\frac{1}{3}+\frac{1}{7}\left(\frac{A-1}{A+1}\right)^2-\frac{2}{5}\left(\frac{A-1}{A+1}\right)\right)v_1^{-1} I(8,4,0)+$$
$$+\frac{4}{3}v_1^3 I(4,2,2)+\frac{8}{15}v_1 I(6,2,2)-\frac{4}{105}v_1^{-1} I(8,2,2)+$$
$$+\left(\frac{1}{3}+\frac{1}{7}\left(\frac{A-1}{A+1}\right)^2+\frac{2}{5}\left(\frac{A-1}{A+1}\right)\right)v_1^6 J(1,4,0)+$$
$$+\left(\frac{1}{3}+3\left(\frac{A-1}{A+1}\right)^2+\frac{18}{5}\left(\frac{A-1}{A+1}\right)\right)v_1^4 J(3,4,0)+$$
$$+\left(-\frac{5}{3}+5\left(\frac{A-1}{A+1}\right)^2-2\left(\frac{A-1}{A+1}\right)\right)v_1^2 J(5,4,0)+$$
$$+\left(1+\left(\frac{A-1}{A+1}\right)^2-2\left(\frac{A-1}{A+1}\right)\right)J(7,4,0)-$$
$$-\frac{4}{105}v_1^6 J(1,2,2)+\frac{8}{15}v_1^4 J(3,2,2)+\frac{4}{3}v_1^2 J(5,2,2)\Bigg], \tag{6}$$

where

$$A = \frac{m_1}{m_2},$$

$$\alpha = \frac{8\pi^2 A}{(1+A)^2}\left(\frac{\lambda}{2}\right)^2 N\left(\frac{k_B T}{2\pi m_2^{1/3}}\right)^{3/2} a^2\,, \tag{7}$$

$$\beta = \frac{8\pi^2 A^2}{(1+A)^2}\left(\frac{\lambda}{2}\right)^2 N\left(\frac{k_B T}{2\pi m_2^{1/3}}\right)^{3/2} k_B T\, a^2\,, \tag{8}$$

and I and J are defined as follows:

$$I(l,m,n) = \int_0^1 B\,\mathrm{d}B \int_0^{v_1} \frac{v_2^l \cos^m \psi \sin^n \psi}{\mathrm{e}^{v_2^2/2}}\,\mathrm{d}v_2\,, \tag{9a}$$

$$J(l,m,n) = \int_0^1 B\,\mathrm{d}B \int_{v_1}^{5} \frac{v_2^l \cos^m \psi \sin^n \psi}{\mathrm{e}^{v_2^2/2}}\,\mathrm{d}v_2\,, \tag{9b}$$

where

$$B = \frac{b}{(\lambda/2)\, a}.$$

We now consider transport coefficients. We know that the test particles suffer many small deflections for the reason that interaction has been assumed to be taking place between heavy ions and electrons. Initially all the test particles have the same velocity but with passage of time as successive encounters take place, they will suffer changes in velocity $\Delta v_{||}$ and $\Delta v_{\perp}$ in directions parallel and perpendicular to the original direction respectively. Hence, the velocity distribution will progressively get broadened. The effect is similar to diffusion of particles in ordinary gas. The quantity $\langle \mathrm{d}v_{||}/\mathrm{d}t \rangle$ (the averages being determined in the same way as for energy loss) represents the rate at which moving test particles are slowed down by interaction with field particles and following Chandrasekhar [12] we call it "the coefficient of dynamical friction". The quantities $\langle \mathrm{d}v_{||}^2/\mathrm{d}t \rangle$ and $\langle \mathrm{d}v_{\perp}^2/\mathrm{d}t \rangle$ give a measure of the rate of velocity dispersion of particles in directions parallel and perpendicular to the original direction of motion of test particles and have been called parallel and perpendicular diffusion coefficients (also known as Fokker-Planck coefficients for a charged assembly).

The expressions for these coefficients can be deduced in the same way as the earlier expressions for energy loss. We obtain

$$\left\langle \frac{\mathrm{d}v_{||}}{\mathrm{d}t} \right\rangle = -\gamma \Big[2v_1^2 I(2, 2, 0) + \frac{4}{3} I(4, 2, 0) - \frac{2}{15} v_1^{-2} I(6, 2, 0) + + \frac{8}{15} v_1^3 J(1, 2, 0) + \frac{8}{3} v_1 J(3, 2, 0) \Big], \tag{10}$$

$$\begin{aligned} \left\langle \frac{\mathrm{d}v_{\perp}^2}{\mathrm{d}t} \right\rangle = \delta \Big[& 4v_1^3 I(2, 2, 0) + 8v_1 I(4, 2, 0) + \frac{4}{5} v_1^{-1} I(6, 2, 0) - \\ & - \frac{8}{105} v_1^{-3} I(8, 2, 0) - 4v_1^3 I(2, 4, 0) - \frac{16}{3} v_1 I(4, 4, 0) + \\ & + \frac{4}{15} v_1^{-1} I(6, 4, 0) - \frac{8}{105} v_1^{-3} I(8, 4, 0) - \frac{4}{3} v_1 I(4, 2, 2) - \\ & - \frac{8}{15} v_1^{-1} I(6, 2, 2) + \frac{4}{105} v_1^{-3} I(8, 2, 2) + \frac{4}{5} v_1^4 J(1, 2, 0) + \\ & + 8v_1^2 J(3, 2, 0) + 4J(5, 2, 0) - \frac{92}{105} v_1^4 J(1, 4, 0) - \\ & - \frac{104}{15} v_1^2 J(3, 4, 0) - \frac{4}{3} J(5, 4, 0) + \frac{4}{105} v_1^4 J(1, 2, 2) - \\ & - \frac{8}{15} v_1^2 J(3, 2, 2) - \frac{4}{3} J(5, 2, 2) \Big], \end{aligned} \tag{11}$$

$$\left\langle\frac{dv_{\|}^2}{dt}\right\rangle = \delta\left[4v_1^3 I(2, 4, 0) + \frac{16}{3} v_1 I(4, 4, 0) - \frac{4}{15} v_1^{-1} I(6, 4, 0) + \right.$$
$$+ \frac{8}{105} v_1^{-3} I(8, 4, 0) + \frac{4}{3} v_1 I(4, 2, 2) + \frac{8}{15} v_1^{-1} I(6, 2, 2) -$$
$$- \frac{4}{105} v_1^{-3} I(8, 2, 2) + \frac{92}{105} v_1^4 J(1, 4, 0) + \frac{104}{15} v_1^2 J(3, 4, 0) +$$
$$\left. + \frac{4}{3} J(5, 4, 0) - \frac{4}{105} v_1^4 J(1, 2, 2) + \frac{8}{15} v_1^2 J(3, 2, 2) + \frac{4}{3} J(5, 2, 2)\right], \quad (12)$$

where

$$\gamma = \frac{8\pi^2}{(1 + A)} \left(\frac{\lambda}{2}\right)^2 N \left(\frac{k_B T}{(2\pi)^{3/2} m_2}\right) a^2 , \quad (13)$$

$$\delta = \frac{8\pi^2}{(1 + A)^2} \left(\frac{\lambda}{2}\right)^2 N \left(\frac{k_B T}{2\pi m_2}\right)^{3/2} a^2 . \quad (14)$$

From the stopping power we can calculate the range R of the incident beam which is

$$R = \frac{E}{-\left\langle\frac{dE}{dx}\right\rangle} . \quad (15)$$

3. Results and Discussion

The numerical values of the rate of energy loss, mean square energy loss, and the Fokker-Planck coefficients have been evaluated for energies of incident beam ranging from 500 to 0.1 keV. For one case, that of Xe^+ in W, we have evaluated the stopping power for various values of the parameter a ranging from 1.0 to 2.0 for the sake of comparison.

All the double integrals involved in these quantities have been computed using Burnside formula on IBM 360 computer. Satisfactory convergence was obtained with $\lambda/2$ as upper limit of integration for impact parameter. The scattering angle ψ has been evaluated using eight point Gauss quadrature formula. Numerical values for different beams are given in Tables 1 to 3 for Debye potential. Energy loss (keV/μm) has been plotted. It is found that the energy loss (keV/μm) and the mean square energy loss increase with increasing energy of the incident beam. All other transport coefficients also show an increase with increasing energies of test particles.

Results for stopping power, i.e. $-\langle dE/dx\rangle$ for Debye potential have been plotted in Fig. 1 and 2. In Fig. 1 curves 1, 2, and 3 refer to singly ionized beams of Cu in Cu, Kr in Cu, and Cu in Ge. For Cu in Ge the energy loss is comparitively much lower. This is due to low electron density in germanium as compared to that of copper. The results based on the Coulomb potential are essentially the same (within 0.1%) as compared with those based on Debye potential.

The plot of stopping power versus energy shows that the stopping power varies as E^n where n takes values from 0.5 to approximately unity as the energy of the incident beam increases from 0.1 to 500 keV. This type of energy depend-

Table 1

Results for a copper ion beam passing through copper

energy (keV)	$\left\langle \frac{dE^2}{dt} \right\rangle$ (10^{-8} erg^2 s^{-1})	$-\left\langle \frac{dv_\parallel}{dt} \right\rangle$ (cm μs^{-2})	$\left\langle \frac{dv_\perp^2}{dt} \right\rangle$ (cm^2 μs^{-3})	$\left\langle \frac{dv_\parallel^2}{dt} \right\rangle$ (cm^2 μs^{-3})	range (μm)
500	0.852×10^{3}	0.243×10^{8}	0.385×10^{4}	0.524×10^{5}	3.06
100	0.185×10^{3}	0.535×10^{7}	0.360×10^{3}	0.992×10^{4}	2.78
40	0.210×10^{1}	0.225×10^{7}	0.357×10^{3}	0.628×10^{4}	2.62
20	0.489	0.141×10^{7}	0.323×10^{3}	0.533×10^{4}	2.08
10	0.103	0.769×10^{6}	0.250×10^{3}	0.353×10^{4}	1.91
5	0.253×10^{-1}	0.465×10^{6}	0.126×10^{3}	0.153×10^{4}	1.59
1	0.154×10^{-2}	0.162×10^{6}	0.527×10^{2}	0.909×10^{2}	0.919
0.8	0.112×10^{-2}	0.142×10^{6}	0.523×10^{2}	0.677×10^{2}	0.840
0.4	0.466×10^{-3}	0.962×10^{5}	0.530×10^{2}	0.385×10^{2}	0.620
0.1	0.105×10^{-3}	0.476×10^{5}	0.564×10^{2}	0.301×10^{2}	0.313

Table 2

Results for a krypton ion beam passing through copper

energy (keV)	$\left\langle \frac{dE^2}{dt} \right\rangle$ (10^{-8} erg^2 s^{-1})	$-\left\langle \frac{dv_\parallel}{dt} \right\rangle$ (cm μs^{-2})	$\left\langle \frac{dv_\perp^2}{dt} \right\rangle$ (cm^2 μs^{-3})	$\left\langle \frac{dv_\parallel^2}{dt} \right\rangle$ (cm^2 μs^{-3})	range (μm)
500	0.453×10^{3}	0.144×10^{8}	0.106×10^{4}	0.219×10^{5}	3.92
100	0.951×10^{1}	0.311×10^{7}	0.200×10^{3}	0.475×10^{4}	3.61
40	0.120×10^{1}	0.145×10^{7}	0.206×10^{3}	0.354×10^{4}	3.08
20	0.259	0.834×10^{6}	0.175×10^{3}	0.267×10^{4}	2.67
10	0.581×10^{-1}	0.472×10^{6}	0.114×10^{3}	0.155×10^{4}	2.36
5	0.149×10^{-1}	0.292×10^{6}	0.542×10^{2}	0.554×10^{3}	1.93
1	0.104×10^{-2}	0.104×10^{6}	0.300×10^{2}	0.366×10^{2}	1.08
0.8	0.776×10^{-3}	0.917×10^{5}	0.301×10^{2}	0.292×10^{2}	0.986
0.4	0.339×10^{-3}	0.630×10^{5}	0.308×10^{2}	0.199×10^{2}	0.717
0.1	0.789×10^{-4}	0.315×10^{5}	0.328×10^{2}	0.172×10^{2}	0.359

Table 3

Results for a copper ion beam passing through germanium

energy (keV)	$\left\langle \frac{dE^2}{dt} \right\rangle$ (10^{-8} erg^2 s^{-1})	$-\left\langle \frac{dv_\parallel}{dt} \right\rangle$ (cm μs^{-2})	$\left\langle \frac{dv_\perp^2}{dt} \right\rangle$ (cm^2 μs^{-3})	$\left\langle \frac{dv_\parallel^2}{dt} \right\rangle$ (cm^2 μs^{-3})	range (μm)
500	0.130	0.763×10^{4}	0.908×10^{1}	0.802×10^{1}	9.76
100	0.261×10^{-2}	0.160×10^{4}	0.944	0.139×10^{1}	9.30
40	0.338×10^{-3}	0.708×10^{3}	0.346	0.100×10^{1}	8.33
20	0.110×10^{-3}	0.500×10^{3}	0.236	0.134×10^{1}	5.86
10	0.438×10^{-4}	0.375×10^{3}	0.167	0.162×10^{1}	3.89
5	0.132×10^{-4}	0.262×10^{3}	0.803×10^{-1}	0.852	2.82
1	0.597×10^{-4}	0.779×10^{2}	0.230×10^{-1}	0.465×10^{-1}	1.91
0.8	0.418×10^{-6}	0.660×10^{2}	0.217×10^{-1}	0.324×10^{-1}	1.81
0.4	0.161×10^{-6}	0.410×10^{2}	0.798×10^{-2}	0.148×10^{-1}	1.45
0.1	0.346×10^{-7}	0.182×10^{2}	0.193×10^{-2}	0.100×10^{-1}	0.818

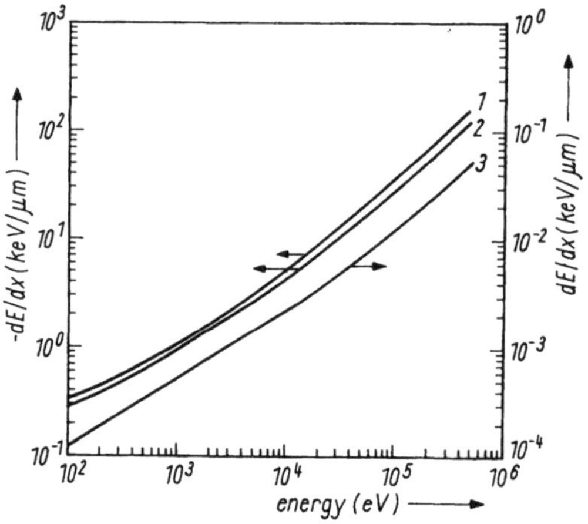

Fig. 1. Variation of stopping power $-(dE/dx)$ with energy. (1) Cu^+ in Cu, (2) Kr^+ in Cu (3) Cu^+ in Ge. These results correspond to a Debye potential

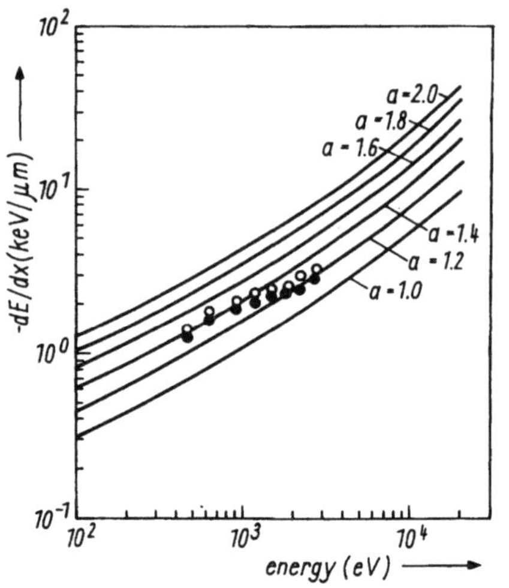

Fig. 2. Variation of stopping power for Xe ions in W. Continuous curves: present results for Xe^+ in W, ● Pathak and Yussouff (RPA), ○ Pathak and Yussouff (Hubbard)

ence is in agreement with earlier studies. Also, the numerical estimates are in fairly good agreement with results obtained previously. For comparison, the stopping power for Xe^+ in W has been plotted in Fig. 2 for various values of the parameter, a, along with the values obtained by Pathak and Yussouff [2]. As expected, the values of stopping power depend upon the parameter, a, and we find that for $a \approx 1.3$ our values agree closely with the values obtained by Pathak and Yussouff [2]. It may also be noted that the energy dependence of stopping power obtained by Pathak and Yussouff is slightly different. It varies approximately as $E^{0.6}$.

The mean square energy loss and other transport coefficients are reported in Tables 1, 2, and 3 for different beams. In general values obtained using Coulomb and Debye potentials are the same within 0.1%. The values based on a Coulomb potential are not reported. All the coefficients increase with increase in energy of the incident beam. The parallel diffusion coefficient varies more rapidly as compared to perpendicular diffusion coefficient.

We thus find that the present method gives much more detailed information about the passage of an ionized beam through crystals. The problem of the passage of ionized beams through dense gases is being considered and the results will be reported shortly.

References

[1] C. ERGINSOY, Interaction of Radiation with Solids, Ed. A. BISHAY, Plenum Press, Inc., New York 1967.

[2] A. P. PATHAK and M. YUSSOUFF, phys. stat. sol. (b) **49**, 431 (1972).

[3] A. B. LIDIARD, Proc. Winter School of Physics of Imperfect Crystalline Solids, Trieste, IAEA, Vienna 1970.

[4] S. Chandrasekhar, Astrophys. J. **93**, 235 (1941).
[5] L. Spitzer, Physics of Fully Ionized Gases, 2nd edition, Interscience Publishers, 1957.
[6] S. T. Butler and M. J. Buckingham, Phys. Rev. **126**, 1 (1962).
[7] I. P. Shkarofsky, T. W. Johnston, and M. P. Bachynski, The Particle Kinetics of Plasmas, Addison Wesley, Reading (Mass.) 1966.
[8] A. A. Husseiny and H. K. Forsen, Phys. Rev. A **1**, 1483 (1970).
[9] M. P. Srivastava and C. Devanathan, Beiträge aus der Plasma-Physik **8**, 247 (1968).
[10] S. Chandrasekhar, Principles of Stellar Dynamics, Dover, New York 1942.
[11] C. Kittel, Introduction to Solid State Physics, Wiley, 1966.
[12] S. Chandrasekhar, Astrophys. J. **97**, 285 (1943).

(Received November 19, 1973)

phys. stat. sol. (b) 62, 491 (1974)

Subject classification: 20.1; 13.1; 14.3; 22.2.3

Institut für Physikalische Elektronik der Technischen Hochschule Wien and Ludwig-Boltzmann-Institut für Festkörperphysik, Wien

Shape of the Cyclotron Resonance Absorption Line of Polar Semiconductors: n-InSb at 337 μm[1]

By

L. Palmetshofer[2]) and D. Kranzer[3])

The shape of the cyclotron resonance absorption line of n-InSb samples is investigated. Experiments were carried out using a HCN laser beam. The temperature range considered being 77 to 150 K. According to the preparation of the sample surface different shapes of the absorption line are observed. Additionally a theory is worked out which is based on an exact numerical solution of the time-dependent Boltzmann equation. The theoretical results are compared with the experiment.

Die Form der Zyklotronresonanzabsorptionslinie von InSb-Proben wird untersucht. Die Experimente werden mit Hilfe eines HCN-Lasers im Temperaturbereich zwischen 77 und 150 K durchgeführt. Entsprechend der Oberflächenbehandlung werden verschiedene Formen der Absorptionslinie gemessen. Die experimentellen Ergebnisse werden mit theoretischen Resultaten verglichen. Letztere werden durch eine numerische exakte Lösung der zeitabhängigen Boltzmanngleichung gewonnen.

1. Introduction

Cyclotron resonance measurements yield information concerning the band structure and the scattering mechanisms. From the position of the absorption peak the cyclotron mass can be obtained [1]. Roughly speaking the width of the cyclotron absorption line is inversely proportional to the relaxation time derived from the dc mobility [2]. This fact holds in the classical limit, i.e., if the thermal energy of the carrier is large compared to the photon energy $\hbar\omega$ (ω is the angular frequency of the incident wave). In this case the Drude theory is usually applied to study the shape of the absorption line [2]. Obviously, statistical effects and nonparabolicity are neglected in the Drude formalism. Both, however, are known to be important in the case of polar compounds with a narrow energy gap [3]. An accurate theory has to start from the time-dependent Boltzmann equation. So far this has been performed with the aid of the relaxation time approximation [4]. This approach, however, fails if polar (PO) scattering by optical modes predominates [5]. Consequently there is a clear need for a theory incorporating the nonparabolicity and PO scattering to investigate the cyclotron absorption line of polar compounds.

In this paper we present a theory which accounts for these complications, and apply it to the free-carrier cyclotron absorption line of n-InSb. We studied different samples using submillimeter waves of a HCN laser. These samples are

[1]) Supported by the "Hochschuljubiläumsstiftung der Stadt Wien" and the "Jubiläumsfonds der Österreichischen Nationalbank".

[2]) Present address: Institut für Physik, Johannes-Kepler-Hochschule Linz, Austria.

[3]) Present address: Marshal Space Flight Center, Huntsville, Alabama 35812, U.S.A.

prepared differently: (i) the surface is polished; (ii) rough surfaces are produced by etching. As a result different shapes of the absorption line have been observed. The experimental procedure is described in Section 2. The theory, developed in Section 3 is based on an exact numerical solution of the time-dependent Boltzmann equation. It accounts for the nonparabolicity of the conduction band and PO, acoustic deformation potential (AC), piezoelectric (PE), and ionized impurity (ION) scattering. Apart from the effective mass m_0^* at the band edge no adjustable parameter is used in the ac calculation. The position of the power transmission minimum is used to fit m_0^*. The ionized impurity concentration N_i is determined from a fit of the dc mobility to experimental data. For a parabolic band and PO scattering a comparison is made between the results of ours and Drude's theory (Section 4). A discussion of the effect of nonparabolicity on the shape of the absorption line is included in Section 4. The experimental and theoretical results are compared in Section 5. A summary of the results is given in Section 6.

2. Experimental Procedure

2.1 Sample preparation and dc measurements

The samples used have been prepared from single crystals of n-type InSb. The electrical properties at 77 K are given in Table 1. The samples were cut

Table 1

Sample properties at 77 K

sample	n (cm^{-3})	μ_{dc} (cm^2/Vs)	N_i (cm^{-3})	d (μm)
1	8×10^{13}	740000	4.3×10^{14}	140
2a	5.1×10^{13}	550000	6×10^{14}	114
2b	5×10^{13}	580000	—	112

to proper dimensions $5 \times 5 \times 0.1$ mm^3. Samples 1 and 2b were etched in CP-4 and sample 2a was polished. The resulting different surface properties are responsible for the different behaviour concerning the Fabry-Perot interference effects. After the ac measurements In (3% Te) contacts were soldered at each corner of the slab. The van der Pauw method was used to measure the dc conductivity and the carrier concentration.

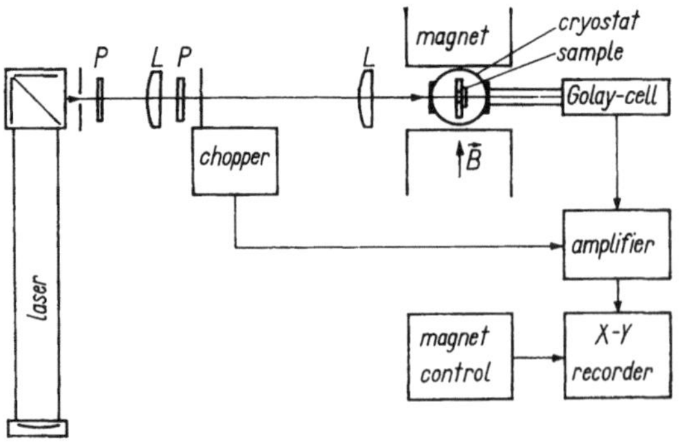

Fig. 1. Experimental set-up: The radiation from the HCN laser is guided by polyethylene lenses (L) through an optical cryostat, a small window in the sample holder, and the sample to a Golay cell. The polarizers (P) provide the correct polarization of the laser beam. The transmission is plotted versus magnetic field strength B on an XY recorder

2.2 Experimental set-up

The samples were mounted in an optical cryostat between the poles of an electromagnet (Fig. 1). All ac measurements were performed using the Voigt configuration [2], the temperature range investigated being 77 to 150 K. A small window in the sample holder limited the cross-section of the CW HCN-laser beam, the wavelength being 337 μm. At a fixed temperature the magnetic field strength $\boldsymbol{B}$ and the power transmission, detected with a Golay cell, are plotted by an XY recorder. A grid polarizer is used to ensure that $\boldsymbol{F}$ is perpendicular to $\boldsymbol{B}$, where $\boldsymbol{F}$ is the electric field strength arising from the electromagnetic wave. To avoid heating of the carriers the incident power is limited to about 10 μW.

3. Theory

3.1 Calculation of the power transmission coefficient [TR]

Apart from the thickness d of the samples the complex refractive index

$$N^* = N - ik \tag{1}$$

is needed to calculate [TR]. N^* is determined by a dispersion relation which can be written in the case of a cubic semiconductor as [2]

$$N^* = \varkappa_\mathrm{r} - i\,\frac{\sigma_\mathrm{eff}}{\omega\varkappa_0}\,, \tag{2}$$

where σ_eff is given in the Voigt configuration by [2]

$$\sigma_\mathrm{eff} = \frac{\omega\varkappa_\mathrm{r}\varkappa_0\sigma_{xx} - i(\sigma_{xx}^2 + \sigma_{xy}^2)}{\omega\varkappa_\mathrm{r}\varkappa_0 - i\sigma_{xx}}\,. \tag{3}$$

The magnetic field is assumed parallel to the z-axis of the coordinate system. The components of the conductivity tensor $\boldsymbol{\sigma}$,

$$\boldsymbol{\sigma} = \begin{pmatrix} \sigma_{xx} & \sigma_{xy} & 0 \\ -\sigma_{xy} & \sigma_{xx} & 0 \\ 0 & 0 & \sigma_{zz} \end{pmatrix}, \tag{4}$$

are complex quantities. In (2) and (3) $\varkappa_\mathrm{r}$ and $\varkappa_0$ denote the relative low-frequency dielectric constant and the vacuum dielectric constant, respectively. The power transmission coefficient is given by [6]

$$[\mathrm{TR}] = \frac{\mathrm{e}^{-Kd}\,[(1 - R)^2 + 4R\sin^2\alpha]}{(1 - R\,\mathrm{e}^{-Kd})^2 + 4R\,\mathrm{e}^{-Kd}\sin^2(\alpha + \beta)} \tag{5}$$

and

$$[\mathrm{TR}] = \frac{(1 - R)^2\,\mathrm{e}^{-Kd}\,(1 + k^2/N^2)}{1 - R^2\,\mathrm{e}^{-2Kd}}\,, \tag{6}$$

respectively. Equations (5) and (6) are valid for wafers with perfect optical and rough surfaces, respectively [6]. The abbrevations appearing in (5) and (6) have the following meaning:

$$K = \frac{2\omega}{c}\,k \text{ (power absorption coefficient)}\,, \tag{7}$$

$$R = \frac{(N-1)^2 + k^2}{(N+1)^2 + k^2} \text{ (power reflection coefficient)} , \tag{8}$$

$$\alpha = \arctan \frac{2k}{k^2 + N^2 - 1}, \tag{9}$$

$$\beta = \frac{\omega}{c} Nd , \tag{10}$$

and c is the velocity of light in vacuum. To calculate [TR] we have to evaluate the components of $\boldsymbol{\sigma}$ which is performed below.

3.2 Conductivity tensor

To calculate $\boldsymbol{\sigma}$ we have to start from the time-dependent Boltzmann equation. For small electric fields oscillating in time as

$$\boldsymbol{F}(t) = \text{Re}\,(\boldsymbol{F}_0 e^{i\omega t}) , \tag{11}$$

the distribution function $f(\boldsymbol{q}, t)$ may be written as

$$f(\boldsymbol{q}, t) = f_0(E) + \boldsymbol{A}(E)\, \boldsymbol{q} e^{i\omega t}, \tag{12}$$

where $\boldsymbol{q}$ is the wave vector and E the energy of the electron, $f_0(E)$ is the thermal equilibrium distribution function, and $\boldsymbol{A}(E)$ is an unknown vector which will be calculated below.

The theory presented in this section is an extension of treatments applied to an oscillating electric field without a dc magnetic field [3, 7] and to dc electric and magnetic fields [8]. Hence a short description of the theory is sufficient.

If $\boldsymbol{B}$ is parallel to the z-direction, two components of $\boldsymbol{A}$, A_x and A_y, do not vanish. Inserting (12) into the time-dependent Boltzmann equation we obtain

$$\left.\begin{aligned} i\omega A_x q - evF_{0x} \frac{\mathrm{d}f_0}{\mathrm{d}E} + vA_y \frac{e}{\hbar} B - \left(\frac{\partial A_x q}{\partial t}\right)_\mathrm{c} &= 0 , \\ i\omega A_y q - evF_{0y} \frac{\mathrm{d}f_0}{\mathrm{d}E} - vA_x \frac{e}{\hbar} B - \left(\frac{\partial A_y q}{\partial t}\right)_\mathrm{c} &= 0 , \end{aligned}\right\} \tag{13}$$

where v is the group velocity of the electrons and $(\partial A_i q/\partial t)_\mathrm{c}$ represents the non-vanishing part of the collision term. We consider two types of scattering mechanisms: a) elastic scattering (AC, PE, ION) and b) inelastic (PO) scattering. The former can be described by a momentum relaxation time while the latter gives rise to a difference equation for $(\partial A_i q/\partial t)_\mathrm{c}$ [7 to 9]. Hence (13) represents a coupled difference equation system.

The components of $\boldsymbol{A}$ are complex quantities:

$$A_l = A_l' + iA_l'' , \tag{14}$$

where $l = x, y$. For low electric fields the real and imaginary part may be written as

$$\left.\begin{aligned} A_x' &= a_{xx}' F_{0x} + a_{xy}' F_{0y} , \\ A_x'' &= a_{xx}'' F_{0x} + a_{xy}'' F_{0y} , \\ A_y' &= a_{yx}' F_{0x} + a_{yy}' F_{0y} , \\ A_y'' &= a_{yx}'' F_{0x} + a_{yy}'' F_{0y} . \end{aligned}\right\} \tag{15}$$

Inserting (14) and (15) into (13) we obtain 8 coupled difference equations which are not linearly independent. Since [2]

$$\left.\begin{aligned} a'_{yy} &= a'_{xx}\,, & a''_{yy} &= a''_{xx}\,, \\ a'_{yx} &= -a'_{xy}\,, & a''_{yx} &= -a''_{xy}\,, \end{aligned}\right\} \tag{16}$$

the following difference equation system is obtained:

$$\left.\begin{aligned} \omega a''_{xx} + \omega_c a'_{xy} + \frac{1}{q}\left(\frac{\partial q a'_{xx}}{\partial t}\right)_c &= \frac{e}{\hbar m^*(E)}\frac{df_0}{dE}, \\ \omega a''_{xy} - \omega_c a'_{xx} + \frac{1}{q}\left(\frac{\partial q a'_{xy}}{\partial t}\right)_c &= 0, \\ -\omega a'_{xx} + \omega_c a''_{xy} + \frac{1}{q}\left(\frac{\partial q a''_{xx}}{\partial t}\right)_c &= 0, \\ -\omega a'_{xy} - \omega_c a''_{xx} + \frac{1}{q}\left(\frac{\partial q a''_{xy}}{\partial t}\right)_c &= 0, \end{aligned}\right\} \tag{17}$$

where

$$\omega_c = \frac{eB}{m^*(E)} \tag{18}$$

and

$$[m^*(E)]^{-1} = \frac{1}{\hbar^2}\frac{1}{q}\frac{dE(q)}{dq}. \tag{19}$$

Equation (19) was obtained by Kołodziejczak [10] for nonparabolic spherical bands.

Kane [11] worked out the energy momentum relation for the electrons in the Γ_1 conduction band. If the spin-orbit split energy Δ is large or small as compared to the energy gap E_g, the relation can be written as [12]

$$\frac{\hbar^2 q^2}{2m_0^*} = E\left(1 + \frac{E}{E_g}\right), \tag{20}$$

where m_0^* is the effective mass at the band edge. Equation (20) applies to n-InSb because $\Delta = 0.8$ eV which is large compared to $E_g = 0.235$ eV. Because of the nonparabolicity the symmetry properties of the electron wave functions are no longer of pure s-type. An admixture of p-type wave functions must be accounted for, which results in a smaller scattering probability [5]. The nonparabolicity and the p-function admixture are taken into account. We obtain the collision term as presented in the literature [5]. Hence all coefficients appearing in (17) are well known.

If B equals zero, $a'_{xy} = a''_{xy} = 0$, and we obtain from (17) the equation system (7) of reference [7] used to determine the high-frequency conductivity of polar semiconductors. On the other hand, if we take $\omega = 0$, a_{xy} and a_{xx} vanish and we obtain the equation system (3) of reference [8] used to investigate the scattering factor. The equation system (17) is solved by a numerical method described in reference [7].

Using the normalization condition

$$\frac{1}{n}\int f_0(E)\frac{d^3q}{4\pi^3} = 1\,, \tag{21}$$

where n is the carrier concentration and f_0 the known distribution function, the components of $\boldsymbol{\sigma}$ can be easily calculated:

$$\left.\begin{aligned}\sigma_{xx} &= -\frac{e}{3\pi^2}\int vq^2\,\mathrm{d}q\,[a'_{xx}(E) + ia''_{xx}(E)]\,,\\ \sigma_{xy} &= -\frac{e}{3\pi^2}\int vq^2\,\mathrm{d}q\,[a'_{xy}(E) + ia''_{xy}(E)]\,.\end{aligned}\right\} \tag{22}$$

Neglecting PO scattering we obtain from (17) and (22) the results of the relaxation time approximation [4] and, in the case of an energy-independent relaxation time, the results of Drude's theory [2].

4. Comparison with Drude's Theory and Influence of Nonparabolicity

4.1 Comparison with the results of Drude's theory

We assume a parabolic energy band and define an energy-independent relaxation time τ_0 by use of the dc conductivity σ_0:

$$\tau_0 = \frac{\sigma_0 m^*}{ne^2}\,. \tag{23}$$

Furthermore we consider PO scattering. To facilitate the discussion we calculate $\gamma'_{ik} = \sigma'_{ik}/\sigma'_{ik\mathrm{D}}$ and $\gamma''_{ik} = \sigma''_{ik}/\sigma''_{ik\mathrm{D}}$ where σ'_{ik} and σ''_{ik} means the real and imaginary part of (22), respectively. The subscript D denotes the corresponding quantity calculated by Drude's theory and (23). The results are plotted in Fig. 2 for $\omega_c\tau_0 = 5$ and $\eta = 3$ (Fig. 2a) and $\eta = 2$ (Fig. 2b), where $\eta = \theta/T$ and θ denotes the characteristic optical phonon temperature. In the limit of large

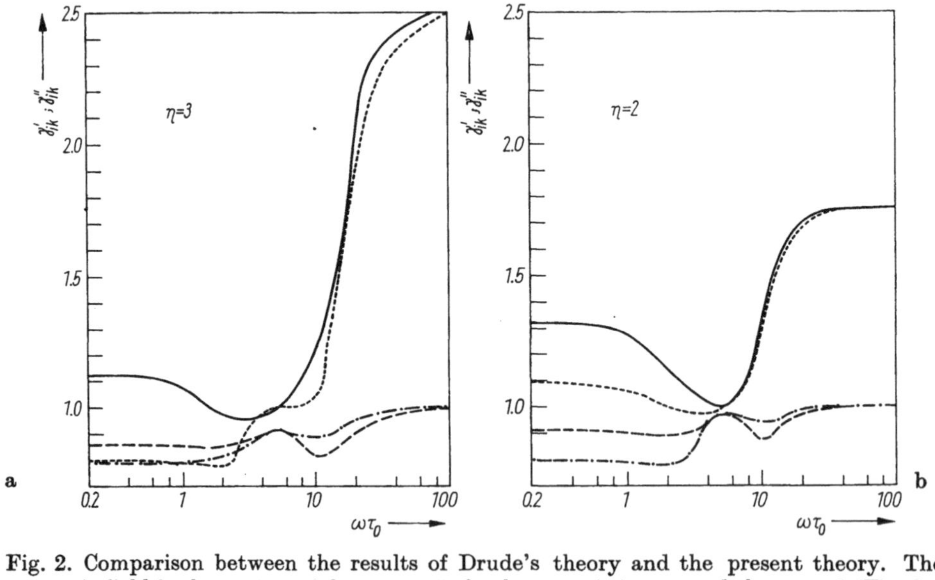

Fig. 2. Comparison between the results of Drude's theory and the present theory. The magnetic field is chosen to satisfy $\omega_c\tau_0 = 5$; furthermore it is assumed that $\eta = 3$ (Fig. 2a) and $\eta = 2$ (Fig. 2b), respectively. The curves correspond to γ'_{xx} (———), γ''_{xx} (–·–·–), γ'_{xy} (– – – –), and γ''_{xy}(······)

frequencies ($\omega\tau_0 \gg 1$, $\omega > \omega_c$) γ'_{xx} equals γ''_{xy} while γ''_{xx} and γ'_{xy} become unity independent of $\omega\tau_0$; the values of γ'_{xx} and γ''_{xy}, however, depend on η. If cyclotron resonance occurs ($\omega = \omega_c$), $\gamma'_{xx} = \gamma''_{xy} = 1$ and $\gamma''_{xx} = \gamma'_{xy}$. As an important result Fig. 2 shows the deviation between both theories to become small, depending on the smallness of $|\omega - \omega_c|\ \tau_0$.

4.2 *Effect of nonparabolicity*

Because of nonparabolicity the effective mass depends on E. Roughly speaking the electrons can be divided in different groups characterized by different values of ω_c. Each group contributes to the absorption, but the corresponding absorption peak appears at different values of B. If the strength of the scattering mechanism depends on E, the width of the absorption line of each group is different. The envelope of these absorption lines is therefore determined by the scattering mechanisms and the shift of the absorption peaks due to nonparabolicity. As a consequence the position of the absorption peak of the envelope depends weakly on the energy dependence of the scattering mechanisms and the absorption line broadens.

5. Experimental Results and Discussion

5.1 *dc measurements*

The temperature dependence of the dc mobility is obtained from van der Pauw measurements. We used a value of $B = 5$ kG. The samples were cut from the same ingots as the corresponding samples of reference [3]. We refer to this paper for a detailed discussion of the dc properties. The dc mobility of the corresponding samples is equal within 10%. The carrier concentration of sample 1 remains constant up to about 110 K, while n of sample 2a and 2b continuously increases with T up to the onset of intrinsic conduction. Samples 1 and 2 of reference [3] show the same behaviour. The total ionized impurity concentration N_i is determined from the dc mobility by treating N_i as adjustable. The results are included in Table 1. Other material parameters needed in the calculation are presented in Table 2.

Table 2

Material parameters used in the calculation

optical Debye temperature	278 K
relative high-frequency dielectric constant	15.7
relative low-frequency dielectric constant	17.9
energy gap at $T = 0$ K	0.235 eV
acoustic deformation potential constant	7.2 eV
piezoelectric tensor component h_{14}	4.47×10^6 V/cm

5.2 *Temperature dependence of the position B_0 of the absorption peak*

The position B_0 of the absorption peak is measured with an accuracy of $\pm 0.5\%$ during the measurements of one sample, while an accuracy of $\pm 1\%$ is estimated if results of different samples are compared. Fig. 3 shows the temperature dependence of B_0 of sample 1 and sample 2. Below, say 110 K,

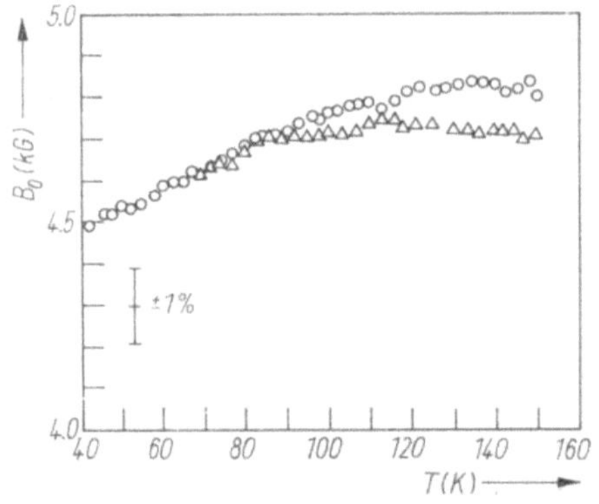

Fig. 3. Experimental results of the position of the transmission minimum B_0 as a function of temperature of sample 1 (○) and sample 2a (△)

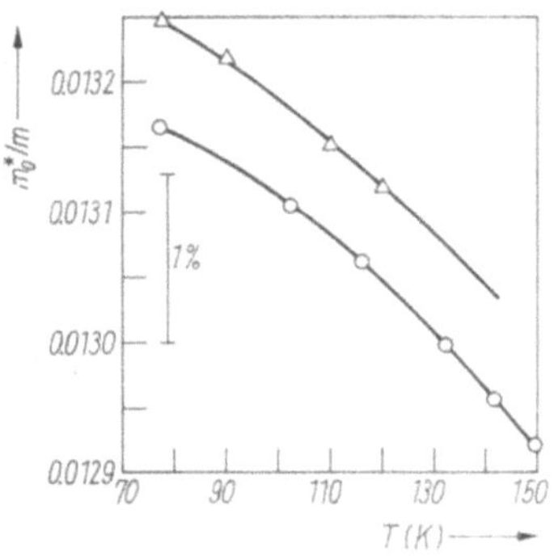

Fig. 4. Temperature dependence of the effective mass at the band edge as determined from the position of the transmission minimum (B_0) of sample 1 (○) and sample 2a (△)

B_0 of sample 1 increases linearly with T and reaches a maximum near 140 K. Because n of sample 1 remains constant up to 110 K, the increase of B_0 with T must be explained considering $m^*(E)$. The increase of n above 110 K causes a Voigt shift [2] which compensates the influence of $m^*(E)$. The flat maximum of B_0 of sample 2 is due to the competing effects of the Voigt shift and m^*.

In the calculation of the shape of the transmission line the effective mass at the band edge, m_0^*, is treated as adjustable and determined from the position of the absorption peak. We take into account the temperature dependence of n. The results for sample 1 and sample 2 agree within 1% and are presented in Fig. 4.

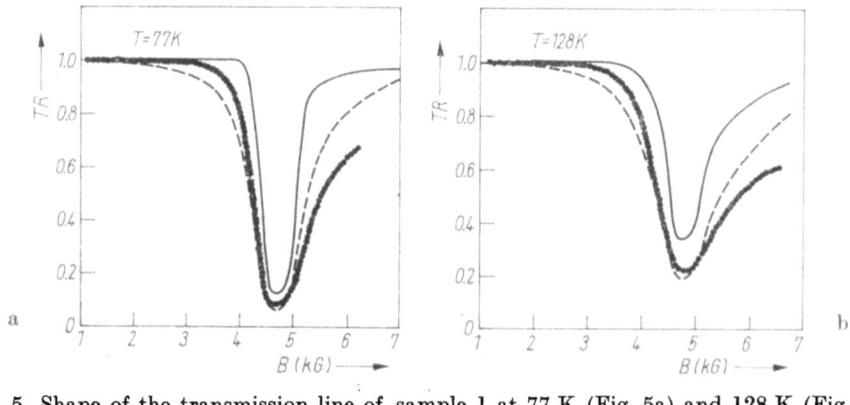

Fig. 5. Shape of the transmission line of sample 1 at 77 K (Fig. 5a) and 128 K (Fig. 5b). The full and dashed curves are calculated using equation (5) and (6), respectively. The full circles represent experimental results. The power transmission coefficient [TR] is normalized to the low values of B

5.3 Shape of the transmission line

The linewidth is usually used to characterize the cyclotron resonance absorption line. In the case of a spherical parabolic energy band the linewidth is determined by the strength of the scattering mechanisms provided multiple reflections are suppressed. As nonparabolicity is important in n-InSb the linewidth loses its physical meaning. Therefore we discuss the whole shape of the absorption line instead of its width.

In Fig. 5 experimental results of sample 1 are presented at $T = 77$ K (Fig. 5a) and $T = 128$ K (Fig. 5b), respectively. Calculated results using equations (5) (full curves) and (6) (dashed curves) are included. The values of [TR] are normalized to the value at low magnetic fields. The full and broken curves represent two limiting cases. They are relevant for samples with perfect optical surfaces leading to interference phenomena (full curves) and samples with optically rough surfaces (broken curves) suppressing interference. The measured results lie well within these two limits for $B < B_0$, while at larger values of B considerable deviations exist.

In general an unsymmetric behaviour of the transmission lines is expected from the following facts: (i) Kramers-Kronig relation, (ii) Fabry-Perot resonance effects, and (iii) nonparabolicity. These facts are included in our treatment. They are, however, insufficient to explain the observed asymmetry. These deviations are presently not understood. It should be noticed that the broken line is in better agreement with experiment, in particular near B_0. In Fig. 6 the results measured at 77 K (Fig. 6a) and 128 K (Fig. 6b) on samples with optically polished surfaces (sample 2a) are presented. A better agreement with theory (full curves) is obtained if [TR] is calculated according to (5). The broken curves are calculated using (6). At 77 K [TR] becomes larger than unity at $B > B_0$, which is due to Fabry-Perot resonance effects. Additional results of sample 2b at 77 K are included in Fig. 6a. Since sample 2a and 2b obey

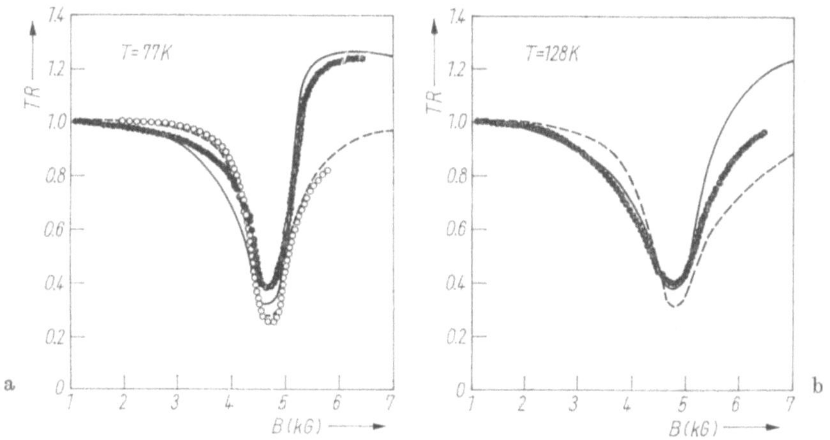

Fig. 6. Shape of the transmission line of sample 2a (●) and sample 2b (○). Fig. 6a and b represent results at 77 and 128 K, respectively. The full and dashed curves are calculated using equation (5) and (6), respectively. The power transmission coefficient [TR] is normalized to the low values of B

almost the same dc properties (Table 1) and thickness, the experimental results (Fig. 6a) demonstrate the sensitivity of Fabry-Perot resonance effects regarding the surface properties.

6. Concluding Remarks

We presented a study of the cyclotron resonance transmission line of n-InSb at HCN laser frequencies. A theory was worked out which includes the essential effect of nonparabolicity and accounts for PO, AC, PE, and ION scattering. The calculation is based on exact numerical solutions of the time-dependent Boltzmann equation. While $B \leqq B_0$, the theoretical results are in a good accord with experiments; deviations exist at $B > B_0$. The asymmetry of the observed transmission lines increases with temperature and is larger than the results predicted by the theory.

Acknowledgements

We wish to thank Prof. H. W. Pötzl for his interest in our work and many valuable discussions. We are very thankful to the "Rechenzentrum der Technischen Hochschule Wien" for being allowed to use the computer.

References

[1] B. LAX, Proc. Internat. School of Physics "Enrico Fermi", Course XXII, Ed. R. A. SMITH, Academic Press, New York 1963 (p. 240).
[2] E. D. PALIK and G. B. WRIGHT, in: Semiconductors and Semimetals, Vol. 3, Ed. R. K. WILLARDSON and A. BEER, Academic Press, New York 1966 (p. 421).
[3] H. LÖSCHNER and D. KRANZER, J. appl. Phys. **44**, 3663 (1973).
[4] J. KOŁODZIEJCZAK and E. KIERZEK-PECOLD, phys. stat. sol. **19**, 623 (1967).
[5] D. L. RODE, in: Semiconductors and Semimetals, Vol. 11, Ed. R. K. WILLARDSON and A. BEER, Academic Press, New York 1973 (in the press).
[6] D. L. GREENAWAY and G. HARBEKE, Optical Properties and Band Structure of Semiconductors, Internat. Series of Monographs in the Science of the Solid State, Vol. 1, Pergamon Press, London 1968.
[7] D. KRANZER, phys. stat. sol. (b) **46**, 591 (1971).
[8] D. KRANZER, J. Phys. Chem. Solids **34**, 9 (1973).
[9] D. KRANZER, H. HILLBRAND, H. PÖTZL, and O. ZIMMERL, Acta phys. Austriaca **35**, 110 (1972).
[10] J. KOŁODZIEJCZAK, Acta phys. Polon. **21**, 637 (1962).
[11] E. O. KANE, J. Phys. Chem. Solids **1**, 249 (1957).
[12] D. MATZ, Phys. Rev. **168**, 843 (1968).

(Received December 6, 1973)

phys. stat. sol. (b) 62, 501 (1974)
Subject classification: 6 and 20.1; 22.8,

Departement de Recherches Physiques, associé au C.N.R.S., Université Paris VI

Raman Spectra of $BaZnF_4$

By

M. QUILICHINI and H. POULET

Raman spectra of a single crystal of $BaZnF_4$ have been obtained. Frequencies and symmetries of the Raman active optical phonons are given. The temperature dependence of the spectra have been examined. The directional dispersion of each of the oblique optical phonons has been measured from the Raman data and calculated. The dimensionless oscillator strengths are calculated.

Es wurden die Ramanspektren von einem $BaZnF_4$-Einkristall beobachtet und die Frequenzen und Symmetrien der Raman-aktiven optischen Phononen bestimmt. Die Temperaturabhängigkeit der Spektren wurde untersucht. Die Richtungsdispersion von jedem der indirekten optischen Phononen wurde aus den Ramanwerten bestimmt und berechnet. Die dimensionslosen Oszillatorstärken wurden berechnet.

1. Introduction

Crystalline ferroelectric $BaZnF_4$ (the ferroelectric Curie point lying probably above the melting point 785 °C) is orthorhombic with lattice constants $a = 5.841$, $b = 14.563$, $c = 4.206$ Å [1]. Keve et al. [2] reported the space group as $A2_1am$ (C_{2v}^{12}). The International Union of Crystallographic convention $c < a < b$ was followed in labeling the axes. The matrix which transforms the space group $A2_1am$ into the conventional setting $Cmc2_1$ is $(001/0\bar{1}0/100)$.

The principal refractive indices are $n_\alpha = 1.498$, $n_\beta = 1.5155$, $n_\gamma = 1.5215$ and lie along b-, a- and c-axes, respectively; a is the polar axis. The primitive unit cell built on lattice constants $(a/2, b/2, 0)$ $(-a/2, b/2, 0)$ $(0, 0, c)$ contains two formula groups. All the atoms belong to the (001) mirror plane.

The resulting 36 vibrational degrees of freedom occur as 33 optical modes with Γ point symmetries:
$11A_1$ (aa, bb, cc): a + $11B_2$ (ab): b + $6A_2$ (bc) + $5B_1$ (ac): c. The nonvanishing Raman tensor elements are given inside the parentheses in short notation, and the polarization direction of the polar modes is indicated by the letter behind the colon. abc is a reference frame such as a-axis parallel to C_2, b- and c-axes normal to symmetry planes.

A_1, B_1, B_2 modes are both Raman and infrared active; so we expect that the associated long range macroscopic electric field can modify the polar phonon spectrum as function of the direction of the phonon wave vector inside the crystal [3].

In the Raman experiments, the LO and TO phonons can sometimes be separated as discrete lines when applying different scattering geometries.

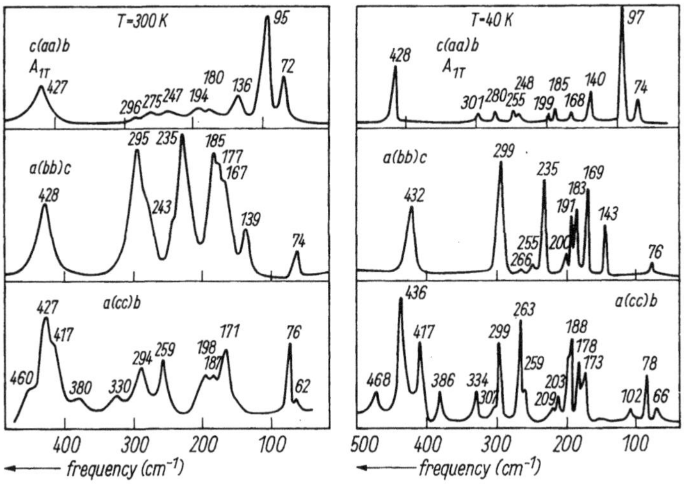

Fig. 1. Raman spectra of A_1 polar modes at 300 and 40 K

2. Experimental

In this paper we present our results on the Raman spectrum of $BaZnF_4$.

The sample was obtained from Scott of Bell Telephone Laboratories. The polished sample faces were oriented normal to the *abc* crystal axes which are defined above. The sample was mounted on the cold finger of a liquid helium Coderg cryostat for the data at 40 °K. The crystal exhibit some impurity fluorescence, which fortunately was beyond the range of the Raman spectrum.

To observe the mixing between A_1 and B_1, A_1 and B_2, B_1 and B_2 polar phonons, the sample was immersed in a cell filled with index matching-oil ($n = 1.49$). The direction of the wave vector $\boldsymbol{k}$ in the scattering plane could be monitored by simply rotating the crystal.

The Raman spectra were recorded using the 5145 Å line of a Spectra Physics

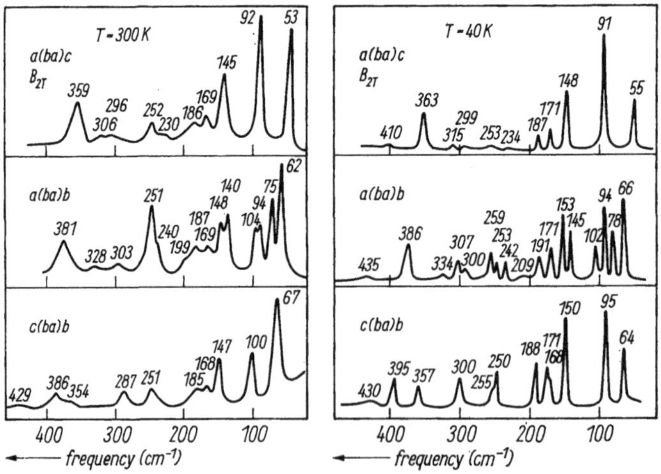

Fig. 2. Raman spectra of B_2 polar modes at 300 and 40 K

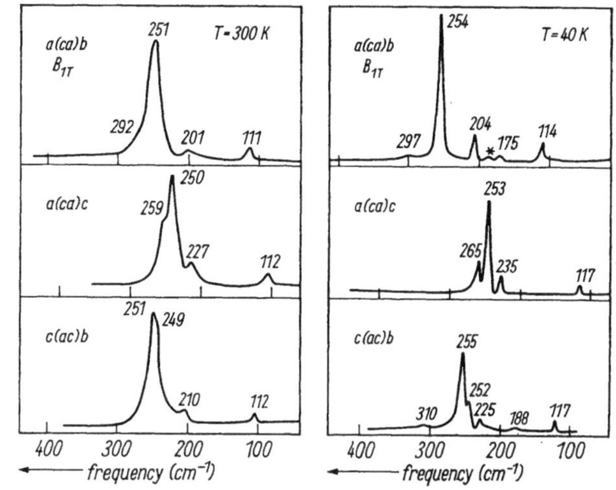

Fig. 3. Raman spectra of B_1 polar modes at 300 and 40 K

Argon ion laser. The light scattered at 90° was spectrally analyzed by a Coderg T 800 triple monochromator and detected by a cooled photomultiplier tube.

Spectral slit width was usually kept at about 1 cm^{-1}. Frequency measurements are accurate to within ± 1 cm^{-1} for sharp bands. Directional dispersion of the phonons was measured in the following way: the propagation directions of incident and scattered radiation were fixed in the right angle geometry. As a result of conservation, the phonon wave vectors created in a first order Raman Stokes process was forced to propagate in the scattering plane at 45° to incident photon. We define a phonon propagation angle θ as the angle between the wave vector $\boldsymbol{k}$ and one chosen a-, b-, c-axis. The angle θ is known with an accuracy of $\pm 2°$, and was varied by jumps of 10° between 0 and 90° at room temperature.

At 295 °K some modes are lost in the background. For several weak lines the linewidth is larger than the instrumental resolution, this makes difficult the identification of some peaks. Cooling the crystal down to 40 °K has revealed a better resolution of the lines, but there remains uncertainty for a few of them. Polarized Raman spectra of $BaZnF_4$ at room temperature and 40 °K are exhibited in Fig. 1 to 4. We collect in Table 1 the mode frequencies and symmetry assignment. We take three rectangular $x\,y\,z$ axes as a reference experimental frame with $\boldsymbol{x} \parallel \boldsymbol{k}_i$, $\boldsymbol{y} \parallel \boldsymbol{k}_s$, and $\boldsymbol{z}$ normal to the plane xy. We define four polarization filters: (zz) when the incident and scattered polarizations are $\parallel$ to $\boldsymbol{z}$, (yx) when the incident and scattered polarizations are $\parallel$ to $\boldsymbol{y}$ and $\boldsymbol{x}$, respectively; and also (zx) and (yz) filters. From the experiments we see that: the 94 cm^{-1} A_1

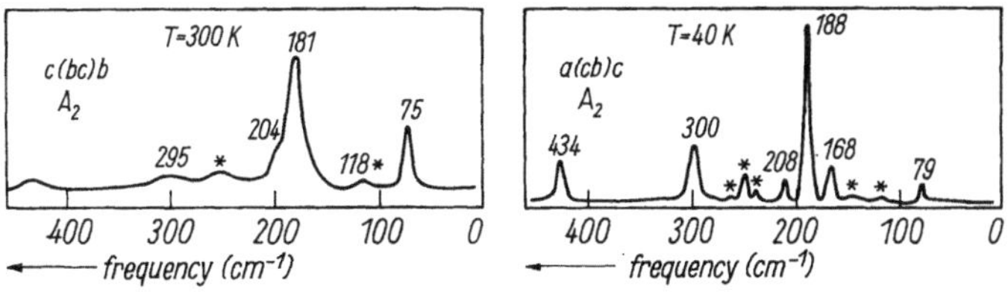

Fig. 4. Raman spectra of A_2 non polar modes at 300 and 40 K

a

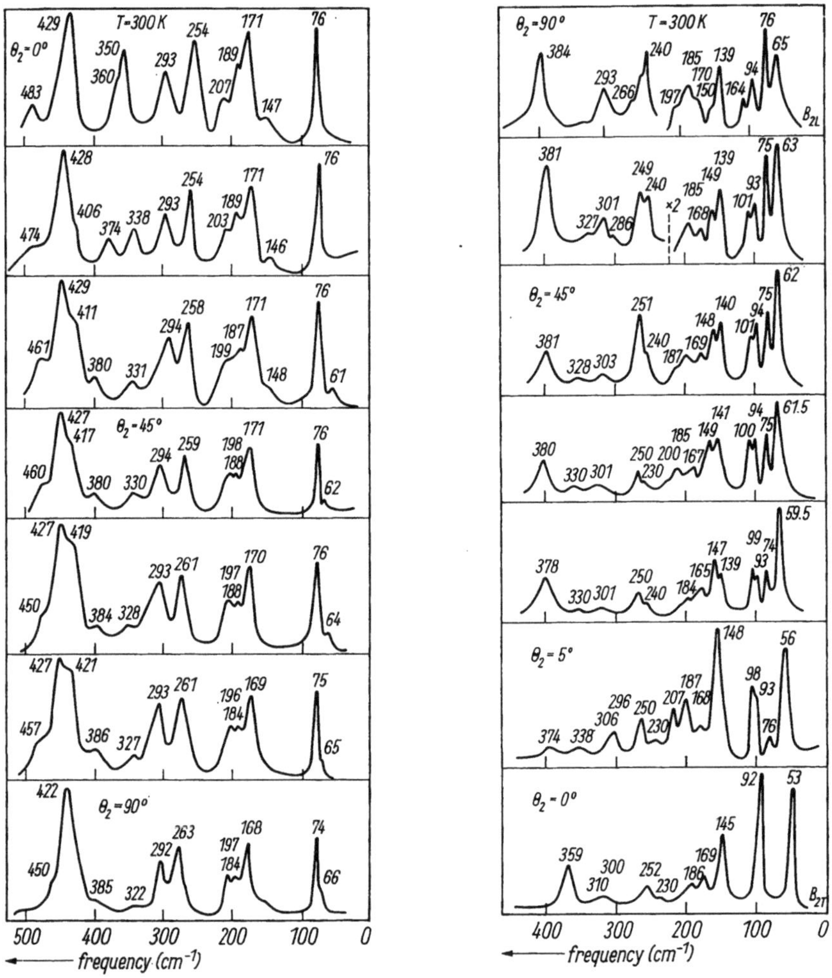

b

Fig. 5. Raman spectra of the oblique phonons in the three *bc*, *ab*, and *ac* planes. a) and b) report the spectra recorded for **k**, in the *ab* plane through *cc* and *yx* polarization filters, respectively. c) and d) report the spectra recorded for **k** in the *ac* plane through (*bb*) and (*yx*) polarization filters, respectively. e) and f) report the spectra recorded for **k** in the *bc* plane through (*yz*) and (*zx*) polarization filters, respectively

phonon has only one (*aa*) non zero tensor coefficient and the 450 cm^{-1} A_1 phonon has only one (*bb*) non zero tensor coefficient. The assignments are difficult around 200, 250, and 300 cm^{-1}. We note that some phonons exhibit a slight frequency shift when the temperature varies from 300 to 40 K.

The angular dispersion of mixed modes "oblique phonon" was measured in the *ab*, *ac*, *bc* planes successively. In *ab* plane θ_2 is the angle (**a**, **k**) and varies from 0 (A_{1L}, B_{2T}) to 90° (A_{1T}, B_{2L}). In *ac* plane θ_3 is the angle (**a**, **k**) and varies from 0 (A_{1L}, B_{1T}) to 90° (A_{1T}, B_{1L}). In *bc* plane θ_1 is the angle (**c**, **k**) and varies from 0 (B_{1L}, B_{2T}) to 90° (B_{1T}, B_{2L}). The corresponding Raman spectra are collected in Fig. 5.

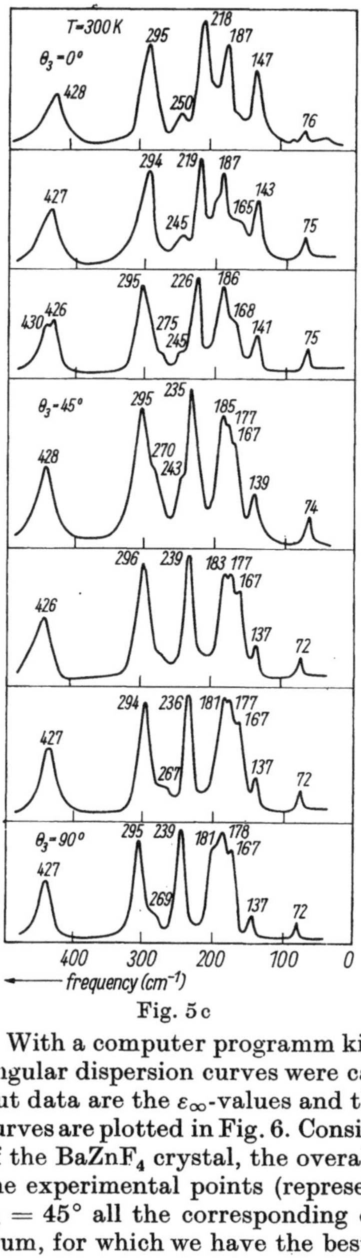

Fig. 5c

3. Discussion

Phenomena associated with electromagnetic interactions among polar modes have been extensively studied and are well understood. From classical electromagnetic theory the dielectric function of an orthorhombic crystal can be written neglecting damping [4]

$$\varepsilon_\alpha(\omega) = \varepsilon_\alpha^\infty + \sum_j \frac{S_j^\alpha \omega_j^2}{\omega_j^2 - \omega^2}\,. \qquad (1)$$

The constants S_j are dimensionless oscillator strengths. The poles of this function occur at the TO frequencies ω_{Tj} while the zeros occur at the LO frequencies ω_{Lj}. The high frequency dielectric constants ε_a^∞, ε_b^∞, ε_c^∞ are obtained from: $\varepsilon_a^\infty = n_\beta^2$, $\varepsilon_b^\infty = n_\alpha^2$, $\varepsilon_c^\infty = n_\gamma^2$. As the TO and LO frequencies are known we are able to calculate the S_j^α constants.

In the crystal the electromagnetic field associated with the phonon follows the Maxwell equations, and we have

$$\boldsymbol{D} = \frac{k^2}{\omega^2/c^2}\left(\boldsymbol{E} - (\boldsymbol{k}\cdot\boldsymbol{E})\boldsymbol{k}\right)\,; \qquad (2)$$

$\boldsymbol{D} = \boldsymbol{\epsilon}\boldsymbol{E}$ [3], where $\boldsymbol{E}$ is the macroscopic field.

The normal modes are purely transverse or purely longitudinal only for phonon propagation along principal axes of the dielectric tensor $\boldsymbol{\epsilon}$.

For any oblique propagation direction $\boldsymbol{k}$ the projection of equations (2) and (3) on the *abc*-axes gives a homogeneous system.

With a computer programm kindly provided by Krauzman [4], the theoretical angular dispersion curves were calculated in the three *ab*, *ac*, *bc* planes. The input data are the ε_∞-values and the experimental TO and LO frequencies. These curves are plotted in Fig. 6. Considering the rather complicated phonon spectrum of the $BaZnF_4$ crystal, the overall agreement between the calculated curves and the experimental points (represented by open circles) is rather satisfactory. For $\theta_i = 45°$ all the corresponding curve points are checked with the 40 °K spectrum, for which we have the best resolution and the best assignment. The only exception occurs for the 353 to 422 cm^{-1} curve in the *ac* plane for which we have no experimental point.

We see that two different types of behaviour exist as $\boldsymbol{k}$ rotates. The vibration can change its character from $T \to L$ but keeps its symmetry at both extremes. On the other hand, the phonon can keep its T or L character but change its symmetry properties.

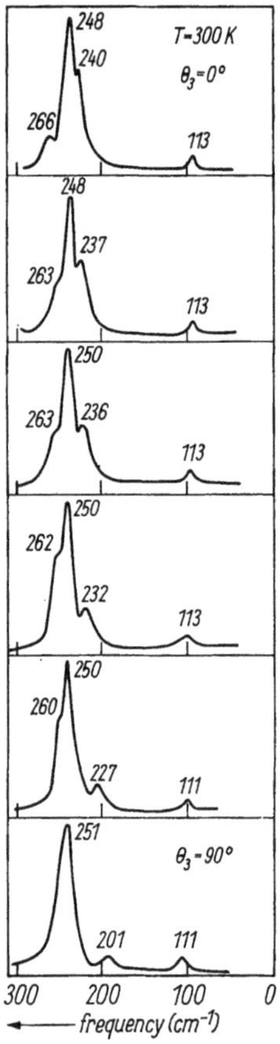

Fig. 5d

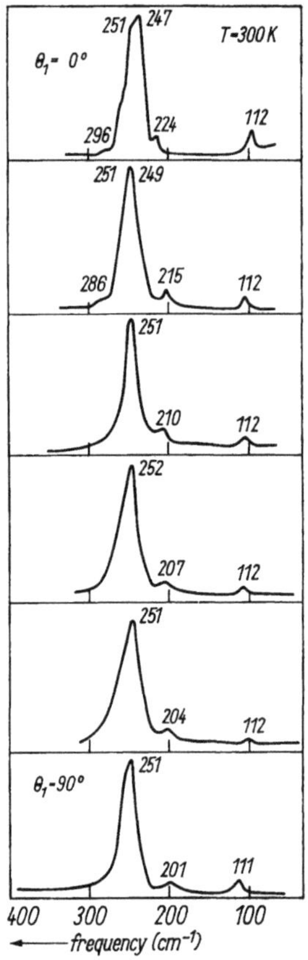

Fig. 5e

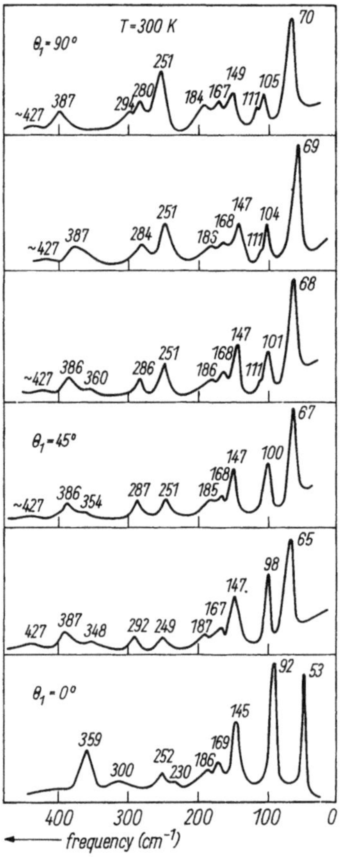

Fig. 5f

In the second case the dominance of the long range electromagnetic forces over the anisotropic short range forces maintains the T or L character, where as in the first case the anisotropic short range forces dominate.

It is worth noting that we observe some net "inverted" transverse longitudinal splitting of phonon modes such has been observed in quartz [6], $BeSO_4 \cdot 4H_2O$ [5] and in trigonal GeO_2 [7].

It is the case for A_1 modes for the branch 263 to 254 cm^{-1} observed with the (*cc*) polarization filter, for the branch 239 to 219 cm^{-1} observed with the (*bb*) polarization filter. For the B_1 modes we observed this behaviour for the 251 to 247 cm^{-1} branch.

From (1) we have for $\omega = 0$

$$\varepsilon_\alpha(0) = \varepsilon_\alpha^\infty + \sum_j S_j^\alpha .$$

We calculated $\varepsilon_a(0)$, $\varepsilon_b(0)$, $\varepsilon_c(0)$ and obtained the values 9.638, 14.703, and 10.683 respectively. These values are in relatively good agreement with the values given in [3] $\varepsilon_a = 11$, $\varepsilon_b = 17$, $\varepsilon_c = 10$, knowing that we have a precision of about one per cent for each measured frequency.

Table 1

Phonon frequencies (cm^{-1}) at Γ and calculated oscillator strengths

calculated oscillator strength' S_j	A_1 species			B_2 species			calculated oscillator strength S_j
	300 K TO	300 K LO	40 K TO	300 K TO	300 K LO	40 K TO	
1.338	72	76	74	52	70	55	7.685
0.817	94	98	97	92	105	91	1.407
1.680	137	147	140	145	150	148	0.502
0.636	167	171	168	168	172	171	0.394
0.791	181	187	185	186	198	187	0.717
1.116	197	219	199	230	252	234	1.268
0.738	239	254	248	253	280	253	0.039
0.463	263	293	280	296	301	299	0.085
0.042	295	350	301	306	320	315	0.078
0.011	426	427	428	359	387	363	0.220
0.174	450	483	?	406	429	410	0.064

calculated oscillator strength S_j	B_1 species			A_2 species	
	300 K TO	300 K LO	40 K TO	300 K	40 K
0.516	111	113	117	75	79
1.720	173	178	175	181	168
5.735	201	247	204	204	188
0.331	251	290	254	295	208
0.068	292	422	297		300
				429	434

We took the dielectric constants determined in the MHz region by Eibschütz et al. [8] to calculate the generalized Lyddane-Sachs-Teller (LST) relation, and we compare the result with the one obtained by

$$\prod_{i=1}^{n_k} \frac{\omega_i^k(\mathrm{LO})^2}{\omega_i^k(\mathrm{TO})^2} .$$

The results are given in Table 2.

Table 2

	A_1	B_2	B_1
LST_{th}	4.793	7.575	4.323
LST_{cal}	4.329	6.393	4.589

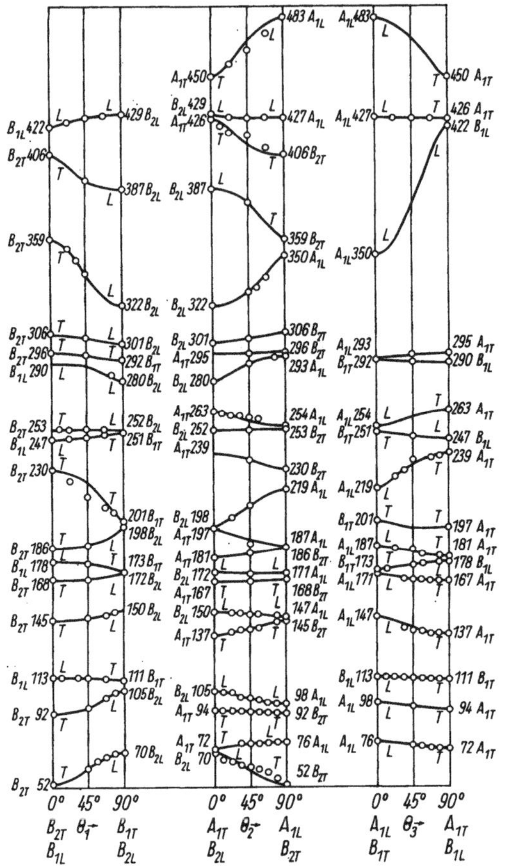

Fig. 6. Calculated dispersional curves in the three *bc*, *ab*, and *ac* planes, θ_1 is the angle (***k***, ***c***) in the *bc* plane; θ_2 is the angle (***k***, ***a***) in the *ab* plane and θ_3 is the angle (***k***, ***a***) in the *ac* plane. We report as experimental points (open circles) only the frequencies known with a good precision. As seen in Fig. 5, it is easier to observe some oblique phonons in one plane instead of the other one. For some broad line or shoulder the accuracy with which the frequency is known is not sufficient to give a point that could be compared with one of a calculated curve

4. Conclusions

The Raman spectra of single crystal of $BaZnF_4$ have been obtained. All group-theoretically predicted modes have been found and assigned. Directional dispersion of polar phonons has been measured and calculated using a programm based on the crystal optics formalism, reasonable agreement is found.

Acknowledgements

The authors thank J. F. Scott who supplied the crystal, Dr. Krauzman for his computer programm, and Mrs. Lenain for X-ray orientation.

References

[1] H. G. V. SCHNERING and P. BLECKMANN, Naturwissenschaften **55**, 342 (1968).
[2] E. T. KEVE, S. C. ABRAHAMS, and J. L. BERNSTEIN, J. chem. Phys. **51**, 4928 (1969).
[3] H. POULET, Ann. Phys. (France) **10**, 908 (1955).
[4] M. KRAUZMAN, C.R. Acad. Sci. (France) B **270**, 856 (1970); private communication.
[5] B. UNGER, phys. stat. sol. (b) **49**, 107 (1972).
[6] J. F. SCOTT and S. P. PORTO, Phys. Rev. **161**, 903 (1967).
[7] J. F. SCOTT, Phys. Rev. B **1**, 3488 (1970).
[8] M. EIBSCHÜTZ, H. J. GUGGENHEIM, S. H. WEMPLE, I. CAMLIBEL, and M. DIDOMENICO, JR., Phys. Letters (Netherlands) **29A**, 409 (1969).

(Received January 11, 1974)

phys. stat. sol. (b) **62**, 511 (1974)

Subject classification: 6 and 20.1; 22.5.3

Physikalisches Institut der Universität Würzburg

Investigation of the Infrared-Active Lattice Vibration in $SrCl_2$

By

R. DROSTE and R. GEICK

$SrCl_2$ single crystals were grown by the Czochralsky method. Their reflectivity at 300 K and transmission at 4.2 K were measured for frequencies from 10 to 930 cm^{-1} and 10 to 150 cm^{-1}, respectively. The reststrahl band found in the reflection spectrum has been analyzed by means of a classical oscillator model. In addition, a Kramers-Kronig analysis was performed since a satisfactory fit could not be achieved with the classical model. The frequencies of the TO-mode and the LO-mode were determined to be 138 and 223 cm^{-1}, respectively. A pronounced shoulder on the reststrahl band was observed as is also the case for CaF_2, BaF_2 etc. Its origin is discussed in terms of multi-phonon processes. The results of the transmission measurements show that $SrCl_2$ is transparent in the far infrared up to 110 cm^{-1} where the lattice absorption starts at 4.2 K.

Es wurden $SrCl_2$-Einkristalle nach dem Czochralsky-Verfahren gezüchtet. Im Bereich 10 bis 930 cm^{-1} wurden Reflexionsmessungen (bei 300 K) und im Bereich 10 bis 150 cm^{-1} Transmissionsmessungen (bei 4,2 K) durchgeführt. Die im Reflexionsspektrum gefundene Reststrahlenbande wurde mittels klassischer Dispersionstheorie analysiert und, da eine befriedigende optimale Anpassung nicht zu erreichen war, zusätzlich einer Kramers-Kronig-Analyse unterworfen. Es ergaben sich die Frequenzen der TO-Eigenschwingung zu 138 cm^{-1} und der LO-Mode zu 223 cm^{-1}. In der Reststrahlenbande taucht die schon aus den Spektren von CaF_2, BaF_2 etc. bekannte Schulter auf, deren Ursache anhand von Mehrphononenprozessen diskutiert wird. Aus den Transmissionsmessungen geht hervor, daß $SrCl_2$ im fernen Infrarot durchlässig ist bis 110 cm^{-1}, wo bei 4,2 K die Gitterschwingungsabsorption einsetzt.

1. Introduction

Recently, there has been considerable interest in $SrCl_2$ as well as in CaF_2, BaF_2, and SrF_2. One of the reasons for this interest is that these crystals are nearly ideal host lattices for paramagnetic ions. A great number of electronic excitation and magnetic resonance experiments has been performed on rare earth ions and other ions in these materials. For some of these investigations, the host lattice was $SrCl_2$ [1 to 6]. Since electronic transitions within the crystal field multiplet have a magnetic dipole character it is useful to know the absorption of the host lattice in the range of the lattice vibrations which is generally rather strong due to its electrical dipole character. Therefore it is of some importance to study the infrared active vibrations of the pure material, namely $SrCl_2$ in our investigation.

The structure of $SrCl_2$ is of the type of CaF_2. The space group is O_h^5, and for the long wavelength lattice vibrations at $q \approx 0$ a group theoretical analysis yields the following classification of these modes (in the notation according to [7]): $2\Gamma_4^- + \Gamma_5^+$. The Γ_5^+-mode is Raman active, and the two triply degenerate Γ_4^--modes are the acoustic and optical lattice vibrations. The long range electrostatic fields cause a splitting of the optical mode into a longitudinal (LO) and a doubly degenerate transverse (TO) mode. The reflection spectra exhibit a pronounced reststrahl band due to the interaction of this TO-mode with electromagnetic radiation.

In the past, the reflection spectra and the Raman effect have been studied for CaF_2, BaF_2, and SrF_2 [8, 9]. In connection with vibronic spectra, theoretical calculations of the lattice vibrations have been performed in CaF_2, EuF_2, and $SrCl_2$ [10]. The Raman effect in $SrCl_2$ has been studied by Srivastava et al. [9]. An investigation of the reflectivity of $SrCl_2$ was first reported by Hisano et al. [11]. However, for reasons having to do with their experimental method an intensive study of the infrared reflection spectra of $SrCl_2$ with more reliable results was believed to be necessary and useful.

2. Experimental

$SrCl_2$ single crystals are strongly hygroscopic. It should be noted that the hydrated forms $SrCl_2 \cdot 6H_2O$ readily comes into being when $SrCl_2$ crystals are exposed to air. Therefore special precautions are necessary when growing the

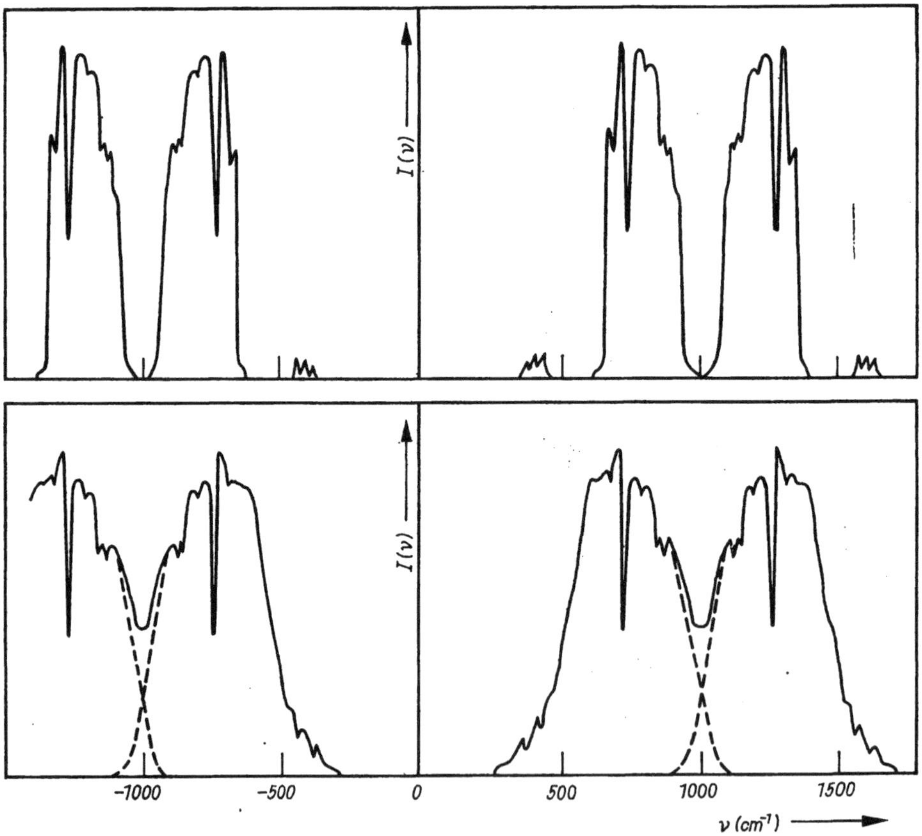

Fig. 1. Spectrum from 0 to 2000 cm^{-1} filtered correctly (upper part) and unsatisfactorily (lower part)

crystals, and preparing and storing the samples. The crystals were grown in an argon atmosphere by a Czochralsky method. Grinding and polishing the sample surfaces was acomplished with an abrasive dispersed in a waterfree organic liquid. Following the recommendations in [9], the samples have been annealed at 650 °C for at least 24 h. The samples were kept immersed in paraffin oil. The organic liquids were only removed immediately before reflection measurements and the sample was brought as quickly as possible into the sample chamber of the spectrometer which was immediately evacuated. The spectrometer used is a Fourier spectrometer FIR 30 (Polytec) with a Fourier transform computer attached to it.

The theory of Fourier transform spectroscopy and the practical details are well known [12 to 14]. However, some comments on the operating range of the FIR 30 and its extension by means of special filters are believed to be worth mentioning. The mechanical precision, smallest sampling distance etc. of the FIR 30 are such that measurements are possible up to a maximum frequency of 1000 cm^{-1}. However, the filters provided with the instrument, limit this range to 700 cm^{-1}. In a Fourier spectrometer filters are used in order to define a frequence ν_{max} with the property that the spectral density of the intensity $I(\nu)$ vanishes for $\nu \geqq \nu_{max}$. This filtering is necessary for a spectrometer with a digital computer, because the interferogram $I(s)$ is measured at discrete points, i.e. at path differences $s = n\,\Delta s$, rather than continuously. For this case, the Fourier transform $I(\nu)$ of the interferogram $I(s)$ is a multivalued function, and therefore, all radiation at frequencies above ν_{max} must be removed. Explicitly, $I(\nu)$ satisfies the relation $I(\nu + m(1/\Delta s)) = I(\nu)$ where $m = 0, \pm 1, \pm 2, \dots$. From basic principles, $I(\nu)$ has to be an even function of frequency, i.e. $I(-\nu) = I(\nu)$. Combining these two relations, the spectrum $I(\nu)$ will be unique only if it is limited to the range $0 \leqq \nu \leqq 1/2\Delta s$, which means that $\nu_{max} = 1000\ cm^{-1}$ for the smallest sampling interval $\Delta s = 5\ \mu m$ of the FIR 30. Recently, filters with a rather sharp cutoff have been developed in the 1000 cm^{-1} range for Raman experiments with a CO_2 laser. Such a filter was employed to extend the range of the spectrometer up to 930 cm^{-1} (see Fig. 1). For comparison, a spectrum when using unsatisfactory filters is shown also. It can be seen, that the intensity $I(\nu)$ in the range 1000 to 2000 cm^{-1} is the intensity $I(-\nu)$ shifted by $1/\Delta s = 2000\ cm^{-1}$ to higher frequencies.

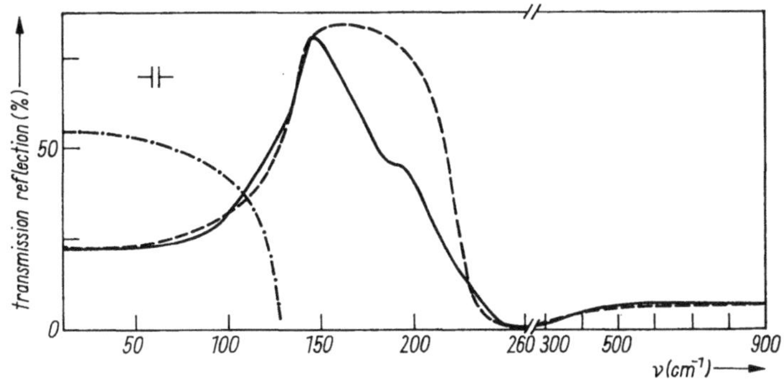

Fig. 2. Reflectivity (———) and transmission (— · — · —) of $SrCl_2$: experimental. Dashed curve (— — —): classical oscillator fit

Under these circumstances, the reflectivity of $SrCl_2$ single crystals was studied over the whole range from 10 to 930 cm^{-1} at 300 K (see Fig. 2). Between 100 and 250 cm^{-1}, the reflection spectra exhibit the reststrahl band expected from group theoretical considerations. For rather low and rather high frequencies the reflectivity is practically constant, from which the static dielectric constant $\varepsilon(0)$ and the high frequency dielectric constant $\varepsilon(\infty)$, respectively, can be determined. It should be noted that the dominant features of the reflection spectra of the investigation by Hisano et al. agree reasonably with our results. In the details however, there are some discrepancies.

The transmission spectrum measured at 4.2 K exhibits an absorption which is nearly zero at 30 cm^{-1} and increases with frequency. Above 110 cm^{-1}, the sample with a thickness of 1.5 mm becomes opaque. This absorption is due to the lattice vibrations of pure $SrCl_2$.

3. Analysis of the Reflection Spectra

An analysis of the reflection spectra was first undertaken by attempting to find an optimum fit with a classical oscillator model:

$$\varepsilon(\nu) = \varepsilon' + i\varepsilon'' = (n + ik)^2 = \varepsilon(\infty) + \frac{\big(\varepsilon(0) - \varepsilon(\infty)\big)\, \nu_{TO}^2}{\nu_{TO}^2 - \nu^2 - i\gamma\nu}, \qquad (1)$$

where ε' and ε'' are the real and imaginary part of ε, respectively, n and k are refractive index and absorption coefficient, respectively, ν_{TO} and γ are eigenfrequency and damping parameter of the TO-mode, respectively, $\varepsilon(0)$ and $\varepsilon(\infty)$ are the static and high frequency dielectric constant, respectively. By means of (1) ε, n, and R were evaluated, and the calculated values of R were compared with the experimental data. Thus by trial and error, the optimum fit values were found to be

$$\varepsilon(0) = 7.55\,, \qquad \varepsilon(\infty) = 2.86\,,$$

$$\nu_{TO} = 138\ \text{cm}^{-1}\,, \quad \gamma = 10\ \text{cm}^{-1}\,,$$

and

$$\nu_{LO} = \sqrt{\varepsilon(0)/\varepsilon(\infty)}\, \nu_{TO} = 223\ \text{cm}^{-1}\,.$$

The reflectivity as calculated with these parameter values agrees quite well with the experimental data with the exception of the frequencies between 145 and 230 cm^{-1} where the shoulder on the reststrahl band occurs. Here the discrepancy with theoretical spectra which do not exhibit this shoulder is considerable. In order to include this feature into our analysis, a Kramers-Kronig analysis of the experimental results was performed.

It may be worthwhile to mention the special form of the Kramers-Kronig relation (KKR) used by us. This relation connects the real and imaginary parts of the response function of a causal linear system by means of an integral transformation. Usually, this response function is the amplitude reflection coefficient $r = \sqrt{R} \exp(i\theta)$ or, more specific, its logarithm $\ln r = \frac{1}{2} \ln R + i\theta$. The data obtained with a Fourier spectrometer sometimes happen to be in error by a constant factor over the whole spectrum. Then, it is advantageous, to take the derivative of the function $\ln r$ and to evaluate the phase angle θ as follows:

$$\theta(\eta) = -\frac{1}{2\pi} \int\limits_0^\infty \frac{\mathrm{d} \ln R(\nu)}{\mathrm{d}\nu} \left[\ln |\eta + \nu| - \ln |\eta - \nu|\right] \mathrm{d}\nu\,. \qquad (2)$$

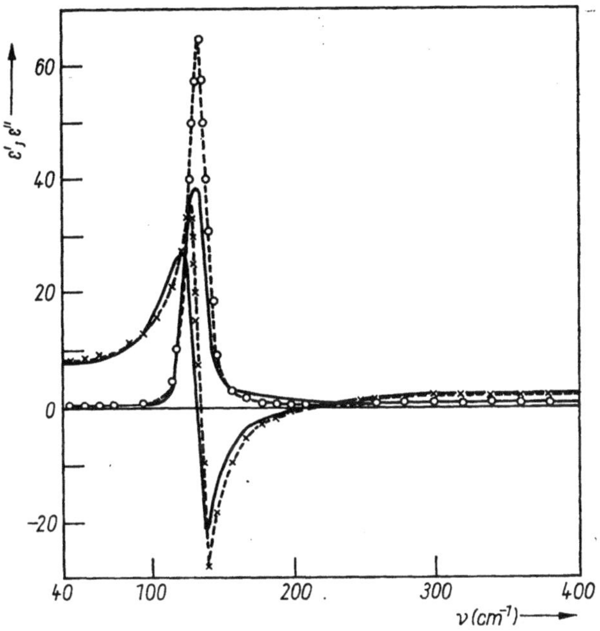

Fig. 3. Real and imaginary part of the dielectric constant: ——— from KKR-analysis, - - - - - from classical oscillator fit

In order to use the KKR, it is necessary to know R for all frequencies from 0 to ∞. Because R was only determined experimentally in a limited range it was extrapolated to frequencies outside of this range. In this formulation, no contributions to the integral arise for sufficiently high frequencies when the derivative of R approaches zero.

After calculating θ by means of (2), the real and imaginary part of ε were evaluated from R and θ. These are shown as functions of frequency in Fig. 3 and are compared to the data obtained by the classical oscillator fit. Near the shoulder in the reflectivity (145 to 230 cm^{-1}), the KKR data show that the imaginary part ε'' of ε is much larger than calculated by means of the classical model with a constant ad hoc damping parameter. Thus it seems that a more correct description of the optical constants of $SrCl_2$ will have to include a frequency dependent damping parameter. This will be discussed in more detail in the next section. As fas as the dominant features are concerned, however, the dielectric constants from KKR and from the classical model agree reasonably well, namely the frequency of the maximum of ε'' (eigenfrequency ν_{TO}), the frequency where $\varepsilon' = 0$ (ν_{LO}), and the area under the ε''-curve (oscillator strength). Therefore, the optimum fit values of the parameters are taken to be reliable values for these quantities except for the damping parameter γ.

4. Discussions and Conclusions

For all crystals with the CaF_2 structure, the frequency ν_{Ra} of the Raman active lattice mode happens to occur between ν_{TO} and ν_{LO}, the frequencies of the infrared active TO-mode and the corresponding LO-mode. This is true also

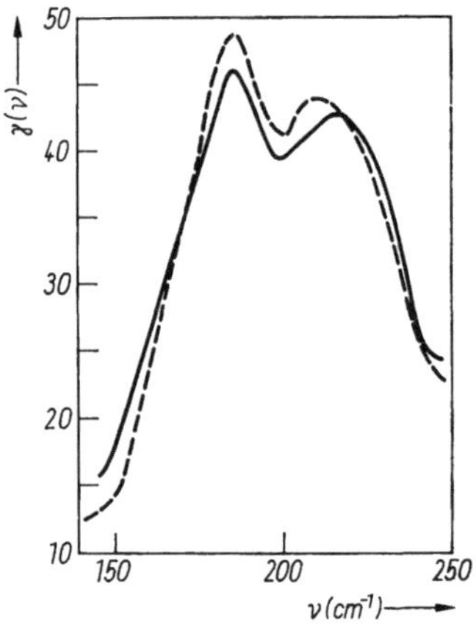

Fig. 4. Damping parameter as a function of frequency, – – – classical oscillator fit, variation of γ, ——— $\gamma(\nu)$ from ε' and ε'' (formula of Bilz et al. [16])

for $SrCl_2$. From the reflection data, we conclude that $\nu_{TO} = 138$ cm^{-1} and $\nu_{LO} = 223$ cm^{-1}, whereas the Raman frequency is $\nu_{Ra} = 183$ cm^{-1} [9].

From the theoretical treatments of anharmonic effects in the infrared lattice absorption it is well known that the damping parameter can vary with frequency due to a frequency dependent probability that an anharmonic decay of the ir-active phonon into two other lattice phonons can occur [15]. Here processes due to the nonlinear dipole moment have been neglected. Thus, any structure in ε'' versus frequency except for the dominant maximum at ν_{TO} is ascribed to two phonon summation or difference processes. For the range 150 to 230 cm^{-1} in $SrCl_2$, i.e. within the reststrahl band, it can easily be shown that only summation processes are important. The peaks in ε'' due to two phonon processes generally are smaller and are dominated by the resonance (in the classical picture) at ν_{TO}. Therefore it may be useful to evaluate a damping parameter $\gamma(\nu)$ as a function of frequency from the experimental data in a simple minded way. Essentially, this is done merely to get rid of the resonant structure at ν_{TO} in order to exhibit more clearly the structure in ε'' due to anharmonic effects [16, 17]. The frequency dependent damping parameter was evaluated in two ways:

a) For every frequency in this range, an optimum fit was obtained by means of (1). In this procedure, only a variation of the damping parameter γ was considered. The optimum value γ is then the value of the function $\gamma(\nu)$ at ν.

b) From the KKR data, $\gamma(\nu)$ was evaluated by means of a formula first introduced by Bilz et al [16].

The results of these calculations show that $\gamma(\nu)$ obtained by the two methods are in reasonable agreement (see Fig. 4). $\gamma(\nu)$ exhibits two maxima at $\nu = 185$ and 210 cm^{-1}. The dispersion relations $\nu_j(q)$ as calculated with a shell model by Kühner et al. [10] show a transverse acoustic branch at the edge of the Brillouin zone with little dispersion and a frequency of about 75 cm^{-1}. From 90 to 150 cm^{-1}, there are several branches with relatively little dispersion. In the density of states, the acoustic branch produces a relatively high maximum at 71 cm^{-1}, while from the other branches pronounced maxima arise at 114, 133, and 147 cm^{-1}. Now, it is assumed that these peaks in the one phonon density of states represent the major contributions to the two phonon density of states. Then, two phonon peaks are expected at 185, 204, and 218 cm^{-1}. These numbers are in unbelievably good agreement with the maxima of $\gamma(\nu)$ which was determined from the experimental data. Therefore, it is believed that the maxima of $\gamma(\nu)$ and, accordingly, the shoulder in the reflectivity is caused by the two phonon processes deduced from the model calculation.

Acknowledgements

Financial support of the Deutsche Forschungsgemeinschaft is gratefully acknowledged. We are indebted to Mrs. Schuhmacher, University of Freiburg, for valuable assistance in growing the samples, and to Mr. Deuling, FU Berlin, for the KKR computer program.

References

[1] J. D. Axe and P. P. Sorokin, Phys. Rev. **30**, 945 (1963).
[2] T. S. Piper, J. P. Brown, and D. S. McClure, J. chem. Phys. **46**, 1353 (1967).
[3] E. Loh, Phys. Rev. B **7**, 1846 (1973).
[4] E. S. Sabisky, Phys. Rev. **141**, 352 (1966).
[5] M. M. Abraham, L. A. Boatner, C. B. Finch, and R. W. Reynolds, Phys. Rev. B **3**, 2864 (1971).
[6] R. W. Mires, D. J. Arnold, and W. K. Dean, Phys. Rev. B **5**, 3654 (1972).
[7] G. F. Koster, J. O. Dimmock, R. G. Wheeler, and H. Statz, Properties of the Thirty-Two Point Groups, M.I.T. Press, Cambridge (Mass.) 1963.
[8] W. Kaiser, W. G. Spitzer, R. H. Kaiser, and L. E. Howarth, Phys. Rev. **127**, 1950 (1962).
[9] R. Srivastava, H. V. Lauer, L. L. Chase, and W. E. Bron, Phys. Letters A **36**, 333 (1971).
[10] D. H. Kühner, H. V. Lauer, and W. E. Bron, Phys. Rev. B **5**, 4112 (1972).
[11] K. Hisano, N. Ohama, and O. Matumura, J. Phys. Soc. Japan **20**, 2294 (1965).
[12] L. Genzel, Plenarvorträge der 33. Physikertagung, Karlsruhe 1968.
[13] R. Geick, Chemie für Labor und Betrieb **23**, 194; **23**, 250; **23**, 300 (1972).
[14] R. J. Bell, Introductory Fourier Transform Spectroscopy, Academic Press, New York 1972.
[15] R. A. Cowley, Adv. Phys. **12**, 421 (1963).
[16] H. Bilz, L. Genzel, and H. Happ, Z. Phys. **160**, 535 (1960); **169**, 53 (1962).
[17] A. A. Maradudin and R. F. Wallis, Phys. Rev. **120**, 442 (1960); **123**, 777 (1961); **125**, 1277 (1962).

(Received December 28, 1973)

phys. stat. sol. (b) 62, 519 (1974)

Subject classification: 20.1 and 20.3; 13.4; 22.8

Laboratoire de Diffraction Neutronique, D.R.F., C.E.N.G., Grenoble (a), Laboratoire de Spectroscopie et de Luminescence, Université Claude Bernard, Lyon I, Villeurbanne (b), and Laboratoire de Magnétisme, C.N.R.S., Grenoble (c)

Crystal Field Determination in Rare-Earth Oxysulfides

II. Optical Experiments for Ho^{3+} and Dy^{3+} Ions

By

J. ROSSAT-MIGNOD (a), J. C. SOUILLAT (b), and G. QUEZEL (c)

The crystal field levels of the $^5I_8$ ground state manifold of Ho^{3+} have been obtained from the luminescence spectrum of Y_2O_2S:5% Ho^{3+}. The ground state of the Ho^{3+} ion is found to be a set of five levels (0, 8, 14, 18, 23 cm^{-1}). An attempt has been made to deduce the parameters of the CEF Hamiltonian from the experimental level energies. For Dy^{3+}, the breaking up of the fundamental $^6H_{15/2}$ multiplet has been obtained by analysing the fluorescence spectrum of Y_2O_2S:1% Dy^{3+} and the absorption spectrum of Dy_2O_2S. The ground state is found to be a doublet relatively well separated (21 cm^{-1}) from the first excited level. In these two compounds, CEF calculations give satisfactory agreement with both the value and direction of the magnetic moment obtained from neutron diffraction experiments.

Les niveaux de champ cristallin du multiplet fondamental $^5I_8$ de l'ion Ho^{3+} ont été obtenues par l'analyse du spectre de luminescence de Y_2O_2S:5% Ho^{3+}. L'état fondamental est constitué d'un ensemble de 5 niveaux très serrés (0, 8, 14, 18, 23 cm^{-1}). Sur la base de ces résultats expérimentaux, nous avons déterminé les six paramètres de l'hamiltonien de champ cristallin. En ce qui concerne Dy^{3+}, la décomposition du multiplet fondamental $^6H_{15/2}$ a été déduite du spectre de fluorescence de Y_2O_2S:1% Dy^{3+} et du spectre d'absorption du composé pur. L'état fondamental est un doublet relativement bien séparé (21 cm^{-1}) du premier niveau excité. Pour ces deux composés, les calculs de champ cristallin sont en accord satisfaisant avec les résultats de diffraction neutronique, en ce qui concerne la direction et la valeur des moments magnétiques.

1. Introduction

A general survey of the work done on the rare-earth oxysulfides is given in paper I [1] of this series together with the optical results on Tb_2O_2S. In this paper we report the optical measurements and the crystal field determination for Ho^{3+} and Dy^{3+} in oxysulfides.

2. CEF Splitting of the $^5I_8$ Manifold of Ho^{3+} Ion

2.1 Experimental results

We have only determined the CEF splitting of the $^5I_8$ ground state multiplet. The Ho^{3+} ion has no excited level with a low quantum number J ($J = 0$ or 1); so no absorption spectra have been recorded because of the difficulty to index the great number of peaks. The CEF levels have therefore been given by an emission spectrum, from a sample of Y_2O_2S:5% Ho^{3+}. Three fluorescence spectra have been recorded which show the transitions originating from the excited

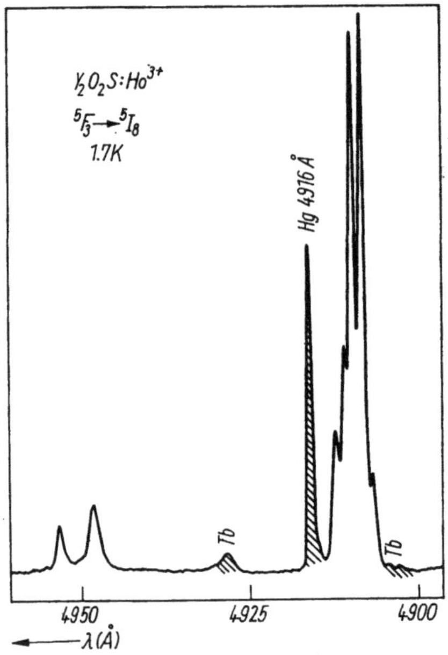

Fig. 1. Fluorescence spectrum of $Y_2O_2S:Ho^{3+}$; $^5F_3 \rightarrow {}^5I_8$ transitions

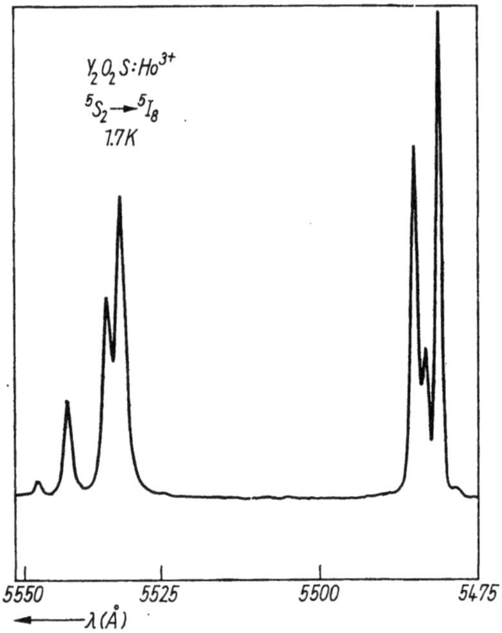

Fig. 2. Fluorescence spectrum of $Y_2O_2S:Ho^{3+}$; $^5S_2 \rightarrow {}^5I_8$ transitions

levels $^5F_3$ (Fig. 1), $^5S_2$ (Fig. 2), and $^5F_5$ (Fig. 3). On the $^5F_3 \rightarrow {}^5I_8$ transition spectrum the hatched lines are due to Tb^{3+} impurities and the line at 4916 Å is a mercury peak. These experiments have been performed at 1.7 K, so only the lowest component of the emitting levels is populated. Thus from the pattern we obtain directly the level scheme of the $^5I_8$ ground state. This manifold is split by the CEF of C_{3v} symmetry in six doublets and five singlets [1], but the eleven expected lines have never been observed on a same spectrum. So the three spectra have been analysed and the energies of the eleven levels have been determined by cross-checking the results. Experimental results are reported in Table 1: Columns 1 and 2 give the peak energies, in wavelength and cm^{-1}, respectively, and column 3 the energy difference between each peak and the original peak.

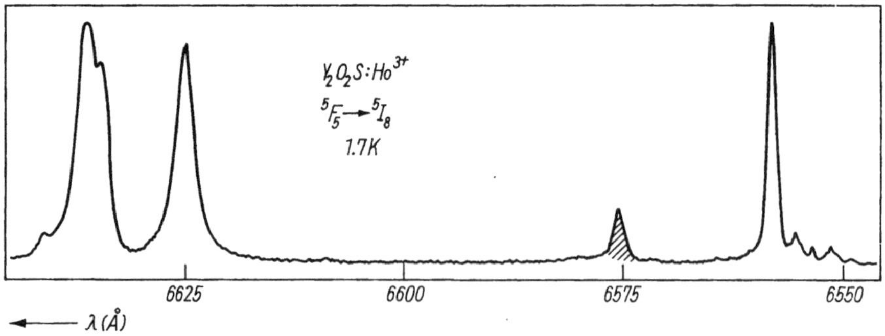

Fig. 3. Fluorescence spectrum of $Y_2O_2S:Ho^{3+}$; $^5F_5 \rightarrow {}^5I_8$ transitions

Table 1

Fluorescence spectra of Y_2O_2S:5% Ho^{3+}; $T = 1.7$ K

line number	λ (Å)	energy (cm^{-1})	$\Delta E = E - E_1$ (cm^{-1})	remarks
		$^5F_3 \rightarrow {}^5I_8$ transition		
1	4906.7	20375	0	
2	4908.5	367	8	
3	4910	361	14	
4	4911	357	18	
5	4912.1	352	23	
6	4914.8	341	34	unassigned line
7	4948.2	204	171	
8	4950.4	192	183	
9	4953.5	182	193	
10	4956	172	203	
		$^5S_2 \rightarrow {}^5I_8$ transition		
1	5478.5	18248	0	
2	5481.5	238	10	
3	5483.5	231	17	
4	5485.3	225	23	
5	5531.8	072	176	
6	5534	065	183	
7	5540	045	203	
8	5544.7	030	218	
		$^5F_5 \rightarrow {}^5I_8$ transition		
1	6548	15268	0	intense at 4.2 K
2	6551.2	260	8	
3	6553.5	255	13	intense at 4.2 K
4	6555.2	251	17	
5	6558	244	24	
6	6575.5	204	64	unassigned line
7	6625	090	178	
8	6635	067	201	
9	6636.5	064	204	
10	6641.5	053	215	

From this table, two sets of levels may be distinguished: a first set of five levels between 0 and 23 cm^{-1} and a second set of six levels, between 171 and 218 cm^{-1}; the following level energies can reasonably be deduced: 0, 8, 13, 17, 23, 171, 176, 183, 193, 203, and 218 cm^{-1}.

2.2 CEF determination

CEF parameters have been adjusted in such a manner to obtain the best fit between calculated and observed energies. With the following values of the parameters $V_2^0 = 30$, $V_4^0 = 76$, $V_4^3 = -2450$, $V_6^0 = 17.5$, $V_6^3 = 176$, and $V_6^6 = 200\,cm^{-1}$, the r.m.s. deviation is $1.8\,cm^{-1}$. The obtained results are reported in Table 2 together with the calculated irreducible representation associated with each level. But in each set, the levels are very close and it is impossible to determine unambiguously the succession of the irreducible representations: Only Zeeman effect experiments on single crystals could give a correct answer

Table 2

Energy (in cm^{-1}) and wave function of CEF levels of the $^5I_8$ multiplet

exp. energy	0	8	13	17	23	171	176	183	193	203	218
calc. energy	0	10	11	15	24.5	170	177	185.5	194.5	201	216
ΔE	0	−2	2	2	−1.5	1	−1	−2.5	−1.5	2	2
irreducible associated representations	Γ_3	Γ_3	Γ_1	Γ_2	Γ_3	Γ_3	Γ_1	Γ_3	Γ_2	Γ_3	Γ_1
$M = 6$			−0.289	−0.264			−0.398		−0.656		0.508
3			+0.456	0.656			0.268		−0.254		0.469
0			−0.645				0.735				0.208
−3			−0.456	0.656			−0.268		−0.254		−0.469
−6			−0.289	0.264			−0.398		0.656		0.508
8	0.772	0.045			−0.633	0.11		−0.25		−0.01	
5	−0.390	0.104			−0.443	−0.22		−0.684		0.415	
2	0.467	0.123			0.585	−0.130		−0.298		0.531	
−1	−0.166	0.626			−0.186	−0.473		0.494		0.278	
−4	0.073	0.724			0.158	0.18		−0.362		−0.530	
−7	0.001	0.235			0.180	0.831		0.260		0.432	

to this question. However, since the Néel temperature is 2.25 K, the ground state is certainly a doublet because the first excited level lies at 8 cm^{-1} (11.6 K) and no magnetic order could appear if the ground state was a singlet. Moreover, we shall see in paper III [2] that this level scheme fits very well the specific heat curve.

Selection rules given in paper I [1] for electric dipole transitions show that eleven lines are allowed if the emitting level is a doublet whereas only nine or eight can be observed if the emitting level is a singlet of Γ_1 or Γ_2 type. For the $^5F_3 \rightarrow {}^5I_8$ transition, the five peaks corresponding to the lowest set of levels are well observed, which indicates that the emitting level is probably a doublet; but the intensities of the lines corresponding to the upper set of levels are too weak to allow a more definite conclusion about the nature of this emitting level. Concerning the $^5F_5 \rightarrow {}^5I_8$ transition, the line intensities vary strongly when the temperature is decreased from 4.2 to 1.7 K: This indicates that the emitting rate is probably made of at least two close levels and therefore we cannot give any conclusion about these levels. On the contrary, in the $^5S_2 \rightarrow {}^5I_8$ transition we have distinctly observed four lines corresponding to the lowest set and four corresponding to the upper set of levels: The emitting level belongs probably to the Γ_2 representation ($\Gamma_2 \rightarrow \Gamma_1$ transitions are forbidden). For the lowest set of levels, the missing line corresponds to the third level, which is in good agreement with CEF calculations indicating that this level belongs to the Γ_1 representation.

3. CEF Splitting of the $^6H_{15/2}$ Manifold of Dy^{3+} Ion

3.1 Experimental results

Luminescence and absorption spectra have been recorded for this compound. Fig. 4 shows the absorption spectrum for the $^6H_{15/2} \rightarrow {}^6F_{5/2}$ transition. At 40 K, the three observed lines give the CEF splitting of the $^6F_{5/2}$ excited level (empirically noted D). At 77 K, additional peaks appear due to the splitting of the $^6H_{15/2}$ ground state manifold (these peaks are noted Z). At this temperature only

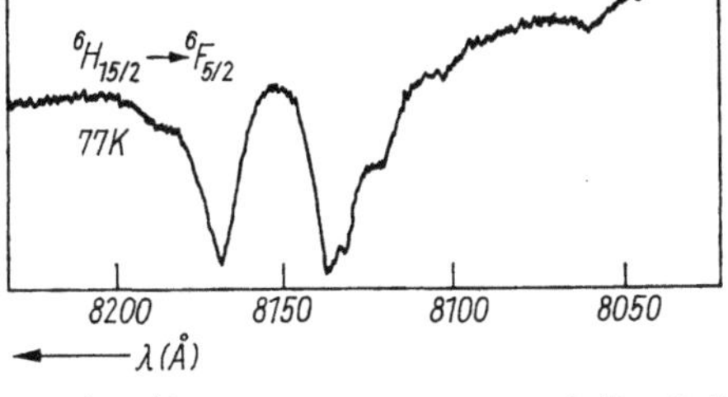

Fig. 4. Absorption spectrum of Dy_2O_2S; $^6H_{15/2} \rightarrow {}^6F_{5/2}$ transitions

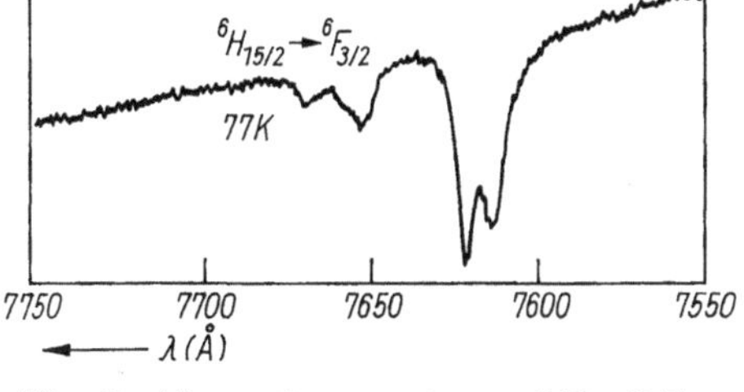

Fig. 5. Absorption spectrum of Dy_2O_2S; $^6H_{15/2} \rightarrow {}^6F_{3/2}$ transitions

Table 3
Absorption spectrum of Dy_2O_2S

wavelength (Å)	energy (cm^{-1})	assignment	remarks
	$^6H_{15/2} \rightarrow {}^6F_{5/2}$ transition		
8055.5	12410	D_3–Z_1	observed at 10 K
8066.5	12393	D_3–Z_2	
8103	12338	D_3–Z_3	
8117.5	12316	D_2–Z_1	observed at 10 K
8122.5	12308	D_1–Z_1	observed at 10 K
8132	12293	D_2–Z_2	
8137.5	12285	D_1–Z_2	
8168.5	12239	D_2–Z_3	
8173	12232	D_1–Z_3	
8181	12203	D_2–Z_4	
8187	12211	D_1–Z_4	
	$^6H_{15/2} \rightarrow {}^6F_{3/2}$ transition		
7610.5	13136	E_2–Z_1	observed at 10 K
7615.5	13127	E_1–Z_1	
7623.5	13113	E_2–Z_2	
7654.5	13060	E_2–Z_3	
7659.5	13052	E_1–Z_3	
7670.5	13033	E_1–Z_4	

four Kramers doublets are populated, with the following energies: 0, 23, 76, and 97 cm^{-1}. These results are confirmed by the $^6H_{15/2} \rightarrow {}^6F_{3/2}$ absorption spectrum (Fig. 5). At 10 K, only one line at 7610.5 Å is observed corresponding to one component of the $^6F_{3/2}$ multiplet; the second component appears only at 77 K. At this temperature, the observed lines, denoted E, give the following splitting for the ground state multiplet: 0, 23, 76, and 95 cm^{-1}.

Table 3 summarizes all these results. Other transitions have been observed which confirm the above values, but they are less accurate because the lines are not well defined. We only indicate that the lowest doublet of the $^4F_{9/2}$ excited multiplet is located at 20866 cm^{-1}.

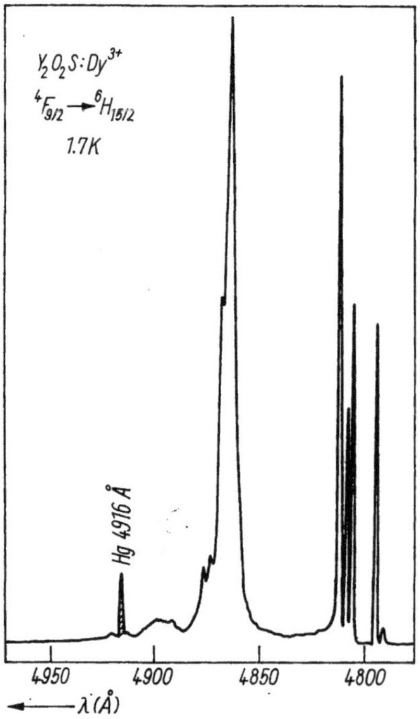

Fig. 6. Fluorescence spectrum of $Y_2O_2S:Dy^{3+}$; $^4F_{9/2} \rightarrow {}^6H_{15/2}$ transitions

In order to get the four upper doublet energies of the $^6H_{15/2}$ manifold, we have recorded luminescence spectra for Y_2O_2S: 1% Dy^{3+}. Fig. 6 shows the transitions originating from the $^4F_{9/2}$ excited manifold at 1.7 K.

Since at this temperature only the $^4F_{9/2}$ low-lying doublet is populated, the observed lines might give directly the splitting of the $^6H_{15/2}$ ground state. At least nine peaks are observed whereas there are only eight $^6H_{15/2}$ doublets; so this spectrum is due to two kinds of transitions and the corresponding lines are mixed. In Table 4 we report the experimental results.

Table 4
Luminescence spectrum of Y_2O_2S:1% Dy^{3+}; $T = 1.7$ K

wavelength (Å)	energy (cm^{-1})	
4794	20854	0
4798	20837	17
4808.5	20791	63
4811	20780	74
4815	20762	92
4866	20545	309
4870	20528	326
4874.3	20510	344
4878	20494	360

The first line (20854 cm^{-1}) is due to a transition from the $^4F_{9/2}$ lowest doublet to the $^6H_{15/2}$ ground state doublet, since absorption experiments give for this line an energy of 20866 cm^{-1}.

We have found five levels between 0 and 92 cm^{-1}, whereas only four doublets are observed on the absorption spectra; thus the line at 20791 cm^{-1} does not belong to the $^4F_{9/2} \rightarrow {}^6H_{15/2}$ transition. Finally, only the energies of the four lowest doublets can be obtained with some certainty: The values obtained from the fluorescence spectrum (0, 17, 74, and 92 cm^{-1}) are in good agreement with those obtained from absorption spectra (0, 23, 76, and 95 cm^{-1}). The slight discrepancies can be explained, first by the experimental accuracy (2 or 3 cm^{-1}) and secondly by the difference of lattice host (Y_2O_2S instead of Dy_2O_2S).

3.2 CEF determination

In order to determine the CEF parameters, the following experimental mean

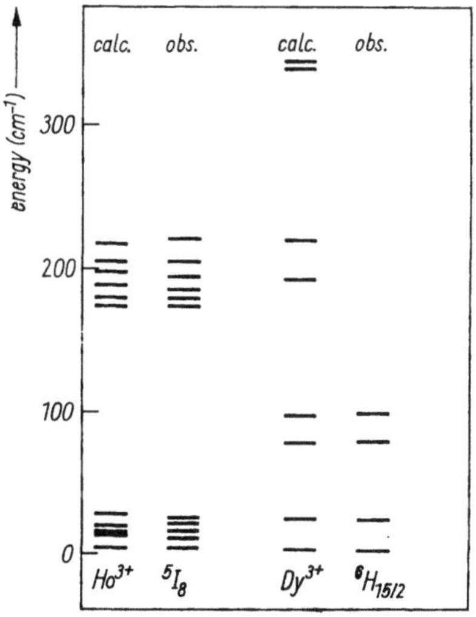

Fig. 7. Calculated and observed CEF splitting of the ground state multiplets of Ho^{3+} and Dy^{3+} in oxysulfides

energy values are used: 0, 21, 76, and 94 cm^{-1}. In this case the determination of six CEF parameters is difficult and not very accurate since we have only three experimental data. However, the order of magnitude of these parameters is known from the results for other oxysulfides (Tb, Ho) so that it is possible to have an approximative determination. The best fit is obtained for the following parameters:

$$V_2^0 = 62\ ,\ V_4^0 = 88\ ,\ V_4^3 = -3200\ ,\ V_6^0 = 23\ ,\ V_6^3 = 310\ ,\ \text{and}\ V_6^6 = 400\ \text{cm}^{-1}\ .$$

The results are reported in Table 5. In Fig. 7, we compare the experimental and calculated CEF levels for Dy^{3+} and Ho^{3+} in oxysulfides.

Table 5

Energy (in cm^{-2}) and wave functions of CEF levels of the $^6H_{15/2}$ multiplet

exp. energy	0	21	76	94				
calc. energy	0	21	76	94	190	218	339	341
ΔE	0	0	0	0				
irreducible associated representations	$\Gamma_5 + \Gamma_6$	Γ_4	Γ_4	$\Gamma_5 + \Gamma_6$	Γ_4	Γ_4	$\Gamma_5 + \Gamma_6$	Γ_4
$M = \frac{15}{2}$	0.815			−0.481			0.311	
$\frac{9}{2}$	−0.478			−0.255			0.836	
$\frac{3}{2}$	0.280			0.666			0.420	
$-\frac{3}{2}$	0.167			0.495			0.168	
$-\frac{9}{2}$	0.280			0.085			0.004	
$-\frac{15}{2}$	0.19			0.086			0.005	
$\frac{13}{2}$		0.613	−0.589		−0.171	−0.008		0.498
$\frac{7}{2}$		−0.578	0.096		−0.054	−0.061		0.806
$\frac{1}{2}$		0.465	0.756		−0.211	0.322		0.253
$-\frac{5}{2}$		0.258	0.192		0.823	−0.429		0.185
$-\frac{11}{2}$		−0.082	−0.190		0.495	0.841		0.060

4. Conclusion

As for Tb^{3+} [1] the ground state of Ho^{3+} in oxysulfide cannot be considered as a doublet but as a set of four levels (two doublets and two singlets) very close to the ground state doublet. However, the first excited level energy (about 12 K) is much larger than the ordering temperature ($T_N = 2.25$ K), so that in the antiferromagnetic state the magnetic anisotropy mainly depends on the nature of the ground state doublet. The wave function of this doublet (Table 2) indicates an important magnetization anisotropy with the easy magnetization direction along the c-axis, and a moment value of about 7.5 μ_B at $T = 0$ K. These results are in good agreement with neutron diffraction data [3].

For Dy^{3+}, the first excited level lies at 30 K so that in the ordered state ($T_N = 5.12$ K) the magnetic anisotropy depends only on the ground state doublet. As this doublet belongs to the ($^1\Gamma_5 + {}^1\Gamma_6$) representation, the magnetic moment has an Ising behaviour imposed by group theory [1]. The only non-zero component of the moment is the *c*-component; its value, calculated from the wave function given in Table 5, is in good agreement with neutron diffraction experiments [4] and susceptibility and magnetization measurements on single crystals [5].

References

[1] J. Rossat-Mignod, J. C. Souillat, and G. Quezel, phys. stat. sol. (b) **62**, 223 (1974).

[2] J. Rossat-Mignod, G. Quezel, A. Berton, and J. Chaussy, phys. stat. sol. (b) **63**, No. 1 (1974).

[3] R. Ballestracci, E. F. Bertaut, and G. Quezel, J. Phys. Chem. Solids **29**, 1001 (1968).

[4] Y. Abbas, J. Rossat-Mignod, G. Quezel, and C. Vettier, Solid State Commun., to be published.

[5] G. Quezel, J. Rossat-Mignod, and Y. Abbas, to be published.

(Received November 26, 1973)

phys. stat. sol. (b) **62**, 529 (1974)

Subject classification: 2 and 13.1; 22.1.1

IBM Thomas J. Watson Research Center, Yorktown Heights, New York

A Simple Tight-Binding Model Applicable to Calculations of the Band Structure of Amorphous Germanium

By

D. HENDERSON[1])

A simple semi-empirical tight-binding model applicable to germanium is presented which takes into account the effects of rotation of the dihedral angle and of bond distortions. The effects of rotation of the dihedral angle are taken into account by including all interactions between nearest-neighbors, the effects of distortions in bond angles by means of a two-center approximation together with the properties of spherical harmonics, and the effects of distortions in bond-length by fitting the matrix elements to the sophisticated band calculations of Herman et al. not only for crystalline Ge at the observed density but also for uniformly compressed and dilated crystalline Ge. Comparison of the results of this model with sophisticated calculations for various polymorphs of Ge shows that this model is, at very least, qualitatively satisfactory and should be useful in band structure calculations for amorphous Ge.

Nous présentons un modèle simple, semi-empirique et adapté au germanium, dans l'approximation de liaison forte, où nous tenons compte des effets de la rotation autour de la liaison entre voisins, et de la modification de ces liaisons. Nous tenons compte de l'effet de la rotation autour de la liaison en considérant toutes les interactions entre les plus proches voisins, et de celui des distorsions dans les angles des liaisons par une approximation à deux centres, complétée par les propriétés des harmoniques sphériques; pour la variation de longueur des liaisons nous ajustons les éléments de matrice aux calculs de bandes précis de Herman et al., non seulement pour Ge cristallin à la densité observée, mais aussi pour Ge cristallin uniformément comprimé ou dilaté. La comparaison des résultats de ce modèle avec des calculs plus complets pour diverses variétés cristallines de Ge montre à tout le moins un accord qualitatif satisfaisant, et suggère son utilité dans les calculs de bandes pour Ge amorphe.

1. Introduction

Possibly the most promising method for calculating the band structure of amorphous solids is through the use of boxes or cells with periodic boundary conditions which contain large numbers of atoms which are disordered within the box. As long as the dimensions of the box are large compared to the bond-length and as long as there are no internal symmetries within the box, the effects of the periodic boundary conditions should not be appreciable.

Obviously, the size of the box represents a compromise. The effects of the periodic boundary conditions diminishes as the size of the box increases. However, the complexity of the calculation of the band structure increases as the size of the box increases. Based in part on the author's experience with periodic boundary conditions in the theory of liquids, a roughly cubic box with about

[1]) Above address applicable until July, 1974. Permanent address: IBM Research Laboratory, San Jose, California.

$N = 60$ atoms seems a reasonable compromise. Some models based on periodic boundary conditions are already available [1].

Unit cells containing 60 atoms are beyond the capacity of most methods of calculating band structures. Thus, it is of interest to consider the tight-binding or LCAO approximation. If one s and three p functions are used for the calculation for the band structure of amorphous Ge, a $4N \times 4N$ (or, in the case $N = 60$, 240×240) matrix must be diagonalized. This is a large calculation but not beyond the capacities of modern computers.

In this paper a semi-empirical tight-binding model for Ge is considered. This model is different from earlier models not only because the effects of rotation of the dihedral angle are considered but because the effects of bond distortions are also considered.

Some results based on this model have been reported earlier [2, 3]. Because of the interest expressed in this model it was decided to publish a short paper about the model itself.

2. Tight-Binding Model

In the tight-binding approximation, the wave function is assumed to be of the form,

$$\psi(\boldsymbol{k}, \boldsymbol{r}) = \sum_d \sum_f \sum_\alpha C_{\alpha f}(\boldsymbol{k})\, \mathrm{e}^{i\boldsymbol{k}(\boldsymbol{d}+\boldsymbol{R}_f)} \varphi_\alpha(\boldsymbol{r} - \boldsymbol{d} - \boldsymbol{R}_f)\,, \tag{1}$$

where $\boldsymbol{d}$ is a translation vector of the lattice, d is a cell index, $f = 1, ..., N$ is the atom number within the cell, $\boldsymbol{R}_f$ is the position of atom f, and $\alpha = 1, ..., 4$ implies that, for the case of Ge or some other tetrahedrally bonded solid, an s, p_x, p_y, or p_z wave function is being used. The Hamiltonian is assumed to be of the form

$$H = -\frac{\hbar^2}{2m} \nabla^2 + \sum_d \sum_f V_f(\boldsymbol{r} - \boldsymbol{d} - \boldsymbol{R}_f) \tag{2}$$

and the energy levels are determined from the equation

$$|H_{ij} - E\delta_{ij}| = 0\,, \tag{3}$$

where

$$H_{\alpha'+f',\alpha+f} = \sum_d \mathrm{e}^{i\boldsymbol{k}(\boldsymbol{d}+\boldsymbol{R}_f-\boldsymbol{R}_{f'})} \int \varphi^*_{\alpha'}(\boldsymbol{r} - \boldsymbol{R}_{f'})\, H\varphi_\alpha(\boldsymbol{r} - \boldsymbol{R}_f)\, \mathrm{d}\tau\,. \tag{4}$$

Following the notation of Slater and Koster [4], write

$$E_{\alpha'\alpha}(x, y, z) = \int \varphi^*_\alpha\, H\varphi_\alpha(\boldsymbol{r} - x\boldsymbol{i}_1 - y\boldsymbol{i}_2 - z\boldsymbol{i}_3)\, \mathrm{d}\tau\,, \tag{5}$$

where $\boldsymbol{i}_1$, $\boldsymbol{i}_2$, and $\boldsymbol{i}_3$ are unit vectors in the x-, y- and z-directions.

The angular dependence of $E_{\alpha'\alpha}$ can be easily obtained from the properties of spherical harmonics if we assume a two-center approximation. In this approximation, only the terms of the sum in (2) located on atoms f' and f are retained and the three-center integrals are discarded. Within the two-center approximation,

$$E_{ss} = \xi(R)\,, \tag{6}$$

$$E_{sx} = l\eta(R)\,, \tag{7}$$

$$E_{xx} = l^2\lambda(R) + (1 - l^2)\,\mu(R)\,, \tag{8}$$

$$E_{xy} = lm\lambda(R) - lm\mu(R)\,, \tag{9}$$

where R is the magnitude of $\boldsymbol{R}$, l, m and n are the direction cosines of $\boldsymbol{R}$ and

$$\eta(0) = 0\,, \tag{10}$$

$$\lambda(0) = \mu(0)\,. \tag{11}$$

Throughout the remainder of this paper it is assumed that only interactions between nearest-neighbors need be included.

The widely used tight-binding model of Weaire [5] results from the further approximations

$$\xi(0) = -3V_1 + V_0\,, \tag{12}$$

$$\lambda(0) = V_1 + V_0\,, \tag{13}$$

$$-\xi(R_0) = -\frac{1}{\sqrt{3}}\,\eta(R_0) = \frac{1}{3}\,\lambda(R_0) = \frac{1}{4}\,V_2\,, \tag{14}$$

$$\mu(R_0) = 0\,. \tag{15}$$

The Weaire model, although extremely enlightening, suffers from some deficiencies. Firstly, the top of the valence band is flat so that there is a delta function in the electronic density of states. Furthermore, the matrix elements do not depend on the dihedral angle of neighboring atoms. As a result, the Weaire model does not distinguish between the band structures obtained using the cubic diamond or the hexagonal wurtzite lattices. This is unfortunate because the difference in the band structures of Ge in these lattices is of the order of magnitude of the differences between crystalline and amorphous Ge. In addition,

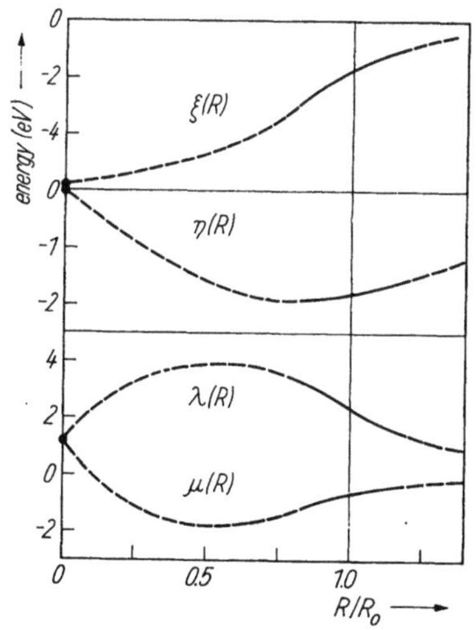

Fig. 1

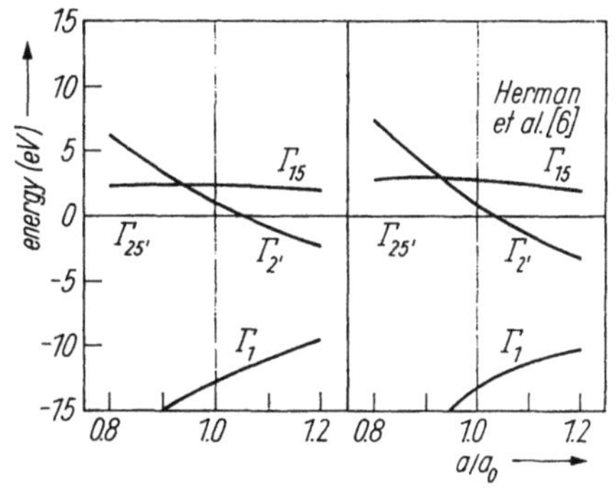

Fig. 2

Fig. 1. Values of $\xi(R)$, $\eta(R)$, $\lambda(R)$, and $\mu(R)$. The circles at $R = 0$ and solid parts of the curves are the parts determined by fitting the results of Herman et al. [6, 7]. The broken parts merely give a reasonable interpolation

Fig. 2. Energy levels of cubic Ge at Γ point. The left-hand part gives the LCAO results and the right-hand part gives the results of Herman et al. [6, 7]. The quantity a is the lattice constant and a_0 is the lattice constant at the observed density

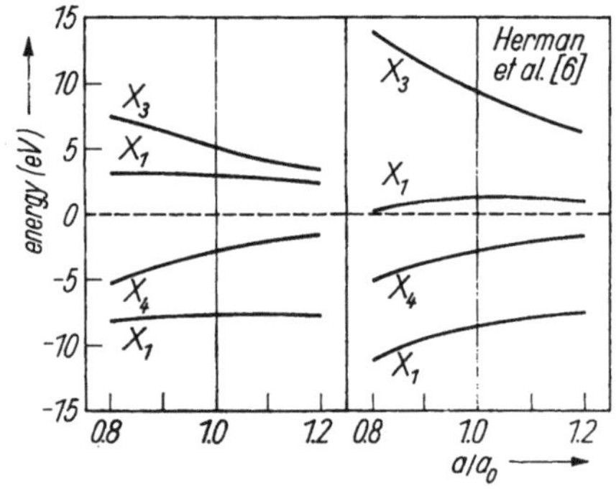

Fig. 3. Energy levels of cubic Ge at X point. The meaning of the curves and symbols are the same as in Fig. 2

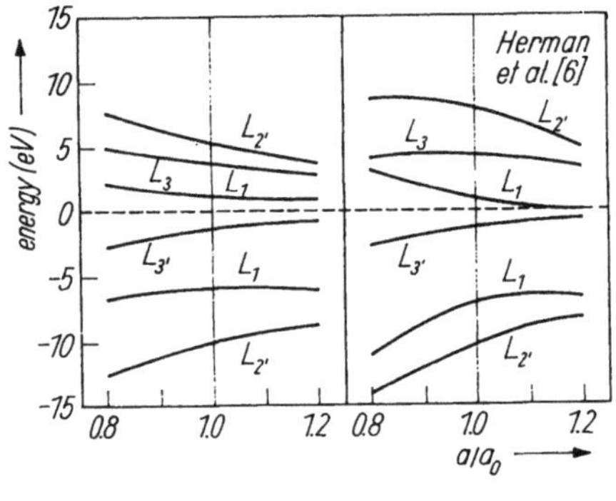

Fig. 4. Energy levels of cubic Ge at L point. The meaning of the curves and symbols are the same as in Fig. 2

changes in the matrix elements due to distortions in the tetrahedral bonding are not included.

The delta function at the top of the valence band results from (15). Relaxation of this condition will remove this feature of Weaire's model. The other deficiencies can be removed by introducing the angle and distance dependence contained in (6) to (9).

3. Parametrization

We have found that if $\xi(0) = -5.8$ eV, $\xi(R_0) = 1.7$ eV, $\eta(0) = 0$, $\eta(R_0) = \sqrt{3}$ eV, $\lambda(0) = \mu(0) = 1.2$ eV, $\lambda(R_0) = 2.3$ eV, and $\mu(R_0) = -0.70$ eV, a fairly reasonable fit of the more elaborate band structure calculations of Herman et al. [6, 7] for Ge in the diamond lattice is obtained. A comparison has been given previously [2, 3]. The agreement is good for the valence band but only partially satisfactory for the conduction band.

The above values for the functions $\xi(R)$, etc., probably bear only a qualitative relation to the values which would be obtained from accurate integration of (4). They should be thought of as effective values, compensating in part for the neglect of non-nearest neighbor interactions.

In amorphous Ge there are distortions in bond length as well as angular distortions. To include these effects $\xi(R)$, $\eta(R)$, $\lambda(R)$, and $\mu(R)$ must be parametrized. Bond distortions in amorphous Ge are small so that it is sufficient to parametrize these functions for $R \approx R_0$. This may be done by making a fit to the band structure results of Herman et al. [6, 7] for Ge in a uniformly compressed or dilated diamond lattice. The fitted values of the functions $\xi(R)$, etc., are shown in Fig. 1. The resulting band structure is compared with the accurate results in Fig. 2 to 4. The agreement is quite pleasing.

4. Summary

A simple tight-binding model has been formulated and has been shown to give a reasonable fit to the properties of Ge in the diamond structure for a wide range of densities. Further, this model takes into account the effects of changes in the dihedral angle. As a result, this model obtains the differences in the band structure of cubic and hexagonal Ge [3]. Comparison with more sophisticated calculations [8] shows that this model correctly predicts the shift to higher energies of the peak at the top of the valence band.

This model also accounts for the effects of bond distortions. Application of this model [2, 3] to polymorphs of Ge which contain bond distortions shows reasonable agreement with the results of more sophisticated methods [9].

References

[1] D. Henderson and F. Herman, J. non-cryst. Solids **8/10**, 359 (1972).
[2] D. Henderson and I. B. Ortenburger, in: Computational Methods for Large Molecules and Localized States in Solids, Ed. F. Herman, A. D. McLean, and R. K. Nesbet, Plenum Press, New York 1973 (p. 318).
[3] D. Henderson, phys. stat. sol. (b) **57**, 661 (1973).
[4] J. C. Slater and G. F. Koster, Phys. Rev. **94**, 1498 (1954).
[5] D. Weaire, Phys. Rev. Letters **26**, 1541 (1971).
[6] F. Herman, R. L. Kortum, C. D. Kuglin, and R. A. Short, in: Quantum Theory of Atoms, Molecules, and the Solid State, Ed. P. O. Löwdin, Academic Press, Inc., London/New York 1966 (p. 381).
[7] F. Herman and J. P. Van Dyke, Phys. Rev. Letters **21**, 1575 (1968).
[8] I. B. Ortenburger, W. E. Rudge, and F. Herman, J. non-cryst. Solids **8/10**, 653 (1972).
[9] D. Henderson and I. B. Ortenburger, J. phys. Chem. **6**, 631 (1973).

(Received December 17, 1973)

phys. stat. sol. (b) 62, 535 (1974)

Subject classification: 2; 21.1; 21.1.1; 21.6

Research Institute of Mineral Dressing and Metallurgy, Tôhoku University, Sendai

Static Structure of Liquid Noble and Transition Metals by X-Ray Diffraction

By

Y. WASEDA and M. OHTANI

X-ray diffraction patterns were obtained from liquid noble and transition metals (Cu, Ag, Au, Fe, Co, Ni) at temperatures between melting point and 1650 °C. After calculating the structure factor (Fourier analysis), the atomic radial distribution function was evaluated from which interatomic distance and coordination number were obtained. Comparing the results with those observed previously by other researchers, adequate agreement was found.

Röntgenbeugungsbilder von flüssigen Edelmetallen und Übergangsmetallen (Cu, Ag, Au, Fe, Co, Ni) wurden bei Temperaturen zwischen dem Schmelzpunkt und 1650 °C erhalten. Durch eine schon mehrfach dargestellte Methode (Fourier-Analyse) wurde der Strukturfaktor in eine radiale Atomverteilungsfunktion transformiert, aus der sich der mittlere Abstand nächster Atome und die Koordinationszahl ergeben. Bei einem Vergleich mit früher von anderen Autoren angegebenen Daten wurde ausreichende Übereinstimmung gefunden.

1. Introduction

In the last few years, there have been a number of papers on the static structure of liquid metals and alloys at temperatures below 1100 °C by X-ray and neutron diffraction. The results of these structural information seem to make a great contribution to the interpretation of various phenomena in liquid metals such as electron-transport and atomic-transport properties [1]. But only a few papers [2 to 7] on the structure factor near the melting point for liquid noble and transition metals have been reported. Moreover the temperature dependence of the structure factor for these liquid metals has not been investigated, because of experimental difficulties arising from their high melting point and active chemical reactivity. As the structure factor is indispensable information for the quantitative discussion of various properties such as electrical resistivity and self-diffusion, it would be desirable to derive the precise knowledge of the structure factor, including temperature dependence, for liquid noble and transition metals.

The main purpose of this paper is not only to show the atomic distribution in liquid noble and transition metals by the common Fourier transformation method but also to give a detailed information of the structure factor (containing numerical values) at several temperatures carefully measured by X-ray diffraction. The reason for the use of X-ray diffraction is that the difficulty of correction for magnetic scattering prevent the light-hearted application of neutron diffraction in the case of liquid transition metals [5, 6].

2. Experimental Procedures

The experimental arrangement for the measurement of the scattered X-ray intensities of liquid sample (stationary horizontal sample goniometer) has been fully discussed [8]. The high-temperature chamber used in this work was identical to the one described in a previous paper [9]. A few additional details are given below.

All data were obtained with the combination of Mo, Cu, and Cr radiations. The intensity in dependence on the scattering angle was stepwise measured in the range θ from 2° to 75°. A graphite monochrometer in the diffracted beam was used to eliminate the white spectrum, the Kβ radiation, most of the Compton scattering and fluorescent radiation from the sample. A high-temperature camera reported previously [9] was slightly modified to maintain temperatures in the sample up to 1660 °C in inert atmosphere. The sample was heated in a boron nitride crucible or alumina crucible (30×20 mm^2) by a tungsten heater element held between two water cooled copper electrodes. A heat shield of thin pyrolitic graphite was also inserted in this chamber. The sample temperature was measured by a Pt–Rh (6/30) thermocouple which was inserted into a hole in the side of crucible.

The liquid metal sample were prepared by 99.99% copper, 99.9% silver, and 99.9% gold. In the case of transition metals, electrolytic iron (99.8%) and nickel (99.9%) and Mond cobalt (99.5%) was previously degassed in vacuum of 10^{-6} Torr by electron beam melting. Before each run the sample was heated to about 1500 °C, in the case of transition metals to about 1650 °C, where the oxide was reduced by the inert atmosphere with hydrogen (Ar–3% H_2) and then brought to the desired temperature within ± 5 deg. As these liquid noble and transition metals show the high value of surface tension, the X-ray beam was collimated and focused to make the best use of the characteristics of the stationary horizontal sample goniometer. Namely, the incident X-rays of 5 mm width and the divergence slit of (1/6)° was used.

3. Analysis of Intensity Patterns

The method of analysing the measured X-ray intensities was the same as that described in the previous paper [9]. For convenience of discussion, the essential features are given below. The distribution of atoms as a function of the radial distance r for a given reference atom, $4\pi r^2 \varrho_0 g(r)$, can be obtained from the X-ray intensity $I(Q)$ scattered coherently by N atoms with atomic scattering factor f as a function of Q by means of the well-known relations

$$4\pi r^2 \varrho_0 \, g(r) = 4\pi r^2 \varrho_0 + \frac{2r}{\pi} \int_0^\infty Q[S(Q) - 1] \sin (Qr) \, dQ \tag{1}$$

and

$$S(Q) = \frac{I(Q)}{Nf^2}, \tag{2}$$

where ϱ_0 is the average number density of atoms, $g(r)$ is the pair correlation function, $S(Q)$ is the structure factor, $Q = 4\pi \sin \theta/\lambda$, where 2θ is the angle between the incident and diffracted X-rays and λ is the wavelength. In order to convert the observed intensities into electron units, the Krogh-Moe-Norman method [10, 11] was used with atomic scattering factors calculated by Cromer and Waber [12] including the anomalous dispersion correction [13].

In this work, the observed intensity data at small Q values less than 0.50 Å^{-1} have been smoothly extrapolated to the value of $Q = 0.0$ Å^{-1} with reference to the value of $S(0)$ which is connected with the coherent small-angle scattering due to the isothermal compressibility. This relation is,

$$\lim_{Q \to 0} S(Q) = S(0) = \varrho_0 \chi_T \, kT \,, \tag{3}$$

where k is the Boltzmann constant, χ_T is the isothermal compressibility and T is the absolute temperature. The isothermal compressibility calculated from the sound velocity observed by Filippov et al. [14] was used. But as far as we know, the sound velocity of liquid Au, Ni, and Co is not reported. We adopted, therefore, the isothermal compressibility of liquid Cu to calculate the value of $S(0)$ for liquid Au and in the case of liquid Ni and Co, that of liquid Fe was used. As shown in a previous paper [5], the effect of this extrapolation on the contribution to the calculation of the radial distribution function is quite small.

The numerical calculation was performed on the NEAC-2200, Model 500, Computer Centre, Tôhoku University.

4. Results and Discussion

As mentioned in the introduction, recent advances in the theory of liquid metals has emphasized the necessity of a more precise determination of the structure factor $S(Q)$ in the small Q region ($Q < 1.2$ Å^{-1}) and the large Q region ($Q > 10.0$ Å^{-1}) [1]. In order to improve this problem of the small Q region, a few attempts were reported in terms of a transmission geometry method [15 to 17]. This method must be one of the useful methods to determine the structure factor of liquid metals. But in this method the absorption correction with respect to sample and cell is not necessarily simple process. From this point of view, the Bragg-Brentano focusing geometry coupled with Mo, Cu, and Cr radiations was performed in this work. In this geometry, it is well-known that the absorption correction is constant value provided that a fairly flat surface

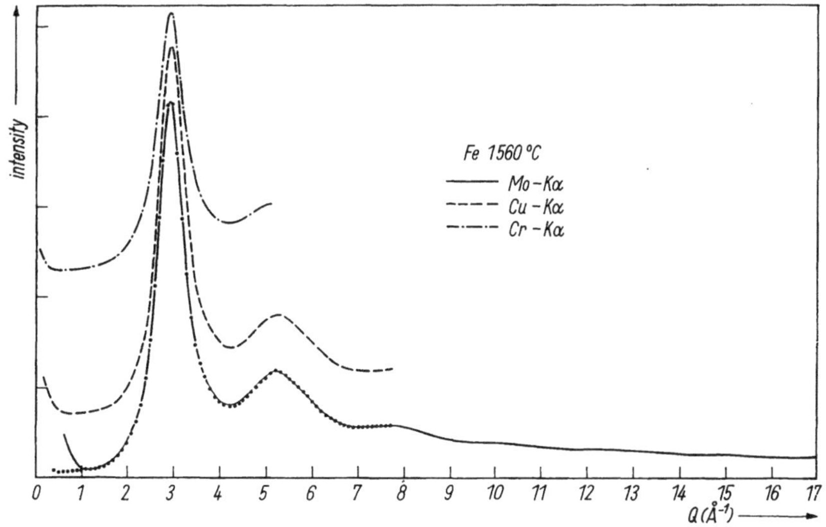

Fig. 1. Intensity patterns of liquid Fe at 1560 °C for three different wavelengths $Q = 4\pi \sin \theta/\lambda$

of liquid sample is maintained. The intensity pattern of liquid Fe at 1560 °C measured by three kinds of radiations is plotted in Fig. 1 as a function of Q. In order to obtain the intensity $I(Q)$ in (2) for calculation of $S(Q)$, we superposed the intensity patterns of Cu and Cr radiations on that of Mo radiation. Solid dots in Fig. 1 show the results of this procedure. For other liquid metals, the same process was performed to determine the structure factor $S(Q)$. Using the data measured by three different wavelengths, we obtained the structure factor with sufficient accuracy in the range from $Q_{min} = 0.50$ Å^{-1} to $Q_{max} = 17.0$ Å^{-1}. This is one of the new information obtained in this work.

In order to determine the structure factor of liquid noble and transition metals with much higher accuracy than heretofore, the accumulated counts, varying from 30000 at low angles to 50000 at high angles, were chosen in this work to hold statistical errors approximately uniform and below 1% because taking an equal number of counts gives equal counting statistics. But the source of systematic error in X-ray diffraction seems to arise from the normalization procedure of measured intensity. For this reason, an independent check of the normalization was performed according to the method proposed by Rahman [18]. This method has been already applied in the previous works [8, 17].

Rahman [18] showed that the structure factor must satisfy the following relation:

$$4\pi\varrho_0 L^3 \frac{j_1(\mu L)}{\mu L} = \frac{L}{\pi\mu} \int_0^\infty Q[S(Q) - 1]\,[j_0\{(Q + \mu)\,L\} - j_0\{(Q - \mu)\,L\}]\,\mathrm{d}Q\,, \quad (4)$$

where $j_i(x)$ is the i-th spherical Bessel function, μ is an arbitrary parameter in dimension of Å^{-1} and L is the value less than the effective hard-core diameter. Equation (4) is the formularized expression of the physical property that the

Table 1

Check of the normalization for X-ray diffraction data of liquid Fe at 1560 °C by the Raman method

L (Å)	μ (Å^{-1})	ε_{theor}	ε_{exp}	correction factor to $S(Q)$
1.00	1.00	0.286	0.283	1.00_3
1.00	1.50	0.251	0.248	1.00_2
1.00	2.00	0.207	0.210	0.99_7
1.00	2.50	0.158	0.156	0.99_5
1.00	3.00	0.109	0.112	1.01_0
1.50	1.00	0.847	0.850	0.99_4
1.50	1.50	0.617	0.607	1.01_1
1.50	2.00	0.369	0.359	1.00_7
1.50	2.50	0.152	0.154	1.00_3
1.50	3.00	-0.001_0	-0.001_1	1.01_1

ε_{theor} are the values calculated from the left-hand side of equation (4). ε_{exp} are the values calculated from the right-hand side of equation (4). The last column lists the correction factor by which the experimental structure factor should be multiplied to obtain perfect agreement between ε_{theor} and ε_{exp}.

atoms do not mutually approach inside of the atomic diameter due to the repulsive forces in the pair potential. In order to check the experimental results in this work, the calculation of the structure factor of liquid Fe at 1560 °C for a variety of values of μ and L was performed in terms of the right-hand side of (4). These values are listed in the fourth column of Table 1 under the notation of ε_{exp}. The corresponding theoretical values evaluated from the left-hand side of (4) are also listed in the third column of this table as $\varepsilon_{\text{theor}}$ for each μ and L. The magnitude of the error in normalization may be estimated at the correction factor by which the experimental structure factor must be multiplied to make the value of ε_{exp} equal to that of $\varepsilon_{\text{theor}}$. For this reason, the last column of Table 1 shows the correction factor by which the experimental structure factor should be multiplied to obtain perfect agreement between $\varepsilon_{\text{theor}}$ and ε_{exp} of (4). As shown in Table 1, the structure factor of liquid Fe at 1560 °C obtained in this work need be changed by less than 1% to satisfy (4) exactly for each μ and L.

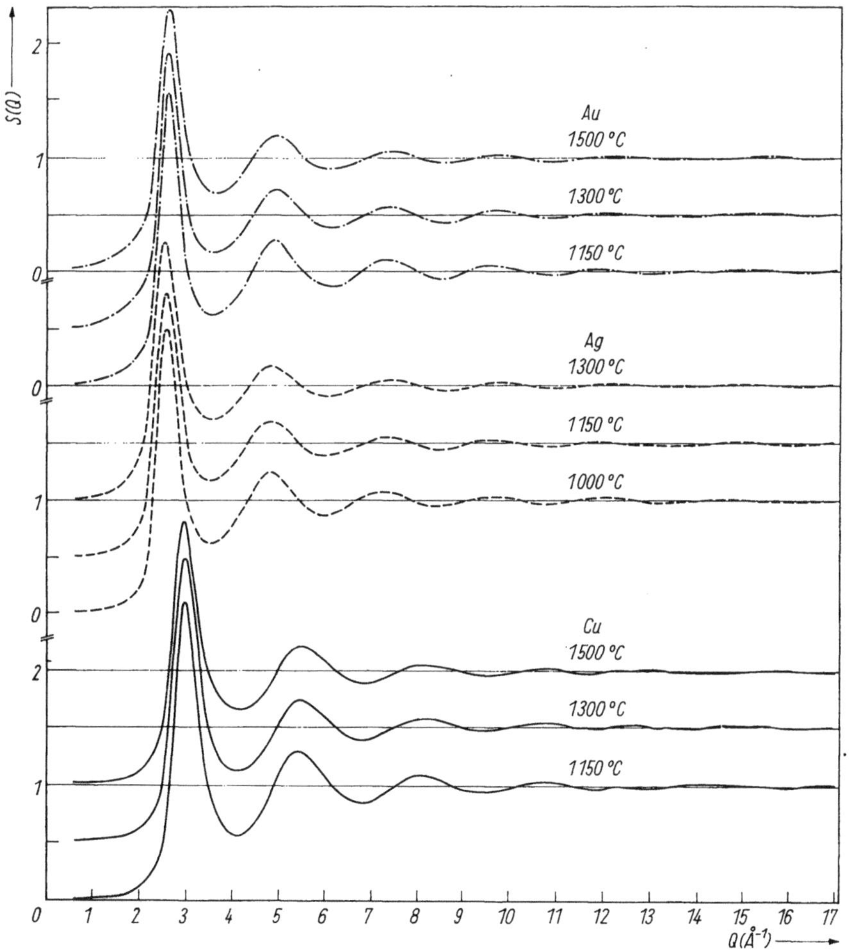

Fig. 2. Temperature dependence of liquid noble metals

Table 2

Summary of diffraction experiments of liquid noble and transition metals

	references	radiation	temp.	number density (atoms/Å^3)	Q_1 (Å^{-1})	Q_2 (Å^{-1})	$S(Q_1)$	$S(Q_2)$	r_1 (Å)	n_1 (atoms)
Cu	Wagner et al. [3]	X-ray	1125	0.0754	2.96	5.45	1.95	1.16	2.61	11.5
	Breuil and Tourand [7]	neutron	1150	—	3.00	5.44	2.41	1.26	—	—
			1300	—	3.00	5.44	2.28	1.24	—	—
			1450	—	2.95	5.44	2.14	1.21	—	—
	this work	X-ray	1150	0.0755	3.00	5.46	2.59	1.28	2.57	11.4
			1300	0.0745	3.00	5.48	2.48	1.24	2.57	11.3
			1500	0.0728	3.00	5.50	2.31	1.22	2.55	11.0
Ag	Pfannenschmid [2]	X-ray	1000	—	2.63	4.80	2.63	1.24	2.86	10.0*)
	Wagner et al. [3]	X-ray	1050	0.0515	2.62	4.87	2.52	1.21	2.87	11.5
	this work	X-ray	1000	0.0517	2.61	4.86	2.48	1.25	2.87	11.3
			1150	0.0509	2.61	4.87	2.31	1.19	2.87	11.0
			1300	0.0501	2.60	4.90	2.24	1.17	2.86	10.8
Au	Pfannenschmid [2]	X-ray	1100	—	2.66	4.89	2.47	1.26	2.85	8.5*)
	this work	X-ray	1150	0.0525	2.66	4.90	2.55	1.27	2.86	10.9
			1300	0.0522	2.66	4.90	2.41	1.23	2.86	10.6
			1500	0.0517	2.65	4.92	2.27	1.20	2.85	10.3
Fe	Ruppersberg and Seemann [4]	X-ray	1550	—	2.98	5.50	2.35	1.22	2.52	8.2*)
	Waseda and Suzuki [6]	neutron	1620	0.0746	2.97	5.48	2.39	1.27	2.58	10.3
	this work	X-ray	1560	0.0756	2.98	5.46	2.39	1.25	2.58	10.8
			1600	0.0749	2.97	5.50	2.29	1.22	2.58	10.8
			1650	0.0742	2.97	5.53	2.20	1.19	2.57	10.4

Table 2 continued

Co	this work	X-ray	1550	0.0786	3.02	5.60	2.46	1.20	2.56	11.5
			1600	0.0780	3.02	5.63	2.41	1.18	2.56	11.3
			1650	0.0774	3.02	5.65	2.32	1.15	2.55	11.2
	Waseda et al. [5]	neutron	1500	0.0792	3.10	5.85	2.30	1.13	2.53	11.6
Ni	this work	X-ray	1500	0.0792	3.10	5.70	2.42	1.21	2.53	11.7
			1600	0.0779	3.10	5.73	2.35	1.19	2.53	11.7
			1650	0.0773	3.10	5.80	2.24	1.16	2.52	11.4

Q_i the position of the i-th peak in the structure factor, $S(Q_i)$ the height of the i-th peak in the structure factor, r_1 the nearest neighbour distance, n_1 the coordination number of atoms.
*) This value is estimated by the method of symmetrical radial distribution function.

This implies that the normalization performed in this work to obtain the structure factor is in error by less than 1%.

The structure factors for liquid Cu, Ag, Au, Fe, Co, and Ni determined in this work are shown in Fig. 2 and 3. In Table 2 the peak position and the peak height of the first two peaks are given together with those found in the previous works [2 to 7].

The results for liquid Cu including the temperature dependence of the structure factor generally agree with those of Breuil and Tourand [7] by neutron diffraction, but differ from that of Wagner et al. [3] by X-ray diffraction. Namely, the structure factor of Wagner et al. shows that the height of the first peak maximum is a fairly small value in comparison with those of Breuil-Tourand and this work as shown in Table 2. This may be related to the error in normalization performed by them [3].

The result obtained for liquid Ag near the melting point is similar to those of Pfannenschmid [2] and Wagner et al. [3]. The structure factor of liquid Au near the melting point is also in general agreement with that of Pfannenschmid by X-ray diffraction.

The general feature of the structure factor for liquid Fe near the melting point is close to the result of Waseda and Suzuki [6] by neutron diffraction, but sightly differs from that obtained by Ruppersberg and Seemann [4] by X-ray diffraction. This difference may be related to the special diffraction procedure (non-focusing stationary sample method) used by them. The structure factor of liquid Co is one of the new data obtained in this work. The structure factor of liquid Ni at 1500 °C shows the sharp peak maxima in comparison with the previous work of Waseda et al. [5] by neutron diffraction. The new data in this work seem to be quantitatively accurate in contrast to the previous result because the correction for the magnetic scattering is one of the serious problems in neutron diffraction [6].

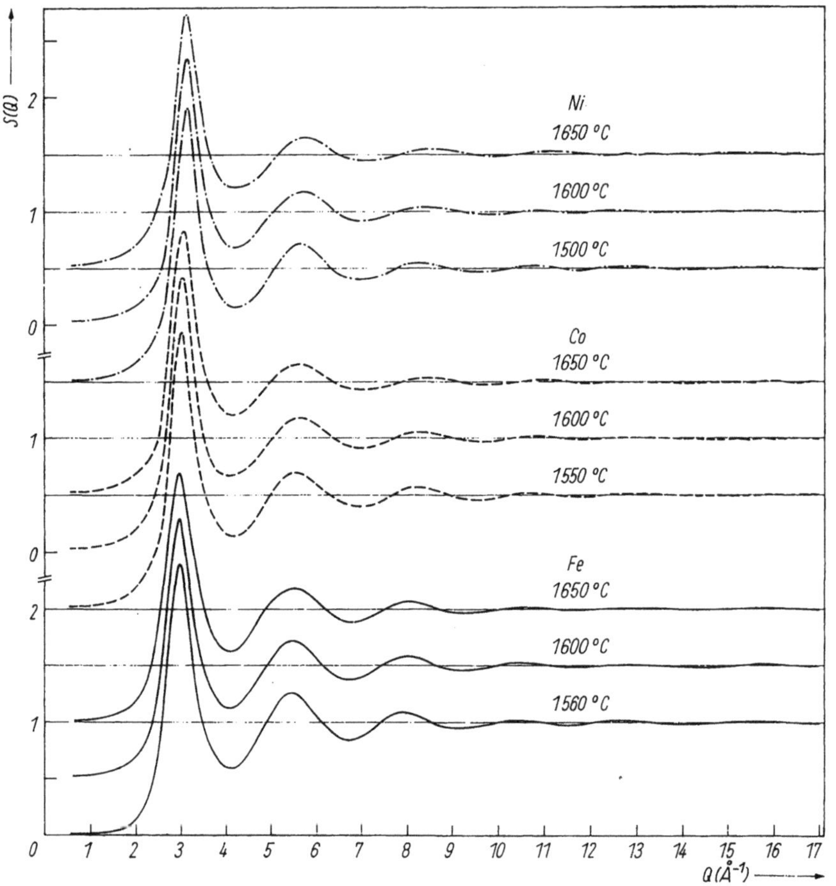

Fig. 3. Temperature dependence of liquid transition metals

The structure factors for liquid noble and transition metals look similar to those of the most simple liquid metals such as the alkali metals in contrast to liquid Hg, Sn, Bi, etc. which show the small hump on the high angle side of the first peak maximum or the large asymmetry of the first peak maximum in the structure factor [19]. Namely the first peak of the structure factor is quite symmetrical as shown in Fig. 2 and 3 and at temperatures just above the melting point the structure factor as the first approximation agrees with that of the hard sphere model using the Percus-Yevick equation with the packing density of 0.45 [20]. But the oscillation of the structure factor for these liquid metals is damped more rapidly than that of the hard sphere model. This behaviour of liquid metals can be considered to be related to the soft repulsive core in the effective interionic potentials [19, 21]. This characteristics may be connected with the overlap of the incomplete d-band with the s-band in these metals. But this problem will be discussed in detail after performing some further experiments.

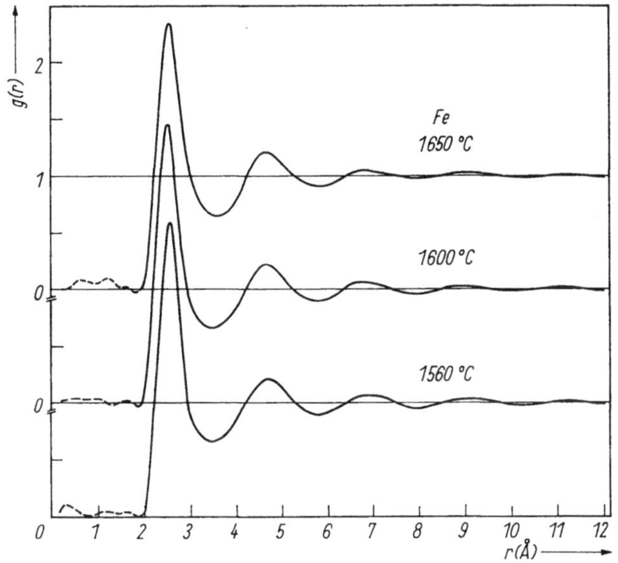

Fig. 4. Pair correlation function of liquid Fe at three different temperatures

Though a new experimental information in this work is the temperature dependence of the structure factor for liquid noble and transition metals, it is not observed a new result which differs from those of liquid Cu observed by Breuil and Tourand [7]. Namely at higher temperatures a broadening of the peak maxima is observed.

The information on the distribution of atoms in real space is given by the common Fourier transform of the structure factor. As an example, the pair correlation function $g(r)$ of liquid Fe calculated in terms of (1) is shown in Fig. 4. As shown in this figure, the spurious ripple in the pair correlation function at small r region is considerably reduced. This is consistent with the physical property that the pair correlation function is essentially zero inside the atomic diameter and support the reliability of the structure factor obtained in this work again. The nearest neighbour distance and the coordination number of the nearest neighbour atoms for six liquid metals evaluated from the atomic radial distribution function $4\pi r^2 \varrho_0 g(r)$ are summarized in Table 2. As we have no unique method of evaluating the coordination number of atoms, only a mean value obtained by three methods [6] is given. But the reasonable coordination number of these liquid metals will be considered to have a value close to that shown in Table 2. These results are almost in agreement with those found previously [2 to 7].

To summarize the results in this work, the static structure of liquid noble and transition metals can be interpreted as a simple fluid in which the arrangement of atoms is random distribution, although these metals have the complicated electronic structure connected with the d-band.

Acknowledgements

The authors are grateful for valuable discussion with the members of our laboratory. Thanks are due to Mr. K. Moriya, for many useful assistance with

Table 3

Numerical values of the structure factor of liquid noble metals in the range from 0.50 to 4.40 Å^{-1}

	Cu			Ag			Au		
temp. (°C)	1150	1300	1500	1000	1150	1300	1150	1300	1500
density (g/cm^3)	7.97	7.86	7.69	9.27	9.12	8.98	17.2	17.1	16.9
Q (Å^{-1})	$S(Q)$	$S(Q)$	$S(Q)$	$S(Q)$	$S(Q)$	$S(Q)$	$S(Q)$	$S(Q)$	$S(Q)$
0.5	0.018_4	0.024_1	0.026_0	0.018_9	0.025_0	0.028_0	0.030_4	0.031_0	0.033_3
0.6	0.019_3	0.025_1	0.025_8	0.018_6	0.028_7	0.030_5	0.033_7	0.036_6	0.042_6
0.7	0.019_6	0.026_3	0.027_9	0.018_6	0.030_1	0.034_3	0.036_8	0.042_9	0.046_4
0.8	0.020_0	0.027_0	0.028_1	0.020_2	0.034_5	0.036_5	0.041_4	0.046_5	0.048_8
0.9	0.020_5	0.028_8	0.029_0	0.024_2	0.036_9	0.038_9	0.046_2	0.054_7	0.052_1
1.0	0.020_9	0.031_0	0.031_7	0.022_6	0.038_8	0.040_9	0.052_4	0.050_7	0.056_7
1.1	0.020_9	0.032_1	0.034_1	0.030_8	0.042_1	0.044_1	0.058_0	0.060_7	0.062_3
1.2	0.020_9	0.035_0	0.036_7	0.039_6	0.051_6	0.053_3	0.063_6	0.072_5	0.078_8
1.3	0.021_1	0.036_8	0.038_9	0.043_9	0.058_2	0.063_5	0.078_7	0.088_9	0.093_3
1.4	0.022_5	0.038_0	0.040_2	0.052_2	0.063_1	0.082_7	0.090_8	0.107	0.118
1.5	0.026_7	0.040_1	0.042_1	0.061_5	0.082_4	0.101	0.104	0.131	0.144
1.6	0.037_2	0.043_9	0.048_5	0.086_4	0.105	0.128	0.121	0.161	0.170
1.7	0.043_2	0.058_9	0.061_4	0.113	0.138	0.161	0.142	0.194	0.203
1.8	0.060_6	0.070_7	0.081_1	0.139	0.174	0.206	0.176	0.229	0.243
1.9	0.075_9	0.089_2	0.093_7	0.186	0.208	0.275	0.217	0.281	0.293
2.0	0.102	0.109	0.115	0.265	0.291	0.357	0.261	0.336	0.352
2.1	0.139	0.140	0.143	0.445	0.456	0.487	0.334	0.410	0.433
2.2	0.181	0.176	0.186	0.815	0.826	0.883	0.439	0.526	0.591
2.3	0.231	0.234	0.252	1.04_9	1.05_3	1.19_7	0.762	0.768	0.953
2.4	0.301	0.307	0.344	1.65_2	1.68_5	1.70_5	1.36_5	1.43_3	1.44_9
2.5	0.479	0.455	0.457	2.21_9	2.08_1	2.10_9	2.05_0	2.03_2	1.95_7
2.6	0.804	0.844	0.850	2.48_5	2.31_6	2.24_1	2.49_3	2.38_1	2.24_9
2.7	1.25_3	1.29_7	1.34_2	2.33_5	2.16_2	2.06_9	2.47_7	2.32_1	2.20_7
2.8	1.83_5	1.83_8	1.86_5	1.88_2	1.87_4	1.70_3	2.06_6	1.93_1	1.89_1
2.9	2.35_4	2.29_3	2.19_4	1.37_2	1.45_7	1.31_8	1.51_1	1.44_7	1.48_1
3.0	2.58_6	2.48_2	2.31_7	1.00_8	1.06_9	1.03_4	1.09_5	1.08_7	1.14_6
3.1	2.43_6	2.33_9	2.17_1	0.836	0.843	0.877	0.901	0.873	0.950
3.2	2.00_5	1.95_4	1.84_8	0.740	0.775	0.802	0.796	0.787	0.858
3.3	1.51_7	1.51_2	1.49_4	0.681	0.717	0.758	0.704	0.726	0.805
3.4	1.15_8	1.17_0	1.21_3	0.644	0.691	0.723	0.654	0.692	0.735
3.5	0.970	0.923	1.03_2	0.635	0.683	0.703	0.623	0.678	0.709
3.6	0.813	0.802	0.916	0.622	0.683	0.703	0.613	0.660	0.688
3.7	0.718	0.725	0.827	0.645	0.692	0.719	0.636	0.669	0.691
3.8	0.635	0.674	0.752	0.681	0.711	0.743	0.675	0.693	0.708
3.9	0.602	0.637	0.701	0.717	0.768	0.774	0.698	0.714	0.730
4.0	0.569	0.617	0.678	0.761	0.808	0.814	0.715	0.739	0.760
4.1	0.578	0.617	0.670	0.824	0.848	0.864	0.754	0.781	0.804
4.2	0.593	0.627	0.666	0.902	0.910	0.918	0.826	0.847	0.862
4.3	0.599	0.633	0.663	0.981	0.989	0.969	0.917	0.924	0.927
4.4	0.611	0.647	0.674	1.05_0	1.06_2	1.01_7	1.00_3	0.998	0.990

Table 4

Numerical values of the structure factor of liquid transition metals in the range from 0.50 to 4.40 Å^{-1}

	Fe			Co			Ni		
temp. (°C)	1560	1600	1650	1550	1600	1650	1500	1600	1650
density (g/cm^3)	7.01	6.95	6.88	7.70	7.64	7.58	7.72	7.60	7.58
Q (Å^{-1})	$S(Q)$	$S(Q)$	$S(Q)$	$S(Q)$	$S(Q)$	$S(Q)$	$S(Q)$	$S(Q)$	$S(Q)$
0.5	0.020_2	0.023_0	0.026_2	0.020_3	0.022_6	0.024_7	0.020_3	0.025_6	0.028_4
0.6	0.022_1	0.024_5	0.028_8	0.021_1	0.022_5	0.025_3	0.020_7	0.028_0	0.030_2
0.7	0.024_7	0.026_7	0.029_8	0.022_0	0.022_6	0.026_7	0.020_4	0.030_9	0.033_1
0.8	0.025_7	0.029_6	0.020_8	0.024_4	0.024_2	0.028_5	0.020_7	0.032_6	0.036_8
0.9	0.026_6	0.030_2	0.032_6	0.025_9	0.026_0	0.030_6	0.021_4	0.033_4	0.038_9
1.0	0.028_6	0.032_7	0.036_2	0.026_7	0.028_5	0.032_5	0.022_7	0.035_3	0.040_1
1.1	0.030_7	0.036_6	0.041_0	0.029_0	0.034_1	0.038_5	0.024_6	0.036_1	0.043_6
1.2	0.030_2	0.041_0	0.045_8	0.035_7	0.041_2	0.044_5	0.028_8	0.052_3	0.058_7
1.3	0.032_5	0.048_4	0.053_5	0.040_5	0.046_5	0.051_4	0.037_4	0.063_6	0.067_8
1.4	0.037_3	0.052_8	0.059_1	0.048_7	0.052_1	0.062_0	0.048_2	0.075_1	0.076_0
1.5	0.039_2	0.063_1	0.066_4	0.061_5	0.069_9	0.068_4	0.063_6	0.086_3	0.088_2
1.6	0.048_6	0.078_3	0.080_9	0.078_3	0.082_6	0.087_0	0.079_0	0.096_0	0.105
1.7	0.071_6	0.096_1	0.104	0.097_1	0.101	0.112	0.099_0	0.101	0.121
1.8	0.085_9	0.113	0.128	0.124	0.127	0.138	0.118	0.130	0.141
1.9	0.117	0.152	0.166	0.151	0.157	0.165	0.142	0.143	0.163
2.0	0.153	0.184	0.211	0.185	0.190	0.201	0.166	0.170	0.198
2.1	0.203	0.236	0.266	0.231	0.234	0.243	0.197	0.203	0.237
2.2	0.264	0.297	0.343	0.282	0.282	0.293	0.223	0.245	0.282
2.3	0.360	0.375	0.465	0.344	0.351	0.285	0.264	0.290	0.344
2.4	0.513	0.535	0.622	0.438	0.453	0.362	0.326	0.349	0.425
2.5	0.732	0.747	0.894	0.560	0.585	0.484	0.397	0.435	0.532
2.6	1.06_0	1.07_2	1.17_6	0.736	0.787	0.717	0.516	0.552	0.663
2.7	1.50_3	1.49_5	1.59_0	1.12_4	1.17_7	1.15_4	0.707	0.777	0.864
2.8	1.97_2	1.92_3	1.96_0	1.66_9	1.71_2	1.70_3	1.13_8	1.14_2	1.16_4
2.9	2.30_9	2.21_7	2.17_3	2.18_2	2.19_5	2.10_6	1.67_4	1.63_4	1.65_4
3.0	2.38_1	2.27_2	2.17_3	2.43_7	2.41_4	2.31_5	2.18_0	2.10_6	2.04_5
3.1	2.17_2	2.08_9	1.98_8	2.33_5	2.28_7	2.28_4	2.41_9	2.34_2	2.24_1
3.2	1.79_4	1.76_4	1.70_9	1.96_8	1.91_4	1.99_6	2.29_2	2.24_4	2.12_3
3.3	1.14_4	1.43_1	1.42_9	1.54_5	1.49_7	1.65_3	1.90_9	1.89_8	1.88_6
3.4	1.14_1	1.17_3	1.19_7	1.23_2	1.19_5	1.36_0	1.48_3	1.49_7	1.47_4
3.5	0.983	1.00_2	1.01_7	1.06_5	1.03_7	1.14_0	1.17_8	1.19_3	1.24_0
3.6	0.884	0.884	0.875	0.973	0.873	1.07_3	0.945	1.02_1	1.03_5
3.7	0.793	0.787	0.765	0.881	0.768	0.903	0.857	0.924	0.917
3.8	0.705	0.707	0.675	0.774	0.722	0.835	0.778	0.841	0.836
3.9	0.639	0.656	0.662	0.687	0.697	0.773	0.733	0.756	0.774
4.0	0.609	0.635	0.652	0.647	0.679	0.736	0.697	0.702	0.738
4.1	0.602	0.633	0.645	0.644	0.669	0.718	0.681	0.684	0.728
4.2	0.605	0.640	0.636	0.648	0.688	0.717	0.681	0.688	0.725
4.3	0.619	0.657	0.641	0.649	0.696	0.721	0.662	0.694	0.719
4.4	0.655	0.692	0.680	0.664	0.711	0.738	0.665	0.696	0.715

experimental work. We would also like to thank the members of Computer Centre, Tôhoku University for allowing the use of their computer. The authors are greatly indebted to Prof. K. Suzuki and Prof. S. Tamaki for their continuous encouragement in this study.

Appendix

Some researchers report only the atomic radial distribution function, but the authors believe that the structure factor $S(Q)$ is more useful information for the discussion of various phenomena in liquid metals. However, it is most common to show the data in the compact form of a figure and this is inconvenient for the quantitative estimation of the properties such as electrical resistivity in liquid metals. From this reason, the numerical values of the structure factors near the first peak maximum for liquid noble and transition metals at several temperatures obtained in this work are listed in Tables 3 and 4.

References

[1] S. TAKEUCHI, The Properties of Liquid Metals, Tokyo Conference, Taylor and Francis, London 1973.
[2] O. PFANNENSCHMID, Z. Naturf. **15a**, 603 (1960).
[3] C. N. J. WAGNER, H. OCKEN, and M. L. JOSHI, Z. Naturf. **20a**, 235 (1965).
[4] H. RUPPERSBERG and H. J. SEEMANN, Z. Naturf. **21a**, 820 (1966).
[5] Y. WASEDA, K. SUZUKI, S. TAMAKI, and S. TAKEUCHI, phys. stat. sol. **39**, 181 (1970).
[6] Y. WASEDA and K. SUZUKI, phys. stat. sol. **39**, 669 (1970).
[7] M. BREUIL and G. TOURAND, J. Phys. Chem. Solids **31**, 549 (1970).
[8] Y. WASEDA and K. MORIYA, Bull. Res. Inst. Min. Dressing and Metallurgy, Tôhoku University, Vol. 29 (1973), in the press.
[9] Y. WASEDA, F. TAKAHASHI, and K. SUZUKI, Sci. Rep. Res. Inst. Tôhoku University **23A**, 127 (1971).
[10] J. KROGH-MOE, Acta cryst. **9**, 351 (1956).
[11] N. NORMAN, Acta cryst. **10**, 370 (1957).
[12] D. T. CROMER and J. T. WABER, Acta cryst. **18**, 104 (1965).
[13] D. T. CROMER, Acta cryst. **18**, 17 (1965).
[14] S. I. FILIPPOV, N. B. KAZAKOV, and L. A. PROUNIN, Izv. vyshikh. chern. Met. **9**, 8 (1966); Chem. Abstracts **65**, 1410d (1966).
[15] P. A. EGELSTAFF, G. DUFFILL, V. RAINEY, J. E. ENDERBY, and D. M. NORTH, Phys. Letters **21**, 286 (1966).
[16] D. M. NORTH and C. N. J. WAGNER, J. appl. Cryst. **2**, 149 (1969).
[17] A. J. GREENFIELD, J. WELLENDORF, and N. WISER, Phys. Rev. A **4**, 1607 (1971).
[18] A. RAHMAN, J. chem. Phys. **42**, 3540 (1965).
[19] Y. WASEDA and K. SUZUKI, The Properties of Liquid Metals, Tokyo Conference, Taylor and Francis, London 1973 (p. 37).
[20] N. W. ASHCROFT and J. LEKNER, Phys. Rev. **145**, 843 (1966).
[21] Y. WASEDA and M. OHTANI, Sci. Rep. Res. Inst. Tôhoku University **24A**, 218 (1973).

(Received January 15, 1974)

phys. stat. sol. (b) 62, 547 (1974)

Subject classification: 13.1; 6; 14.3; 22.5.1

Physics Department, Purdue University, West Lafayette, Indiana

Temperature Dependence of the Cyclotron Resonance Absorption of Electrons in AgBr[1])

By

J. E. BAXTER and G. ASCARELLI

The temperature dependence of the cyclotron mass and mobility of electrons in zone-refined AgBr has been measured from 5 to 22 °K. The electron mobility, measured at 55 GHz, approaches 6×10^5 cm²/Vs at 5 °K ($\omega_c \tau \approx 30$). The measurements indicate that there are two types of zone-refined AgBr the difference arising from the relative electron and hole trap densities. The bulk of our experiments were performed on material where the electron traps are dominant. In this material the mass change with temperature is exponential with an activation energy (95 °K) nearly equal to the binding energy of the $n = 2$ bound state of the electron traps. The same 95 °K activation energy is also observed in the temperature dependence of the mobility. The exponential temperature dependence of both the mass and the mobility can be explained as arising from a resonant scattering of the electron by the traps with the aid of a virtual phonon. The AgBr where the hole traps dominate, on the other hand, shows a linear temperature dependence of the effective mass that is in agreement with what is expected of an electron moving in a polarizable continuum.

Nous avons mesuré la variation avec la température de la masse cyclotronique et de la mobilité des électrons dans le AgBr purifié par zones. Nous avons couvert l'interval de températures de 5 à 22 °K. A 55 GHz et à 5 °K la mobilité des électrons approche 6×10^5 cm²/Vs ($\omega_c \tau \approx 30$). Nos mesures indiquent qu'il y a deux types de AgBr: cibles où le nombres de pièges pour électrons est supérieur à celui pour les trous et vice versa. La plupart de nos mesures a été faite sur des cibles du premier type. Dans ces cibles la variation de la masse effective avec la température est exponentielle avec une énergie d'activation (95 °K) qui est presque égale à l'énergie de l'état $n = 2$ des électrons pièges. La même énergie est aussi observée dans la mobilité. La dependence de la température de la masse et de la mobilité peut être expliquée en supposant une diffusion resonante des électrons par les impuretés avec échange d'un phonon virtuel. Par contre dans le AgBr où les pièges pour trous sont dominants la masse effective des électrons dépend linéarement de la température comme prévu par la théorie du polaron.

1. Introduction

Few transport experiments [1] clearly demonstrate the existence of polarons; most evidence for a strong electron–phonon coupling arises from optical measurements [2 to 6].

To first order in the ionic displacements the electron–phonon interaction is described by the Fröhlich Hamiltonian [7]:

$$H = \frac{(\hbar k)^2}{2m_b} + \sum_q \hbar\omega_L a_q^+ a_q + \sum_q U_q a_q e^{i\mathbf{q}\cdot\mathbf{r}} + \text{c.c.}\,, \qquad (1)$$

where $U_q = \{i\hbar\omega_L(\hbar/2m_b\omega_L)^{1/4}\,(4\pi\alpha/v)^{1/2}\,(1/q)\}$ and the unit of length is $1/u = (\hbar/2m_b\omega_L)^{1/2}$. In the limit where perturbation theory is applicable the eigen-

[1]) Supported in part by the National Science Foundation.

values of equation (1) are [8, 9] to first order in the electron–phonon coupling constant α:

$$\left.\begin{aligned} E(k) &= \frac{(\hbar k)^2}{2m_b} + \alpha(n+1)\,\hbar\omega_L\left(\frac{u}{k}\right)\arcsin\frac{u}{k}\,,\\ E(k) &\approx \alpha(n+1)\,\hbar\omega_L + \frac{(\hbar k)^2}{2m}\left(1-\frac{\alpha(n+1)}{6}\right) - \frac{3}{40}\,\frac{(\hbar k)^4}{4m_b^2}\,\frac{\alpha(n+1)}{\hbar\omega_L}\,, \end{aligned}\right\} \quad (2)$$

where n is the thermal population of phonons whose energy is $\hbar\omega_L$.

It is easily seen that the last term in equation (2) gives rise to a non-parabolicity of the polaron band. Both the perturbation result as well as other approaches [11, 12] predict a linear change of the polaron mass with energy when $[(\hbar k)^2/2m_b] \ll \hbar\omega_L$. In case of electrons in equilibrium with the lattice this implies that the mass will change linearly with temperature.

The polaron mobility has been calculated as well [13, 14] and the results can be applied to both the silver and the alkali halides where the conduction band is spherical [15 to 18]. At low temperatures, in both the silver [13] and alkali halides [15], like in semiconductors [20], an electric field of a few V/cm gives rise to important hot electron effects. This phenomenon has been beautifully exploited [15] to detect cyclotron resonance (C.R.). Unfortunately the carrier distribution is unknown and from the experimental results it is extremely difficult to reach conclusions about the polaron band non-parabolicity.

The present communication reports cyclotron resonance measurements in the ohmic regime. The unexpected result is that scattering of the carriers by impurities contributes to the mass renormalization [21].[2]) These bound states have been studied previously by others [23, 24].

2. Experimental

Measurements were done at 55 GHz with a superheterodyne spectrometer whose sensitivity at 4.2 °K is $\approx 10^{11}$ spins/cm^3 G with an incident power of 1 μW [22]. From this we estimate that we can detect changes of electron density $10^3\ \text{cm}^{-3} \lesssim n \lesssim 10^8\ \text{cm}^{-3}$ the lower limit being for $\omega_c\tau \approx 20$ the upper limit being determined by the dynamic range of the spectrometer. On account of the small magnetic field modulation (MFM) depth available and its nearly unavoidable associated microphonic noise, MFM of the carrier mobility is a much less sensitive technique for detecting C.R.

As a result of both the relatively low electron densities used in the measurements ($n \lesssim 10^8\ \text{cm}^{-3}$) and the fact that the r.f. electric field is parallel to the sample surface plasma effects are negligible in these measurements.

Light sources used were a combination of a 500 W high pressure mercury arc whose IR had been blocked by a water filter and an incandescent lamp whose visible and UV emission were blocked by appropriate filters.

At temperatures $T \gtrsim 18$ °K where $\omega\tau \lesssim 5$ errors in the effective mass arise if one detects a mixture of the real and imaginary part of the signal reflected by the microwave cavity. CR measurements in rapidly quenched Si samples whose electron mobilities are comparable to those in AgBr for 5 °K $\lesssim T \lesssim$ $\lesssim$ 22 °K allow us to measure such phase errors as $\approx 5°$ when $\omega_c\tau \lesssim 3$; in Si this gives to an error in m^* of 1.5%.

[2]) For more details see paper [22].

By contrast for $T \lesssim 12$ °K errors in the electron effective mass in AgBr have a completely different origin: the electron CR signal decreases in magnitude and is superposed on a much larger non-resonant signal due to free holes. A small fluctuation of their density, produced e.g. by changes of light intensity, and the necessity of using microwave powers $P \lesssim 10^{-7}$ W to avoid hot electron effects were the main sources of errors.

Samples of two different sources were used; most originated from two ingots of zone refined AgBr kindly given to us by J. E. Luvalle previously of Fairchild Corporation while some measurements were taken on zone refined samples given to us by F. Moser of the Kodak Research Laboratories. The latter are similar to material used by Tamura [25].

In both cases the samples were cut with a SiC saw, etched in HBr and annealed for 36 h at 390 °C in a purified flowing He atmosphere. Annealing removes strains detectable by crossed polaroids and increases the electron lifetime below 16 °K. Electron lifetimes where longer in the Fairchild samples than in the Kodak samples. At the very low powers used we had difficulties measuring C.R. in the latter below ≈16 °K on account of the small sample sizes available. In the range 16 °K $\lesssim T \lesssim$ 22 °K our results agree with those of [25].

3. Experimental Results

3.1 Thermal equilibrium results

Experimental traces (not to scale) of the electron cyclotron resonance signal at different temperatures for Fairchild AgBr are given in Fig. 1. Two aspects of the results are striking: the asymmetry of the cyclotron resonance line and the shift of the resonance peak with temperature. Neither of these effects are observed in a Si sample that was thermally quenched from 1200 °C in LN. Its electron mobility is comparable to that of electrons in AgBr. Such asymmetry of the cyclotron resonance line is present even when cyclotron resonance is carried out in the dark (above ≈14 °K) using electrons thermally excited from previously filled shallow traps [23, 24] (using MFM). An indication of the asymmetry of the cyclotron resonance line is given in Fig. 2 by the width at half maximum of the cyclotron resonance line on the side in which the magnetic field is respectively higher or lower than that corresponding to the maximum cyclotron resonance absorption. Finally, in Fig. 3 we show the observed change of mass, mobility, and carrier conductivity (number of photocarriers at constant

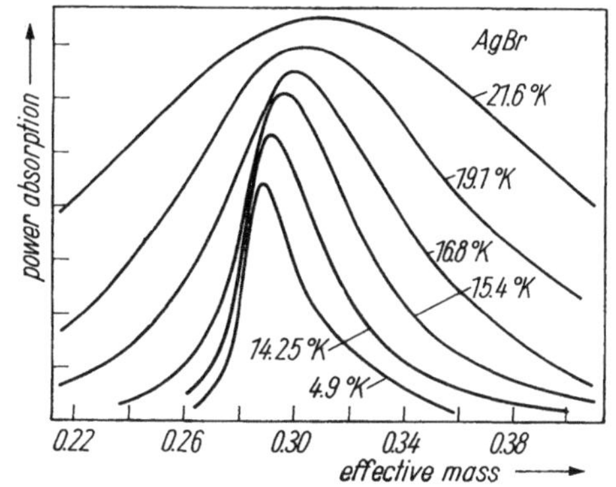

Fig. 1. The temperature dependence of the cyclotron resonance absorption in AgBr. The relative magnitudes of the lines are not to scale but have been chosen for drafting convenience

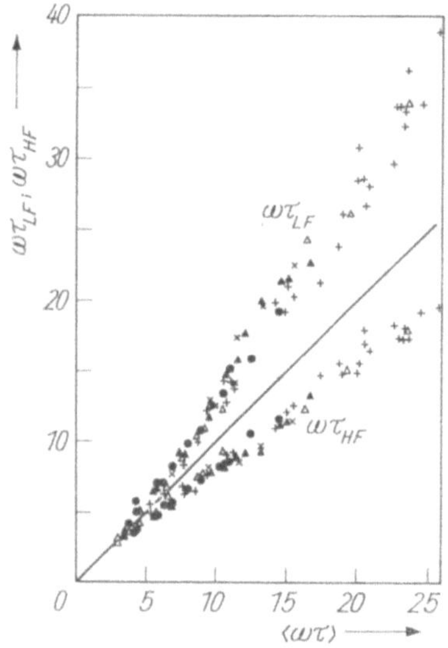

Fig. 2. The half widths on the high and low magnetic field side of the cyclotron resonance line plotted as a function of the mean half width of the observed cyclotron resonance line. The half width is varied by changing temperature.
× No. 3; ● No. 4; ▲ No. 5; △ No. 6; + No. 8

light intensity times their mobility) as a function of temperature. Contrary to what is expected from polaron theory in perfect crystals (equation (2)) the mass does not change linearly with temperature but exponentially with an activation energy near 95 °K. At low temperatures ($T \lessapprox 12$ °K) both the effective mass and the mobility are strongly dependent on the magnitude of the microwave electric field used in the measurements. The values representative of the material can only be obtained by an extrapolation to zero microwave electric field. It is interesting to notice that both the mobility (as measured by the change of width of the resonance) and the change of mass are dependent on the individual sample and its history. The numbers 3 through 8 in Fig. 3 indicate different Fairchild samples. The Kodak samples give rise to a much smaller change in mass. Their change of mass [25] is close to the line obtained from equation (2) or some of its theoretical variants. Our observation that the changes of mass of the Fairchild samples are sample dependent indicates an extrinsic effect representative of either impurities or intrinsic defects.

The relative electron density as a function of temperature is calculated by measuring the area under the cyclotron resonance absorption obtained when the sample is either subject to constant illumination or when the sample is in the dark (after intense illumination) and the carriers are thermally excited from the

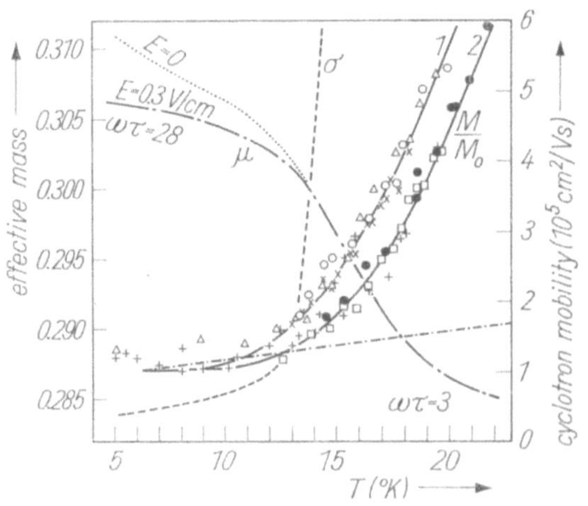

Fig. 3. Temperature dependence of the cyclotron resonance parameters in AgBr. The electron conductivity (σ) scale is in arbitrary units while the mobility (μ) is obtained from the width of the cyclotron resonance line. AgBr: ○ No. 2; × No. 3, ● No. 4, △ No. 6, □ No. 7, + No. 8.
(–·–·– after equation (2))

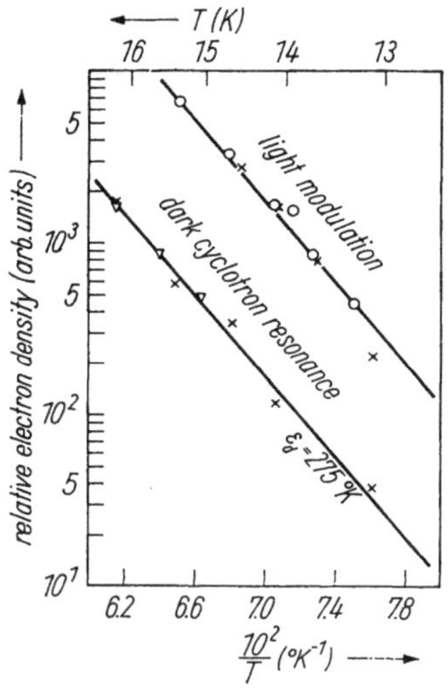

Fig. 4

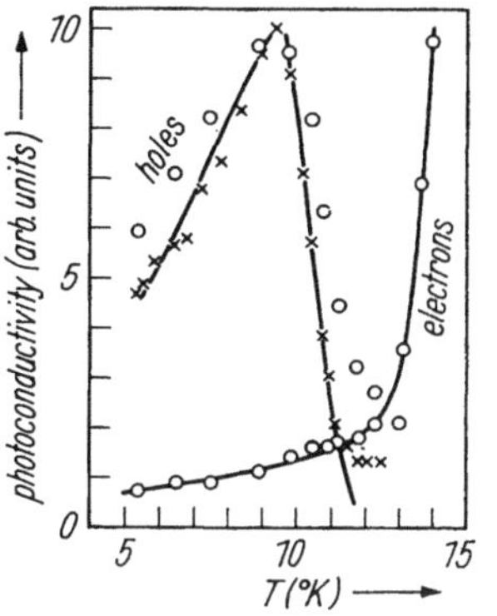

Fig. 5

Fig. 4. Temperature dependence of the electron density as measured from the area of the cyclotron resonance absorption. Dark cyclotron resonance indicates the data obtained with magnetic field modulation of the cyclotron absorption of the electrons thermally excited out of the traps. AgBr: × No. 1; ○ No. 2; ▽ No. 8

Fig. 5. Temperature variation of the photosignal due to both electrons (C.R. absorption) and holes. The light intensity is kept constant. AgBr: ○ No. 3; × No. 8

traps. In both cases (Fig. 4) the results indicate the same activation energy of (275 ± 50) °K a result that is consistent with the binding energy of the shallow traps measured by Brandt and Brown [23]. The temperature range over which the measurement can be carried out is limited on the low temperature side by small signals and by the critical importance of the density of holes. At high temperatures phase errors, temperature stability and the accuracy of the temperature measurement were the limiting parameters. The results of Fig. 4 indicate that the traps are initially filled by bandgap illumination to the point where all N_A hole traps and a large fraction of the N_D electron traps (whose number is larger than the hole traps) are filled. In this situation the sample behaves like a partially compensated n-type semiconductor. When all the N_A hole traps are filled (i.e. $T \lessapprox 10$ °K) any other photo-hole must be free and contribute, to a non-resonant background whose magnitude is determined primarily by the free hole–bound electron recombination cross section and the value of $(N_D - N_A)$. The experimental evidence is indeed that such a non-resonant background exists for $T \lessapprox 12$ °K. It disappears when the temperature is increased with a de-activation energy that is equal to the activation energy of the electrons (Fig. 5). We interpret this result as an indication that the electron trap density is larger than the hole trap density but that most of the photo-holes remain trapped. We accept other authors estimates [23, 24] for $N_D \approx 10^{16}$ cm^{-3}, and estimate that with the optical fluxes available the number of free electrons are $\lessapprox 10^7$ cm^{-3} at $T \approx 12$ °K.

Another consequence of the presence of N_A hole traps is that the measured thermal activation energy of the electron traps should be equal to the optically

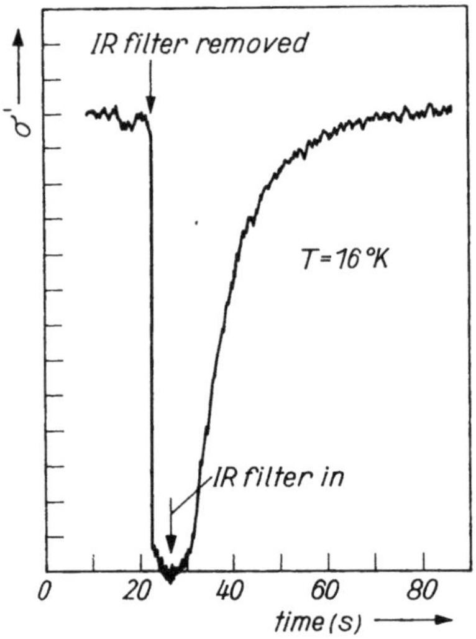

Fig. 6. Quenching of the photoconductivity by near IR light

measured [23] series limit (23.8 meV) of their excitation spectrum just like in a compensated semiconductor. This is experimentally verified. In the absence of compensating traps the thermal activation energy of the electron traps would instead be half that of the optical result.

From the above considerations we conclude that the recombination cross-sections of an electron (e) and a hole (h) are such that $\sigma_{\text{fh-fe}} > \sigma_{\text{fh-be}}$, where the subscripts f and b stand respectively for free and bound. We speculate that the complex kinetics of the decay of the population of the shallow traps [23] reflects the different possible fates of the electron i.e. it is excited out of a trap and either recombines with a hole (free or bound) or it is retrapped in another trap. Tunneling between electron and hole traps is also possible.

Preliminary measurements of the quenching of the electron photoconductivity by infrared light indicate that the hole traps have a depth of $\sim$0.8 eV. The quenching of the photoconductivity spectrum is complex and we postpone its discussion to a successive publication. Near-infrared light ($h\nu \lessapprox 1$ eV) also enhances the low temperature ($T \lessapprox 10$ °K) non-resonant background assigned to free holes. We have not observed this non resonant background in the Kodak samples. In the Fairchild samples the background is sample dependent and it is sensitive to the stress state of the sample. An indication of the efficiency of the infrared quenching of the photoconductivity is given in Fig. 6. An optical absorption at 0.88 eV was observed in V^{3+} or Cr^{3+} doped AgBr by Ulrici [26] and was attributed to hole traps. Sadykova et al. observed that the illumination of AgBr emulsions with near IR enhanced the radiative electron hole recombination. Both effects are different manifestations of the presence of N_A hole traps.

3.2 Hot carrier effects

The original observation of mass changes of the electron polaron with electron temperature were observed in connection with high field effects [15]. High

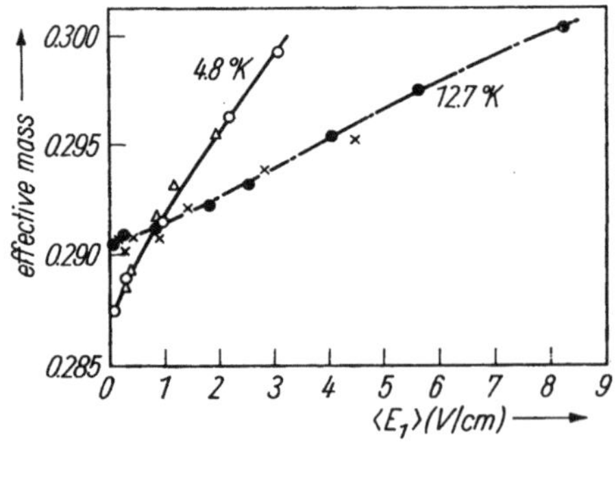

Fig. 7. Microwave electric field dependence of the effective mass of electrons in AgBr, No. 8.

Fig. 8. Comparison of the relation of mobility vs. mass for the case when the carriers are heated by increasing the lattice temperature (○) or by absorbing microwave power (△). The corresponding data of quenched Si are given in order to give an indication of the errors

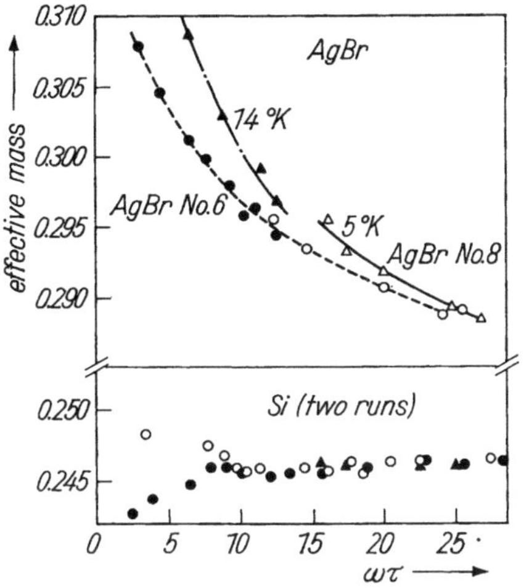

electric fields that accellerate the electrons to a kinetic energy above thermal equilibrium change both the electron mass [15] and the electron mobility [13]. Such hot carrier effects are in fact used in detecting cyclotron resonance in cross modulation experiments [15, 20].

Data indicating such an effect with relatively modest fields is shown in Fig. 7. A given effective mass may be reached either by heating the carriers or by heating the lattice or a combination of both. The electron distribution is not the same in these cases. In order to emphasize this difference, we plot in Fig. 8 the measured effective mass as a function of the corresponding mobility (obtained from the half width of the cyclotron resonance line) when the carrier temperature is changed by either heating the lattice or by heating the carrier. No corrections are made to the experimental data to take into account the different power absorbed at the cyclotron resonance maximum and at the half power point. The principal observation to be made is that while the thermal equilibrium data follow a smooth curve the high field data follow a curve (or possibly two curves if the data are taken at two lattice temperatures) that is different from that obtained in conditions of thermal equilibrium. It is to be noticed that no mass changes are apparent for electrons in Si with either high fields or changes in lattice temperature although the mobilities were similar to those of electrons in AgBr. This indicates that the observed changes in AgBr are not instrumental. As is usual in the case of hot electron effects, when the energy loss mechanism is primarily due to L A phonons, the mobility approaches a value that varies [28] as the inverse of the square root of the electric field ($E^{-1/2}$). This is indeed the case for electrons in AgBr (as well as Si) at rather moderate electric fields (Fig. 9). The same observation had previously been made in Hall measurements [19] in AgCl.

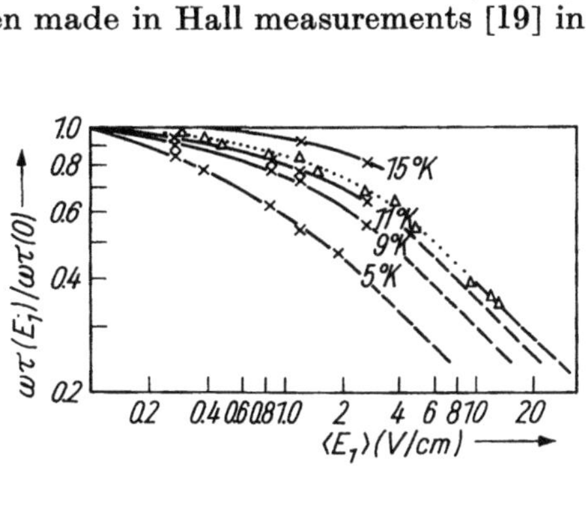

Fig. 9. Microwave electric field dependence of the mobility at different lattice temperatures. × AgBr, △ Si, 4.8 °K; - - - - $E^{1/2}$ dependence

We conclude that at our present level of knowledge it is difficult to obtain detailed information on the polaron dispersion from hot carrier measurements insofar as the electron distribution is in all likelihood not Maxwellian and at present unknown.

4. Interpretation of the Experimental Mass Shift and Mobility

The activation energy for the mass change (95 °K) coincides with the energy of the acoustic phonon (8 meV) necessary for momentum conservation in the indirect optical absorption in AgBr [29]. This phonon has a $\boldsymbol{k}$-vector that corresponds to the zone bondary along $\langle 111 \rangle$.

Although the effective mass approximation is not precise on account of central cell corrections [30] even for the ground state of group V impurities in homopolar semiconductors, it gives excellent results for the excited states. Similar results are expected for the shallow traps in AgBr provided that the binding energy of the state is smaller than a LO phonon energy [31]. Using a dielectric constant [32] of 10.6 and a polaron effective mass (m_p) equal to $0.287 m_0$ we calculate a binding energy of the $n = 2$ state of the shallow electron trap equal to 8.7 meV. A dielectric constant of 11.5 (at 55 GHz we measure $K_0 = 11.5 \pm 0.3$) predicts a binding energy of 7.4 meV. In either case the difference of energy between the ground state of the trap (23.8 meV) and the first excited state is extremely close to the energy of a LO phonon (17.2 meV). Such a near resonance is expected to slightly shift the energy of the excited state and decrease the lifetime of any electron populating it [33].

In the case in which a conduction band polaron in state $|k\rangle$ is resonantly trapped into a bound state $|b\rangle$ of an isolated impurity with the emission of a phonon $(\hbar\omega(q\mu))$ and successively re-emitted into the conduction band by reabsorbing the same virtual phonon, the energy of the perturbed state is [21]

$$W = \frac{(\hbar k)^2}{2m_p} + \sum_{q,\mu} \frac{|\beta_{q\mu}|^2 \, (n_{q\mu} + 1)}{W + E_b - \hbar\omega(q\mu) + \frac{1}{2} i\Gamma} \tag{3}$$

from which one calculates [21]

$$(m^*(T) - m(0))/m(0) = \left(\frac{A}{T}\right) \exp\left(-\frac{E_b}{kT}\right), \tag{4}$$

where

$$\langle b, n_{q\mu} + 1 | \, H_1 \, | k, n_{q\mu} \rangle = \beta_{q\mu} (n_{q\mu} + 1)^{1/2} \, . \tag{5}$$

H_1 is the perturbation Hamiltonian that describes the trapping of free polarons in the bound state $|b\rangle$, $n_{q\mu}$ is the population of phonons whose energy is $\hbar\omega(q, \mu)$, E_b the energy of the bound state of the impurity and $\hbar/\Gamma$ its lifetime.

Both the case of a free electron whose kinetic energy is $\hbar\omega_0$ above the lowest free polaron state, and the case of a free electron whose energy differs from that of a bound state by the energy of a phonon are similar, provided that in the latter case the electron is sufficiently strongly coupled to the phonon and that the electron is sufficiently close to the impurity to permit interaction.

A longitudinal phonon at the zone boundary along $\langle 111 \rangle$ describes a motion in which either the positive (or the negative) ions oscillate around their position of equilibrium while the negative (or positive) ions remain fixed [17]. The dipole moment created by such a motion is similar to that created by LO phonons and we expect that the coupling of an electron to this vibration will be

Fig. 10. Temperature dependence of the cyclotron resonance mobility. AgBr: △ No. 8, × No. 4

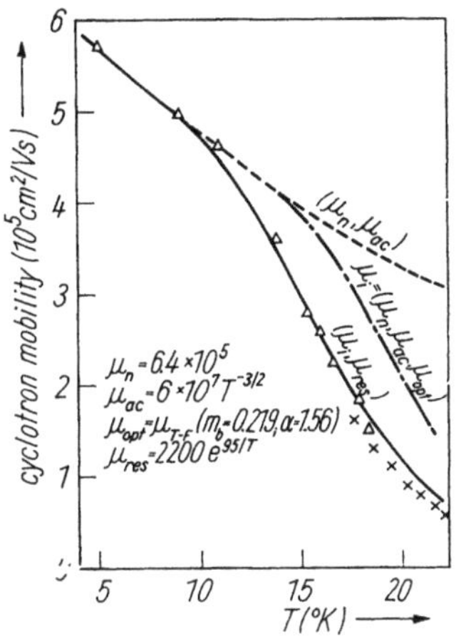

similar to that when a LO phonon is involved. This interaction is not important in the case of free electrons since it would not be possible to conserve crystal momentum if such a phonon would be emitted. This difficulty does not exist near an impurity due to the destruction of the lattice periodicity.

A slow electron near an impurity with an empty bound state can emit a phonon just like an energetic electron in free space can emit an LO phonon [34, 35]. In the former case like in the latter, the energy momentum relation is expected to be modified. In the present model the resonant interaction is turned on only when the electron is sufficiently near the impurity.

The effect expected in the present case at 0 °K should be negligible compared with the large effect [34, 35] that is expected in the case of an electron whose kinetic energy is $\hbar\omega_{LO}$ that propagates in a lattice at 0 °K since the number of impurity ions is $\approx 10^{-6}$ smaller than the number of ions of the lattice. The calculation shows [21] that the temperature independent resonant renormalization is small compared with the temperature dependent term. The latter increases with temperature much more rapidly than what is expected for cold polarons in a perfect lattice because $\omega_{q\mu} < \omega_{LO}$ and accordingly $n_{q\mu} \gg n_{LO}$.

It can be easily shown that the imaginary part of the self energy [21] obtained from equation (3) will vary exponentially with temperature as $\exp(-E_b/kT)$. This term describes an additional contribution to the mobility, i.e.

$$\mu_{res} \approx B \exp\left(\frac{E_b}{kT}\right). \tag{6}$$

In Fig. 10 we present an analysis of the mobility obtained from the half width of the cyclotron resonance curve. We assume that at very low temperature the mobility is limited by both neutral impurity scattering [36] and acoustical phonon scattering assumed to be the same for all Fairchild samples. The absence of a predominant ionized impurity scattering mechanism is supported by the fact that at all the temperatures studied (4 °K $< T <$ 22 °K) the mobility monotonically *decreases* with the application of progressively higher microwave fields (Fig. 8). If ionized impurity scattering would be dominant we would expect the mobility to initially increase with increasing microwave electric field. This is contrary to experimental evidence. From Hall measurements [24] it is known that LO phonon scattering determines the mobility at high temperature ($T \gtrsim 50$ °K). The high temperature scattering is well described by the Thornber-Feynman formula [13].

The total mobility should therefore be

$$\frac{1}{\mu} = \frac{1}{\mu_n} + \frac{1}{\mu_{ac}} + \frac{1}{\mu_{LO}} + \frac{1}{\mu_{res}}. \tag{7}$$

The last two terms in equation (7) should be negligible for $T \lessapprox 10$ °K. The mobility obtained from the first three terms in equation (7) is shown in Fig. 10, where it is obvious that it does not completely describe the experimental results. Introduction of the resonant scattering term whose exponent is the same as that necessary to describe the changes of mass gives good agreement between experiment and the phenomenological fit.

Further support of the resonant scattering mechanism is given by a comparison of the experimental results for coefficients B in equation (6) and A in (4). If the mechanism determining the mass renormalization and the mobility are the same the product AB should be sample independent. Although there is an appreciable scattering in the experimental data the weighted average of the experimental results for various samples is $AB = (55 \pm 14)\,10^3$ indicating a notable correlation between resonant scattering mobility and mass changes.

Table 1

Value of α and m_b for AgBr deduced from the extrapolated polaron mass, $m_p(0) = 0.2875 m_e$, $\varepsilon_s = 10.6$, $\varepsilon_\infty = 4.68$

polaron	α	m_b (units of m_e)
perturbation solution to Fröhlich Hamiltonian	1.54	0.214
LLP solution to Fröhlich Hamiltonian [8]	1.59	0.227
LLP solution (Larsen) [12]	1.56	0.218
Feynman solution (Langreth) [13, 39]	1.56	0.219

Finally in Table 1 the calculated coupling constants and band masses are given for different polaron theories. The values of α obtained in this fashion are in excellent agreement with those obtained from the direct measurement of the self energy and its temperature dependence [29].

We may speculate on the origin of the shallow traps that appear to exist in all the AgBr samples that have been studied both by us and by others [23, 24]. Similar traps appear in AgCl and possibly in the alkali halides [23, 15]. In the latter they seem to be related to excited states of the F-center [15].

We were not able to detect an EPR signal associated with these traps when they were either empty or filled. Their density is supposedly large ($\approx 10^{16}$ cm^{-3}) in comparison with our sensitivity limits. We must conclude that their EPR line is either extremely broad or extremely sensitive to saturation. If the negative result would be due to saturation it appears strange that no trace of an EPR signal was observed in either absorption or dispersion over a temperature range from 2 to ≈ 30 °K. We are therefore tempted to assign the negative result to an exceedingly broad line of the kind that is expected for shallow acceptors in Si or Ge [30]. Based on this hypothesis we believe that a likely model for the shallow traps would be intestitial Ag that precipitates in dislocations. Such a model would explain the sensitivity of the samples to the state of stress and the importance of careful annealing. Possible hole traps could arise on the basis of the model proposed by Palma et al. [37] i.e. Ag^+ vacancies. The relative importance of these two traps would then be sensitive to the concentration of impurities (e.g. divalent ions) relative to the number of silver interstitials that are frozen in during the annealing process. Since both numbers are comparable it would be easy to have n- or p-type samples depending on the details of fabrication. A mass spectrometric analysis of both the Kodak and the Fairchild samples did not permit any conclusion.

5. Conclusions

We have measured the cyclotron mass of electrons in AgBr and found that the temperature dependence of the mass is sample dependent. From a comparison of the activation energy of the mass with that of the electron traps found in AgBr samples from different origins [23, 24] we have identified these traps as the source of the resonant electron–phonon interaction responsible for the extrinsic exponential temperature dependence of the mass. The electron mobility approaches 6×10^5 cm^2/Vs in some of our samples and is affected by the same scattering mechanism that gives rise to mass renormalization in the same temperature interval (10 to 22 °K). Free holes have been observed for $T \lessapprox 10$ °K indicating that in AgBr the hole is not self trapped[3]) but that it can be found in extrinsic traps whose binding energy is $\approx$0.8 eV. This binding energy is close to the peak of the optical absorption observed by Ulrici [26] and assigned to hole traps.

Acknowledgements

The authors are grateful to Dr. J. E. Luvalle presently at the S.C.M. Corporation and Dr. F. Moser of the Kodak Research Laboratory that provided them with the samples that made this research possible. The authors are grateful to Prof. S. Rodriguez who made the calculations of the temperature dependence of the mass renormalization with which experiments were compared. The authors thank Prof. F. C. Brown, Prof. M. U. Palma, and Prof. H. Tamura for communication of their results before publication.

References

[1] J. Appel, Solid State Phys. **21**, 193 (1968).
[2] R. S. Knox, Solid State Phys., Suppl. **5**, 152 (1962).
[3] J. J. Markham, Solid State Phys., Suppl. **8** (1966).
[4] C. J. Summers, R. B. Dennis, B. S. Wherret, P. G. Harper, and S. D. Smith, Phys. Rev. **170**, 755 (1968).
[5] B. D. McCombe and R. Kaplan, Phys. Rev. Letters **21**, 756 (1968).
[6] E. I. Johnson and D. M. Larsen, Phys. Rev. Letters **16**, 655 (1966).
[7] H. Fröhlich, in: Polarons and Excitons, Ed. G. C. Kuper and G. D. Whitfield, Oliver and Boyd, London 1963 (p. 1).
[8] D. Pines, in: Polarons and Excitons, Ed. G. C. Kuper and G. D. Whitfield, Oliver and Boyd, London 1963 (p. 33).
[9] T. Fulton, Phys. Rev. **103**, 1712 (1956).
[10] Y. Osaka, J. Phys. Soc. Japan **21**, 423 (1966).
[11] R. P. Feynman, Phys. Rev. **97**, 660 (1955).
[12] D. M. Larsen, Phys. Rev. **137**, A760 (1965).
[13] F. C. Brown, in: Defects in Solids, Ed. J. E. Crawford, to be published.
[14] K. K. Thornber and R. P. Feynman, Phys. Rev. B **1**, 4099 (1970).

[3]) Tamura and Masumi [38] detected cyclotron resonance of holes in Kodak material.

[15] J. W. HODBY, in: Polarons in Ionic Crystals and Polar Semiconductors, Ed. J. T. DEVREESE, North-Holland Publ. Co., Amsterdam 1972.
[16] H. H. TIPPENS and F. C. BROWN, Phys. Rev. **129**, 254 (1963).
[17] F. BASSANI, R. S. KNOX, and W. BEAL FOWLER, Phys. Rev. **137**, A1217 (1965). P. M. SCOP, Phys. Rev. **139**, A934 (1965).
[18] A. B. KUNZ, phys. stat. sol. **29**, 115 (1968); Phys. Rev. **175**, 1147 (1968).
[19] T. MASUMI, Phys. Rev. **129**, 2564 (1963).
[20] B. LAX and J. G. MAVROIDES, Solid State Phys. **11**, 261 (1960).
[21] J. E. BAXTER, G. ASCARELLI, and S. RODRIGUEZ, Phys. Rev. Letters **27**, 100 (1971). S. RODRIGUEZ, in: Polarons in Ionic Crystals and Polar Semiconductors, Ed. J. T. DEVREESE, North-Holland Publ. Co., Amsterdam 1972.
[22] J. E. BAXTER, Thesis, Purdue University, 1972 (unpublished).
[23] R. C. BRANDT and F. C. BROWN, Phys. Rev. **181**, 1242 (1969).
[24] D. C. BURNHAM, F. C. BROWN, and R. S. KNOX, Phys. Rev. **119**, 1560 (1960). D. C. BURNHAM, Thesis, University of Illinois, 1959 (unpublished).
[25] H. TAMURA, Solid State Commun. **10**, 297 (1972).
[26] W. ULRICI, phys. stat. sol. **40**, 557 (1970).
[27] A. A. SADYKOVA, L. N. ITSKOVICH, and P. V. MEIKLYAR, Optika i Spektroskopiya **30**, 54 (1971).
[28] W. SHOCKLEY, Bell Syst. tech. J. **30**, 990 (1951).
[29] G. ASCARELLI and J. E. BAXTER, Solid State Commun. **10**, 315 (1972).
[30] W. KOHN, Solid State Phys. **5**, 257 (1957).
[31] D. M. LARSEN, Phys. Rev. **180**, 319 (1969).
[32] R. P. LOWNDES, Phys. Letters **21**, 26 (1966).
[33] S. RODRIGUEZ and T. D. SHULTZ, Phys. Rev. **178**, 1252 (1969).
[34] D. M. LARSEN, Phys. Rev. **144**, 697 (1966).
[35] G. D. WHITFIELD and R. PUFF, Phys. Rev. **139**, A338 (1965).
[36] C. ERGINSOY, Phys. Rev. **79**, 1013 (1950).
[37] L. CORDONE, S. L. FORNILI, S. MICCIONCIO, and M. U. PALMA, Phys. Rev. Letters **26**, 135 (1971) and private communication.
[38] H. TAMURA and T. MASUMI, Solid State Commun. **12**, 1183 (1972).
[39] D. M. LARSEN, Phys. Rev. **137**, A760 (1965).

(Received January 15, 1974)

phys. stat. sol. (b) 62, 559 (1974)

Subject classification: 13.4; 21.1; 21.1.1

Department of Metallurgical Engineering, Tokyo Institute of Technology, Ookayama, Meguro, Tokyo

Interaction between a Vacancy and Impurity Atoms

By

K. MASUDA

The vacancy–impurity interaction energy is calculated on the Wolff-Clogston model using the Hartree-Fock approximation. It is suggested that if a bound state exists below the band bottom and if most of the excess electrons are localized at the impurity sites, the interaction energy tends to become negative. There is a qualitative agreement between the theory and the experimental results in Fe base alloys with 3d-transition metal impurities. The interaction due to the magnetic origin inherent in the host metal structure is also taken into account using the simple Stoner model.

Mit der Hartree-Fock-Näherung wird auf der Grundlage des Wolff-Clogston-Modells die Leerstellen–Fremdstörstellen-Wechselwirkungsenergie berechnet. Es wird gezeigt, daß die Wechselwirkungsenergie zu negativen Werten tendiert, wenn unterhalb der Bandkante ein gebundener Zustand existiert und wenn der größte Teil der überschüssigen Elektronen an den Störstellenplätzen lokalisiert ist. Es existiert eine qualitative Übereinstimmung zwischen der Theorie und experimentellen Ergebnissen in Eisenlegierungen mit 3d-Übergangsmetallstörstellen. Die Wechselwirkung magnetischen Ursprungs, die von der Wirtskristallstruktur herrührt, wird mit dem einfachen Stonerschen Modell ebenfalls berücksichtigt.

1. Introduction

Two main types of theoretical approaches have been used for the study of point defects in metals. One is based on the pseudo-potential theory which can deal with metals with spherical energy surfaces [1 to 4]. The other uses the Green's function method, which seems to offer the appropriate basis for treatment of metals with arbitrary energy surfaces [5 to 8].

However, the work reported so far for the transition metals and belonging to the latter category has been limited to the study of impurities [9 to 10]. Thus, the present study was conducted to calculate the interaction energy between a vacancy and an impurity atom in transition metals with the Green's function method. The interaction due to the magnetic origin was also evaluated. In the following calculations, we will use the Hartree-Fock approximation starting from the Wolff-Moriya Hamiltonian [11] to obtain the Green's function for the system containing two defects. Our discussion will refer explicitly to the numerical calculations made on a tight-binding band of a simple cubic lattice, in order to see the general features of the interaction, which result more or less independently from the initial models. As a possible application of the present theory, we will estimate the binding energies between a vacancy and impurity atoms in the Fe base alloys, with 3d transition metal impurities. For simplicity, we will consider only the s-symmetry solution concerning vacant sites. One can expect, however, that the obtained results are general with respect to the shape of the energy surface. Some experiments concerning the impurity diffusion in the iron base alloys will be examined qualitatively in the light of the present calculation of the vacancy–impurity by assuming that the difference in the activation energy between the self-diffusion and the impurity diffusion [12] merely reflects the binding energy between the vacancy and the impurity.

2. Hamiltonian and Green's Function

We consider two impurities or defects, at positions $\boldsymbol{R}_1$ and $\boldsymbol{R}_2$. The point defect in metals is described by a perturbing potential $V(r)$, which is considered to be both distance- and spin-dependent (magnetic case).

However, if the Fermi surface has a large enough cross-section or the density of states at the Fermi level is large enough, it is expected that the long-range oscillatory fringes of $V(r)$ are small compared to the central atomic part. This condition applies for a transition metal impurity in a transition metal. Therefore it is reasonably assumed in this case that the matrix element of $V(r)$ to be considered is only the diagonal one, connecting the two Wannier functions localized at the impurity [13].

The perturbing potential due to a vacancy is assumed to extend to the nearest-neighbour sites of the vacant site, so that Friedel's condition may be satisfied. For even an infinitely high repulsive potential, localized at the defect site, is not able to repel enough electrons to satisfy the charge neutrality condition. Thus, we can generally write the Hamiltonian for the two-defect case as[1])

$$\begin{aligned}\mathcal{H} = {}& \sum_\sigma \sum_j \sum_l T_{jl} c_{j\sigma}^+ c_{l\sigma} + \sum_j U_j^{(0)} c_{j\uparrow}^+ c_{j\uparrow} c_{j\downarrow}^+ c_{j\downarrow} + E_{00}^{(1)} \sum_\sigma c_{1\sigma}^+ c_{1\sigma} + \\ & + U^{(1)} c_{1\uparrow}^+ c_{1\uparrow} c_{1\downarrow}^+ c_{1\downarrow} + E_{11}^{(1)} \sum_\sigma \sum_\varrho c_{1+\varrho,\sigma}^+ c_{1+\varrho,\sigma} + E_{00}^{(2)} \sum_\sigma c_{2\sigma}^+ c_{2\sigma} + \\ & + U^{(2)} c_{2\uparrow}^+ c_{2\uparrow} c_{2\downarrow}^+ c_{2\downarrow} + E_{11}^{(2)} \sum_\sigma \sum_\varrho c_{2+\varrho,\sigma}^+ c_{2+\varrho,\sigma}\,, \end{aligned} \tag{2.1}$$

where a single-band tight-binding model is assumed to represent the 3d bands. The sum of the first and the second term is the Hamiltonian, H_0, for the unperturbed host metal. $c_{j\sigma}^+$ and $c_{l\sigma}$ are the creation and annihilation operators for an electron with spin σ on the j-th and l-th sites, and $c_{d+\varrho,\sigma}^+$ and $c_{d+\varrho,\sigma}$ $(d = 1, 2)$ are those for the one of the nearest-neighbour sites of the defect site. T_{jl} is the transfer integral between the Wannier orbitals on the j-th and l-th sites and is connected with the energy eigenvalues, $\varepsilon_{\boldsymbol{k}}$, by the relation

$$\varepsilon_{\boldsymbol{k}} = \sum_j T_{jl} \exp\left[i\boldsymbol{k} \cdot (\boldsymbol{R}_j - \boldsymbol{R}_l)\right].$$

Table 1

The assumed values of the Coulomb integral

	Ti	V	Cr	Mn	Fe	Co	Ni
U	0.86	0.96	1.06	1.16	1.31	1.41	1.46
$U^{(2)}$	−0.45	−0.35	−0.25	−0.15	0.00	0.10	0.15

$U = (U^{(0)} + U^{(2)})$, and $U^{(2)}$ are shown in the unit of a half the band width of the host metal.

[1]) The Hamiltonian in (2.1) can be used for the divacancy situation in transition metals. Therefore, the last term can be ignored if defect 1 and defect 2 are, respectively, a vacancy and a transition metal impurity atom from the previously mentioned reason.

The rest of (2.1) is the perturbation, V^σ, due to the two defects situated on $\boldsymbol{R}_1$ and $\boldsymbol{R}_2$. $E_{00}^{(d)}$ and $E_{11}^{(d)}$ ($d = 1, 2$) stand for the perturbing potentials of the defect centre and its nearest-neighbour site, respectively. $U_j^{(0)}$ is the intra-atomic Coulomb integral for the host metal (present case: Fe) and $U^{(2)}(U^{(1)})$ represents the change of its value for the impurity atom (defect site). Consequently, the intra-atomic Coulomb integral for the impurity atoms is given by $U^{(\mathrm{im})} = U^{(0)} + U^{(2)}$. It is expected that $U^{(2)}$ is small compared with $U^{(0)}$ in the case of the impurities in the same transition series as shown in Table 1. The symbol ϱ includes the sum of all the nearest-neighbour lattice sites.

In the tight-binding limit for a simple cubic lattice, the eigenvalue of the unperturbed Hamiltonian with the wave vector $\boldsymbol{k}$ is given by

$$\varepsilon_{\boldsymbol{k}} = -2T(\cos k_x a + \cos k_y a + \cos k_z a) , \tag{2.2}$$

where we have omitted the term which does not depend on the wave vector. In order to investigate the interaction between the defects, we start with the resolvent form of the one-particle Green's function given by

$$G^\sigma(\varepsilon + is) = (\varepsilon - H_0^\sigma + is)^{-1} + (\varepsilon - H_0^\sigma + is)^{-1} V^\sigma(\varepsilon - \mathcal{H}_0^\sigma + is)^{-1} . \tag{2.3}$$

The matrix element of the Green's function is obtained from (2.1) and (2.3) in the localized state representation within the Hartree-Fock approximation as

$$G_{jl}^\sigma = g_{jl} + g_{j1} V_{00}^{(1)\sigma} G_{1l}^\sigma + g_{j2} V_{00}^{(2)\sigma} G_{2l}^\sigma + \sum_\varrho g_{j,1+\varrho} V_{11}^{(1)\sigma} G_{1+\varrho,j}^\sigma + \\ + \sum_\varrho g_{j,2+\varrho} V_{11}^{(2)\sigma} G_{2+\varrho,l}^\sigma . \tag{2.4}$$

Here $V_{00}^{(i)\sigma}$ and $V_{11}^{(i)\sigma}$ are the Hartree-Fock potential of the i-th impurity and its nearest-neighbour sites, and are given by

$$\left.\begin{aligned} V_{00}^{(1)\sigma} &= E_{00}^{(1)} + (U^{(0)} + U^{(1)})\langle n_{1,-\sigma}\rangle , & V_{11}^{(1)\sigma} &= E_{11}^{(1)} + U^{(0)}\langle n_{1+\varrho,-\sigma}\rangle , \\ V_{00}^{(2)\sigma} &= E_{00}^{(2)} + (U^{(0)} + U^{(2)})\langle n_{2,-\sigma}\rangle , & V_{11}^{(2)\sigma} &= E_{11}^{(2)} + U^{(0)}\langle n_{2+\varrho,-\sigma}\rangle \end{aligned}\right\} \tag{2.5}$$

and g_{jl} is the Green's function for the unperturbed lattice:

$$g_{jl}(\varepsilon + is) = N^{-1} \sum_{\boldsymbol{k}} (\varepsilon + is - \varepsilon_{\boldsymbol{k}} - U^{(0)}\langle n_{-\sigma}\rangle)^{-1} \exp[i\boldsymbol{k}\cdot(\boldsymbol{R}_j - \boldsymbol{R}_l)] . \tag{2.6}$$

In (2.5) $\langle n_{i,\sigma}\rangle$ is the number of localized electrons with spin σ of the i-th atom.

The calculation of the diagonal element of the Green's function from (2.4) is tedious, but it is needed in order to discuss the excess magnetic moments, spatial distribution of the electrons around impurity atoms, and the interaction between the two defects. As shown in the following section, therefore, the approximated solution will be used for the diagonal component for the two-defect case.

For the sake of later convenience we will use the following notations interchangeably; i.e.,

$$g_{jl}(\varepsilon + is) = g_{j_1-l_1,j_2-l_2,j_3-l_3}(\varepsilon + is) , \tag{2.7}$$

where j_1, j_2, and j_3 (and also l_1, l_2, and l_3) are integers indicating the position vector $\boldsymbol{R}_j$ by the relation $\boldsymbol{R}_j = a(j_1\boldsymbol{e}_1 + j_2\boldsymbol{e}_2 + j_3\boldsymbol{e}_3)$ in which $\boldsymbol{e}_1$, $\boldsymbol{e}_2$, and $\boldsymbol{e}_3$ are the unit vectors in the x-, y-, and z-directions, respectively.

3. Calculation of the Green's Function

3.1 Single-defect case

3.1.1 Isolated vacancy

Since the highly-localized-potential approximation has the unphysical feature in the case of the vacancy as mentioned before, we consider that the perturbing potential due to the vacancy is extended to the nearest-neighbour sites of the vacant site. Therefore, we take into account the perturbing potentials, both $V_{00}^{(1)\sigma}$ and $V_{11}^{(1)\sigma}$.

The diagonal parts of the Green's function are obtained simply from (2.3) as follows. Owing to the symmetry of the lattice, the Green's function $g_{l_1l_2l_3}$ is an even function of l_1, l_2, and l_3, and it remains unchanged with the mutual interchange of them. Therefore, it can be shown that there exist three types of defect levels, s-like level, p-like level with threefold degeneracies, and d-like level with twofold degeneracies. For simplicity, only the s-like level, which alone does not have a node at the centre of the perturbation, is taken into account here. The Green's functions, $G^{\sigma}_{1+\varrho,j}$, for all the nearest neighbours become equal. By using this property, the following equation is obtained from (2.4):

$$\begin{aligned}\tilde{G}^{\sigma}_{jj} = g_{jj} + [V_{00}^{(1)\sigma} g_{j0}g_{0j} + V_{11}^{(1)\sigma} g_{\varrho j}g_{js} - V_{00}^{(1)\sigma}V_{11}^{(1)\sigma}\,(g_s g_{0j}g_{j0} + g_{000}g_{\varrho j}g_{js}) +\\ + 12 g_{100} V_{00}^{(1)\sigma}V_{11}^{(1)\sigma} g_{j0}g_{\varrho j}]\,[1 - g_{000}V_{00}^{(1)\sigma} - g_s V_{11}^{(1)\sigma} +\\ + V_{00}^{(1)\sigma}V_{11}^{(1)\sigma} g_{000}g_s - 6 V_{00}^{(1)\sigma}V_{11}^{(1)\sigma} g_{100}^2]^{-1}\,,\end{aligned} \tag{3.1}$$

by introducing a symbol, $\sim$, which represents the respective quantity in the presence of a single defect, impurity atom or vacancy. Here, g_s and g_{js} are defined by

$$g_s = g_{000} + g_{200} + 4 g_{110}\,, \tag{3.2}$$

$$g_{js} = \sum_{\varrho} g_{j,1+\varrho}\,. \tag{3.3}$$

The total screening charge due to a single vacancy is given by

$$\begin{aligned}Z &= -\frac{1}{\pi}\,\mathrm{Im}\sum_{\sigma}\sum_{j}\int^{\epsilon_F} (\tilde{G}^{\sigma}_{jj} - g_{jj})\,d\epsilon =\\ &= -\frac{1}{\pi}\,\mathrm{Im}\sum_{\sigma}\log\{1 - g_{000}(\epsilon_F)\,V_{00}^{(1)\sigma} - g_s(\epsilon_F)\,V_{11}^{(1)\sigma} + V_{00}^{(1)\sigma}V_{11}^{(1)\sigma} g_{000}(\epsilon_F)\,g_s(\epsilon_F) -\\ &\quad - 6 V_{00}^{(1)\sigma}V_{11}^{(1)\sigma}[g_{100}(\epsilon_F)]^2\}\,.\end{aligned} \tag{3.4}$$

In reality, a procedure to introduce a vacancy consists in removing an atom inside and depositing this atom on the surface. Thus, the spin dependence of the perturbing potential due to a vacancy should be determined by the charge neutrality condition (Friedel's sum rule) for each spin band. The charge neutrality condition sets $Z^{\uparrow} = -0.92$, $Z^{\downarrow} = -0.48$ (paramagnetic case: $Z^{\uparrow} = Z^{\downarrow} = -0.70$). These figures are close to the commonly accepted values of the numbers of the d electrons of Fe metal if multiplied by five. The potential parameters, $V_{00}^{(1)\sigma}$ and $V_{11}^{(1)\sigma}$, allowed by this set of $Z^{\uparrow}$ and $Z^{\downarrow}$ are shown in Table 2.

Table 2

Perturbing potential due to a vacancy ($\lambda = V_{00}^{(1)}/V_{11}^{(1)}$)

spin state	potential parameter	λ 2	4	6	8	10	12	14	16	18	20
ferro	$V_{00}^{(1)}$	12.518	20.992	29.384	37.755	46.101	54.449	62.802	71.140	79.459	87.815
(majority)	$V_{11}^{(1)}$	6.259	5.248	4.897	4.719	4.610	4.537	4.486	4.446	4.414	4.390
ferro	$V_{00}^{(1)}$	1.656	2.312	2.804	3.212	3.566	3.882	4.169	4.433	4.678	4.908
(minority)	$V_{11}^{(1)}$	0.828	0.578	0.467	0.401	0.356	0.323	0.297	0.277	0.259	0.245
para	$V_{00}^{(1)}$	2.286	3.555	4.695	5.775	6.821	7.844	8.851	9.847	10.834	11.815
	$V_{11}^{(1)}$	1.143	0.888	0.782	0.772	0.682	0.653	0.632	0.615	0.602	0.591

The energy unit is the one-sixth bandwidth of the host metal.

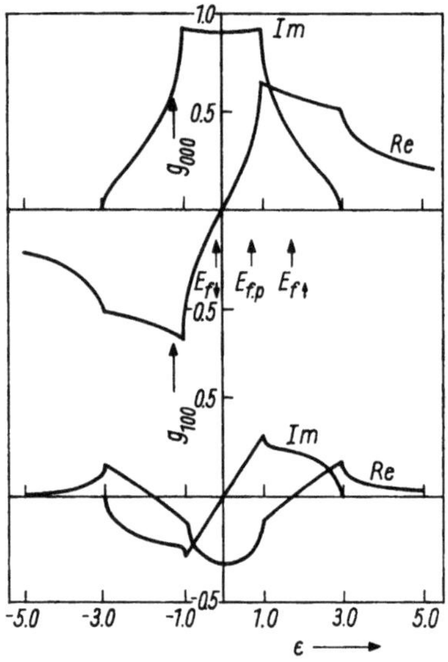

Fig. 1. The unperturbed Green's function g_{l00} for the simple cubic lattice. The abscissa represents the energy parameter ϵ. Re and Im indicate the real and imaginary parts, respectively

A change in the Hartree-Fock energy due to the presence of a single vacancy is evaluated by

$$\delta E(v) = -\frac{1}{\pi}\sum_{\sigma}\sum_{j}\int^{\epsilon_F}(\epsilon-\epsilon_F)\,\mathrm{Im}\,[\tilde{G}^{\sigma}_{jj}(\epsilon)-g_{jj}(\epsilon)]\,d\epsilon - [(U^{(0)}+U^{(1)})\,\tilde{N}_{1\uparrow}\tilde{N}_{1\downarrow} - {} - U^{(0)}N^{(0)}_{1\uparrow}N^{(0)}_{1\downarrow}] - \sum_{j\neq 1} U^{(0)}(N^{(0)}_{j\uparrow}\delta\tilde{N}_{j\downarrow} + N^{(0)}_{j\downarrow}\delta\tilde{N}_{j\uparrow} + \delta\tilde{N}_{j\uparrow}\,\delta\tilde{N}_{j\downarrow})\,, \quad (3.5)$$

where the second and third term are introduced because the interaction energies are counted twice in the integral, and $\delta\tilde{N}_{j\sigma}$ is the deviation in the number of localized electrons with spin σ of the j-th atom from its unperturbed value, $N^{(0)}_{j\sigma}$, and given by

$$\delta\tilde{N}_{j\sigma} = -\frac{1}{\pi}\,\mathrm{Im}\int^{\epsilon_F}[\tilde{G}^{\sigma}_{jj}(\epsilon)-g_{jj}(\epsilon)]\,d\epsilon\,. \quad (3.6)$$

Table 3

Unperturbed Green's functions at the Fermi energy (dimension: 1/unit energy)

spin state	$g^{\mathrm{Re}}_{000}(\epsilon_F)$	$g^{\mathrm{Im}}_{000}(\epsilon_F)$	$g^{\mathrm{Re}}_{100}(\epsilon_F)$	$g^{\mathrm{Im}}_{100}(\epsilon_F)$
ferro (majority)	0.575	0.377	0.000	0.217
ferro (minority)	−0.023	0.896	−0.339	−0.019
para	0.294	0.897	−0.279	0.207

The energy unit is the one-sixth bandwith of the host metal.

The calculations involved for (3.4) to (3.6) require the analytic expression for the unperturbed Green's function given by (2.6). The unperturbed Green's function for the simple cubic lattice was expressed by Horiguchi and Morita [14 to 15] as a sum of simple integrals of the complete elliptic integral of the first kind. From their expressions we can numerically evaluate the unperturbed Green's functions at general positions, as shown in Fig. 1. The values of the unperturbed Green's functions at the Fermi energy are given in Table 3.

The results of the numerical calculations of $\delta E(v)$ are tabulated in Table 4. We see from Table 4 that the parameter dependent part of $\delta E(v)$ is very small and is within a few per cent of the total $\delta E(v)$, while the changes in the potential parameters are very large. These tendencies are considered to be general, since the potential parameters must satisfy Friedel's condition. Therefore, in the following we will use the potential parameters, $V_{00}^{(1)\sigma}$ and $V_{11}^{(1)\sigma}$, with the condition $V_{00}^{(1)\sigma}/V_{11}^{(1)\sigma} = 10$, which is the widely adopted one [5].[2])

3.1.2 Single impurity atom

The impurity states in transition metals have been successfully discussed by several authors with the strongly localized perturbation theory based on the Koster-Slater approach [20]. Therefore, we put $V_{11}^{(2)\sigma} = 0$ for the impurity potentials of the transition metal impurity atoms. In this case, the change in the Hartree-Fock energy due to the presence of a single impurity atom, $\delta E(i)$, is evaluated by (3.5), changing the index 1 to 2 and equating the Green's function parts to the following expression

$$\sum_j [\tilde{G}_{jj}^\sigma(\epsilon) - g_{jj}(\epsilon)] = \frac{\frac{\partial}{\partial\epsilon}\left(1 - g_{000}(\epsilon)\, V_{00}^{(2)\sigma}\right)}{\left(1 - g_{000}(\epsilon)\, V_{00}^{(2)\sigma}\right)},$$

where $V_{00}^{(2)\sigma}$ is the impurity potential parameter and obtained as follows.

The total screening charge due to a single impurity atom is given by

$$Z = -\frac{1}{\pi}\sum_\sigma \operatorname{arctg}\frac{g_{000}^{\mathrm{Im}}(\epsilon_\mathrm{F})\, V_{00}^{(2)\sigma}}{1 - g_{000}^{\mathrm{Re}}(\epsilon_\mathrm{F})\, V_{00}^{(2)\sigma}}.$$

Putting $Z = -4/5$, $-3/5$, $-2/5$, $-1/5$, $1/5$, $2/5$, $3/5$, which represent the impurity atom, Ti, V, Cr, Mn, Co, Ni, and Cu, respectively, into the above equation, we can obtain the potential parameter with the Hartree-Fock relation between majority and minority spin potentials [20]. The results are given in Table 5.

3.2 Two-defect case

Our problem is now to obtain an expression of the interaction between a vacancy and a transition metal solute atom in ferromagnetic metals (including the paramagnetic state). As mentioned before, we consider the impurity potential localized only at the impurity site and put $V_{11}^{(2)\sigma} = 0$. The matrix elements

[2]) Here, we do not undertake an elaborate procedure of solving (2.5) and (3.4) self-consistently. The self-consistent solution apparently gives λ much smaller and larger than 10 in the majority and minority spin band, respectively, since $U^{(1)} \approx -U^{(0)}$ in the vacant site. However, this does not cause any essential changes in final results as long as Friedel's condition is satisfied.

Table 4

$\delta E^{\sigma}(v)$ for each spin band ($\lambda = V_{00}^{(1)}/V_{11}^{(1)}$)

spin state	λ									
	2	4	6	8	10	12	14	16	18	20
ferro (majority)	2.400	2.398	2.382	2.381	2.373	2.368	2.364	2.361	2.358	2.360
ferro (minority)	0.542	0.554	0.564	0.571	0.577	0.582	0.586	0.589	0.592	0.594
para	1.110	1.134	1.149	1.159	1.167	1.173	1.178	1.182	1.185	1.187

The energy unit is the one-sixth bandwidth of the host metal.

Table 5
Impurity potentials for each spin band

spin state	Ti ($Z = -4/5$)	V ($Z = -3/5$)	Cr ($Z = -2/5$)	Mn ($Z = -1/5$)	Co ($Z = 1/5$)	Ni ($Z = 2/5$)	Cu ($Z = 3/5$)
ferro (majority)	2.85	3.36	3.20	1.90	0.33	−0.55	−5.88
ferro (minority)	0.50	−0.36	−1.45	−2.00	−1.25	−2.10	−6.17
para	1.71	1.05	0.65	0.32	−0.41	−1.06	−2.80

The energy unit is the one-sixth bandwidth of the host metal.

of the Green's function in the presence of a vacancy and the impurity atom are given by

$$G_{1j}^{\sigma} = g_{1j} + g_{000} V_{00}^{(1)\sigma} G_{1j}^{\sigma} + g_{12} V_{00}^{(2)\sigma} G_{2j}^{\sigma} + 6 g_{100} V_{11}^{(1)\sigma} G_{1+\varrho,j}^{\sigma} , \tag{3.7}$$

$$G_{2j}^{\sigma} = g_{2j} + g_{21} V_{00}^{(1)\sigma} G_{1j}^{\sigma} + g_{000} V_{00}^{(2)\sigma} G_{2j}^{\sigma} + g_{2,1s} V_{11}^{(1)\sigma} G_{1+\varrho,j}^{\sigma} , \tag{3.8}$$

$$G_{1+\varrho,j}^{\sigma} = g_{1+\varrho,j} + g_{100} V_{00}^{(1)\sigma} G_{1j}^{\sigma} + g_{1+\varrho,2} V_{00}^{(2)\sigma} G_{2j}^{\sigma} + g_{s} V_{11}^{(1)\sigma} G_{1+\varrho,j}^{\sigma} , \tag{3.9}$$

$$g_{2,1s} = \sum_{\varrho} g_{2,1+\varrho} ,$$

where the s-symmetry around the vacancy is assumed. From (3.7) to (3.9) we can obtain the Green's function for each lattice site. By inserting the Green's functions, thus obtained into (2.4), and by some manipulation, we can obtain the trace of the diagonal elements as follows:

$$\sum_{j} (G_{jj}^{\sigma} - g_{jj}) = \frac{\frac{\partial \Delta^{\sigma}}{\partial \epsilon}}{\Delta^{\sigma}} \tag{3.10}$$

with

$$\begin{aligned} \Delta^{\sigma} = {} & (1 - g_{000} V_{00}^{(1)\sigma}) [\tfrac{1}{6} g_{2,1s}^{2} V_{11}^{(1)\sigma} - (1 - g_{000} V_{00}^{(2)\sigma}) (1 - g_{s} V_{11}^{(1)\sigma})] + \\ & + g_{21} V_{00}^{(1)\sigma} [g_{12} V_{00}^{(2)\sigma} (1 - g_{s} V_{11}^{(1)\sigma}) - g_{100} g_{2,1s} V_{11}^{(1)\sigma} V_{00}^{(2)\sigma}] + \\ & + g_{100} V_{00}^{(1)\sigma} [g_{12} g_{2,1s} V_{11}^{(1)\sigma} V_{00}^{(2)\sigma} + 6 g_{100} V_{11}^{(1)\sigma} (1 - g_{000} V_{00}^{(2)\sigma})] . \end{aligned} \tag{3.11}$$

A change in the Hartree-Fock energy due to the presence of two defects is evaluated by

$$\begin{aligned} \delta E(v, i) = {} & \frac{1}{\pi} \sum_{\sigma} \int^{\epsilon_F} \operatorname{arctg} \frac{\operatorname{Im} \Delta^{\sigma}}{\operatorname{Re} \Delta^{\sigma}} \, d\epsilon - [(U^{(0)} + U^{(1)}) N_{1\uparrow} N_{1\downarrow} - U^{(0)} N_{1\uparrow}^{(0)} N_{1\downarrow}^{(0)}] - \\ & - [(U^{(0)} + U^{(2)}) N_{2\uparrow} N_{2\downarrow} - U^{(0)} N_{2\uparrow}^{(0)} N_{2\downarrow}^{(0)}] - \\ & - \sum_{j \neq 1,2} U^{(0)} (N_{j\uparrow}^{(0)} \delta N_{j\downarrow} + N_{j\downarrow}^{(0)} \delta N_{j\uparrow} + \delta N_{j\uparrow} \delta N_{j\downarrow}) . \end{aligned} \tag{3.12}$$

If a vacancy exists at one of the nearest-neighbour sites of the impurity atom, the real and imaginary parts of Δ^σ are given as

$$\begin{aligned}
\operatorname{Re} \Delta^\sigma = {} & (1 - f_{000} V_{00}^{(1)\sigma}) [(1 - f_{000} V_{00}^{(2)\sigma}) (1 - f_s V_{11}^{(1)\sigma}) + j_{000} j_s V_{00}^{(2)\sigma} V_{11}^{(1)\sigma} + \\
& + \tfrac{1}{6} V_{11}^{(1)\sigma} V_{00}^{(2)\sigma} (f_s^2 - j_s^2)] + j_{000} V_{00}^{(1)\sigma} [j_s V_{11}^{(1)\sigma} (1 - f_{000} V_{00}^{(2)\sigma}) + \\
& + j_{000} V_{00}^{(2)\sigma} (1 - f_s V_{11}^{(1)\sigma}) + \tfrac{1}{3} f_s j_s V_{11}^{(1)\sigma} V_{00}^{(2)\sigma}] + f_{100} V_{00}^{(1)\sigma} [f_{100} V_{00}^{(2)\sigma} + \\
& + V_{11}^{(1)\sigma} V_{00}^{(2)\sigma} (f_{100} f_s - j_{100} j_s) + 6 f_{100} V_{11}^{(1)\sigma} (1 - f_{000} V_{00}^{(2)\sigma}) + 6 j_{000} j_{100} \times \\
& \times V_{11}^{(1)\sigma} V_{00}^{(2)\sigma}] - j_{100} V_{00}^{(1)\sigma} [j_{100} V_{00}^{(2)\sigma} (f_{100} j_s + j_{100} f_s) V_{11}^{(1)\sigma} V_{00}^{(2)\sigma} + \\
& + 6 j_{100} V_{11}^{(1)\sigma} (1 - f_{000} V_{00}^{(2)\sigma}) - 6 f_{100} j_{000} V_{11}^{(1)\sigma} V_{00}^{(2)\sigma}] ,
\end{aligned} \tag{3.13}$$

$$\begin{aligned}
\operatorname{Im} \Delta^\sigma = {} & j_{000} V_{00}^{(1)\sigma} [(1 - f_{000} V_{00}^{(2)\sigma}) (1 - f_s V_{11}^{(1)\sigma}) - j_{100} j_s V_{00}^{(2)\sigma} V_{11}^{(1)\sigma} - \tfrac{1}{6} V_{11}^{(1)\sigma} V_{00}^{(2)\sigma} \times \\
& \times (f_s^2 - j_s^2)] + (1 - f_{000} V_{00}^{(1)\sigma}) [j_s V_{11}^{(1)\sigma} (1 - f_{000} V_{00}^{(2)\sigma}) + j_{000} V_{00}^{(2)\sigma} \times \\
& \times (1 - f_s V_{11}^{(1)\sigma}) + \tfrac{1}{3} f_s j_s V_{11}^{(1)\sigma} V_{00}^{(2)\sigma}] + j_{100} V_{00}^{(1)\sigma} [f_{100} V_{00}^{(2)\sigma} + V_{11}^{(1)\sigma} V_{00}^{(2)\sigma} \times \\
& \times (f_{100} f_s - j_{100} j_s) + 6 f_{100} V_{11}^{(1)\sigma} (1 - f_{100} V_{00}^{(2)\sigma}) + 6 j_{100} j_{000} V_{11}^{(1)\sigma} V_{00}^{(2)\sigma}] + \\
& + f_{100} V_{00}^{(1)\sigma} [j_{100} V_{00}^{(2)\sigma} + V_{11}^{(1)\sigma} V_{00}^{(2)\sigma} (f_{100} j_s + j_{100} f_s) + 6 j_{100} V_{11}^{(1)\sigma} (1 - f_{000} V_{00}^{(2)\sigma}) - \\
& - 6 f_{100} j_{000} V_{11}^{(1)\sigma} V_{00}^{(2)\sigma}] .
\end{aligned} \tag{3.14}$$

Here, the functions denoted by f and j are the real and imaginary part of the unperturbed Green's functions, respectively. The results of the numerical calculations of the integrand of the first term, $Z(\epsilon)$[3]), in (3.12) in the paramagnetic state are shown in Fig. 2. The interaction energy between a vacancy and an impurity atom is defined by

$$E_{\text{int}} = \delta E(v, i) - \delta E(v) - \delta E(i) . \tag{3.15}$$

$\delta E(i)$ is the change in the Hartree-Fock energy due to the presence of the single impurity atom and is obtained in the manner shown in Section 3.1.2. For simplicity, we put here $U^{(2)} = 0$, since the intra-atomic Coulomb integrals are expect-

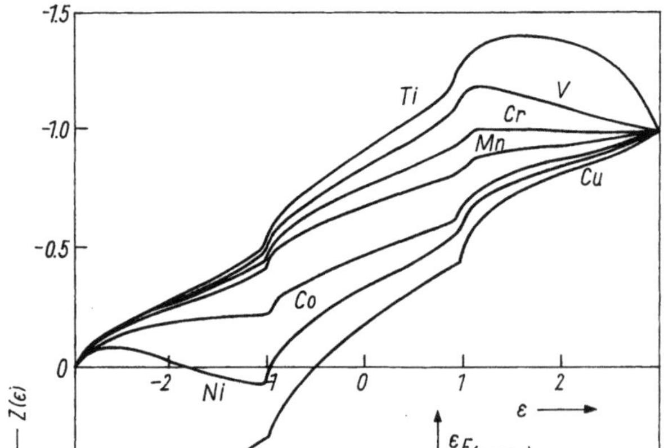

Fig. 2. The charge displaced up to the energy ϵ is shown against the energy in the paramagnetic state when a vacancy and an impurity atom are added to the host lattice

[3]) This value gives the charge displaced up to the energy ϵ, as well as the scattering phase shift devided by π (single defect case).

ed to have almost the same value for the atoms belonging to the same transition metal group. Even if we take $U^{(2)} \neq 0$, the qualitative feature of the interaction is scarcely changed since the potential parameters used in the present calculations must always satisfy Friedel's condition and the value of the quantity, $U^{(2)}$, is itself small compared with $U^{(0)}$ as shown in Table 1.

In order to investigate the effects of ferromagnetism of the host metal, we extensively made the following assumption. The majority spin bands and the minority ones are shifted to each other so that the correct magneton number of Fe(2.2) may be given, i.e. the majority spin bands and the minority spin bands are occupied by 4.6 and 2.4 electrons per atom, respectively. From these figures, we can obtain the effective exchange integral. Here the spin dependence of the band shape is completely ignored.

4. Results and Discussion

We have examined the feature of the interaction between a vacancy and an impurity atom in transition metals using the Wolff-Clogston tight-binding model within the non-degenerate band scheme. The results of the numerical calculations on the interaction energy are shown in Fig. 3. It may appear that the numerical calculations in the present paper need some arbitrary parameters such as $V_{00}^{(1)\sigma}$ and $V_{11}^{(1)\sigma}$. However, the obtained results are almost independent of the values assumed for these parameters. This is due to the fact that the parameters must always satisfy Friedel's condition. Therefore, the results of the present numerical calculations are considered to be mainly determined by the model adopted. It might be further argued that the present model is extremely simple. However, it is our contention that this model can still qualitatively represent the situation for the Fe base alloys with the first transition-metal impurities, as indicated by the following discussions.

The present calculations indicate that the binding energy of a vacancy with an impurity in Fe base alloys is positive for all the impurities except Cu which alone has negative binding with a vacancy. As shown in Fig. 3, this indication is in complete agreement with the experimental studies except those of Mn and Ti. This tendency of the binding energy seems to arise from the fact that the impurity atom provides the bound state below the band bottom, and most of

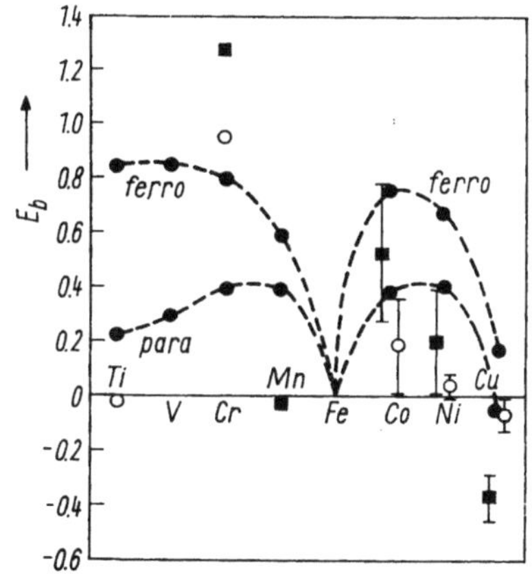

Fig. 3. The binding energy between a vacancy and the impurity atom is shown in the para- and ferromagnetic state. ○ and ■ indicate the values derived from the impurity diffusion experiments in α- and γ-iron base alloys, respectively. E_b is measured in 1/6 band width

the excess electrons of the impurity atom are localized at the impurity site. In fact, in the case of Cu impurity in Fe base alloys, most of the excess electrons are found to be confined within the impurity site. Specifically, 99% of the excess electrons are localized at this site. This fact seems to more or less support Blandin and Deplante's treatment [16], in which the interaction energies are obtained by a somewhat phenomenological approach with the electrostatic potential theory based on the model of virtual bound states.

The present study also includes the evaluation of the binding energies in the ferro- and the paramagnetic state by using the Stoner model [17], and it has been shown that the binding energies in the ferromagnetic state are considerably larger than those in the paramagnetic state. This is not due to the effects of the magnetic moments, localized at the impurity sites which is demagnetized in the paramagnetic state, but is introduced by the almost full occupation in the majority spin band. The result of the diffusion study of Ni in Fe reported by Hirano et al. [18], seems to support the above tendency of the binding energy in the ferro- and the paramagnetic state. That is, the change in the diffusion constant of Ni in Fe, introduced by the magnetic transformation at 1043 K, has been considered to have come mainly from the difference of the frequency factor [19]. In addition to this interpretation on the experimental finding, it must also be considered based on the above discussion that the change of the activation energy coming from the electronic structure change due to the magnetic transformation is never small enough to be ignored.

It is hoped that the present theory will be extended to more realistic situations such that the effects of interaction due to the band degeneracy, the s–d interaction, and the more rigorous treatment of two defects, including the symmetrical consideration, may be taken into account.

Acknowledgements

The author wishes to express his gratitude to Prof. R. Abe for many helpful discussions, and to Prof. T. Mori for suggesting the problem.

Appendix

In this appendix, the unperturbed Green's function, g_s, is shown to be expressed in a simple form by using a recurrence formula. $g_{l,m,n}$ for the simple cubic lattice satisfies the following recurrence formula:

$$g_{l+1,m,n} + g_{l-1,m,n} + g_{l,m+1,n} + g_{l,m-1,n} + g_{l,m,n+1} + g_{l,m,n-1} = \\ = -2\delta_{l0}\delta_{m0}\delta_{n0} + 2(\epsilon + is)\, g_{l,m,n}\,, \tag{A1}$$

where δ's are the usual Kronecker symbols. This leads to

$$g_{110} = -\tfrac{1}{4}\, g_{000} + \tfrac{1}{2}\,(\epsilon + is)\, g_{100} - \tfrac{1}{4}\, g_{200}\,. \tag{A2}$$

Inserting (A2) into (3.2), the unperturbed Green's function, g_s, is expressed as

$$g_s = g_{000} + g_{200} + 4g_{110} = 2(\epsilon + is)\, g_{100} \tag{A3}$$

which simplifies numerical calculations involved in the present study.

References

[1] W. A. Harrison, Pseudopotentials in the Theory of Metals, W. A. Benjamin, Inc., New York 1966.
[2] Y. Fukai, Phys. Rev. **186**, 697 (1969).
[3] R. Benedek and A. Baratoff, J. Phys. Chem. Solids **32**, 1015 (1971).
[4] R. Yamamoto, M. Doyama, and O. Takai, Phys. Letters A **43**, 247 (1973).
[5] A. Seeger, J. Phys. Radium **23**, 616 (1962).
[6] A. Seeger, E. Mann, and G. Thierer, Z. Phys. **193**, 295 (1966).
E. Mann, phys. stat. sol. **11**, 753 (1965); **11**, 767 (1965); **13**, 293 (1966); **32**, 867 (1969).
[7] W. Hehl and E. Mann, phys. stat. sol. (b) **48**, 271 (1971).
[8] E. Mann, Phys. kondens. Materie **12**, 210 (1971).
[9] J. Kanamori, J. appl. Phys. **36**, 929 (1965).
[10] A. A. Gomes and I. A. Campbell, J. Phys. C **1**, 253 (1968).
[11] P. A. Wolff, Phys. Rev. **124**, 1030 (1961).
[12] N. L. Peterson, Solid State Phys. **22**, 409 (1968).
[13] A. A. Gomes, J. Phys. Chem. Solids. **27**, 451 (1966).
[14] T. Horiguchi, J. Phys. Soc. Japan **30**, 1261 (1971).
[15] T. Morita and T. Horiguchi, J. Phys. A **5**, 67 (1972); J. Phys. Soc. Japan **30**, 957 (1971); J. Math. Phys. **12**, 981 (1971).
[16] A. Blandin and J. L. Deplante, J. Phys. Chem. Solids **26**, 381 (1965).
[17] E. C. Stoner, Proc. Roy. Soc. **A154**, 656 (1936).
[18] K. Hirano, B. L. Averbach, and M. Cohen, Acta metall. **9**, 440 (1962).
[19] W. Hume-Rothery, Acta metall. **11**, 630 (1963).
[20] I. A. Cambell and A. A. Gomes, Proc. Phys. Soc. **91**, 319 (1967).

(Received November 15, 1973)

phys. stat. sol. (b) 62, 573 (1974)

Subject classification: 14.2; 21

Department of Physics, Massachusetts Institute of Technology, Cambridge (a) and Department of Physics, Georgetown University, Washington, D.C. (b)

Obtaining a Consistent Set of Thermodynamic Variables for Real Superconductors[1)] [2)]

By

J. Bostock (a), W. N. Mathews, Jr. (b), and W. D. Gregory (b)

A method is presented for obtaining a consistent set of thermodynamic parameters and temperature-dependent thermodynamic variables for a real superconductor. The method depends upon determining a temperature-dependent effective interaction strength for the particular material. Conversely, given the effective interaction strength for the superconductor, the thermodynamic variables as functions of temperature can be obtained. The method is illustrated for Al, V, Ta, Nb, and Pb. The results of this approach and the McMillan theory of the critical temperature are compared for Nb, Pb, and the In–Tl alloy system.

Eine Methode wird angegeben, die es möglich macht, für einen realen Supraleiter einen konsistenten Satz von thermodynamischen Parametern und temperaturabhängigen Variablen zu bestimmen. Zuerst wird ein temperaturabhängiger effektiver Wechselwirkungsparameter für ein Material bestimmt. Wenn andererseits dieser Parameter bekannt ist, kann man die thermodynamischen Variablen als Funktionen der Temperatur erhalten. Diese Methode wird für Al, V, Ta, Nb und Pb angewendet. Die Ergebnisse werden mit der McMillan-Theorie der kritischen Temperatur für Nb, Pb und In–Tl-Legierungen verglichen.

1. Introduction

In this paper a method for obtaining the temperature-dependent thermodynamic variables and a consistent set of thermodynamic parameters characterizing actual superconducting materials is presented. The method is based on Sheahen's reformulation [2] of the BCS model [3] to account for the deviations of the thermodynamic variables of actual superconducting materials from the ideal "law of corresponding states" behaviour predicted by BCS. Sheahen has shown that these deviations can be accounted for within the framework of the BCS model by taking the effective interaction strength of the superconductor to be temperature-dependent. This point of view can be justified on three grounds: 1. The theoretical basis has been discussed by Rothwarf [4] using the Eliashberg formulation of the microscopic theory of superconductivity [5]. 2. The temperature dependence of the effective electron mass, which has its origins in the electron–phonon interaction, has been discussed in several papers [6 to 8]. 3. It has been shown explicitly that the transverse electron–phonon coupling strength for the A-15 compounds varies with temperature [9, 10].

In Section 2 a description of the calculation of the temperature-dependent effective interaction strength NOV(t) for actual superconducting materials is

---

1) Supported in part by the Air Force Office of Scientific Research, Project Themis, and the U.S. Atomic Energy Commission.

2) A preliminary account of this work is abstracted in [1].

given. Here $t \equiv T/T_c$ is the reduced temperature, T the absolute temperature, and T_c the critical temperature for superconductivity. In Section 3 the results for Al, V, Ta, Nb, and Pb are discussed, and the procedure for inverting the determination of NOV(t) to obtain the dependence of the other thermodynamic quantities on temperature is described. A specific illustration is given for the gap parameter $\Delta(T)$.

An equation for T_c, which is formally similar to the BCS weak-coupling relation and which is in good agreement with experimental data on a variety of superconductors, emerges from the work of Sections 2 and 3. In Section 4 this relation and the equation for T_c which has been derived on the basis of the strong-coupling theory [11] by McMillan [12] are compared for Nb, Pb, and the In–Tl alloy system recently investigated by Dynes [13]. Section 5 is a summary.

2. Calculational Procedure

The necessary extension of the BCS model consists of replacing the BCS gap equation by

$$\frac{1}{\text{NOV}(t)} = \int\limits_0^{\omega_c} \mathrm{d}\varepsilon \frac{\tanh \{[\varepsilon^2 + \Delta(T)^2]^{1/2}/2k_\text{B}T\}}{[\varepsilon^2 + \Delta(T)^2]^{1/2}} . \tag{1}$$

This differs from the original BCS gap equation in that the interaction strength is allowed to be temperature-dependent and the upper limit, ω_c, is no longer simply some average phonon energy such as the Debye energy.

Equation (1) is used to calculate NOV(t) for a specific superconducting material as follows. A value for NOV is first obtained with the use of the experimentally measured value of $2\Delta(0)/k_\text{B}T_c$ and the equation

$$\frac{2\Delta(0)}{k_\text{B}T_c} = \frac{3.528}{1 - \exp\left(\frac{-2}{\text{NOV}}\right)} . \tag{2a}$$

This value of NOV, which is henceforth denoted as "the Sheahen value", is used to evaluate[3])

$$\frac{\Delta C}{\gamma T_c} = \frac{1.48}{\left[1 - \exp\left(\frac{-2}{\text{NOV}}\right)\right]^3} , \tag{2b}$$

$$\frac{\gamma T_c^2}{V_\text{m} H_0^2} = 0.1683 \left[1 - \exp\left(\frac{-2}{\text{NOV}}\right)\right] , \tag{2c}$$

and

$$\left(\frac{\mathrm{d}h}{\mathrm{d}t}\right)_1 = - \frac{1.7725}{1 - \exp\left(\frac{-2}{\text{NOV}}\right)} . \tag{2d}$$

---

[3]) As Sheahen [2] points out, equations (2) follow from the BCS model in the weak-coupling limit. ΔC is the specific heat jump at the critical temperature; γ is the electronic specific heat coefficient; $2\Delta(0)$ is the energy gap at absolute zero; k_B is the Boltzmann constant; V_m is the molar volume; H_0 is the critical field at absolute zero; $(\mathrm{d}h/\mathrm{d}t)_1$ is the slope of the reduced critical field, $h \equiv H_c(T)/H_0$, at the critical temperature.

If the values so computed for these thermodynamic parameters are in agreement with the experimental values, then ω_c is determined from[4])

$$\omega_c = \frac{k_B T_c}{1.134} \exp\left(\frac{1}{\text{NOV}}\right). \tag{3}$$

Next the values of $\Delta(0)$, T_c, and ω_c are used in (1) to determine TOP $\equiv$ NOV(0) and BOT $\equiv$ NOV(1). Finally, NOV(t) for $0 < t < 1$ can be calculated using some set of data for $\Delta(T)$ in (1).

As Sheahen [2] points out, TOP and "the Sheahen value" of NOV must be very close to one another. (They may differ slightly, however, since "the Sheahen value" was calculated with the weak-coupling limit of the BCS equations.) Then NOV(t) determined, as discussed, from (1) is necessarily a monotonically non-increasing function of t for $0 \leqq t \leqq 1$.

For weak-coupling superconductors, "the Sheahen value" of NOV satisfies (1) for $0 \leqq t \leqq 1$. This is because the quantities on the left hand sides of (2) are much more sensitive to the form and strength of the pairing interaction than is the reduced energy gap $\delta(t) \equiv \Delta(T)/\Delta(0)$ [14].

The significance of ω_c determined in this way is that $2\omega_c$ is the minimum width (at $t = 0$) of the energy shell which gives a net attractive pairing interaction. That is, when the upper limit on the integral in (1) is less than ω_c, there is no solution to (1) consistent with $\delta(0) = 1$; when the upper limit is equal to or greater than ω_c, there is such a solution, given that the coupling strength may not be larger than TOP. Moreover, ω_c is the smallest value of the upper limit which gives $\delta(1) = 0$ for the given value of NOV. For weak-coupling superconductors this is a restatement of the fact that TOP (= BOT) is the largest possible value of the coupling strength. For strong-coupling materials this is equivalent to requiring that BOT be the largest possible value of the coupling strength consistent with $\delta(1) = 0$. If the value of TOP is not approximately the same as "the Sheahen value" of NOV, these conditions will not be satisfied.[5])

Of course, any one of (2) can be employed to determine "the Sheahen value" of NOV. The remaining three of (2) are then used to check this value.

In carrying out these calculations, (1) is used to determine which values of the "data point" $(\delta(t) \pm \Delta\delta,\ t \pm \Delta t)$, where $\Delta\delta$ and Δt are the experimental uncertainties in the reduced gap and temperature, respectively, are consistent with TOP $\geqq$ NOV $\geqq$ BOT. That is, each "data point" generates a set of points to be tested, (δ_n, t_m) where, for example

$$\delta_n = \delta - \Delta\delta + 10^{-4}n\,, \qquad t_m = t - \Delta t + 10^{-3}m$$

with

$$n = 0, 1, \ldots, 2(\Delta\delta)\,10^4 \quad \text{and} \quad m = 0, 1, \ldots, 2(\Delta t)\,10^3\,.$$

[4]) Equation (3) follows from (1) in the weak-coupling limit, provided we follow BCS in assuming that NOV(t) is in fact independent of t in this limit. Equation (3) is of course just equation (3.29) of BCS except for the replacements of $\hbar\omega$ by ω_c and N(O)V by NOV.

[5]) Failure of the calculated and measured values of the thermodynamic parameters on the left sides of (2) to agree, failure of the condition TOP $\geqq$ BOT to hold, and failure of ω_c to have the correct properties are interpreted to mean that the values of $\Delta(0)$ and T_c used do not reflect the bulk properties of the material. Our experience indicates that this situation would not occur if all of the necessary parameters were carefully measured for the same sample. The measured values of the thermodynamic parameters are compiled in [2] and [15].

Clearly the smaller the experimental errors, the more precisely NOV(t) can be determined. Generally the constraint TOP $\geqq$ NOV(t) $\geqq$ BOT merely delineates a "shell" within which the values of NOV(t) must lie. In the case of weak-coupling superconductors, for which NOV must be independent of t, the "shell" reduces to a straight line. The "shell" also reduces to a precise set of values for NOV(t) if theoretical or otherwise precisely chosen values of $\delta(t)$ are used. The resultant values of NOV(t) are, of course, characteristic of the given material only to the extent that the gap values used are. For strong-coupling superconductors, the change from TOP to BOT can be of the order of 10%, so that the "shell" is quite wide and the data set used must be very accurate if the character of the NOV(t) curve is to be evident.

This method can also be extended to include the effects of anisotropy.[6]) The appropriate generalizations of (1) and (3), and (2), have been given by Clem [17] and Sheahen [2], respectively. For example, (2a) becomes

$$\frac{2\Delta(0)}{k_B T_c} = \frac{3.528\,[1 - \frac{3}{2}\langle a^2\rangle]}{1 - \exp\left(\frac{-2}{\text{NOV}}\right)}, \tag{4}$$

where $\langle a^2\rangle$ is the mean squared anisotropy. In this instance, one uses any two of the generalized version of (2) to determine values for NOV and $\langle a^2\rangle$, and then checks the consistency of these values in the remaining equations. For weak-coupling materials, for which a non-zero $\langle a^2\rangle$ results in a gap ratio of less than the BCS value, (4) alone yields a lower bound on $\langle a^2\rangle$. A full application of the method proposed in this paper is capable of yielding a unique value of $\langle a^2\rangle$, as well as NOV(t).

We must emphasize that the method just described is basically designed to account for deviations from the "law of corresponding states" results predicted by the BCS weak-coupling *model* within the context of the BCS *theory* [3] in an *internally self-consistent* fashion. The absolute validity of the method rests upon the concept of an effective temperature-dependent pairing interaction. Thus, in the following sections, we make no claims for the *accuracy*, i.e., the absolute validity of our results; rather, we comment on the *precision*, i.e. the internal consistency of the results obtained. We have found the method to be useful. Moreover, the technique presented in this paper is a workable method for separating anisotropy and strong-coupling effects. With the renewed interest and activity in tunneling into bulk single-crystal superconductors [18, 19], this may well turn out to be a significant use of our method.

3. Results and Discussion

This method of determining NOV(t) has been used for Al, V, Ta, Nb, and Pb. The results for $t = 0$ and $t = 1$ are given in Table 1.[7]) Al, V, and Ta are seen

[6]) For a different approach to the problem of anisotropy which is, however, also based on Sheahen's ideas, see [16].

[7]) For detailed discussion of the numerical calculations, see [20]. For the materials considered, once a consistent set of thermodynamic parameters was found, the relative *precision* of the calculations of TOP and BOT was not limited by the scatter in the experimental data used, but rather by the accuracy of the atomic constants employed. (Admittedly, energy gap data with this little scatter are not yet readily available.) The *precision* of the calculations is thus about five parts in 10,000.

Table 1

Thermodynamic parameters

material	$\frac{2\Delta(0)}{k_B T_c}$	T_c (K)	$\Delta(0)$ (meV)	θ_c *) (K)	$\frac{\theta_c}{\theta_D}$ *)	$\frac{\Delta C}{\gamma T_c}$	$\left(\frac{dh}{dt}\right)_1$	$\frac{\gamma T_c^2}{V_m H_0^2}$	NOV Sheahen	TOP	BOT	(TOP–BOT)
aluminium	3.53 [21, 22]	1.20 [21, 22]	0.183	41.93	0.098	1.47	1.79	0.173	0.2718	0.2718	0.2718	$<1.0 \times 10^{-4}$
vanadium	3.56 [23]	5.38 [23]	0.825	43.00	0.108	1.52	1.79	0.168	0.4252	0.4251	0.4251	$<1.0 \times 10^{-4}$
tantalum	3.60 [24 to 26]	4.39 [24 to 26]	0.681	27.34	0.106	1.56	1.83	0.170	0.5117	0.5116	0.5115	1.5×10^{-4}
niobium-s	3.81**) [27]	9.20**) [27]	1.511	29.82	0.108	1.85	1.91	0.156	0.7685	0.7683	0.7573	1.1×10^{-2}
niobium-p	3.86***) [28, 29]	9.23 [28]	1.535	27.75	0.100	1.93	1.96	0.160	0.8155	0.8153	0.7989	1.7×10^{-2}
lead	4.32 [30, 31]	7.20 [30, 31]	1.340	14.82	0.140	2.67	2.19	0.141	1.1792	1.1789	1.0725	1.1×10^{-1}

*) $\theta_c \equiv \omega_c/k_B$, and $\theta_D \equiv \omega_D/k_B$ is the Debye temperature, where ω_D is the Debye energy.
**) The data used are for the ⟨111⟩ direction of tunneling in single-crystal specimens.
***) This is an averaged value determined from tunneling measurements on single crystal Nb of low residual resistivity ratio.

to be weak-coupling superconductors in the BCS sense, as expected, and the large differences between TOP and BOT for Nb and Pb show that they are indeed strong-coupling superconductors.

Fig. 1 shows NOV(t) for Pb determined with the data of [31]. The error flags of this data set, $\Delta\delta = 10^{-4}$ and $\Delta t = 5\times10^{-3}$, are so small that a distinctive curve for NOV(t) is traced out. For purposes of comparison, the NOV(t) curve determined from Mühlschlegel's [32] theoretical $\delta(t)$ is also shown.

Since precise energy gap versus temperature data for Nb were unavailable, experimental values of $\delta(t)$ for weak-coupling [21]) and for strong-coupling [31] superconductors were used to determine the extreme values of NOV(t) for this material. The results are shown in Fig. 2. The effect of the size of the error flags is clearly seen.

In order to obtain a distinctive curve for NOV(t), the uncertainty in $\delta(t)$ must beat least an order of magnitude smaller than the relative change between TOP and BOT (assuming $\Delta t = 5\times10^{-3}$). Thus the experimental data for Ta were not sufficiently accurate to permit a unique determination of NOV(t).

In the calculations for Al and V, the weak-coupling data [21] resulted in a constant NOV to within the precision of the computation.

For comparison, the calculations were also carried out with ω_D, rather than ω_c, as the upper limit in (1). The resultant values of NOV(0) are 0.3820, 0.2226, and 0.2718 for Pb, V, and Al, respectively. These values are much more nearly BCS-like (≈ 0.25) than those designated as TOP in Table 1. However, the resultant values of NOV(t) are not consistent with the BCS model. In the case of Pb, for example, as t varies from zero to unity, NOV(t) decreases by $\approx 2.6\times\times10^{-2}$, rather than remaining constant. For Al the values of NOV(t) for $t > 0.48$ are higher than NOV(0) by as much as 10^{-2}. Thus, it is evident that the straightforward use of the original BCS gap equation together with real material properties leads to inconsistencies. The method proposed in this paper entails no such difficulties.

These results indicate the manner in which the method for determining NOV(t) can be inverted for use as a tool in determining the precision of some given set of data $(\delta(t) \pm \Delta\delta, t \pm \Delta t)$, once $\Delta(0)$ and T_c are known. If NOV(t) is known, the

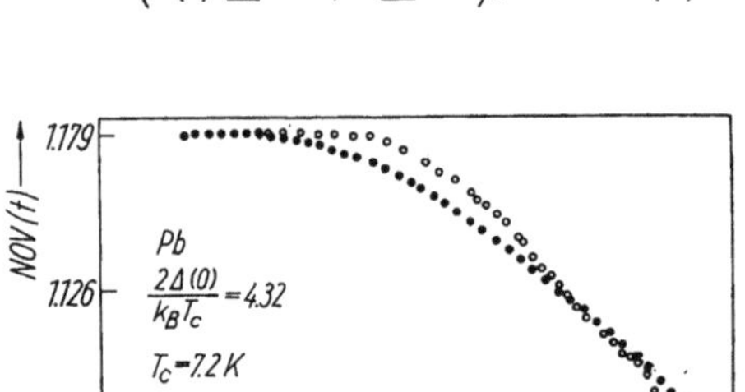

Fig. 1

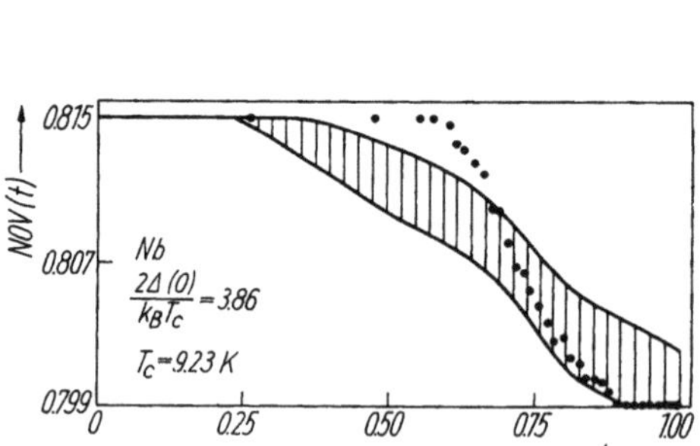

Fig. 2

Fig. 1. The effective interaction parameter for Pb determined with the experimental gap vs. temperature data of Gasparovic et al. [31] and with Mühlschlegel's [32] theoretical values for $\delta(t)$ vs. t. (●) Mühlschlegels $\delta(t)$, (○) experimental $\delta(t)$

Fig. 2. The extreme limits for the effective interaction parameter of Nb determined from weak-coupling [21] and strong-coupling [31] data for the energy gap vs. temperature. Closed points: strong-coupling, hatched area: weak-coupling

Fig. 3. An illustration of the way in which the method of determining NOV(t) can be inverted to reduce the uncertainties in an experimental [33] data set, $(\delta(t) \pm \Delta\delta, t \pm \Delta t)$

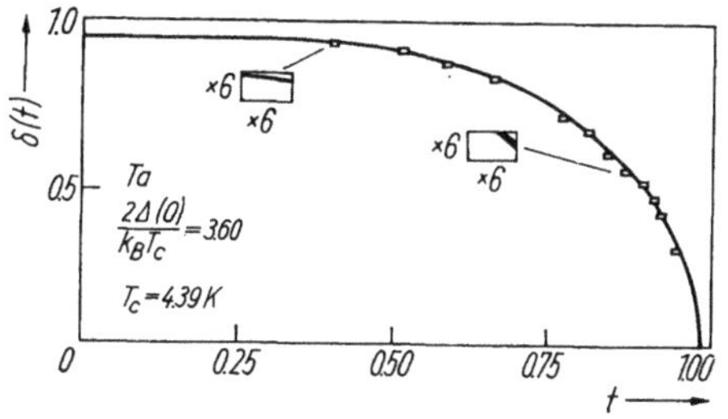

values of $\delta(t)$ can be computed from (1). If NOV(t) is not known, then $\Delta(0)$ and T_c will determine TOP and BOT, from which the values of $(\delta \pm \Delta\delta, t \pm \Delta t)$ consistent with the "shell" of values between TOP and BOT can be determined. In this way the method can be used to check the consistency of and perhaps even improve the statistics on any given set of experimental data. In Fig. 3 an example of this type of calculation is given for data taken on Ta film tunneling junctions [33]. The experimental errors were taken to be $\Delta\delta = 0.007$ and $\Delta t = 0.006$. It should be noted that one data point failed to meet the criterion TOP $\geqq$ NOV(t) $\geqq$ BOT.

A possible use of this procedure for checking the consistency and improving the statistics of an experimental data set for the energy gap would be to apply it to data which is to be used in conjunction with the McMillan-Rowell inversion program [34] to obtain the complex energy gap and phonon density of states. Considering the complexity and operational cost of the McMillan-Rowell program, this might be a significant use of the method proposed in this paper.

Once the values of ω_c and $\delta(t)$ have been determined, then it is possible to obtain the temperature dependence of the other thermodynamic variables of the BCS theory, and even to improve the statistics of experimental data sets for these quantities. However, one must use the general equations of the BCS theory before the weak-coupling approximations have been assumed, e.g., the equations which lead to equation (3.38) of BCS [3].

Additional support for this approach to obtaining a consistent set of thermodynamic parameters can be found in the determination of $H_c(t)$ from the measured specific heat data for Nb by Leupold and Boorse [35]. The deviation of the critical field from parabolic dependence on temperature, which is a very sensitive test of model calculations, is found to be positive for $0 < t \lessapprox 0.3$ and negative (smaller in magnitude than BCS) for $0.3 \lessapprox t < 1$. This is to be contrasted with the wholly positive deviations obtained when the strong-coupling theory is applied to strong-coupling materials)[8], and the totally negative deviations resulting from the BCS theory with or without anisotropy. Calculations of $H_c(t)$ were made by Swihart[9]) using a cutoff, ω_c, of $13.7\Delta(0)$, for

[8]) The so-called strong-coupling theory is, of course, applicable for the entire range of coupling strengths, and might thus more appropriately be referred to as the "general-coupling theory". When applied to a strong-coupling case, this theory yields a wholly positive deviation. (See [11].) By contrast, in the weak-coupling limit, the general theory reproduces the totally negative deviation obtained from the BCS theory. One might speculate that for an intermediate-coupling case, such as Nb, the calculated deviation would indeed be sometimes positive and sometimes negative. Unfortunately, no such calculation, employing a realistic phonon spectrum, for an intermediate-coupling material is as yet available.

[9]) See Fig. 5 and the accompanying discussion in [35].

$2\Delta(0)/k_B T_c$ equal to 3.75 and 3.84. The results of these calculations show the same qualitative behavior as observed experimentally for Nb.

4. Comparison with the McMillan Theory

The results of Section 3 suggest that T_c and NOV are related according to the equation

$$T_c = 1.134\,\theta_c \exp\left(\frac{-1}{\text{NOV}}\right) \tag{5}$$

which is identical in form to the BCS weak-coupling result. However, in contrast to the BCS relation, θ_c is not the Debye temperature, θ_D; rather, $\theta_c \approx \theta_D/10$.[10]) Equation (5) is in good agreement with experimental data on a variety of superconductors [36 to 38], provided NOV is identified with λ, the electron–phonon coupling parameter of the Eliashberg theory. The successful use of (5) in calculating T_c prompts a comparison of this result with the expression for T_c derived from the strong-coupling theory by McMillan [12].

The strong-coupling theory explicitly takes into account quasiparticle damping and the retarded character of the electron–phonon interaction, while retaining the pairing hypothesis of the BCS theory.[11]) The resulting equations relate the superconducting properties of a material to its normal state properties, namely the electron–phonon coupling parameter λ, the Coulomb pseudopotential μ^*, and the Debye temperature θ_D. Once these normal state parameters are given, the equations can be solved for the gap parameter, critical temperature, and electronic density of states, permitting one, in principle, to predict whether a particular material is a superconductor.

Using the strong-coupling theory and assuming a simplified Nb phonon spectrum, McMillan [12] has derived the following equation for T_c:

$$T_c = \frac{\theta_D}{1.45} \exp\left[-\frac{1.04\,(1+\lambda)}{\lambda - \mu^*\,(1 + 0.62\,\lambda)}\right]. \tag{6}$$

The assumption of a Nb phonon spectrum introduces important errors only for those materials with a phonon spectrum significantly different from that of Nb. Otherwise (6) should be applicable to those materials well described by an isotropic, one-band, electron–phonon model of superconductivity.

To compare the results of the present work with the McMillan theory, consider Nb and Pb.[12]) McMillan's values of λ for these two materials are 0.82 and 1.12, respectively. From Table 1 we see that it is reasonable to identify NOV in (5) with λ. However, in making such an identification it is assumed that all screening and renormalization effects on T_c can be accounted for in this model

---

[10]) Calculations for Hg, which must be regarded as preliminary because energy gap data sufficiently precise for a complete calculation are presently unavailable, indicate TOP ≈ 1.33 and $\theta_c/\theta_D \approx 0.13$. Thus θ_c/θ_D increases from ≈ 0.10 for weak-coupling materials to ≈ 0.14 for strong-coupling materials.

[11]) It has been shown (see [39]) that the Eliashberg theory reduces to the BCS theory in the absence of lifetime effects, so that the retarded nature of the pairing interaction is in effect "built into" the BCS theory.

[12]) Comparison for Al, V, and Ta is not possible because the critical data used in the two calculations are different.

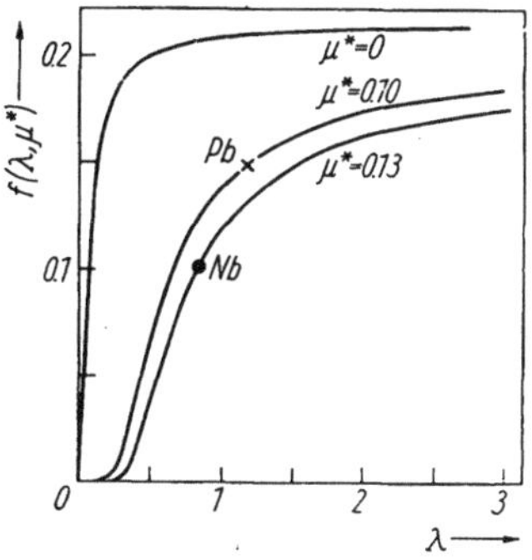

Fig. 4. The variation of Debye coupling strength $f(\lambda, \mu^*)$ as a function of λ and μ^* in the McMillan theory [12]. The values for Nb and Pb are indicated in the figure, assuming the values of λ calculated by McMillan

by a single number, θ_c/θ_D. That is, the McMillan equation, (6), can be written in the form of (5), but with θ_c replaced by $f(\lambda, \mu^*)\,\theta_D$:

$$T_c = f(\lambda, \mu^*)\left[1.134\,\theta_D \exp\left(\frac{-1}{\lambda}\right)\right]. \tag{7}$$

The Debye strength $f(\lambda, \mu^*)$, which is defined by (6) and (7) and which contains the effects of screening and renormalization on T_c, is plotted in Fig. 4 as a function of λ for different values of μ^*. The values of λ and μ^* obtained from experiment are such that the values of $f(\lambda, \mu^*)$ for actual materials are very close to the values of θ_c/θ_D calculated by the present method. For example, the Debye strengths of Pb and Nb determined with McMillan's values of λ and μ^*, which are shown in Fig. 4, are essentially the same as the values of θ_c/θ_D given in Table 1.

A further comparison of the values of NOV with λ is provided by the tunneling data recently obtained by Dynes [13] for the In–Tl alloy system. This alloy system exhibits four distinct crystal structures, each having its own characteristic phonon density of states. The values of T_c calculated from (5) and (6) are plotted as functions of Tl concentration in Fig. 5. The Debye temperature is

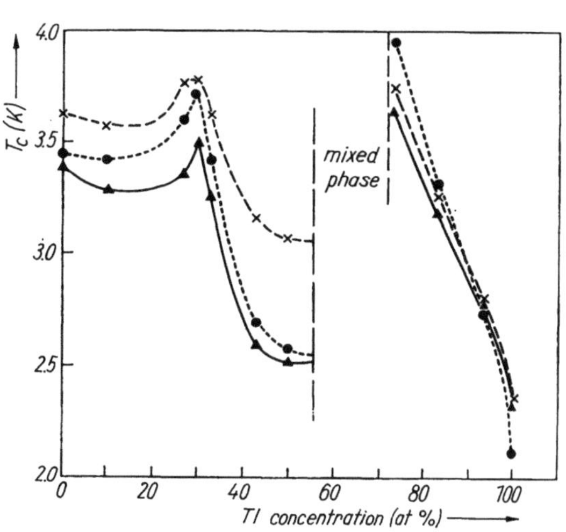

Fig. 5. Comparison of theoretical and experimental [13] values of T_c for the alloy system $In_{1-x}Tl_x$. (▲) T_c experimental, (●) T_c McMillan, (×) T_c present work

calculated from [40]

$$\theta_{\mathrm{D}} = 1.2\left[0.38 + 0.66\,\frac{\langle\omega^2\rangle}{\omega_0^2}\right]\frac{\omega_0}{k_{\mathrm{B}}}\,, \tag{8}$$

where $\langle\omega^2\rangle$ is the mean squared energy of the phonon spectrum and ω_0 is the maximum energy in the phonon spectrum. The values of λ, μ^*, $\langle\omega^2\rangle$, and ω_0 used are those given by Dynes [13]. The results from (5), taking $\theta_c/\theta_D = 0.1$, are well within 10% of the experimental values except for the two alloy compositions immediately preceding the "mixed phase" of the system. Better agreement would be expected if measured values of θ_D were used and if the Debye strength were varied as a function of alloy composition. As noted by Dynes, the values of T_c calculated with the McMillan equation are in good agreement with the experimental values throughout the entire alloy series. Both equations for T_c show the correct qualitative concentration dependence.

Although (5) fits the data of Dynes on the In–Tl system reasonably well, it should be remembered that this equation cannot be used to predict critical temperatures. For a given material the effective interaction strength and Debye strength can be determined only by the entire procedure described in Section 2. However, in addition to giving T_c, the present method also results in a complete set of thermodynamic parameters, and can be used to obtain the temperature-dependent thermodynamic variables as well.[13])

5. Conclusions

A procedure is presented whereby the deviations of the properties of various superconducting materials from the "law of corresponding states" behavior predicted by BCS can be reconciled within the BCS framework. The central feature of this procedure is that the effective interaction strength is permitted to be a function of temperature. This effective interaction strength NOV(t) is computed in two steps. First, the width of that energy shell centered on the Fermi surface throughout which there is an effective pairing interaction is determined. It is required that this width be such that the equations for the thermodynamic parameters of the superconductor, in the weak-coupling limit, yield the experimental values. Second, using this shell as the domain of integration in the BCS gap equation, NOV(t) can be calculated from the experimentally measured temperature dependence of one of the thermodynamic variables of the material, most conveniently the energy gap. The temperature dependence of the other thermodynamic variables can then be determined using NOV(t) and the equations of the BCS theory. The effects of anisotropy can be included from the outset. This procedure can also be inverted to check the consistency of and improve the statistics on experimental data for the temperature dependence of the thermodynamic properties of the material.

This procedure suggests an equation for the critical temperature in terms of the Debye temperature and the electron–phonon coupling constant. This relation is compared with the McMillan equation for T_c, derived on the basis of the strong-coupling theory of superconductivity. There is good agreement with that theory in cases where this method is applicable: a bulk material which is well described by an isotropic, one band, electron–phonon model of superconductivity.

[13]) Similar calculations are being carried out for the additional materials considered by Leavens [41].

Acknowledgements

We would like to thank the authors of [21, 25, 27, 31] for furnishing us their data on the energy gap as a function of temperature, without which this method could not have been tested. We are grateful to the Georgetown University Computation Center for the use of their facilities. One of us (J. B.) would also like to thank the members of the Institut für Theoretische Festkörperphysik of the Universität des Saarlandes, Germany, for suggestions on the manuscript and discussions of strong-coupling superconductivity.

References

[1] J. Bostock, W. N. Mathews Jr., and W. D. Gregory, Bull. Amer. Phys. Soc. **16**, 498 (1971).
[2] T. P. Sheahen, Phys. Rev. **149**, 370 (1966).
[3] J. Bardeen, L. N. Cooper, and J. R. Schrieffer, Phys. Rev. **108**, 1175 (1957).
[4] A. Rothwarf, Phys. Letters **26A**, 43 (1967); **28A**, 430 (1968).
[5] G. M. Eliashberg, Soviet Phys. — J. exper. theor. Phys. **11**, 696 (1960).
[6] P. B. Allen and M. L. Cohen, Phys. Rev. B **1**, 1329 (1970).
[7] G. Grimvall, Phys. kondens. Materie **9**, 283 (1969).
[8] J. Appel, Phys. Rev. Letters **21**, 1164 (1968).
[9] W. Klose and H. Schuster, Z. Phys. **241**, 348 (1971).
[10] W. Dietrich and H. Schuster, Phys. Letters **35A**, 48 (1971).
[11] D. J. Scalapino, in: Superconductivity, Ed. R. D. Parks, Marcel Dekker, Inc., New York 1969 (Chap. 10).
[12] W. L. McMillan, Phys. Rev. **167**, 331 (1968).
[13] R. C. Dynes, Phys. Rev. B **2**, 644 (1970); **4**, 3255 (1971).
[14] J. C. Swihart, IBM J. Res. Developm. **6**, 14 (1962).
[15] E. Bucher, F. Heiniger, J. Muller, and J. L. Olsen, in: Proc. 9th Internat. Conference Low Temperature Phys., Ed. J. G. Daunt, D. O. Edwards, F. J. Milford, and M. Yaqub, Plenum Press, New York 1965 (p. 616).
[16] D. U. Gubser, Phys. Rev. B**6**, 827 (1972).
[17] J. R. Clem, Ann. Phys. (U.S.A.) **40**, 268 (1966).
[18] W. D. Gregory, R. F. Averill, and L. S. Straus, Phys. Rev. Letters **27**, 1503 (1971).
[19] W. D. Gregory, L. S. Straus, R. F. Averill, J. C. Keister, and C. Chapman, in: Proc. 13th Internat. Conference Low Temperature Phys., Ed. K. D. Timmerhaus and W. O'Sullivan, Plenum Press, New York 1974 (Vol. 3, p. 316).
[20] J. Bostock, W. D. Gregory, L. Leopold, and W. N. Mathews, Jr., Self-Consistent Threshold Calculations for the Superconducting Injection Laser, AEC Report No. ORO-3665-24, 1970 (Appendix A).
[21] B. L. Blackford and R. H. March, Canad. J. Phys. **46**, 141 (1968).
[22] G. L. Wells, J. E. Jackson, and E. N. Mitchell, Phys. Rev. B **1**, 3636 (1970).
[23] R. Radebaugh and P. H. Keesom, Phys. Rev. **149**, 209 (1966).
[24] I. Dietrich, Z. Naturf. **17a**, 94 (1962).
[25] L. J. Barnes, Phys. Rev. **184**, 434 (1969).
[26] P. L. Richards and M. Tinkham, Phys. Rev. **119**, 575 (1960).
[27] R. B. Laibowitz and J. J. Cuomo, J. appl. Phys. **41**, 2748 (1970).
[28] P. Townsend and J. Sutton, Phys. Rev. **128**, 591 (1962).
[29] M. L. A. MacVicar and R. M. Rose, J. appl. Phys. **39**, 1721 (1968).
[30] I. Giaever and K. Mergerle, Phys. Rev. **122**, 1101 (1961).
[31] R. F. Gasparovic, B. N. Taylor, and R. E. Eck, Solid State Commun. **4**, 59 (1966).
[32] B. Mühlschlegel, Z. Phys. **155**, 313 (1959).
[33] I. Giaever, in: Proc. 8th Internat. Conference Low Temperature Phys., Ed. R. O. Davies, Butterworths, Washington, D.C. 1963 (p. 171).

[34] W. L. McMILLAN and J. M. ROWELL, in: Superconductivity, Ed. R. D. PARKS, Marcel Dekker, Inc., New York 1969 (Chapter 11).
[35] H. A. LEUPOLD and H. A. BOORSE, Phys. Rev. **134**, A1322 (1964).
[36] F. J. MORIN and J. P. MAITA, Phys. Rev. **129**, 1115 (1963).
[37] P. SPITZLI, Chaleur Specifique D'Alliages de Structure A-15, unpublished thesis, University of Geneva 1970.
[38] R. FLÜKIGER, F. HEINIGER, A. JUNOD, J. MULLER, P. SPITZLI, and J. L. STAUDENMANN, J. Phys. Chem. Solids 32, 459 (1970).
[39] A. BARDASIS and J. R. SCHRIEFFER, Phys. Rev. Letters **7**, 79 (1961).
[40] J. W. GARLAND, in: Proc. 1st Symposium 1969 Spring Superconducting Symposia, Naval Res. Lab. Rep. 6292, Washington, D. C. 1969 (p. 42).
[41] C. R. LEAVENS, Solid State Commun. **13**, 1607 (1973).

(Received January 25, 1974)

phys. stat. sol. (b) 62, 585 (1974)

Subject classification: 13.4; 22.1.2

Department of Electronics, Faculty of Technology, Kanazawa University

LCAO MO Treatment for Electronic Structure of Deep Levels in Semiconductors — S^+ Donor in Silicon

By

T. SHIMIZU and K. MINAMI[1])

The energy and wave function of the ground state of the S^+ donor in silicon are calculated by empirical LCAO MO method. The donor state is simulated by a cluster of a S^+ ion and 274 silicon atoms surrounding it. The interaction and overlap integrals determined by Alstrup and Johansen are used which fit the bands of silicon calculated by LCAO method to those of OPW calculation. The ground state is known to be totally symmetric, so that only the totally symmetric energy states are calculated. The fact enables to take the large cluster. The calculated MO coefficients agree rather well with those estimated from ENDOR data. Extended Hückel theory is found to be unsuccessful in describing the donor state in this case.

Energie und Wellenfunktion des Grundzustands des S^+-Donators in Silizium werden durch eine empirische LCAO-MO-Methode berechnet. Der Donatorzustand wird durch ein Cluster, das ein S^+-Ion und 274 Siliziumatome enthält, simuliert. Es werden Wechselwirkungs- und Überlappungsintegrale benutzt, die von Alstrup und Johansen durch Anpassung der LCAO-Bänder des Siliziums an OPW-Berechnungen bestimmt wurden. Es ist bekannt, daß der Grundzustand totalsymmetrisch ist, deshalb werden die totalsymmetrischen Energiezustände allein berechnet. Diese Tatsache gestattet die Einbeziehung des großen Clusters. Die berechneten MO-Koeffizienten stimmen befriedigend mit denen, die aus ENDOR-Daten bestimmt werden, überein. Es wird gefunden, daß eine erweiterte Hückel-Theorie in diesem Fall für eine Beschreibung des Donatorzustands erfolglos ist.

1. Introduction

The effective mass theory [1] which has been successful in describing the electronic structure of shallow impurity states in semiconductors is not appropriate to describe the electronic structure of deep impurity states in semiconductors. Instead, deep impurity states are known to be successfully described by the LCAO MO treatment. Recent work by Messmer and Watkins [2 to 4] shows that the nitrogen donor in diamond is successfully described by an extended Hückel theory (EHT) [5] which is a simple one-electron LCAO MO theory. They simulated the system by a large molecule made up of the nitrogen atom and 70 carbon atoms surrounding it. The similar treatment is applied to defect centres in SiO_2 by Bennett and Roth [6, 7]. Larkins discussed the validity of the application of the EHT method to impurity problems in solids [8].

The S^+ donor in silicon has been investigated experimentally in detail by optical measurements [9] and ENDOR [10], and the ground state level is known to situate 0.61 eV below the bottom of a conduction band. Several works have been done to describe the electronic structure of the S^+ donor in the framework of the effective mass formalism corrected in several points [11 to 14]. However, those treatments seem not to be satisfactory. For example, Müller's [11] treat-

---

[1]) Present address: Kanazawa Women's college, Kanazawa.

ment has an arbitrariness in determining the radius of a central cell and Ning and Sah [14] did not explain the ENDOR data. Previously, the present authors [15] attempted to treat the electronic structure of the S^+ donor in silicon by utilizing the method of Stocker [16]. The donor wave function was expressed by the linear combination of the antibonding orbitals (made up of sp^3 hybridized orbital) which constitute a conduction band. The interaction integrals between those orbitals were determined so as to fit the conduction band of silicon and the overlap integrals between them were neglected. The effect of substituting a silicon atom by a S^+ ion was expressed by one parameter which was determined by equating the calculated energy of the ground state of the S^+ donor to the observed one. The resulting wave function was compared with that estimated from ENDOR data. The treatment has several shortcomings. 1. The use of a parameter whose value is determined using the observed ionization energy of the donor. 2. The use of the antibonding orbitals made up of sp^3 hybridized orbitals. 3. The neglect of the overlap integrals between antibonding orbitals. In order to improve those shortcomings, a LCAO MO cluster treatment similar to that of Messmer and Watkins is carried out for the S^+ donor in silicon in the present article. A preliminary result has already been published [17].

In our LCAO MO cluster treatment, (i) the values determined by Alstrup and Johansen (hereafter refer to AJ) [18] so as to fit the bands of silicon calculated by the LCAO method to those of OPW calculation are used for the interaction and overlap integrals between silicon atoms instead of those by EHT and (ii) the energy difference between the donor level calculated for the cluster of a finite size and the conduction band minimum calculated by AJ for a perfect silicon crystal of an infinite size by ultilizing a periodicity is compared with the observed ionization energy of the donor. The S^+ ion is assumed to be substitutional in the present calculation. The ground state is known to be totally symmetric (A_1 representation of T_d group) [10], which makes a numerical calculation simple. The calculation by using the interaction and overlap integrals determined by EHT method proves to be unsuccessful.

2. Method of Calculation

The donor state is simulated by a cluster of a S^+ ion and silicon atoms surrounding it. The wave function is taken to be a linear combination of atomic orbitals (LCAO) of them. In order to save computing time, the secular equation is factorized by taking advantage of symmetry and only the molecular orbitals (MO) belonging to A_1 representation are calculated, because the ground state is known to be A_1.

The MO's, ψ_i, belonging to A_1 are expressed by

$$\psi_i = \sum_\nu C_{\nu i} \varphi_\nu \,, \tag{1}$$

where φ_ν is a totally symmetric linear combination of equivalent atomic orbitals $\chi_\alpha^{(\nu)}$:

$$\varphi_\nu = \frac{1}{\sqrt{N_\nu}} \sum_\alpha \chi_\alpha^{(\nu)} \,, \tag{2}$$

where N_ν is the number of equivalent atomic orbitals. The secular equations

$$\sum_\nu (H_{\mu\nu} - E_i G_{\mu\nu}) \, C_{\nu i} = 0 \tag{3}$$

give the energies E_i and MO coefficients $C_{\nu i}$. Here $H_{\mu\nu}$ and $G_{\mu\nu}$ are given as follows:

$$H_{\mu\nu} = \frac{1}{\sqrt{N_\mu N_\nu}} \sum_{\alpha,\beta} \langle \chi_\alpha^{(\mu)} | H | \chi_\beta^{(\nu)} \rangle \tag{4}$$

and

$$G_{\mu\nu} = \frac{1}{\sqrt{N_\mu N_\nu}} \sum_{\alpha,\beta} \langle \chi_\alpha^{(\mu)} | \chi_\beta^{(\nu)} \rangle \,. \tag{5}$$

H is a one-electron Hamiltonian and separated into two parts, H_0 and H', where H_0 is a Hamiltonian for a perfect crystal and H' is a perturbation caused by substituting a silicon atom by a S^+ ion. In the present calculation, 3s orbital of the S^+ ion χ_0 and 3s and 3p orbitals of each silicon atom are taken as atomic orbitals. It is already known that the wave function of the ground state is A_1 from ENDOR, so 3p orbitals of the S^+ ion should not be admixed to stabilize the ground state by a lattice distortion as in the case of a nitrogen donor in diamond [2 to 4].

The present method is different from the EHT method in the point that the interaction and overlap integrals between silicon atoms determined by AJ are used instead of those calculated from the following formulae (EHT method):

$$\langle \chi_\alpha^{(\mu)} | H_0 | \chi_\alpha^{(\mu)} \rangle = - I_\mu \tag{6}$$

and

$$\langle \chi_\alpha^{(\mu)} | H_0 | \chi_\beta^{(\nu)} \rangle = - \frac{1}{2} K(I_\mu + I_\nu) \langle \chi_\alpha^{(\mu)} | \chi_\beta^{(\nu)} \rangle \tag{7}$$

for $(\alpha, \mu) \neq (\beta, \nu)$, where I_μ is the valence state ionization potential, $\langle \chi_\alpha^{(\mu)} | \chi_\beta^{(\nu)} \rangle$ is an overlap integral between Slater-type atomic orbitals and K is a constant equal to 1.75.

AJ calculated the conduction and valence bands of silicon by a linear combination of 3s and 3p atomic orbitals by taking into account the interaction and overlap integrals up to third neighbours, and good overall agreement with OPW calculation is attained by adjusting those integrals so as to make the bands fit to the OPW calculation. The right energy gap 1.1 eV is attained by this way. The overlap integrals were calculated from Slater-type atomic orbitals and the orbital exponent was adjusted. The adjusted value is larger by the factor 1.11 compared with that determined by the Slater rule.

The interaction and overlap integrals are determined as follows using the calculated results by AJ:

1. When μ and ν do not refer to χ_0, $\langle \chi_\alpha^{(\mu)} | H_0 | \chi_\beta^{(\nu)} \rangle$ and $\langle \chi_\alpha^{(\mu)} | \chi_\beta^{(\nu)} \rangle$ are taken as those given by AJ, and as for the matrix elements of H', one takes

$$\langle \chi_\alpha^{(\mu)} | H' | \chi_\beta^{(\nu)} \rangle = - \frac{e^2}{\varkappa(\bar{r})\ \bar{r}} \langle \chi_\alpha^{(\mu)} | \chi_\beta^{(\nu)} \rangle \tag{8}$$

provided that the overlap integral $\langle \chi_\alpha^{(\mu)} | \chi_\beta^{(\nu)} \rangle$ is taken to be that determined by AJ. Here $\varkappa(\bar{r})$ is an r-dependent dielectric constant of silicon given by Azuma and Shindo [19] and $\bar{r}$ is defined by

$$\frac{1}{\bar{r}} = \frac{(1/r_\mu + 1/r_\nu)}{2}, \tag{9}$$

where r_μ and r_ν denote the distance between the S nucleus and silicon nuclei belonging to groups μ and ν, respectively.

2. As for the evaluation of the interaction and overlap integrals relating to the atomic orbital of the S^+ ion, the following method is adopted. First they are calculated by the EHT method, and then they are multiplied by reduction factors which are determined so as for the corresponding parameters between Si atoms calculated by the EHT method to agree with those by AJ. These reduction factors signify that the atoms in consideration exist in the Si crystal.

$$\langle\chi_0|\, H_0 + H'\,|\chi_0\rangle = -\, I_{3s}(S^+)\,\frac{H_0(3s,\, 3s)}{-\, I_{3s}(Si)}\,, \tag{10}$$

where $I_{3s}(Si)$ and $I_{3s}(S^+)$ are I_μ's for 3s orbitals of neutral silicon atom and S^+ ion (15.0 and 34.0 eV, respectively) and $H_0(3s, 3s)$ is a diagonal element of H_0 between the 3s orbital of silicon given by AJ (-12.6 eV). When μ does not refer to χ_0, $\langle\chi_\alpha^{(\mu)}|\, H_0 + H'|\chi_0\rangle$ are approximated by

$$\frac{1}{2} K_\mu\langle\chi_\alpha^{(\mu)}|\chi_0\rangle \left(\langle\chi_0|\, H_0 + H'|\chi_0\rangle + \langle\chi_\alpha^{(\mu)}|\, H_0\,|\chi_\alpha^{(\mu)}\rangle - \frac{e^2}{\varkappa(r_\mu)\, r_\mu}\right) \tag{11}$$

based on the EHT method. The values of K_μ are determined by equating $\langle\chi_\alpha^{(\mu)}|\, H_0\,|\chi_\beta^{(\nu)}\rangle$ given by AJ to

$$\frac{1}{2} K_{\alpha\beta}^{(\mu\nu)}\langle\chi_\alpha^{(\mu)}|\chi_\beta^{(\nu)}\rangle\, (\langle\chi_\alpha^{(\mu)}|\, H_0\,|\chi_\alpha^{(\mu)}\rangle + \langle\chi_\beta^{(\nu)}|\, H_0\,|\chi_\beta^{(\nu)}\rangle)\,, \tag{12}$$

where the interaction and overlap integrals are those given by AJ. The values of K_μ determined in this way are 1.38, 1.83, and 1.75, respectively, when μ is nearest, second, and third neighbours. K_μ depends not only on the relative configuration between the two atoms but also on the types of the atomic orbitals, but the dependence on the latter is averaged. The values of $\langle\chi_\alpha^{(\mu)}|\chi_0\rangle$ are taken to be those calculated by using Slater orbitals multiplied by the ratios of the calculated $\langle\chi_\alpha^{(\mu)}|\chi_\beta^{(\nu)}\rangle$ to those given by AJ. The ratios are 0.78, 0.52, 0.83, 0.46, 0.54, and 0.50 for overlaps between s–s for nearest, second, third neighbours, between s–p for nearest, second, and third neighbours, respectively.

3. Results and Discussions

It is found that the conventional EHT method does not reproduce the right energy bands of silicon and also that it does not work well in describing a S^+ donor state in silicon. The obtained results show that the width of the conduction band is as large as 146 eV while it is 8.7 eV by OPW calculation and that the energy gap is 3.89 eV (the true energy gap is 1.1 eV). Consequently, the calculation is carried out by using the overlap and interaction integrals determined by AJ.

We will mention about the size of the cluster and the band edge from which the donor level is measured. Since we are interested in the donor level, the criterion determining the size of the cluster is the convergence of the calculated position of the donor level by enlarging the size of the cluster. The energy levels corresponding to a conduction and valence band need not be approached to the right energy levels of a perfect crystal of an infinite size. The energy bands of a perfect crystal of an infinite size are already calculated by AJ in agreement with the right energy bands by utilizing the periodicity. Therefore

Fig. 1. The dependences on the size of the cluster of the calculated energies of the lowest level belonging to a conduction band (○), the donor level (●), and the highest level belonging to a valence band (△). The abscissa is the radius of the cluster. Two solid lines show the bottom of the conduction band and the top of the valence band of a perfect crystal and the dashed line shows the observed S^+ donor level

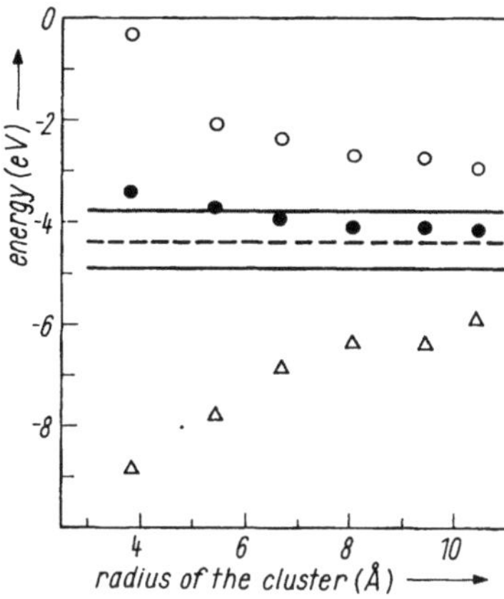

it is desirable to measure the donor level from the conduction band minimum calculated by AJ in contrast with most of previous works in which the impurity level was measured from the highest or lowest level in the group of levels corresponding to a valence or conduction band, respectively. This problem is discussed by Messmer and Watkins [4].

In order to see how the size of the cluster makes the calculated results vary, the energy levels belonging to A_1 representation for the clusters of a S^+ ion and surrounding silicon atoms are calculated for six clusters, i.e., those of a S^+ ion and surrounding 16, 34, 70, 122, 166, and 274 silicon atoms which correspond to clusters with radii 3.84, 5.43, 6.65, 8.03, 9.41, and 10.43 Å, respectively. The dependences on the size of the cluster of the calculated energies of the lowest level belonging to a conduction band, the highest level belonging to a valence band and the donor level are shown in Fig. 1. The size of the largest cluster is found to be sufficient for the convergence of the position of the donor level. Table 5 in [4] shows that the size of the largest cluster taken by Messmer and Watkins seems to be insufficient.

The calculated energy levels for the largest cluster are shown in Fig. 2. The ranges of the conduction and valence bands calculated by AJ are shown together. The groups of levels belonging to a conduction and valence band are seen and the gap between them is about 3 eV which is larger than the true

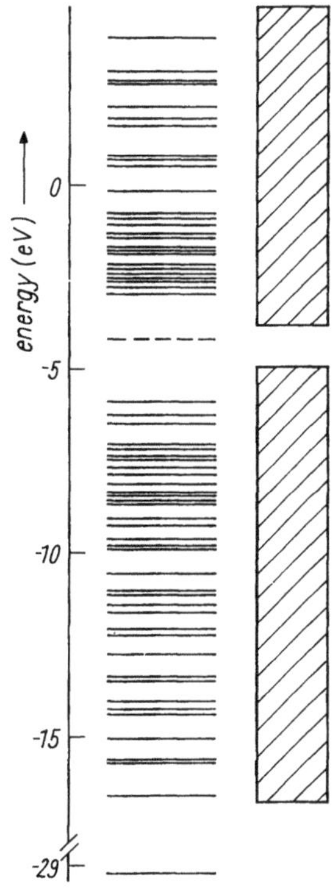

Fig. 2. The calculated MO energies of totally symmetric states for a cluster of a S^+ ion and surrounding 274 silicon atoms. The dashed line shows the S^+ donor level. The ranges of the conduction and valence bands calculated by AJ are shown together by hatched columns

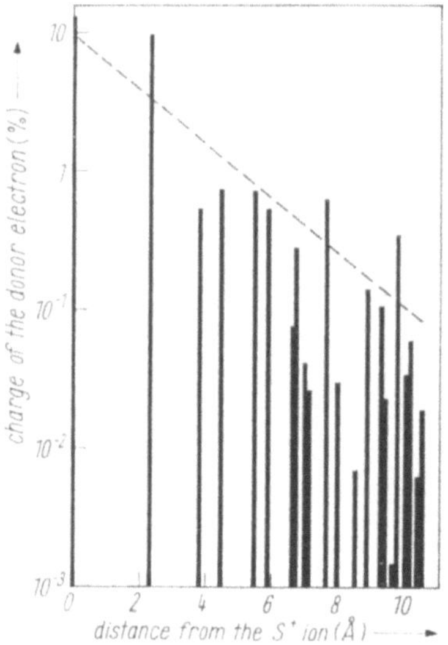

Fig. 3. Percentage of the charge of the donor electron on each of the atom sites vs. the distance from the S^+ ion. The square of the envelope function by an effective mass theory normalized in the cluster is shown by a dashed line for comparison

energy gap of silicon (1.1 eV). The level denoted by a dashed line shows the ground state of the S^+ donor which situates 0.36 eV below the bottom of the conduction band of the perfect lattice. The value is not so different from the observed value 0.61 eV. The MO coefficients C_0, C_s, and C_p are 0.49, −0.47, and −0.71. Here C_0 is the coefficient of χ_0 and C_s and C_p are, respectively, those of the sums of 3s and 3p$_\sigma$ orbital of four nearest neighbour silicon atoms, $\{\chi_s(1) + \chi_s(2) + \chi_s(3) + \chi_s(4)\}/2$ and $\{\chi_p(1) + \chi_p(2)+\chi_p(3) + \chi_p(4)\}/2$. p$_\sigma$ means the p-orbital directed toward the S^+ ion. The percentage of the charge of the donor electron on each of the atom sites is plotted in Fig. 3 versus the distance from the S^+ ion. The square of the envelope function by an effective mass theory of binding energy 0.61 eV (Bohr radius 4.5 Å) normalized in the cluster is also plotted by a dashed line.

Here the factors determining the position of the donor level are worth mentioning. Roughly speaking, this level is an antibonding orbital made up of 3s orbital of the S^+ ion and 3s and 3p orbitals of nearest neighbour silicon atoms. The energy is determined mainly by two factors, i.e., the energy of 3s orbital of the S^+ ion $\langle\chi_0|H|\chi_0\rangle$ and the splitting between the bonding and antibonding state. The magnitude of the splitting is determined by the off-diagonal interaction matrix elements which depend on the overlap integrals between the S^+ ion and silicon atoms and the value of $\langle\chi_0|H|\chi_0\rangle$ in the present approximation. The lowering of the value of $\langle\chi_0|H|\chi_0\rangle$ lowers the average value of the bonding and antibonding state and enlarges the splitting between them. As a result, it is found that this makes the energy of the antibonding state higher. $\langle\chi_0|H|\chi_0\rangle$ is far lower than that of silicon, so that the donor level should situate higher than the average of conduction band levels (antibonding states between silicon atoms) if the difference of the overlap integrals between silicon atoms and that between silicon atom and the S^+ ion might be neglected. In fact, the overlap between silicon atom and S^+ ion is smaller than that between silicon atoms and the fact makes the splitting smaller. Numerical calculation shows that this effect is important and as a result the level situates below the conduction band.

The Coulomb energy $-e^2/\varkappa r$ was found to make only a small effect on the energy level.

In order to compare the calculated MO coefficients with the experimental values, they are estimated from the ENDOR data by Ludwig [10]. The observed hfs constants can be calculated by the following formulae:

$$A_0 = \frac{8\pi}{3} g_e\beta_e g_n\beta_n\, |\chi_0(0)|^2\, C_0^2\,, \tag{13}$$

$$A_1 = \frac{8\pi}{3} g_e\beta_e g'_n\beta_n\, |\chi_s(0)|^2\, \frac{C_s^2}{4}\,, \tag{14}$$

and

$$B_1 = \frac{2}{5} g_e\beta_e g'_n\beta_n\, \langle 1/r^3\rangle_p\, \frac{C_p^2}{4}\,. \tag{15}$$

Here A_0 is the hfs constant due to Fermi contact interaction with S^{33} nucleus and A_1 and B_1 are those due to Fermi contact interaction and dipolar interaction with a nearest neighbour Si^{29} nucleus, respectively. g_e and β_e denote the g-value and Bohr magneton of electron, respectively, g_n and g'_n the nuclear g-value of S^{33} and Si^{29} nuclei, respectively, and β_n the nuclear magneton. If we use the calculated atomic values for $|\chi_0(0)|^2$, $|\chi_s(0)|^2$, and $\langle 1/r^3\rangle_p$, i.e., 10.0, 5.15, and 2.69 at. units, respectively [21], $|C_0| = 0.30$, $|C_s| = 0.17$, and $|C_p| = 0.65$ are obtained from the observed values of A_0, A_1, and B_1. They agree rather well with those of the MO calculation. In the crystal, the values of $|\chi_0(0)|^2$, $|\chi_s(0)|^2$, and $\langle 1/r^3\rangle_p$ may be different from those of free atoms. From the energy band calculation by AJ, the square of the amplitude of the Bloch function of the conduction band minimum normalized in unit volume can be calculated to be $159|\chi_s(0)|^2$ (in at. units). On the other hand its value has been obtained by Shulman and Wyluda [22] from the nuclear spin lattice relaxation time of Si^{29} to be 186. The comparison between them gives $|\chi_s(0)|^2 = 1.17$ at. units which is smaller by a factor 0.23 than the atomic value. $|\chi_0(0)|^2$ and $\langle 1/r^3\rangle_p$ cannot be known in this way, so that they are assumed to be reduced tentatively by the same factor as $|\chi_s(0)|^2$. When those reduced values are used for $|\chi_0(0)|^2$, $|\chi_s(0)|^2$, and $\langle 1/r^3\rangle_p$, $|C_0| = 0.63$, $|C_s| = 0.36$, and $|C_p| = 1.4$ are obtained. There is a possibility that the MO coefficients become larger than unity because of the non-orthogonality of φ_ν's, but 1.4 seems to be too large. Probably, $\langle 1/r^3\rangle_p$ may be smaller than the present estimation. For the S^+ ion $|\chi_0(0)|^2$ is expected to be larger than that for the neutral atom, i.e., 10.0 at. units, then the predicted value of $|C_0|$ should be smaller.

Finally, it seems to be useful to summarize the approximations made in the present treatment which should be improved in future work.

1. The interaction and overlap integrals involving the sulphur orbital are evaluated using a modified EHT method as shown in Section 2. At present, there is no idea how to improve this.

2. Charge distributions are not calculated self-consistently, and the valence state ionization energies are taken to be that of the S^+ ion and those of the neutral silicon atom. In accordance with it, a Coulomb potential around the donor is taken to be the one due to plus one elementary charge screened by an r-dependent dielectric constant. In order to determine the charge distribution self-consistently, all the energy states besides totally symmetric one should also be calculated.

3. The effect of the lattice distortion around the S^+ ion is neglected. If all the energy states were calculated, the magnitude of the lattice distortion could be determined by minimizing the total energy.

In conclusion it may be said that the present method of LCAO MO calculation gives rather good results not only for the ionization energy but also for the wave function of the S^+ donor in silicon.

Acknowledgement

The numerical calculations were carried out by an electronic computer of Kanazawa University, FACOM 230-35.

References

[1] W. KOHN, Solid State Phys. **5**, 257 (1957).
[2] R. P. MESSMER and G. D. WATKINS, Phys. Rev. Letters **25**, 656 (1970).
[3] G. D. WATKINS and R. P. MESSMER, in: Proc. 10. Internat. Conf. Phys. Semicond., Cambridge, Mass., 1970, Ed. S. P. KELLER, J. C. HENSEL, and F. STERN, U.S.AEC, Oak Ridge, Tenn., 1970 (p. 623)
[4] R. P. MESSMER and G. D. WATKINS, Phys. Rev. B **7**, 2568 (1973).
[5] R. HOFFMAN, J. chem. Phys. **39**, 1397 (1963).
[6] L. M. ROTH and A. J. BENNETT, in: Proc. 10. Internat. Conf. Phys. Semicond., Cambridge, Mass., 1970, Ed. S. P. KELLER, J. C. HENSEL, and F. STERN, U.S.AEC, Oak Ridge, Tenn., 1970 (p. 619).
[7] A. J. BENNETT and L. M. ROTH, J. Phys. Chem. Solids **32**, 1251 (1971).
[8] F. P. LARKINS, J. Phys. C **4**, 3065 (1971).
[9] W. E. KRAG, W. H. KLEINER, H. J. ZEIGER, and S. FISCHLER, J. Phys. Soc. Japan, Suppl. **21**, 230 (1966).
[10] G. W. LUDWIG, Phys. Rev. **137**, A1520 (1965).
[11] A. M. K. MÜLLER, Z. Naturf. **20a**, 1476 (1965).
[12] A. GLODEANU, Rev. Roum. Phys. **14**, 139 (1969).
[13] T. CASTNER, JR., Phys. Rev. B **2**, 4911 (1970).
[14] T. H. NING and C. T. SAH, Phys. Rev. B **4**, 3482 (1971).
[15] T. SHIMIZU and K. MINAMI, Memoirs of the Faculty of Technology, Kanazawa University **6**, 303 (1971) (in Japanese).
[16] D. STOCKER, Proc. Roy. Soc. **A270**, 397 (1962).
[17] T. SHIMIZU and K. MINAMI, phys. stat. sol. (b) **48**, K181 (1971).
[18] I. ALSTRUP and K. JOHANSEN, phys. stat. sol. **28**, 555 (1968).
[19] M. AZUMA and K. SHINDO, J. Phys. Soc. Japan **19**, 424 (1964).
[20] C. E. MOORE, Atomic Energy Levels, Vol. 1, National Bureau of Standards (U.S.), Circular **467**, U.S. Government Printing Office, Washington (D.C.) 1949.
[21] C. M. HURD and P. COODIN, J. Phys. Chem. Solids **28**, 523 (1966).
[22] R. SHULMAN and B. WYLUDA, Phys. Rev. **103**, 1127 (1956).

(Received December 28, 1973)

phys. stat. sol. (b) 62, 593 (1974)

Subject classification: 13.5.1

Ukrainian Agricultural Academy, Kiev

Graphical Technique Applied to the Calculation of Excitonic Processes

By

V. D. ISKRA

A modified diagrammatic technique is developed which is convenient to excitons in presence of external fields. The exciton Green's function is calculated as an illustration of the method.

Es wird eine modifizierte Diagrammtechnik entwickelt, die für Exzitonen bei äußeren Feldern geeignet ist. Die Greensche Funktion der Exzitonen wird als Beispiel für diese Methode berechnet.

1. Introduction

In some problems the necessity appears to calculate the two-particle Green's function in a random field. This may be done with the help of the standard diagrammatic technique, representing the one-particle Green's function of an electron (a hole) by means of the directed line describing the propagation of the particle in question [1, 2]. This technique may be used as well in the problems where the electron–hole bound states are of importance.

However, an explicit calculation of the two-particle Green's function using this technique appears to be rather complicated in such problems. In this case it might be more convenient to modify the diagrammatic technique and to describe the motion of the interacting electron and hole in terms of the centre-of-mass motion ("particle" with the total mass M) and the inner motion ("particle" with the reduced mass μ).

In this paper the modified diagrammatic technique is described, the relevant diagrams are compared with those of the conventional technique and the exciton Green's function in weak random field is calculated.

2. The Basic Equations

We shall start from the simplest model of two isotropic and parabolic bands having their extrema in the centre of the Brillouin zone. The Bloch functions are chosen as the basic system, the electron and hole creation and annihilation operators $a_{\boldsymbol{k}}^+$, $a_{\boldsymbol{k}}$ and $b_{\boldsymbol{k}}^+$, $b_{\boldsymbol{k}}$ referring to the states with the crystal momentum $\boldsymbol{k}$. The perturbed Hamiltonian includes just the terms describing the electron–hole pair interaction and their interaction with the static random field (the rest of the Coulomb interaction is taken into account only through the screened potential). Then the two-particle Green's function

$$K_e(\boldsymbol{k}_4, \boldsymbol{k}_3, \boldsymbol{k}_2, \boldsymbol{k}_1; t) = -i \lim_{\substack{t_4 \to t_1 \\ t_3 \to t_2}} \langle \Psi_0| \; T \; \{b_{\boldsymbol{k}_4}^+(t_4) \; b_{\boldsymbol{k}_3}(t_3) \; a_{\boldsymbol{k}_2}(t_2) \; a_{\boldsymbol{k}_1}^+(t_1)\} \; |\Psi_0\rangle \quad (2.1)$$

describes the propagation of the exciton in the random field (here Ψ_0 is the ground state wave function of the system with interaction, T the chronological operator, $t = t_2 - t_1$). The Fourier transform of the exciton Green's function, $K_e(\omega)$, satisfies the equation

$$\{\omega - \varDelta - T - V - U\}\, K_e = -I\,. \tag{2.2}$$

Here T is the kinetic energy operator of an electron and a hole (we use the "hole" energy normalization in the valence band), V the electron–hole interaction potential, U the potential energy of the exciton in the random field

$$U = U_c + U_v\,,$$

U_c and U_v being electron and hole potential energies in the random field, I is the unit operator, $\varDelta$ the band gap, and ω the energy parameter.

In the coordinate representation there are two equivalent ways to describe the system: one may use the electron and hole coordinates, $\boldsymbol{r}^c$ and $\boldsymbol{r}^v$, or the centre-of-mass and relative coordinates,

$$\boldsymbol{\varrho} = \boldsymbol{r}^c - \boldsymbol{r}^v \quad \text{and} \quad \boldsymbol{R} = \frac{m_c \boldsymbol{r}^c + m_v \boldsymbol{r}^v}{m_c + m_v}\,. \tag{2.3}$$

Here m_c and m_v are the electron and hole effective masses: $\hbar = 2m_0 = 1$, m_0 being the free-electron mass.

Let K_0 be the two-particle Green's function satisfying the equation

$$\{\omega - \varDelta - T\}\, K_0 = -I\,. \tag{2.4}$$

In the first case (we call it the m_c, m_v representation)

$$T = T_c + T_v\,, \qquad T_c = -\frac{1}{m_c}\nabla^2_{\boldsymbol{r}^c}\,, \qquad T_v = -\frac{1}{m_v}\nabla^2_{\boldsymbol{r}^v}\,. \tag{2.5}$$

In this case the function K_0 describes the independent propagation of two particles possessing the masses m_c and m_v, respectively. The corresponding diagram has the form

$K_0 =$  , (2.6)

the upper and the lower lines corresponding to the unperturbed electron and hole Green's functions, G_0^+ and G_0^-, respectively.

In the second case (we call it the M, μ representation)

$$T = T_M + T_\mu\,, \qquad T_M = -\frac{1}{M}\nabla^2_{\boldsymbol{R}}\,, \qquad T_\mu = -\frac{1}{\mu}\nabla^2_{\boldsymbol{\varrho}} \tag{2.5a}$$

and the same Green's function describes the independent propagation of two particles with the masses $M = m_v + m_c$ and $\mu = m_c m_v/(m_c + m_v)$. The corresponding diagram has the form

$K_0 =$  . (2.7)

The upper and lower lines correspond here to the unperturbed Green's functions of the particles possessing masses $\mu(G_0^\mu)$ and $M(G_0^M)$, respectively.

Let K be the two-particle Green's function, satisfying the equation

$$\{\omega - \varDelta - T - U\}\, K = -I\,. \tag{2.8}$$

In the m_c, m_v representation it may be represented in the form

$$K = \frac{i}{2\pi} \int\limits_{-\infty}^{+\infty} d\omega_1 \, d\omega_2 \, G^+(\omega_1) \, G^-(\omega_2) \, \delta(\omega - \omega_1 + \omega_2) \, , \tag{2.9}$$

G^+ and G^- being the electron and the hole one-particle Green's functions in the random field

(2.10)

(2.11)

Here the point at the electron (hole) line corresponds to the matrix element of the potential $U_c(U_v)$. Hence in the m_c, m_v representation the Green's function has the standard form

(2.12)

In the M, μ representation the matrix element of the random field has to join the lines of the particles M and μ:

(2.13)

The equation (2.13) may also be written in the symbolic form

$$K(\omega) = K_0(\omega) + K_0(\omega) \, UK(\omega) \, , \tag{2.14}$$

the function K_0 being defined by the equation

$$K_0(\omega) = \frac{i}{2\pi} \int\limits_{-\infty}^{+\infty} d\omega_1 \, d\omega_2 \, G_0^M(\omega_2) \, G_0^\mu(\omega_1) \, \delta(\omega - \omega_1 + \omega_2) \, . \tag{2.15}$$

Note that since the vertical line in (2.13) corresponds to the matrix element of the potential $U = U_c + U_v$, we have in fact

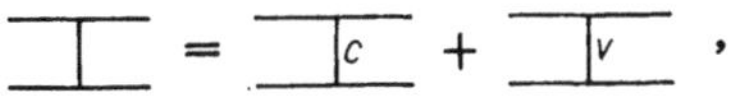
(2.16)

the vertical lines bearing the indices "c" and "v" corresponding to the matrix elements of U_c and U_v, respectively.

In the m_c, m_v representation the exciton Green's function K_e is represented by the diagrammatic equation

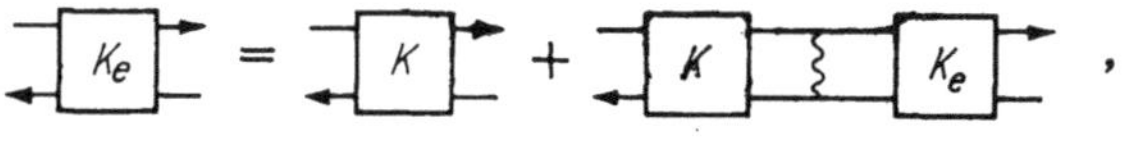
(2.17)

the wavy line corresponding to the matrix element of the interaction potential V.

In the M, μ representation the exciton Green's function K_e may be obtained if one "dresses" the particle μ-line in (2.13) by the potential V. Thus

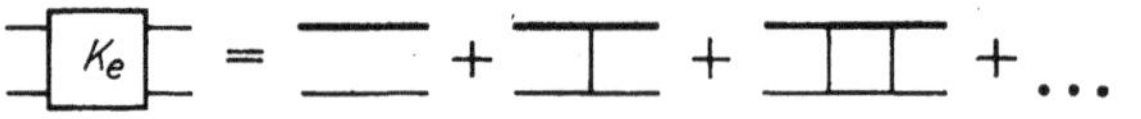

(2.18)

the full line being that of the particle μ "dressed" by the potential V:

(2.19)

Here the cross means the matrix element of the potential V.

Thus the difference between the two representations consists in the following: in the first case each particle, m_c as well as m_v interacts with the random field independently of each other, but both of them interact with each other via the potential V; in the second case both particles "interact" with each other via the random field, but only one of them, the particle μ, is scattered by the potential V.

3. The Calculation Rules

Now we formulate the rules of calculating the contribution of each diagram. We shall use the coordinate representation.

In the m_c, m_v representation these rules are conventional: we merely repeat them here for convenience.

1. Every electron line corresponds to the Green's function $G_0^+(\boldsymbol{r}_1^c - \boldsymbol{r}_2^c; \omega_1)$, $\boldsymbol{r}_1^c$ and $\boldsymbol{r}_2^c$ being the initial and final coordinates respectively:

$$G_0^+(\boldsymbol{r}_1^c - \boldsymbol{r}_2^c; \omega_1) = \sum_{\boldsymbol{k}} \frac{\exp\left[i\boldsymbol{k}(\boldsymbol{r}_1^c - \boldsymbol{r}_2^c)\right]}{\omega_1 - \dfrac{k^2}{m_c} + i\varepsilon}, \qquad \varepsilon \to +0 . \tag{3.1}$$

2. Every hole line corresponds to the Green's function $G_0^-(\boldsymbol{r}_1^v - \boldsymbol{r}_2^v; \omega_2)$:

$$G_0^-(\boldsymbol{r}_1^v - \boldsymbol{r}_2^v; \omega_2) = \sum_{\boldsymbol{k}} \frac{\exp\left[i\boldsymbol{k}(\boldsymbol{r}_1^v - \boldsymbol{r}_2^v)\right]}{\omega_2 + \Delta + \dfrac{k^2}{m} - i\varepsilon}, \qquad \varepsilon \to +0 . \tag{3.2}$$

3. The points correspond to the random potentials $U_c(\boldsymbol{r}^c)$, $U_v(\boldsymbol{r}^v)$.
4. The wavy line corresponds to the interaction potential $V(\boldsymbol{r}^c - \boldsymbol{r}^v)$.
5. Integration over all the intrinsic coordinates of the diagram is implied.
6. Every pair of the electron and hole lines corresponds to the integration over the variables ω_1 and ω_2 within the interval $(-\infty, +\infty)$, ω_1 and ω_2 being the energy parameters of the electron and hole lines: they are related by $\omega_1 - \omega_2 = \omega$, ω being the energy parameter of the exciton Green's function.
7. Every diagram bears the factor $(-1)^n\,(i/2\pi)^{n+1}$, n being the number of the wavy lines in the diagram.

In the M, μ representation the rules are the following:

1. Every line of the particle μ corresponds to the Green's function $G_0^\mu(\boldsymbol{\varrho}_1 - \boldsymbol{\varrho}_2; \omega_1)$:

$$G_0^\mu(\boldsymbol{\varrho}_1 - \boldsymbol{\varrho}_2; \omega_1) = \sum_{\boldsymbol{p}} \frac{\exp\left[i\boldsymbol{p}(\boldsymbol{\varrho}_1 - \boldsymbol{\varrho}_2)\right]}{\omega_1 - \dfrac{p^2}{\mu} + i\varepsilon}. \tag{3.3}$$

2. Every line of the particle M corresponds to the Green's function $G_0^M(\boldsymbol{R}_1 - \boldsymbol{R}_2; \omega_2)$:

$$G_0^M(\boldsymbol{R}_1 - \boldsymbol{R}_2; \omega_2) = \sum_{\boldsymbol{P}} \frac{\exp\,[i\boldsymbol{P}(\boldsymbol{R}_1 - \boldsymbol{R}_2)]}{\omega_2 + \Delta + \dfrac{P^2}{M} - i\varepsilon} . \tag{3.4}$$

3. The cross on the μ-line corresponds to the potential $V(\boldsymbol{\varrho})$.
4. The vertical line corresponds to the potential

$$U(\boldsymbol{R}, \boldsymbol{\varrho}) = U_c\left(\boldsymbol{R} + \frac{m_v}{M}\,\boldsymbol{\varrho}\right) + U_v\left(\boldsymbol{R} - \frac{m_c}{M}\,\boldsymbol{\varrho}\right)$$

(the vertical line possessing index "c" or "v" corresponds to the potential U_c or U_v, respectively).
5. The integration is implied over all the intrinsic coordinates.
6. Every pair of the M- and μ-lines, which are not connected by the vertical line implies the integration over the variables ω_1 and ω_2 within the interval $(-\infty, +\infty)$. Here ω_1 and ω_2 are the energy parameters of the μ- and M-lines, they are related by $\omega_1 - \omega_2 = \omega$.
7. Every diagram bears the factor $(-1)^n\,(i/2\pi)^{n+1}$, n being the number of the vertical lines contained in the diagram.

For example, the contributions, carried into the exciton Green's function $K_e(\boldsymbol{R}_1\boldsymbol{\varrho}_1; \boldsymbol{R}_2\boldsymbol{\varrho}_2; \omega)$ by the diagrams

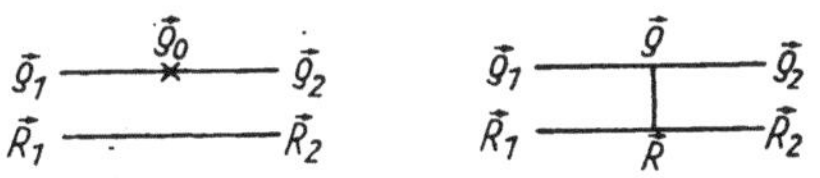

(3.5)

are

$$\frac{i}{2\pi} \int d\omega_1\, d\omega_2\, \delta(\omega - \omega_1 + \omega_2) \times$$
$$\times \int d\boldsymbol{\varrho}_0\, G_0^\mu(\boldsymbol{\varrho}_1 - \boldsymbol{\varrho}_0; \omega_1)\, V(\boldsymbol{\varrho}_0)\, G_0^\mu(\boldsymbol{\varrho}_0 - \boldsymbol{\varrho}_2; \omega_1)\, G_0^M(\boldsymbol{R}_1 - \boldsymbol{R}_2; \omega_2) \tag{3.6a}$$

and

$$-\frac{i}{2\pi} \int d\omega_1\, d\omega_2\, d\omega_3\, d\omega_4\, \delta(\omega - \omega_1 + \omega_2)\, \delta(\omega - \omega_3 + \omega_4) \times$$
$$\times \int d\boldsymbol{R}\, d\boldsymbol{\varrho}\, G_0^\mu(\boldsymbol{\varrho}_1 - \boldsymbol{\varrho}; \omega_1)\, U(\boldsymbol{R}, \boldsymbol{\varrho})\, G_0^M(\boldsymbol{R}_1 - \boldsymbol{R}; \omega_2)\, G_0^\mu(\boldsymbol{\varrho} - \boldsymbol{\varrho}_2; \omega_3)\, G_0^M(\boldsymbol{R} - \boldsymbol{R}_2; \omega_4) , \tag{3.6b}$$

respectively.

Note that the contribution of the free-exciton Green's function K_e^0 is described by the diagram

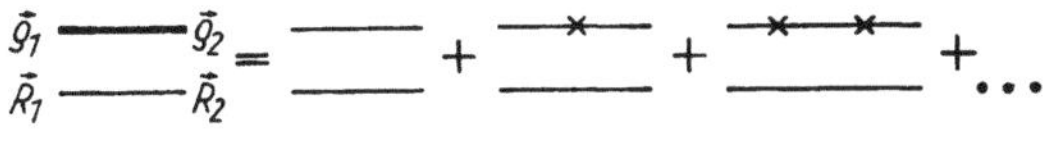

(3.7)

and has the form

$$K_e^0(\boldsymbol{R}_1\boldsymbol{\varrho}_1; \boldsymbol{R}_2\boldsymbol{\varrho}_2; \omega) =$$
$$= \frac{i}{2\pi} \sum_N \int d\omega_1\, d\omega_2\, \delta(\omega - \omega_1 + \omega_2)\, \frac{\Phi_N(\boldsymbol{\varrho}_1)\, \Phi_N^*(\boldsymbol{\varrho}_2)}{\omega_1 - E_N + i\varepsilon}\, G_0^M(\boldsymbol{R}_1 - \boldsymbol{R}_2; \omega_2) , \tag{3.8}$$

Φ_N and E_N being the eigenfunctions and eigenvalues of the operator $T + V$.

4. The Correspondence Rules

It is convenient to compare the graphs of the m_c, m_v, and M, μ representations using the momentum representation. Thus we supply every particle line with the momentum variable and calculate the matrix element of the random potential in the plane wave representation.

Let us call the two-particle Green's function, satisfying (2.8) as $K(\boldsymbol{k}_1, \boldsymbol{k}_2; \boldsymbol{k}_3, \boldsymbol{k}_4; \omega)$ if one uses the m_c, m_v representation, and as $\mathcal{K}(\boldsymbol{p}, \boldsymbol{P}; \boldsymbol{p}', \boldsymbol{P}'; \omega)$ if one uses the M, μ representation, $\boldsymbol{k}_1, \boldsymbol{k}_2, \boldsymbol{k}_3, \boldsymbol{k}_4$ and $\boldsymbol{p}, \boldsymbol{P}, \boldsymbol{p}', \boldsymbol{P}'$ being the momenta of the free ends of the corresponding diagram:

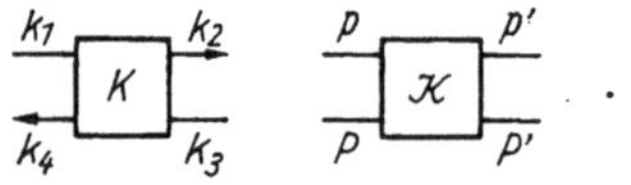

(4.1)

Then using (2.4), (2.9), (2.14), (2.15), and the rules of the previous paragraph one may formulate the correspondence rules connecting the diagrams of the m_c, m_v and M, μ representations. For example it may be shown by the direct calculation that

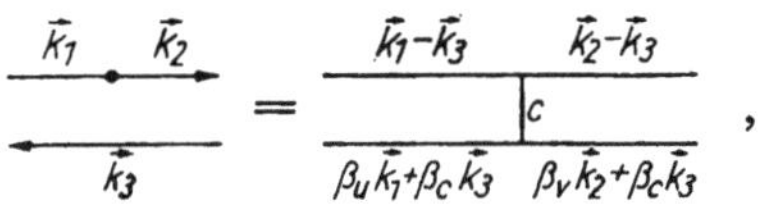

(4.2)

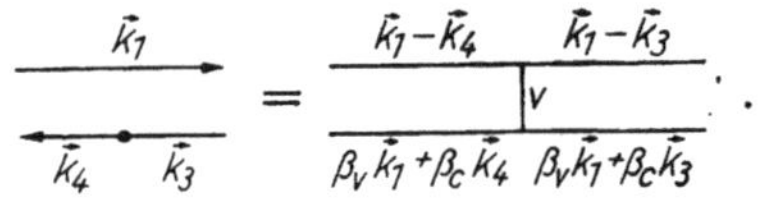

(4.3)

Here

$$\beta_v = \frac{m_v}{M}, \qquad \beta_c = \frac{m_c}{M}.$$

It may be shown as well that

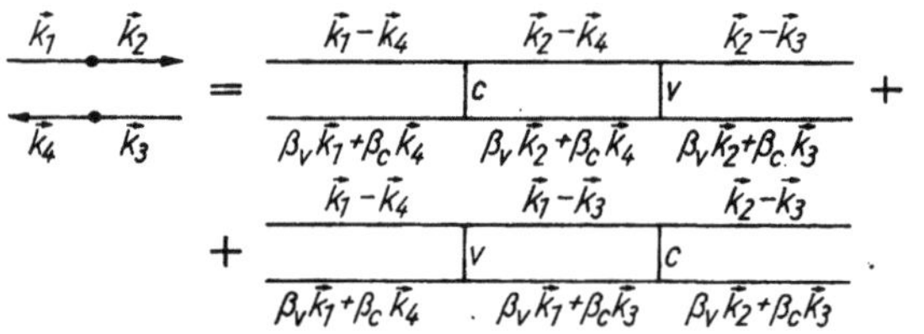

(4.4)

In general case the sum of blocks of the m_c, m_v representation, containing n points, and the block of the M, μ representation, having n vertical lines, give exactly equal contributions, if the momenta of the free ends of the blocks are related by

$$\boldsymbol{p} = \boldsymbol{k}_1 - \boldsymbol{k}_4\,, \;\; \boldsymbol{p}' = \boldsymbol{k}_2 - \boldsymbol{k}_3\,, \quad \boldsymbol{P} = \beta_v \boldsymbol{k}_1 + \beta_c \boldsymbol{k}_4\,, \quad \boldsymbol{P}' = \beta_v \boldsymbol{k}_2 + \beta_c \boldsymbol{k}_3\,. \tag{4.5}$$

If the condition (4.5) is satisfied, the following equality takes place:

$$K(\boldsymbol{k}_1, \boldsymbol{k}_2; \boldsymbol{k}_3, \boldsymbol{k}_4; \omega) = \mathcal{K}\,(\boldsymbol{p}, \boldsymbol{P}; \boldsymbol{p}', \boldsymbol{P}'; \omega)\,. \tag{4.6}$$

5. Averaging over the Random Field

Let us represent the exciton Green's function K_e in the form of the expansion over the eigenfunctions of the unperturbed equation

$$\{T_M + T_\mu + V\}\,\Psi = E\Psi \tag{5.1}$$

and turn to the energy representation. Consider the diagrams of the type

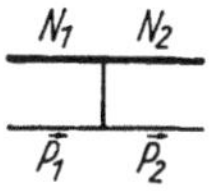

, (5.2)

N being the set of quantum numbers of the relative motion and $\boldsymbol{P}$ being the centre-of-mass momentum. It is easily seen that such diagrams describe second-kind collisions: the exciton state is changed by the interaction with the random field which connects the coordinates of the M- and μ-particles. If the state N_2 belongs to the continuous spectrum, the diagram (5.2) describes the dissociation of the exciton in the random field on cost of the centre of mass kinetic energy.

Let (2.18) be averaged over the random field and let $\langle K_e \rangle$ be the exciton Green's function averaged over the random field.

Some typical blocks, appearing after averaging of (2.18) is done, are represented in the following figure:

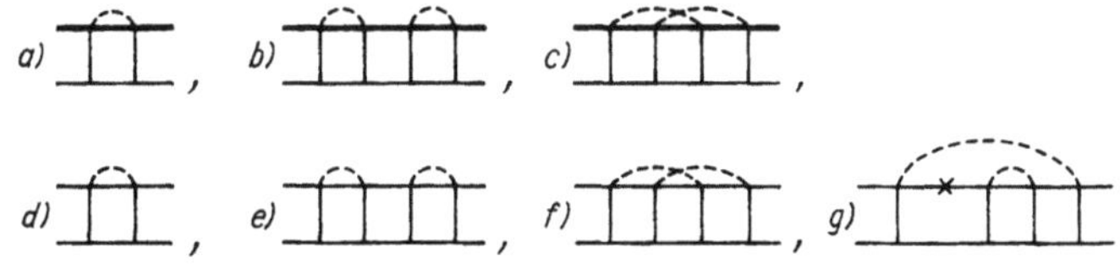

(5.3)

(the dashed line in (5.3) is the symbol of averaging over the random field).

Let us formally construct the function $\langle K_e \rangle_1$ which would take into account only the blocks of the types a) to c). It may be written in the form

$$\langle K_e \rangle_1 = K_e^0 + K_e^0 Q_1 K_e^0\,, \tag{5.4}$$

Q_1 being the operator, given by the following series:

(5.5)

Let $\mathcal{M}_1$ be the first contribution in (5.5):

(5.6)

in the energy representation $\mathcal{M}_1$ has the form

$$\mathcal{M}_1(N_1\boldsymbol{P}_1;\, N_2\boldsymbol{P}_2;\, \omega) = \\ = \sum_{N,\boldsymbol{P}} \langle U(N_1\boldsymbol{P}_1;\, N\boldsymbol{P})\,[U(N\boldsymbol{P};\, N_2\boldsymbol{P}_2)]^* \rangle\, K_e^0(N, \boldsymbol{P}, \omega)\, \delta_{\boldsymbol{P}_1\boldsymbol{P}_2}\,. \tag{5.7}$$

Here

$$K_e^0 = -\frac{1}{\omega - \Delta - E(N, \boldsymbol{P}) + i\gamma}\,, \qquad \gamma \to +0 \tag{5.8}$$

is the diagonal matrix element of the free-exciton Green's function, $E(N, \boldsymbol{P})$ being the eigenvalues of the operator $T_M + T_\mu + V$, $U(N_1\boldsymbol{P}_1; N\boldsymbol{P})$ being the matrix element of the random potential calculated with the eigenfunctions of the operator $T_M + T_\mu + V$.

Note that $\langle K_e \rangle_1$ is the exciton Green's function averaged over the random field but taking account of the second-kind collisions only; it takes no account of the scattering of the particles.

Taking account of the scattering processes, i.e. of the blocks which do not contain the matrix elements of the potential V (for example, the blocks d) to f) and the blocks of the type g)) produces the following result: everywhere in (5.4), including the expression for the operator Q_1, the free-exciton Green's function K_e^0 is to be replaced by the function $\tilde{K}_e^0$ satisfying the equation

$$\{\omega - \Delta - T_M - T_\mu - V + \tilde{\mathcal{M}}\} \tilde{K}_e^0 = - I . \tag{5.9}$$

Here $\tilde{\mathcal{M}}$ is the mass operator corresponding to the averaged two-particle Green's function $\langle \mathcal{K} \rangle$ [3]. Then (5.4) takes the form

$$\langle K_e \rangle = \tilde{K}_e^0 + \tilde{K}_e^0 Q \tilde{K}_e^0 , \tag{5.10}$$

Q being the operator obtained by the replacement of $K_e^0 \to \tilde{K}_e^0$ in the expression (5.5).

Consider now the weak-field approximation using the condition

$$\Psi_1^{1/2} \ll W_B .$$

Here W_B is the Bohr energy in the crystal and Ψ_1 is the bigger of the electron and hole mean-squared potential energy fluctuations.

Using the calculation rules of Section 3 it may be shown that the contribution given by the block c) is essentially smaller than that of the block b) if the condition $\omega \approx \Delta - W_B$ is satisfied. Note that if the scattering is taken account of the ratio of these contributions is given approximately by the parameter $\eta = \Psi_1^{1/2}/W_B$.

Thus, omitting the blocks of the type c) the functions $\langle K_e \rangle_1$ and $\langle K_e \rangle$ satisfy the equations

$$\langle K_e \rangle_1 = K_e^0 + K_e^0 \mathcal{M}_1 \langle K_e^0 \rangle_1 , \tag{5.4a}$$

$$\langle K_e \rangle = \tilde{K}_e^0 + \tilde{K}_e^0 \mathcal{M} \langle K_e \rangle , \tag{5.10a}$$

$\mathcal{M}$ being the mass operator obtained from $\mathcal{M}_1$ by the replacement $K_e^0 \to \tilde{K}_e^0$.

In the first approximation of the perturbation theory for the effective wave equation one obtains

$$\langle K_e(\boldsymbol{R}\boldsymbol{\varrho}; \boldsymbol{R}'\boldsymbol{\varrho}'; \omega) \rangle = \\ = \sum_{N,\boldsymbol{P}} \frac{\Phi_N^*(\boldsymbol{\varrho})\, \Phi_N(\boldsymbol{\varrho}') \exp\,[i\boldsymbol{P}(\boldsymbol{R} - \boldsymbol{R}')]}{\omega - \Delta - E(N, \boldsymbol{P}) - \delta\tilde{E}(N, \boldsymbol{P}, \omega) - \delta E(N, \boldsymbol{P}, \omega)}, \tag{5.11}$$

$$\delta\tilde{E}(N, \boldsymbol{P}, \omega) = \int \Phi_N^*(\boldsymbol{\varrho})\, e^{i\boldsymbol{P}\boldsymbol{R}}\, \tilde{\mathcal{M}}(\boldsymbol{R} - \boldsymbol{R}';\ \boldsymbol{\varrho}\boldsymbol{\varrho}'; \omega)\, \Phi_N(\boldsymbol{\varrho}')\, e^{-i\boldsymbol{P}\boldsymbol{R}'}\, d\boldsymbol{\varrho}\, d\boldsymbol{\varrho}'\, d\boldsymbol{R}\, d\boldsymbol{R}' , \tag{5.12}$$

$$\delta E(N, \boldsymbol{P}, \omega) = \mathcal{M}(N\boldsymbol{P}; N\boldsymbol{P}; \omega) . \tag{5.13}$$

The quantity

$$\tau^{-1} = \operatorname{Im} \delta\tilde{E}(N, \boldsymbol{P}, \omega) + \operatorname{Im} \delta E(N, \boldsymbol{P}, \omega) \tag{5.14}$$

is the inverse relaxation time of the exciton state N, $\boldsymbol{P}$. The second contribution to the r.h.s. of (5.14) comes from the second-kind collisions. It contains the probability of the exciton dissociation via to such collisions (a similar value was calculated in [4]).

Acknowledgement

The author is grateful to Prof. V. L. Bonch-Bruevich for useful discussion on the subject.

References

[1] D. A. Kirznits, Field Methods of the Many-Body Theory, Atomizdat, Moscow 1963 (in Russian).
[2] S. F. Edwards, Phil. Mag. **33**, 1020 (1958).
[3] V. D. Iskra, Ukr. fiz. Zh. **17**, 1483 (1972).
[4] V. L. Bonch-Bruevich and V. D. Iskra, Fiz. Tekh. Poluprov. **5**, 1948 (1971).

(Received January 15, 1974)

phys. stat. sol. (b) 62, 603 (1974)

Subject classification: 18.4; 13.1

Sektion Physik der Karl-Marx-Universität Leipzig (a) and Department of Mathematics, Imperial College London (b)

On the Existence of Antiferromagnetism in Half-Filled Narrow Energy Bands

By

U. LINDNER (a) and A. C. HEWSON (b)

It is shown that the spin-dependent shift of the subbands considered by Roth for the treatment of ferromagnetic solutions of the Hubbard model plays a decisive role in making antiferromagnetism possible on this model. A relation is given which allows to calculate the Néel temperature T_N for a simple cubic lattice and a half-filled band in dependence on the parameters of the Hubbard model and which includes in the limit a criterion formerly obtained by the simple Hubbard decoupling.

Es wird gezeigt, daß die von Roth bei der Behandlung ferromagnetischer Lösungen des Hubbardmodells berücksichtigte spinabhängige Energieverschiebung der Unterbänder auch für antiferromagnetische Lösungen eine entscheidende Rolle spielt. Es wird eine Beziehung angegeben, die die Néeltemperatur T_N für ein einfach kubisches Gitter bei halbgefülltem Band in Abhängigkeit von den Parametern des Hubbardmodells zu berechnen gestattet und die ein früher mit der einfachen Hubbardentkopplung abgeleitetes Kriterium als Grenzfall enthält.

1. Introduction

As a basis for understanding the magnetic and electric behaviour of transition metals and transition metal compounds, e.g. transition metal oxides and sulphides, a simple theoretical model seems to be suitable which considers the d-electron band as an s-band and only takes into account interactions between electrons in Wannier states of the same atom. This model is often called Hubbard model [1].

Of special interest is the question, wether this model has antiferromagnetic solutions. Transition metal oxides and sulphides are insulators or substances with insulator– or semiconductor–metal transitions or metals and the most of them, belonging to the first two groups, show below a critical temperature T_N antiferromagnetic order [2].

In the recent years numerous papers have been published on the antiferromagnetic properties of the Hubbard model. As long ago as in 1964 Herring [3] pointed out the fact that for a half-filled band, if the Coulomb interaction between electrons in Wannier states of the same atom is sufficiently large, the ground state should be antiferromagnetic. Nagaoka's considerations in 1966 [4] for the ground state of different lattices for an exactly half-filled band and nearest neighbour overlapping gave the same results, whereas in several lattices, for a nearly half-filled band the ferromagnetic state can be the ground state, too, provided that the Coulomb interaction is large enough.

Nagaoka's results for the ground state of a half-filled band were verified by Sokoloff [5]. Caron and Kemeny [6] considered in 1970 the ground state of a simple cubic lattice using the t-matrix formalism and found out that the transitions between paramagnetic, ferromagnetic and antiferromagnetic states occur for a smaller ratio of band width W to Coulomb interaction energy U as calculated by the Hartree-Fock decoupling of Green's functions by Penn [7]. For a half-filled band Caron and Kemeny obtain an antiferromagnetic ground state, too. However it is necessary to take in consideration that the application of the t-matrix formalism is justified only for a small concentration of electrons or for a Coulomb interaction energy U small compared with the band width W.

Investigations of the ground state in the limiting case of vanishing band width demand particular attention because of the problems connected with the degeneracy of the energy of the different spin configurations [8].

Of special practical importance is the question of the existence of antiferromagnetic solutions for temperatures $T \neq 0$. Recently there have been several papers on this problem and also discussions of the relationship between antiferromagnetism and metal–insulator transition. We believe that except for limiting cases the dependence of the transition temperatures T_M between insulating and metallic phase on the parameters of the Hubbard model is not yet solved definitly. In this paper we regard only the transition between antiferromagnetic and paramagnetic phase characterized by the Néel temperature T_N.

We restrict us to simple antiferromagnetica (see e.g. [3] and [9]). Expecting antiferromagnetic order we divide the lattice into two equivalent ferromagnetic sublattices A and B such that the resulting magnetic moment is zero without an external magnetic field.

The Néel temperature T_N may be calculated in different ways (e.g. [9]). Following Langer et al. [10] we evaluate T_N from the vanishing of the sublattice magnetization.

Langer et al. calculated T_N in Hartree-Fock approximation, but this approximation will be valid, if at all, only in the small U limit. Using Hubbard's simple decoupling scheme [1] Penn [11], Hewson and Lindner [12], Bari and Kaplan [13], Jelitto [14], Mehrotra and Viswanathan [15] have shown, that the Hubbard model for a half-filled band has no antiferromagnetic solutions in disagreement with a result of Arai [16]. In the limiting case of vanishing band width Hubbard's decoupling scheme yields the exact result but otherwise it overestimates the influence of the correlation and so it is necessary to look for other approximations. Moreover Hewson and Lindner [12] and Bari and Kaplan [13] showed, that for $U \ll W$ the Hubbard decoupling does not give the same results for T_N as those obtained by the Hartree-Fock decoupling.

Here we give a solution for T_N (cp. [17]) using a method published by Roth [18] for the decoupling of Green's functions. This method is essentially identical with the decoupling scheme proposed by Linderberg and Öhrn in 1967 and with the work of Rowe in 1968 and it is strongly related to a paper of Suhl and Werthamer from the year 1961. Sawada [19] showed in 1970 that Roth's method corresponds to a variational method for the free energy at finite temperatures yielding stationary solutions. Investigating the ferromagnetic order in the Hubbard model Roth obtained in the limit case $W \ll U$ the results of Harris and Lange [20], who showed by the moment method of Kubo and Tomita that the existence of an additional spin-dependent shift of the centre of gravity of the bands leads to a preference of the ferromagnetic state compared with the simple Hubbard decoupling.

Kaplan and Bari [21] investigated the stability of the different phases calculating the free energy by variational methods in the limit cases $W \ll U$ and

$W \gg U$, but for $U \approx W$ the validity of their solutions is questionable. Discussing the free energy by functional integration recently Cyrot and Lacour-Gayet [22] got results similar to those obtained by us by Roth's method.

2. Model and Method

The Hubbard Hamiltonian for a cubic lattice with nearest neighbour overlap can be written as

$$H = -\sum_{\substack{i\in A\\ j\in B\\ \sigma}} t_{ij}\left(c_{i\sigma}^{+A} c_{j\sigma}^{B} + c_{j\sigma}^{+B} c_{i\sigma}^{A}\right) + \frac{1}{2} U \sum_{\substack{i\in A\vee B\\ \sigma}} n_{i\sigma} n_{i-\sigma} - \mu \sum_{\substack{i\in A\vee B\\ \sigma}} n_{i\sigma}\,. \tag{1}$$

t_{ij} is the kinetic energy in the band and U the repulsive interaction between electrons of the same atoms. μ denotes the chemical potential. $c_{i\sigma}^{+}$ and $c_{i\sigma}$ are the creation and annihilation operators for an electron with spin σ in a Wannier state of the i-th atom. A and B denote the sublattices to which the lattice point under consideration belongs. If for the mean occupation numbers $\langle n_{i\uparrow}^{A}\rangle$ and $\langle n_{i\downarrow}^{A}\rangle$ of a Wannier state for nearest neighbour atoms i and j the relation

$$\langle n_{i\sigma}^{A}\rangle = \langle n_{j-\sigma}^{B}\rangle \tag{2}$$

holds with $\sigma = \uparrow,\downarrow$ the solutions are of antiferromagnetic type. For abbreviation we put

$$\langle n_{i\sigma}^{A}\rangle = \alpha\,, \qquad \langle n_{i-\sigma}^{A}\rangle = \gamma\,. \tag{3}$$

Then we get

$$n = \alpha + \gamma = \langle n_{i\sigma}^{A}\rangle + \langle n_{i-\sigma}^{A}\rangle = \langle n_{j-\sigma}^{B}\rangle + \langle n_{j\sigma}^{B}\rangle \tag{4}$$

for the average number of electrons per atom and

$$x = \alpha - \gamma = \langle n_{i\sigma}^{A}\rangle - \langle n_{i-\sigma}^{A}\rangle = \langle n_{j-\sigma}^{B}\rangle - \langle n_{j\sigma}^{B}\rangle\,, \tag{5}$$

where x is a quantity proportional to the sublattice magnetization.

Before we start the calculation we want to give a short exposition of Roth's method.

For the present Hamiltonian H we look for quasi-particle operators A_i such that the following relation holds:

$$[A_i, H] = \omega_i A_i\,,$$

or more generally

$$[A_i, H] = \sum_j K_{ij} A_j\,; \tag{6}$$

diagonalizing the matrix $\boldsymbol{K}$ we get the desired quasi-particle energies.

For interacting systems no simple set of operators $\{A_i\}$ exists. It is usual to replace some operators on the right-hand side of (6) approximately by their expectation values. Roth gives the following prescription for this replacement: Take the anticommutator or commutator, respectively, of both sides of (6) with A_l^{+}, where A_l is a member of the set $\{A_i\}$, and take either the expection value in the ground state or the thermal average of both sides:

$$\langle [[A_i, H], A_l^{+}]_{\pm}\rangle = \sum_j K_{ij} \langle [A_j, A_l^{+}]_{\pm}\rangle\,. \tag{7}$$

For abbreviation we put

$$\langle [A_j, A_l^+]_\pm \rangle = N_{jl} \tag{8}$$

and

$$\langle [[A_i, H], A_l^+]_\pm \rangle = E_{il} \,. \tag{9}$$

In matrix notation (7) has the form

$$\boldsymbol{E} = \boldsymbol{K}\boldsymbol{N} \,. \tag{10}$$

If the matrix $\boldsymbol{N}$ is nonsingular the matrix $\boldsymbol{K}$ is uniquely determined and we can use (7) for decoupling the equations of motion of Green's functions.

We restrict us on Fermi operators and choose $\eta = -1$ (cp. e.g. [9]). Then we get

$$\begin{aligned} \omega \langle\langle A_i; B \rangle\rangle_\omega &= \langle [A_i, B]_+ \rangle + \langle\langle [A_i, H]; B \rangle\rangle_\omega \\ &= \langle [A_i, B]_+ \rangle + \sum_l K_{il} \langle\langle A_l; B \rangle\rangle_\omega \,. \end{aligned} \tag{11}$$

If

$$B = A_j^+ \,,$$

where A_j is a member of the set $\{A_i\}$, then we have

$$\sum_l (\omega \delta_{il} - K_{il}) \langle\langle A_l; A_j^+ \rangle\rangle_\omega = \langle [A_i, A_j^+]_+ \rangle \,. \tag{12}$$

Now we can define a Green's function matrix $\boldsymbol{G}$, such that

$$G_{lj}(\omega) = \langle\langle A_l; A_j^+ \rangle\rangle_\omega \tag{13}$$

and then we get from (12) and (10)

$$\boldsymbol{G} = (\omega - \boldsymbol{K})^{-1} \boldsymbol{N} = \boldsymbol{N}(\omega \boldsymbol{N} - \boldsymbol{E})^{-1} \boldsymbol{N} \,. \tag{14}$$

The eigenvalues of the matrix $\boldsymbol{K}$ correspond to the singularities of the Green's functions. If the matrices $\boldsymbol{E}$ and $\boldsymbol{N}$ are Hermitian and if furthermore $\boldsymbol{N}$ is positive definite then the eigenvalues of $\boldsymbol{K}$ are real, this means in this case the approximation yields undamped quasi-particles.

In order to get the set $\{A_i\}$ we start with one-particle operators and complete the set by combinations of these operators such that the bilinear form

$$H_0 = \sum_{ij} \omega_{ij} A_i^+ A_j$$

is of same structure as the Hamiltonian H under consideration. It is possible that H_0 contains more terms than H. These terms ought to yield parameters ω_{ij} which are small compared with the considered ones and may be neglected. This can cause difficulties if one of the terms in the Hamiltonian H is very small. In our opinion this is the reason for the fact that Roth's first calculations [18] for ferromagnetic solutions in the Hubbard model in the low-density limit does not lead to Kanamori's T-matrix result. Later this was obtained by Roth by a suitable extension of the set $\{A_i\}$.

The domain of validity of the approximation is essentially determined by the choice of the set $\{A_i\}$. For a fixed set Roth's method ought to yield the best possible approximation [19]; for a base set of one-particle operators we get the Hartree-Fock approximation, e.g. the calculation of Langer et al. [10] corresponds to the base set

$$A_1^\sigma = c_{\boldsymbol{k}\sigma}^{+\mathrm{A}} = \frac{1}{\sqrt{N}} \sum_{i\in\mathrm{A}} c_{i\sigma}^{+\mathrm{A}} \mathrm{e}^{-i\boldsymbol{k}\boldsymbol{R}_i},$$

$$A_2^\sigma = c_{\boldsymbol{k}\sigma}^{+\mathrm{B}} = \frac{1}{\sqrt{N}} \sum_{j\in\mathrm{B}} c_{j\sigma}^{+\mathrm{B}} \mathrm{e}^{-i\boldsymbol{k}\boldsymbol{R}_j}$$

of Roth's method. ($\boldsymbol{k}$ runs over a Brillouin zone of a sublattice and N is the number of the sublattice atoms.)

3. Extension of Roth's Approximation to Solutions of Antiferromagnetic Type

Like Roth for investigation of ferromagnetic solutions, we include into the base set besides one-particle operators a combination of these operators of the type $c_{i\sigma}^+ n_{i-\sigma}$. This choice is suggested by the simple Hubbard decoupling and by the fact that this set allows to construct a bilinear form H_0 of the same structure as the Hamiltonian H.

We write the set $\{A_i\}$ in such a form, that $\boldsymbol{N}^\sigma$ is a diagonal matrix. (This means, we start from "Hubbard's alloy analogy".)

$$\left.\begin{aligned}
A_1^\sigma &= c_{\boldsymbol{k}\sigma}^{+\mathrm{A}} - d_{\boldsymbol{k}\sigma}^{+\mathrm{A}} = \frac{1}{\sqrt{N}} \sum_{i\in\mathrm{A}} (1 - n_{i-\sigma}^{\mathrm{A}})\, c_{i\sigma}^{+\mathrm{A}} \mathrm{e}^{-i\boldsymbol{k}\boldsymbol{R}_i}\,,\\
A_2^\sigma &= d_{\boldsymbol{k}\sigma}^{+\mathrm{A}} = \frac{1}{\sqrt{N}} \sum_{i\in\mathrm{A}} n_{i-\sigma}^{\mathrm{A}} c_{i\sigma}^{+\mathrm{A}} \mathrm{e}^{-i\boldsymbol{k}\boldsymbol{R}_i}\,,\\
A_3^\sigma &= c_{\boldsymbol{k}\sigma}^{+\mathrm{B}} - d_{\boldsymbol{k}\sigma}^{+\mathrm{B}}\,,\\
A_4^\sigma &= d_{\boldsymbol{k}\sigma}^{+\mathrm{B}}\,.
\end{aligned}\right\} \tag{15}$$

$\boldsymbol{k}$ runs over the first Brillouin zone of the sublattice A or B and N is the number of atoms in a sublattice.

$\boldsymbol{N}^\sigma$, $\boldsymbol{E}^\sigma$ and $\boldsymbol{G}^\sigma$, according to (8), (9), and (13), are now 4×4 matrices for each spin, because we do not take into account anomalous mean values.

We get

$$\boldsymbol{N}^\sigma = \begin{pmatrix} 1-\gamma & 0 & 0 & 0\\ 0 & \gamma & 0 & 0\\ 0 & 0 & 1-\alpha & 0\\ 0 & 0 & 0 & \alpha \end{pmatrix} \tag{16}$$

and

$$\boldsymbol{E}^\sigma = \begin{pmatrix} \mu(1-\gamma) + W_{\mathrm{AA}}^\sigma & -W_{\mathrm{AA}}^\sigma & \varepsilon_{\boldsymbol{k}}(1-\alpha-\gamma) + W_{\mathrm{AB}}^\sigma & \varepsilon_{\boldsymbol{k}}\alpha - W_{\mathrm{AB}}^\sigma\\ -W_{\mathrm{AA}}^\sigma & (\mu - U)\,\gamma + W_{\mathrm{AA}}^\sigma & \varepsilon_{\boldsymbol{k}}\gamma - W_{\mathrm{AB}}^\sigma & W_{\mathrm{AB}}^\sigma\\ \varepsilon_{\boldsymbol{k}}(1-\alpha-\gamma) + W_{\mathrm{BA}}^\sigma & \varepsilon_{\boldsymbol{k}}\gamma - W_{\mathrm{BA}}^\sigma & \mu(1-\alpha) + W_{\mathrm{BB}}^\sigma & -W_{\mathrm{BB}}^\sigma\\ \varepsilon_{\boldsymbol{k}}\,\alpha - W_{\mathrm{BA}}^\sigma & W_{\mathrm{BA}}^\sigma & -W_{\mathrm{BB}}^\sigma & (\mu - U)\,\alpha + W_{\mathrm{BB}}^\sigma \end{pmatrix} \tag{17}$$

with α and γ defined by (3) and with

$$W^{\sigma}_{\mathrm{AA}} = \frac{1}{N} \sum_{\substack{i\in \mathrm{A}\\ j\in \mathrm{B}}} t_{ij} \langle \overset{\mathrm{A}}{n}_{i\sigma} \overset{+\mathrm{B}}{c}_{j-\sigma} \overset{\mathrm{A}}{c}_{i-\sigma} - (1 - \overset{\mathrm{A}}{n}_{i\sigma})\, \overset{+\mathrm{A}}{c}_{i-\sigma} \overset{\mathrm{B}}{c}_{j-\sigma} \rangle \tag{18}$$

and

$$W^{\sigma}_{\mathrm{AB}} = \frac{1}{N} \sum_{\substack{i\in A\\ j\in B}} t_{ij} \mathrm{e}^{-i\boldsymbol{k}(\boldsymbol{R}_i - \boldsymbol{R}_j)} \langle \overset{\mathrm{A}}{n}_{i-\sigma} \overset{\mathrm{B}}{n}_{j-\sigma} - \overset{+\mathrm{A}}{c}_{i\sigma} \overset{\mathrm{B}}{c}_{j\sigma} (\overset{+\mathrm{A}}{c}_{i-\sigma} \overset{\mathrm{B}}{c}_{j-\sigma} + \overset{+\mathrm{B}}{c}_{j-\sigma} \overset{\mathrm{A}}{c}_{i-\sigma}) \rangle \, . \tag{19}$$

We can obtain W^{σ}_{BB} from W^{σ}_{AA} and W^{σ}_{BA} from W^{σ}_{AB} replacing $i \in \mathrm{A}$ by $j \in \mathrm{B}$ and $j \in \mathrm{B}$ by $i \in \mathrm{A}$.

Since all the used operators are invariant against time inversion we have

$$W^{\sigma}_{\mathrm{AA}} = (W^{\sigma}_{\mathrm{AA}})^* , \quad W^{\sigma}_{\mathrm{BB}} = (W^{\sigma}_{\mathrm{BB}})^*, \quad \text{and} \quad W^{\sigma}_{\mathrm{AB}} = (W^{\sigma}_{\mathrm{BA}})^*$$

and the matrices $\boldsymbol{N}^{\sigma}$ and $\boldsymbol{E}^{\sigma}$ are Hermitian, so the quasi-particles are undamped. Using (15) and

$$t_{ij} = \frac{1}{N} \sum_{\boldsymbol{k}} \varepsilon_{\boldsymbol{k}} \mathrm{e}^{i\boldsymbol{k}(\boldsymbol{R}_i - \boldsymbol{R}_j)} , \tag{20}$$

we can write for W^{σ}_{AA}:

$$\begin{aligned} W^{\sigma}_{\mathrm{AA}} &= \frac{1}{N} \sum_{\substack{i\in \mathrm{A}\\ j\in \mathrm{B}}} t_{ij} \langle (1 - \overset{\mathrm{A}}{n}_{i\sigma})\, \overset{\mathrm{B}}{c}_{j-\sigma} \overset{+\mathrm{A}}{c}_{i-\sigma} - \overset{\mathrm{A}}{n}_{i\sigma} \overset{\mathrm{B}}{c}_{j-\sigma} \overset{+\mathrm{A}}{c}_{i-\sigma} \rangle = \\ &= \frac{1}{N} \sum_{\boldsymbol{k}} \varepsilon_{\boldsymbol{k}} (\langle \overset{\mathrm{B}}{c}_{\boldsymbol{k}-\sigma} \overset{+\mathrm{A}}{c}_{\boldsymbol{k}-\sigma} \rangle - 2 \langle \overset{\mathrm{B}}{c}_{\boldsymbol{k}-\sigma} \overset{+\mathrm{A}}{d}_{\boldsymbol{k}-\sigma} \rangle) \, . \end{aligned} \tag{21}$$

The elements of the Green's functions matrix $\boldsymbol{G}^{\sigma}$ contain besides α and γ correlation functions W^{σ}_{AA}, W^{σ}_{BB}, W^{σ}_{AB}, W^{σ}_{BA}, too.

Especially for $\langle\langle \overset{+\mathrm{A}}{c}_{\boldsymbol{k}\sigma} ; \overset{\mathrm{A}}{c}_{\boldsymbol{k}\sigma} \rangle\rangle_{\omega}$ we obtain

$$\begin{aligned} &\langle\langle \overset{+\mathrm{A}}{c}_{\boldsymbol{k}\sigma} ; \overset{\mathrm{A}}{c}_{\boldsymbol{k}\sigma} \rangle\rangle_{\omega} \det |\omega \boldsymbol{N}^{\sigma} - \boldsymbol{E}^{\sigma}| = \\ &= [\alpha (1 - \alpha) (\omega - \mu) (\omega - \mu + U) - W^{\sigma}_{\mathrm{BB}} (\omega - \mu + U\alpha)] \times \\ &\times [\gamma (1 - \gamma) (\omega - \mu + U (1 - \gamma) - W^{\sigma}_{\mathrm{AA}})] - \\ &- (\varepsilon_{\boldsymbol{k}} \alpha\gamma - W^{\sigma}_{\mathrm{BA}}) (\varepsilon_{\boldsymbol{k}} \alpha\gamma - W^{\sigma}_{\mathrm{AB}}) (\omega - \mu + U\alpha) \end{aligned} \tag{22}$$

with

$$\begin{aligned} &\det |\omega \boldsymbol{N}^{\sigma} - \boldsymbol{E}^{\sigma}| = \\ &= W^{\sigma}_{\mathrm{AA}} [- \alpha (1 - \alpha) (\omega - \mu) (\omega - \mu + U) (\omega - \mu + U\gamma) + \\ &+ \varepsilon_{\boldsymbol{k}}^2 \alpha\gamma (\omega - \mu + U (1 - \alpha))] + \\ &+ W^{\sigma}_{\mathrm{BB}} [- \gamma (1 - \gamma) (\omega - \mu) (\omega - \mu + U) (\omega - \mu + U\alpha) + \\ &+ \varepsilon_{\boldsymbol{k}}^2 \alpha\gamma (\omega - \mu + U (1 - \gamma))] + \\ &+ (W^{\sigma}_{\mathrm{AB}} + W^{\sigma}_{\mathrm{BA}}) \varepsilon_{\boldsymbol{k}} \alpha\gamma [(\omega - \mu + U) (\omega - \mu + U (1 - \alpha - \gamma)) - \varepsilon_{\boldsymbol{k}}^2] + \\ &+ (W^{\sigma}_{\mathrm{AA}} W^{\sigma}_{\mathrm{BB}} - W^{\sigma}_{\mathrm{AB}} W^{\sigma}_{\mathrm{BA}}) [(\omega - \mu + U\alpha) (\omega - \mu + U\gamma) - \varepsilon_{\boldsymbol{k}}^2] + \\ &+ \alpha\gamma (1 - \alpha) (1 - \gamma) (\omega - \mu)^2 (\omega - \mu + U)^2 + \varepsilon_{\boldsymbol{k}}^4 \alpha^2 \gamma^2 - \\ &- \varepsilon_{\boldsymbol{k}}^2 \alpha\gamma (\omega - \mu) (\omega - \mu + U) [\gamma (1 - \gamma) + \alpha (1 - \alpha)] - \\ &- \varepsilon_{\boldsymbol{k}}^2 \alpha\gamma (\omega - \mu + U)^2 (1 - \alpha - \gamma)^2 \, . \end{aligned} \tag{23}$$

By the quantities W^{σ}_{AA}, W^{σ}_{AB}, W^{σ}_{BB}, and W^{σ}_{BA} additional correlations are taken into account, compared with Hubbard's simple decoupling; by W^{σ}_{AA} e.g. the influence of motion of a $-\sigma$-electron on the motion of a σ-electron according as an electron with spin σ at the lattice point $\boldsymbol{R}_i$ exists or not. The quantities W^σ_{AA} and W^σ_{BB} do not depend on $\boldsymbol{k}$, this means they cause a spin-dependent shift of the centre of gravity of the bands. By the simple Hubbard decoupling, which yields no antiferromagnetic solutions, except in the limiting case of band width $W \to 0$, the $-\sigma$-electron is fixed and so the spin-dependent shift is neglected. By W^σ_{AA} and W^σ_{BB} a correction of the overestimation of U by Hubbard's simple decoupling is caused, too.

By a straight-forward calculation we get

$$\begin{aligned} W^\sigma_{\mathrm{AA}} = \frac{1}{N} \sum_{\boldsymbol{k}} \varepsilon_{\boldsymbol{k}}\, \mathsf{F}_T\{ \big(W^{-\sigma}_{\mathrm{AA}}(1-2\gamma)\, \varepsilon_{\boldsymbol{k}}\, [W^{-\sigma}_{\mathrm{BB}} - \gamma\,(1-\gamma)\,(\omega - \mu + U(1-\alpha))] + \\ + W^{-\sigma}_{\mathrm{BB}} \varepsilon_{\boldsymbol{k}} \alpha\,(1-\alpha)\,[\gamma(\omega-\mu) - (1-\gamma)\,(\omega - \mu + U)] + \\ + (\omega-\mu)\,(\omega-\mu+U)\,(1-\gamma)^2\,\alpha^2\,(\varepsilon_{\boldsymbol{k}}\gamma - W^{-\sigma}_{\mathrm{AB}}) + \\ + (\omega-\mu)\,(\omega-\mu+U)\,\gamma^2(1-\alpha)^2\,(W^{-\sigma}_{\mathrm{AB}} - \varepsilon_{\boldsymbol{k}}\alpha) + \\ + W^{-\sigma}_{\mathrm{AB}}\alpha\gamma\,(1-\alpha)\,(1-\gamma)\,[-(\omega-\mu)^2 + (\omega-\mu+U)^2] + \\ + \varepsilon_{\boldsymbol{k}}\alpha\gamma\,(1-\alpha)\,(1-\gamma)\,(1-\alpha-\gamma)\,(\omega-\mu+U)^2 + \\ + \varepsilon_{\boldsymbol{k}}(W^{-\sigma}_{\mathrm{AB}} - \varepsilon_{\boldsymbol{k}}\alpha\gamma)\,[\varepsilon_{\boldsymbol{k}}(\alpha^2(1-2\gamma) + \gamma(\alpha-\gamma)) - W^{-\sigma}_{\mathrm{BA}}(1-2\gamma)]\big) \times \\ \times (\det|\omega\boldsymbol{N}^{-\sigma} - \boldsymbol{E}^{-\sigma}|)^{-1}\}\,, \end{aligned} \tag{24}$$

where $\det|\omega\boldsymbol{N}^{-\sigma} - \boldsymbol{E}^{-\sigma}|$ may be obtained by (23) if one replaces everywhere σ by $-\sigma$.

F_T stands for the operation

$$\mathsf{F}_T\{G\} = \frac{i}{2\pi} \lim_{\varepsilon\to 0^+} \int_{-\infty}^{+\infty} \mathrm{d}\omega\, \frac{G(\omega+i\varepsilon) - G(\omega-i\varepsilon)}{\mathrm{e}^{\omega/k_{\mathrm{B}}T} + 1}\,. \tag{25}$$

W^σ_{BB} can be written in an analogous way. Replacing σ by $-\sigma$ we get $W^{-\sigma}_{\mathrm{AA}}$ and $W^{-\sigma}_{\mathrm{BB}}$.

For $x = 0$ we have

$$W^\sigma_{\mathrm{AA}}(x=0) = W^{-\sigma}_{\mathrm{AA}}(x=0) = W^\sigma_{\mathrm{BB}}(x=0) = W^{-\sigma}_{\mathrm{BB}}(x=0) = W_0\,. \tag{26}$$

The quantities $W^\sigma_{\mathrm{AB}} = (W^\sigma_{\mathrm{AB}})^*$, given by (19), depend on $\boldsymbol{k}$ and therefore influence the band. They cannot be constructed by the set (15). A calculation is possible in a way given by Roth [18]. We write W^σ_{AB} in the following form:

$$\begin{aligned} W^\sigma_{\mathrm{AB}} = \frac{1}{N} \sum_{\substack{i\in \mathrm{A}\\ j\in \mathrm{B}}} t_{ij} \mathrm{e}^{-i\boldsymbol{k}(\boldsymbol{R}_i - \boldsymbol{R}_j)}\, (\langle n^{\mathrm{A}}_{i-\sigma} n^{\mathrm{B}}_{j-\sigma}\rangle + \\ + \langle c^{\mathrm{B}}_{j\sigma} c^{+\mathrm{B}}_{j-\sigma} c^{\mathrm{A}}_{i-\sigma} c^{+\mathrm{A}}_{i\sigma}\rangle + \langle c^{\mathrm{B}}_{j\sigma} c^{\mathrm{B}}_{j-\sigma} c^{+\mathrm{A}}_{i\sigma} c^{+\mathrm{A}}_{i-\sigma}\rangle)\,. \end{aligned} \tag{27}$$

The first two terms in the parentheses are the density correlation function on two sites and the spin correlation function $\langle S^{+\mathrm{A}}_i S^{-\mathrm{B}}_j\rangle$ and the third term is a "double-hop" correlation function, that decreases to zero for U goes to infinity. The second and the third term in (27) are invariant against $\sigma \to -\sigma$ and we expect that their influence on the quantity x is small.

The simple Hubbard decoupling comprises W^{σ}_{AB} only in the limiting case of vanishing correlation, this means

$$W^{\sigma}_{AB} \to \overline{W}^{\sigma}_{AB} = \frac{1}{N} \sum_{\substack{i \in A \\ j \in B}} t_{ij} e^{-i\boldsymbol{k}(\boldsymbol{R}_i - \boldsymbol{R}_j)} \langle n^{A}_{i-\sigma} \rangle \langle n^{B}_{j-\sigma} \rangle = \varepsilon_{\boldsymbol{k}} \alpha \gamma \,. \tag{28}$$

Besides density fluctuations, spin and "double-hop" correlation functions are neglected. Compared with the band energy the difference $W^{\sigma}_{AB} - \overline{W}^{\sigma}_{AB}$ ought to be small. In the following we use the approximation $\overline{W}^{\sigma}_{AB}$ for W^{σ}_{AB}.

4. Criterion for Antiferromagnetism in a Half-Filled Band

If an antiferromagnetic solution exists the quantity x, proportional to the sublattice magnetization, has to be unequal zero:

$$x = \alpha - \gamma = \frac{1}{N} \sum_{\boldsymbol{k}} \mathsf{F}_T \{\langle\langle c^{+A}_{\boldsymbol{k}-\sigma}; c^{A}_{\boldsymbol{k}-\sigma} \rangle\rangle_\omega - \langle\langle c^{+A}_{\boldsymbol{k}\sigma}; c^{A}_{\boldsymbol{k}\sigma} \rangle\rangle_\omega\} \neq 0 \,. \tag{29}$$

Differentiating (29) at the Néel point ($x = 0$) we get a criterion for the Néel temperature T_N:

$$1 = \frac{1}{N} \sum_{\boldsymbol{k}} \mathsf{F}_{T_N} \left\{ \frac{\partial}{\partial x} (\langle\langle c^{+A}_{\boldsymbol{k}-\sigma}; c^{A}_{\boldsymbol{k}-\sigma} \rangle\rangle_\omega - \langle\langle c^{+A}_{\boldsymbol{k}\sigma}; c^{A}_{\boldsymbol{k}\sigma} \rangle\rangle_\omega)|_{x=0} \right\}, \tag{30}$$

where F_{T_N} is defined analogously to (25).

We restrict us on a half-filled band:

$$\alpha + \gamma = n = 1 \,. \tag{31}$$

For the chemical potential μ we obtain then

$$\mu = \frac{U}{2} \tag{32}$$

and using (23) to (26) and (31), (32), we get for the quantity W_0 (see (26)):

$$W_0 = 0 \,. \tag{33}$$

With (22) to (26), (28) and (31) to (33) we get for the relation (30)

$$\begin{aligned} 1 = \frac{1}{N} \sum_{\boldsymbol{k}} \mathsf{F}_{T_N} \Bigg\{ & \frac{1}{D} \omega^2 \frac{\partial}{\partial x} (W^{\sigma}_{B} - W^{-\sigma}_{B})|_{x=0} + \\ & + \frac{1}{D} \left(\omega^2 - \frac{1}{4} U^2\right) \left[\frac{\partial}{\partial x} (W^{\sigma}_{A} - W^{-\sigma}_{A})|_{x=0} - U\right] + \\ & + \frac{1}{D^2} \omega^2 \left(\omega^2 - \frac{1}{4} U^2\right) \left[\varepsilon^2_{\boldsymbol{k}} - \left(\omega^2 - \frac{1}{4} U^2\right)\right] \left[\frac{\partial}{\partial x} (W^{\sigma}_{A} - W^{-\sigma}_{A})|_{x=0} + \right. \\ & \left. + \frac{\partial}{\partial x} (W^{\sigma}_{B} - W^{-\sigma}_{B})|_{x=0} \right] \Bigg\} \end{aligned} \tag{34}$$

with

$$W^{\sigma}_{AA} = \alpha\gamma W^{\sigma}_{A}, \qquad W^{\sigma}_{BB} = \alpha\gamma W^{\sigma}_{B}, \tag{35}$$

and

$$D = \left(\omega^2 - \frac{1}{4} U^2\right)^2 - \varepsilon^2_{\boldsymbol{k}} \omega^2 \,. \tag{36}$$

Furthermore it follows

$$\frac{\partial}{\partial x}(W_A^\sigma - W_A^{-\sigma})|_{x=0} = -\frac{\partial}{\partial x}(W_B^\sigma - W_B^{-\sigma})|_{x=0} \doteq w' =$$

$$= \left[-\frac{1}{N}\sum_k \frac{4\varepsilon_k^2}{\Delta_k}\left(\text{th}\,\frac{\Delta_k + \varepsilon_k}{4k_B T_N} + \text{th}\,\frac{\Delta_k - \varepsilon_k}{4k_B T_N}\right)\right] \times$$

$$\times \left[1 + \frac{1}{N}\sum_k \frac{2}{U\Delta_k}\left(-\varepsilon_k(\Delta_k - \varepsilon_k)\,\text{th}\,\frac{\Delta_k + \varepsilon_k}{4k_B T_N} + \varepsilon_k(\Delta_k + \varepsilon_k)\,\text{th}\,\frac{\Delta_k - \varepsilon_k}{4k_B T_N}\right)\right]^{-1} \tag{37}$$

with

$$\Delta_k^2 = U^2 + \varepsilon_k^2. \tag{38}$$

$\boldsymbol{k}$ runs over the Brillouin zone of a sublattice.

Since the expressions in the parentheses are positive, we can immediately see that is $w' \leqq 0$. The equality sign holds for the following cases:

1. for $U \to \infty$ (split-band limit)
2. for band width $W \to 0$ (atomic limit)
3. for $U < \infty$ and $T_N \to \infty$.

Finally we get for (34)

$$1 = \frac{1}{N}\sum_k \mathsf{F}_{T_N}\left\{-\frac{1}{D}U\left(\omega^2 - \frac{1}{4}U^2\right) - \frac{1}{4}w'\frac{U^2}{D}\right\} =$$

$$= \frac{1}{N}\sum_k \frac{U}{2\Delta_k}\left[\text{th}\,\frac{\Delta_k + \varepsilon_k}{4k_B T_N} + \text{th}\,\frac{\Delta_k - \varepsilon_k}{4k_B T_N}\right] + K \tag{39}$$

with

$$K = -\frac{1}{N}\sum_k \mathsf{F}_{T_N}\left\{\frac{1}{4}w'\frac{U^2}{D}\right\} =$$

$$= -\frac{1}{4}w'\frac{1}{N}\sum_k \frac{1}{\varepsilon_k^2\Delta_k}\left(-\varepsilon_k(\Delta_k - \varepsilon_k)\,\text{th}\,\frac{\Delta_k + \varepsilon_k}{4k_B T_N} + \right.$$

$$\left. + \varepsilon_k(\Delta_k + \varepsilon_k)\,\text{th}\,\frac{\Delta_k - \varepsilon_k}{4k_B T_N}\right). \tag{40}$$

Compared with the criterion for the Néel temperature T_N obtained by the use of the simple Hubbard decoupling [12] we have now obtained the correction term K taking into account the spin-depending shift of the subbands. The expression in parentheses is the same as in (37) and therefore it follows

$$K \geqq 0\,. \tag{41}$$

The equality sign is valid for the cases 1, 2, and 3. For the case 3 (39) does not hold, this means T_N cannot be infinite.

We can bring (39) for T_N in a simpler form, if we let $\boldsymbol{k}$ run over the Brillouin zone of the whole lattice (not of the sublattice). Then we obtain

$$1 = \frac{1}{N_g}\sum_k \frac{U}{\Delta_k}\,\text{th}\,\frac{\Delta_k + \varepsilon_k}{4k_B T_N} + \left[\frac{1}{N_g^2}\sum_{k,k'} \frac{4\,\varepsilon_k^2}{U\Delta_k}\left(\frac{U}{\Delta_{k'}} - \frac{U}{\varepsilon_{k'}}\right)\text{th}\,\frac{\Delta_k + \varepsilon_k}{4k_B T_N} \times\right.$$

$$\left.\times\,\text{th}\,\frac{\Delta_{k'} + \varepsilon_{k'}}{4k_B T_N}\right]\left[1 + \frac{4}{N_g}\sum_k \left(\frac{\varepsilon_k^2}{U\Delta_k} - \frac{\varepsilon_k}{U}\right)\text{th}\,\frac{\Delta_k + \varepsilon_k}{4k_B T_N}\right]^{-1}, \tag{42}$$

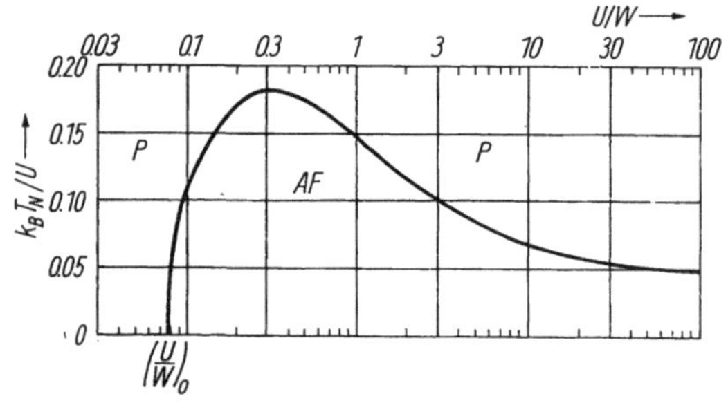

Fig. 1. For a half-filled band the ratio of the Néel temperature $k_B T_N$ to the repulsive interaction U between electrons on the same atom is plotted as function of U/W, where W is the band width

where N_g is the number of the atoms in the whole lattice. Fig. 1 shows the results of the numerical calculation of the Néel temperature T_N for a simple cubic lattice, using the density of state function $\varrho = \varrho_{sc}(\varepsilon)$ stated by Jelitto [23].

The calculations have been done for fixed values of U ($U = 0.1$; 0.5; 1; 2; 5 eV) and we found that all values of U lead for fixed U/W to the same values for $k_B T_N/U$ (the deviations amount to 1% maximally) and so we get only one curve for $k_B T_N/U$ as a function of U/W.

5. Discussion

The curve for $k_B T_N/U$ as a function of U/W, which we have calculated by Roth's method, using the set (15), starts at a certain $(U/W)_0$. For all $U/W < < (U/W)_0$ the formula (42) is not solvable for any T_N; this means that in the present approximation for $U/W < (U/W)_0$ no antiferromagnetic order exists. The exact value for $(U/W)_0$ depends on the used density of states $\varrho(\varepsilon)$. For a simple cubic lattice we get $0.078 < (U/W)_0 < 0.079$. For a flat band with

$$\varrho(\varepsilon) = \begin{cases} \dfrac{1}{D} & \text{for} \quad -\dfrac{D}{2} \leqq \varepsilon \leqq \dfrac{D}{2} \\ 0 & \text{otherwise} \end{cases}$$

we get $(U/D)_0 = 0.14$.

The existence of such a critical value $(U/W)_0$ is in agreement with results of Caron and Kemeny [24] for $T = 0$°K and with CPA-results of Cyrot and Lacour-Gayet [22] and in disagreement with CPA-results of Plischke [25] and results which have been obtained by Hartree-Fock approximation [10], because a Hartree-Fock approximation gives an antiferromagnetic order for arbitrarily small U/W.

For a flat band we can expand the right-hand side of (42) for $U/D \gg 1$. We find that $k_B T_N/U$ is proportional to $1/\log U/D$. Langer [26], using a more general first order decoupling, got the same result. However, this behaviour does not agree with perturbation expansion calculations about the atomic limit that give an antiferromagnetic coupling between local moments is proportional to W^2/U. Kaplan and Bari [21], using the thermal single determinant approximation, found for the Néel temperature $k_B T_N \propto W^2/U$; Plischke [25] gets the same results, using a model density of states given by Hubbard [27]. In the atomic limit $W \to 0$ (42) yields $T_N = 0$ as we have expected.

Above of a maximal value $(k_B T_N/U)_{max}$ no antiferromagnetic order exists, independent of the band width W. For $\varrho = \varrho_{sc}(\varepsilon)$ we get $(k_B T_N/U)_{max} \approx 0.18$, this correspond e.g. for $U = 1$ eV to $T_N \approx 2096$ °K. Always two different values of U/W yield the same value $k_B T_N/U$, e.g. for $\varrho = \varrho_{sc}(\varepsilon)$ we get $k_B T_N/U = = 0.1$ for $(U/W)_1 \approx 0.1$ and $(U/W)_2 \approx 3$, this corresponds e.g. for $U = 1$ eV and band width $W_1 = 10$ eV and $W_2 = 1/3$ eV to the Néel temperature $T_N \approx \approx 1170$ °K. This suggests the conjecture that we could have on the right-hand side of the maximum a transition from antiferromagnetic insulator to paramagnetic insulator and on the left-hand side one from antiferromagnetic to paramagnetic metal.

Calculations for $n \neq 1$ are in preparation.

Acknowledgements

The authors wish to thank for the carrying-out of the numerical calculations Dr. E. Heiner, Sektion Physik der Technischen Universität Dresden, and for stimulating and useful conversations Prof. E. P. Wohlfarth, Prof. M. J. Zuckermann, Dr. D. M. Edwards, and Dr. D. E. Mackenzie.

References

[1] J. Hubbard, Proc. Roy. Soc. **A 276**, 238 (1963).
[2] D. Adler, Solid State Phys. **21**, 1 (1968).
D. Adler, Rev. mod. Phys. **40**, 714 (1968).
[3] C. Herring, in: Magnetism — Treatise on Modern Theory and Materials, Ed. G. T. Rado and H. Suhl, Vol. IV, Academic Press, New York/London **1966** (p. 273).
[4] J. Nagaoka, Phys. Rev. **147**, 392 (1966).
[5] J. B. Sokoloff, Phys. Rev. B **2**, 3707 (1970).
[6] L. G. Caron, G. Kemeny, Phys. Rev. B **4**, 150 (1971).
[7] D. R. Penn, Phys. Rev. **142**, 350 (1966).
[8] D. M. Esterling and R. V. Lange, Phys. Rev. B **1**, 2231 (1970); .
D. M. Esterling, Phys. Rev. B **2**, 4686 (1970).
R. A. Bari, Phys. Rev. B **2**, 2260 (1970).
[9] S. W. Tjablikow, Quantentheoretische Methoden des Magnetismus, B. G. Teubner Verlagsgesellschaft, Leipzig **1968** (Transl. from Russian).
[10] W. Langer, M. Plischke, and D. Mattis, Phys. Rev. Letters **23**, 1448 (1969) and Errata, Phys. Rev. Letters **24**, 635 (1970).
[11] D. R. Penn, Phys. Letters **26A**, 509 (1968).
[12] A. C. Hewson and U. Lindner, phys. stat. sol. **42**, K185 (1970).
[13] R. A. Bari and T. A. Kaplan, Phys. Letters **33A**, 400 (1970).
[14] R. J. Jelitto, Z. Naturf. **27a**, 1394 (1972).
[15] C. Mehrotra, K. S. Viswanathan, Phys. Rev. B **7**, 559 (1973).
[16] T. Arai, Phys. Rev. B **4**, 216 (1971).
[17] U. Lindner, Habilitationsschrift, Leipzig 1971.
[18] L. M. Roth, Phys. Rev. **184**, 451 (1969).
[19] K. Sawada, Progr. theor. Phys. (Kyoto) **43**, 1199 (1970).
[20] A. B. Harris and R. V. Lange, Phys. Rev. **157**, 295 (1967).
[21] T. K. Kaplan and B. A. Bari, J. appl. Phys. **41**, 875 (1970).
[22] M. Cyrot and P. Lacour-Gayet, Solid State Commun. **11**, 1762 (1972).
[23] R. J. Jelitto, J. Phys. Chem. Solids **30**, 609 (1969).
[24] L. G. Caron and G. Kemeny, Phys. Rev. B **3**, 3007 (1971).
[25] M. Plischke, Solid State Commun. **13**, 393 (1973).
[26] W. Langer, J. Phys. C **4**, L56 (1971).
[27] J. Hubbard, Proc. Roy. Soc. **A281**, 401 (1964).

(Received January 7, 1974)

phys. stat. sol. (b) 62, 615 (1974)

Subject classification: 14.3 and 20.1; 6; 22.9

Laboratoire de Biophysique, U.E.R.D.M., Université de Nice (a), and Brown Boveri Research Center, Baden, Switzerland (b)

Polarized Optical Reflection Spectra of the Highly Anisotropic Organic Semiconductor $TEA(TCNQ)_2$ [1])

By

A. BRAU (a), P. BRÜESCH (b), J. P. FARGES (a), W. HINZ (b), and D. KUSE (b)

From polarized reflection spectra of $TEA(TCNQ)_2$ single crystals at room temperature, quantitative values of the complex dielectric function $\varepsilon(\omega)$ and the optical conductivity $\sigma(\omega)$ in the range from the far infrared to the ultraviolet have been derived. The polarization of the vibrational bands almost exclusively parallel to the highly conducting direction suggests strong interaction of conduction electrons with intramolecular TCNQ vibrations.

An Einkristallen des hoch anisotropen organischen Halbleiters $TEA(TCNQ)_2$ wurden bei Zimmertemperatur die polarisierten Reflexionsspektren vom fernen Infrarot bis ins Ultraviolett gemessen. Daraus wurden die komplexe Dielektrizitätskonstante $\varepsilon(\omega)$ sowie die optische Leitfähigkeit $\sigma(\omega)$ bestimmt. Die beobachteten Vibrationsbanden sind fast ausschließlich parallel zur Richtung hoher Leitfähigkeit polarisiert. Das ist ein Hinweis dafür, daß die Leitungselektronen mit den intramolekularen TCNQ-Schwingungen stark gekoppelt sind.

1. Introduction

Triethylammonium-tetracyanoquinodimethane ($TEA(TCNQ)_2$) belongs to a class of charge transfer organic salts based on the molecule TCNQ that are electrical conductors [1].

The crystal structure shows the planar TCNQ molecules stacked on top of each other to form parallel linear chains, the chains being arranged in such a way that layers made up of TCNQ chains alternate with layers of TEA molecules [2]. With the TEA acting as cation there is formally half an unpaired electron per TCNQ site available for conduction. Unlike the highly conducting TCNQ salts such as NMP(TCNQ) [3] or TTF(TCNQ) [4] which are metallic at room temperature and have all the TCNQ molecules in a chain equidistant and equivalent, $TEA(TCNQ)_2$ has a superstructure which prevents a metallic band structure. The repeat unit in the chain consists of four TCNQ molecules with spacings of 3.24 and 3.30 to 3.32 Å alternatively. The molecules are stacked slightly offset with two different modes of overlap. As inferred from bond lengths, the TCNQ appears to be present as $TCNQ^\circ$ and $TCNQ^{-}$, the unpaired electrons residing preferentially on pairs of neighbouring $TCNQ^{-}$ units [2].

The dc conductivity [5 to 7] is highly anisotropic. For the direction of highest conductivity — coinciding with the chain direction — its value at room temperature is of the order of some Ω^{-1} cm^{-1}. It is two to three orders of magnitude

[1]) Work supported in part by the Direction des Recherches et Moyens d'Essais, Paris.

smaller for directions perpendicular to the chains. The conductivity is thermally activated and the activation energy (E_g) varies with direction of measurement and temperature between 0.10 and 0.20 eV. The transport properties of the material are not well understood. The temperature dependence of the paramagnetic susceptibility [8] and of an observed EPR line [9] have been interpreted in terms of a localized electron pair model with a singlet ground state and a triplet state 0.03 to 0.04 eV above the singlet state [8, 9]. This and the structural evidence for $(TCNQ)_2^{\bar{2}}$ dimers both suggest electron localization on the TCNQ sites in the ground state. Suezaki [10] considered that the activation energy of the electrical conduction is due to electron Coulomb correlation. Brau and Farges, on the other hand, tried to reconcile the structural, magnetic, and electrical properties of the material by assuming a band model in which band tails of localized states due to structural disorder effects extend into the band gap [7].

Previous optical measurements on powder samples of $TEA(TCNQ)_2$ and some related compounds [11, 12] have revealed a broad infrared absorption band which overlaps with the vibrational spectra of these compounds. This band has been considered as due to interband transitions or charge transfer transitions involving the exchange of electrons between neighbouring $TCNQ^\circ$ and $TCNQ^-$ units within chains [12].

We have studied the optical properties of $TEA(TCNQ)_2$ in more detail using single crystals. Reflection spectra have been measured with polarized light, at nearly normal incidence, at room temperature, in the spectral range from 20 to 50000 cm^{-1}. For polarization parallel to the TCNQ chains, the optical constants have been determined from the spectrum by means of a Kramers-Kronig analysis. The complexity of the spectrum prevents a correlation of the results with the dc transport properties in terms of a simple model.

2. Experiments

The crystals are triclinic, consequently the axes of the optical indicatrix and of the conductivity tensor can in principle be arbitrarily oriented in the crystal. In our measurements the light was polarized along three mutually orthogonal directions, labelled 1, 2 and 3, which coincide with the directions in which the dc conductivity σ was measured previously [7]. Direction 1 is parallel to the TCNQ chains, directions 2 and 3 are perpendicular to the chains, 2 being within and 3 perpendicular to the alternating layers of TCNQ and TEA molecules. As expected from the structure, the electrical anisotropy is such that $\sigma_1 \gg \sigma_2 \gg$

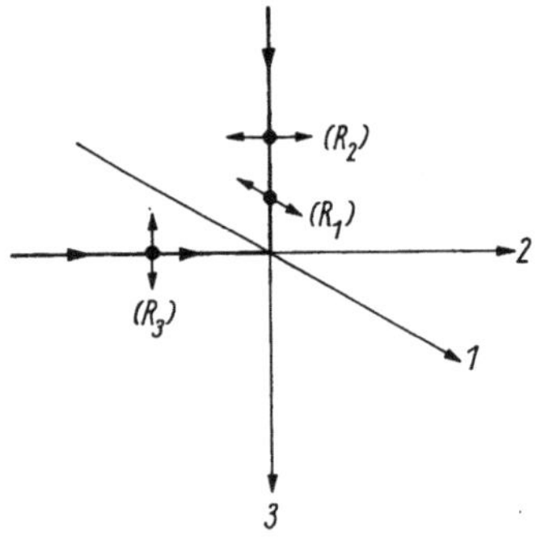

Fig. 1. Definition of the different configurations for the polarized near-normal-incidence reflection spectra of $TEA(TCNQ)_2$. The three mutually orthogonal directions, labelled 1, 2, and 3, coincide with the directions in which the dc conductivity was measured previously [7]. Direction 1 is parallel to the TCNQ chains, direction 2 and 3 perpendicular to the chains, 2 being within and 3 perpendicular to the alternating layers of TCNQ and TEA molecules. In the R_1-configuration, for example, the incident beam was nearly parallel to the direction 3 and polarized parallel to the direction 1

$\gg \sigma_3$. The reflectances R_1 and R_2 (Fig. 1) could be measured on a natural surface of the crystal platelets, which is parallel to the TCNQ layers [13]. Unpolished surfaces gave higher and more reproducible R-values than tentatively polished surfaces. In two different samples the reflectivity agreed to within $\pm 1\%$ throughout the spectra. For measurements of the R_3-spectrum one of the narrow side faces of the crystals had to be dressed at right angle to direction 2 and polished.

Two different spectrometers were used: one equipped with a Spex Model 1702 grating monochromator with 75 cm focal length in the range 50000 to 750 cm^{-1}, and a Fourier photospectrometer (Beckman FS 720) from 1000 to 20 cm^{-1}. A freshly aluminized mirror was used as a reference in the far infrared region, and a rhodium mirror in the range 50000 to 750 cm^{-1}. The reflectance of the rhodium mirror was measured over this range using an absolute reflectometer similar to the one described in reference [14].

3. Results

The reflection spectra are shown in Fig. 2. For the polarization directions 2 and 3, perpendicular to the chains, the reflectance never exceeds 10% and shows only little structure, even in the infrared region of the lattice vibrations. In contrast, the R_1-spectrum, polarized along the chains, shows numerous strong infrared bands with reflection peaks of about 50% and a broad maximum near 3000 cm^{-1} followed by a minimum of less than 1% near 7000 cm^{-1} indicating a plasma resonance. At higher photon energies up to 50000 cm^{-1} R_1 varies only little and remains smaller than 5%. A "plasma" reflection minimum observed by Vlasova and Gutman [15] on the much less conducting salt $Cs_2(TCNQ)_2$ is also shown for comparison in Fig. 2.

The real and imaginary parts of the optical dielectric function $\varepsilon = \varepsilon_1 + i\varepsilon_2$ and the optical conductivity σ, determined from the R_1-spectrum by means of a Kramers-Kronig analysis, are shown in Fig. 3, 4 and 5, respectively. We have performed the Kramers-Kronig analysis with the following extrapolations: At high frequencies ($\omega > 50000$ cm^{-1}) the reflectivity was represented by a constant value of 1.6%. At the low-frequency end of the spectrum ($\omega < 20$ cm^{-1}) a constant value of 56.5% was assumed for the reflectivity. The latter extrapolation, of course, implies a vanishing dc conductivity in contrast to the measured value of $\sigma_{dc} = 7.4\ (\Omega\text{cm})^{-1}$. However, this extrapolation does not introduce serious errors of the calculated optical constants in the measured frequency range.

The optical constants exhibit considerable structure at energies below 5000 cm^{-1} where the large anisotropic structure is observed in the reflection spectrum.

Both the imaginary part ε_2 of the dielectric function and the optical conductivity σ show numerous narrow bands, some of which overlap with a broad band which extends from ≈ 1600 to ≈ 5000 cm^{-1}. The real part ε_1 of the dielectric function passes through zero at several frequencies which correspond to transverse optical (for $d\varepsilon_1/d\omega < 0$) and longitudinal optical (for $d\varepsilon_1/d\omega > 0$) resonances. The highest of the longitudinal resonances which occurs at ≈ 4800 cm^{-1} gives rise to the pronounced "plasma" reflection edge.

Our results may be compared with the powder "absorption" spectra reported by Kondow and Sakata [11] and by Iida [12]. The close resemblance of our

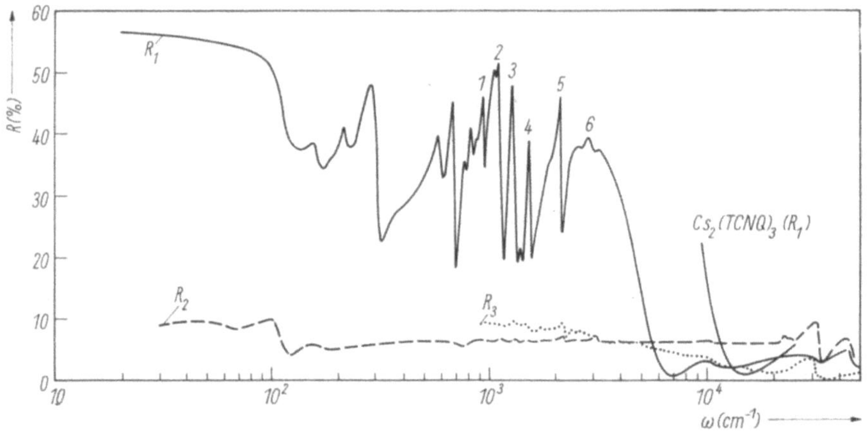

Fig. 2a. Polarized near-normal-incidence reflection spectra of TEA$(TCNQ)_2$ at room temperature in the whole frequency range from the far infrared to the ultraviolet. Note the logarithmic energy scale. The configurations R_1, R_2, and R_3 are explained in Fig. 1. The results obtained by Vlasova and Gutman [15] on $Cs_2(TCNQ)_3$ (R_1-configuration) are also shown for comparison

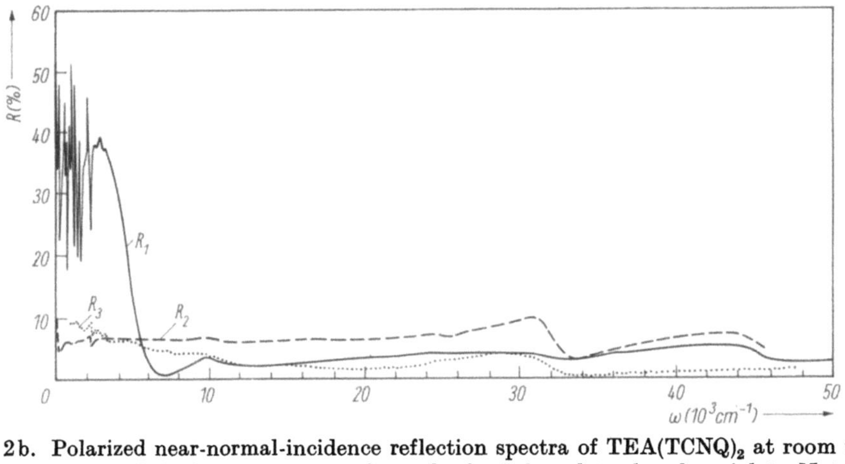

Fig. 2b. Polarized near-normal-incidence reflection spectra of TEA$(TCNQ)_2$ at room temperature in the whole frequency range from the far infrared to the ultraviolet. Note the linear energy scale. The spectra show the details of the electronic transitions at higher energies

R_1-spectrum with the infrared spectrum of reference [11] indicates that the powder spectrum in this particular case was governed by reflection rather than absorption processes. At higher energies our reflection spectra indicate transitions at 10 to 11 kK (polarized preferentially along the chains), at 28 to 32 kK (polarized preferentially perpendicular to the chains), and at 42 kK. These bands also appear in the spectra of ε_2 (Fig. 4) and σ (Fig. 5). Iida [12] reported bands at 2.9 to 3.3, 10.3 to 11.4, 16.4, and 27.2 kK. From a comparison of the spectra of various TCNQ compounds in solution and in powder form he ascribed the 10 to 11 kK band to either electronic transitions of the $(TCNQ)^{\overline{\cdot}}$ monomer

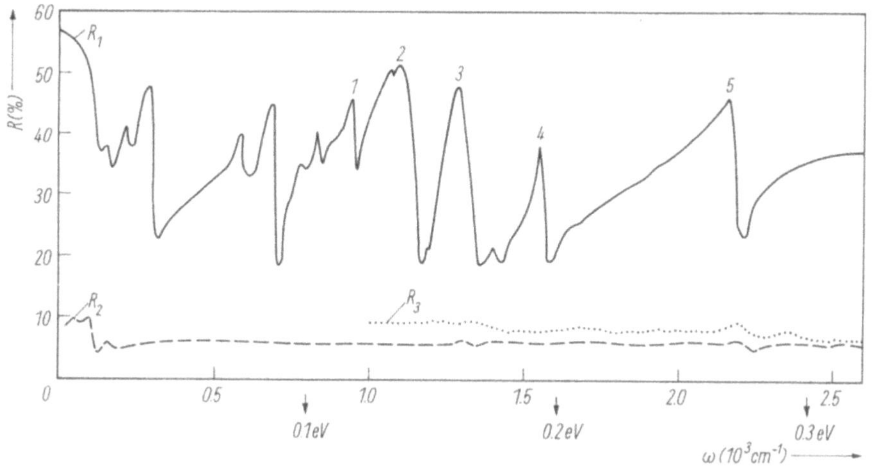

Fig. 2c. Polarized near-normal-incidence reflection spectra of TEA(TCNQ)$_2$ at room temperature in the far infrared and infrared range. The spectra show the details of the vibrational transitions observed in the R_1-configuration

or to charge transfer between (TCNQ)$^{\bar{\cdot}}$ radicals. The latter interpretation agrees with the above-indicated polarization of the band parallel to the chain direction, but is questionable for the following reason: The spectra of TEA(TCNQ)$_2$ and other intermediate conductivity TCNQ compounds closely resemble those of the high-conductivity compounds [12]. The band at 10 to 12 kK may therefore reasonably be expected to have the same origin in all these compounds. In the low-conductivity compounds, on the other hand, a band ascribed to charge transfer between TCNQ$^{\bar{\cdot}}$ radicals [12] was observed below 10 kK. It is

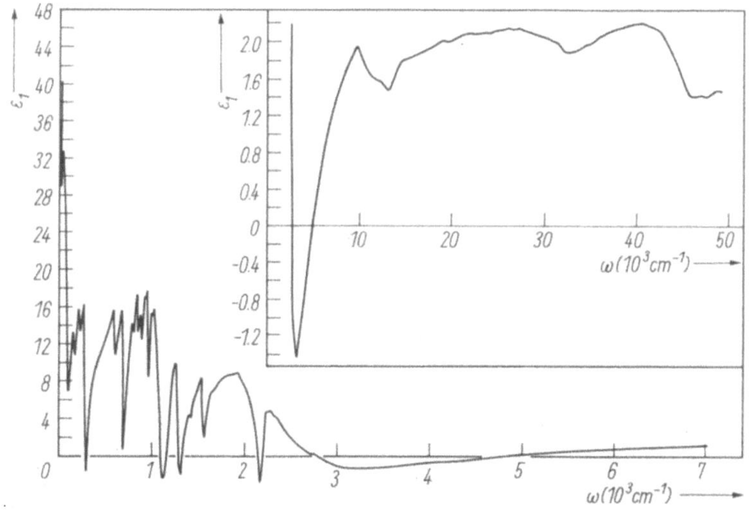

Fig. 3. Real part $\varepsilon_1(\omega)$ of the dielectric constant $\varepsilon = \varepsilon_1 + i\varepsilon_2$ of TEA(TCNQ)$_2$ for light polarized parallel to the TCNQ chains (configuration R_1 of Fig. 1)

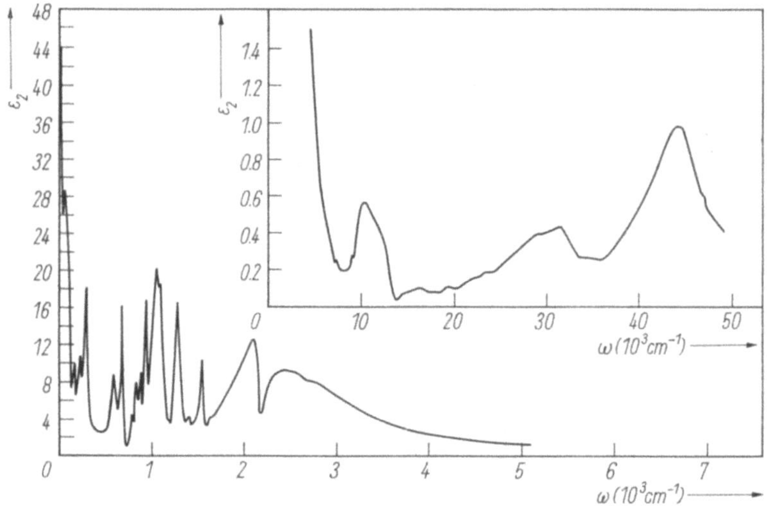

Fig. 4. Imaginary part $\varepsilon_2(\omega)$ of the dielectric constant $\varepsilon = \varepsilon_1 + i\varepsilon_2$ of TEA(TCNQ)$_2$ for light polarized parallel to the TCNQ chains (configuration R_1 of Fig. 1)

unlikely — although not impossible — that the TCNQ$^{\bar{}}$–TCNQ$^{\bar{}}$ charge transfer requires equal energies in the intermediate and high-conductivity compounds and even less in the compounds with low conductivity. The 16.4 kK band was attributed by Iida to dimers, and the 27.2 kK band was assumed to consist of overlapping dimer and monomer bands. Our spectra do not show the 16.4 kK band. Apparently it is too weak for observation in reflection.

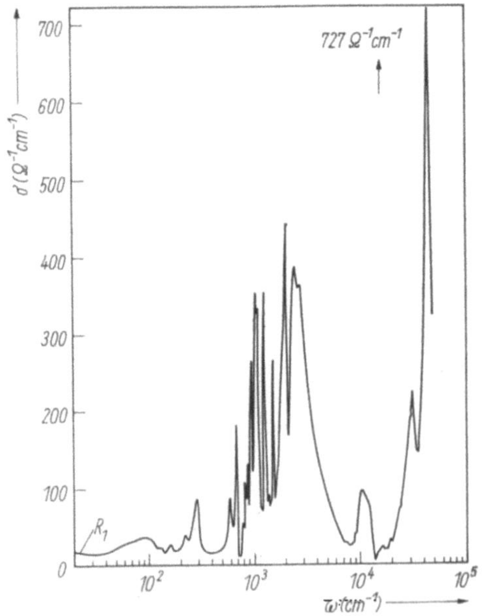

Fig. 5. Optical conductivity $\sigma(\omega)$ of TEA(TCNQ)$_2$ for light polarized parallel to the TCNQ chains (configuration R_1 of Fig. 1)

4. Discussion

From a comparison of the reflection spectrum polarized along the chains (Fig. 2) with the powder absorption spectra of various other TCNQ compounds [11, 16] and with the spectra of TCNQ alone [17] most of the discrete structure in the infrared appears to be associated with vibrations of the TCNQ molecules. For instance, the peaks at 155 and 218 cm^{-1} (Fig. 2) can be identified with those reported by Lunelli and Pecile [17] at 143 to 146 and 220 to 224 cm^{-1} and by Kondow et al. [16] at 143 and 219 cm^{-1}, in the spectra of TCNQ alone. A peak at 296 cm^{-1}, possibly corresponding to the strong band at 290 cm^{-1} in $TEA(TCNQ)_2$ was observed by Kondow et al. [16] in several other TCNQ salts, but appears to be absent in the spectrum of TCNQ alone [16, 17]. The pattern of the bands labelled 1 to 6 in Fig. 2 is also found in the powder spectra of other TCNQ compounds [11], and also correlates with the absorption spectrum of TCNQ [17] (Table 1).

Table 1

Comparison of the vibrational frequencies of TCNQ in $TEA(TCNQ)_2$ with the frequencies of the absorption peaks of largest intensities of pure TCNQ as given by Lunelli and Pecile [17] (frequency of the peaks in cm^{-1})

observed from reflection on $TEA(TCNQ)_2$		observed from absorption in pure TCNQ
1050–1100	(2)	1047 1093 1114 1125 1136
1290	(3)	1354
1550	(4)	1540 1545
2160	(5)	2228
2900	(6)	3053 3065

Numbers in parentheses refer to the peaks in Fig. 2a and 2c.

While the frequencies of these characteristic vibrational bands vary only little between the various TCNQ compounds, the position and width of the broad IR band vary markedly [11]. In the case of $TEA(TCNQ)_2$ the excitation energy near the band maximum (Fig. 5) is comparable to twice the activation energy of the dc conductivity. This and the polarization of the band parallel to the chains support the assumption that the band is of electronic origin [12] and is closely related to the dc conduction mechanism. The situation is, however, complicated by the overlap of the electronic transitions with lattice vibration bands, and the spectrum therefore does not help to distinguish between the proposed electronic models [7 to 10]. In contrast, in view of the mutual overlap of electronic and vibrational excitations it seems possible — although, of course, not necessary — that the separability of electronic and vibrational excitations

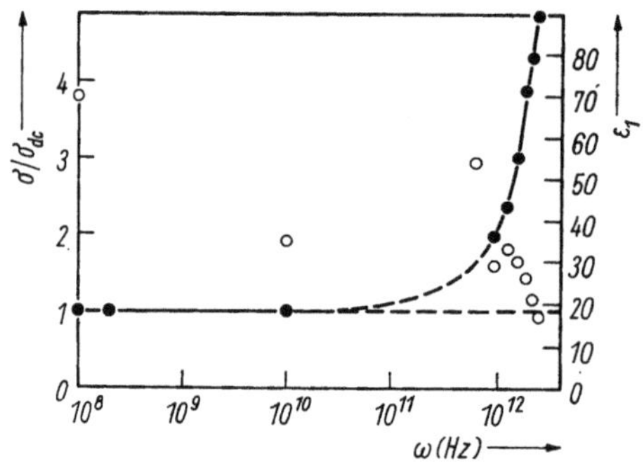

Fig. 6. Comparison of $\varepsilon_1(\omega)$ and $\sigma(\omega)$ at the low-frequency end of the spectrum, as measured along the direction of TCNQ chains, with data in dc field [7], at 10^8 and 2×10^8 Hz [20], and at 10^{10} Hz (Shchegolev) [1]. ● Conductivity ratio σ/σ_{dc}; ○ permittivity ε_1

breaks down. Kondow and Sakata noticed that the vibrational bands in various TCNQ compounds broaden as the overlap with the electronic band increases, and took this as evidence for strong electron–phonon interactions [11]. $TCNQ^{\overline{\cdot}}$ radicals and TCNQ molecules have different internal bond lengths; consequently the transition of an electron from a $TCNQ^{\overline{\cdot}}$ to TCNQ may well excite intense molecular vibrations [2]. A second piece of evidence for vibronic couplings might be the polarization of the vibrational bands almost exclusively parallel to the highly conducting direction [18]. This is just opposite to the case of the chain compound $K_2[Pt(CN)_4]Br_{0.3} \cdot 3\ H_2O$ where only phonon bands polarized perpendicular to the complex chains are observable in the reflection spectra at room temperature [19].

In the case of $TEA(TCNQ)_2$ the observed polarization of the vibrational spectrum is the more surprising as most of the molecular vibrations of TCNQ are in-plane vibrations [17]. Since the molecular planes are only slightly tilted with respect to the chain axis, one would expect the vibrational spectrum to be preferentially polarized perpendicular to the chains. In this direction, however, virtually no structure is observed in the reflection spectrum. The assumption seems plausible that this unexpected anisotropy is caused somehow by a coupling of the phonons to the highly anisotropic electron system.

In Fig. 6 we compare our values for ε_1 and σ at the low-frequency end of the spectrum as measured along the direction of TCNQ chains, with data obtained elsewhere in dc field [7], at 10^8 and 2×10^8 Hz [20] and (by Shchegolev) at 10^{10} Hz [1]. The conductivity values for the various frequencies result in a reasonably smooth curve which remains constant between dc and $\approx 10^{11}$ Hz. The comparison of the ε_1-data is less clear, possibly because the values in the range 10^8 to 10^{10} Hz were determined at a temperature of ≈ 77 K while the optical measurements were made at room temperature.

5. Conclusions

From polarized reflection spectra of $TEA(TCNQ)_2$ single crystals we have derived quantitative values of the complex dielectric function ε and the optical conductivity σ as functions of photon energy, covering the energy range from 20 to 50000 cm^{-1}. However, the complexity of the spectra and the overlap between intense electronic and vibrational transitions which possibly gives rise to strong vibronic interactions, prevent a close analysis of the spectrum in relation to the dc transport properties and a description in terms of a simple model. In particular, the results obtained at room temperature only, do not distinguish between the models that have already been proposed to describe the properties of TEA(TCNQ)$_2$ and other TCNQ compounds with intermediate conductivities [1, 7 to 10].

Acknowledgements

It is a pleasure to thank Dr. P. Dupuis in Nancy for the crystals and Mr. F. Lehmann for his help during the measurements. Two of us (A. B. and J.P.F.) wish to record their appreciation of the hospitality received from Dr. H. R. Zeller and colleagues at B.B.C. One of us (J.P.F.) also is indebted to Dr. H. Launois in Orsay for a useful discussion on the subject.

References

[1] For reviews see: I. F. Shchegolev, phys. stat. sol. (a) **12**, 9 (1972); H. R. Zeller, Festkörperprobleme XIII, 31 (1973).
[2] H. Kobayashi, Y. Ohashi, F. Marumo, and Y. Saito, Acta cryst. **B29**, 459 (1970).
[3] A. J. Epstein, S. Etemad, A. F. Garito, and A. J. Heeger, Phys. Rev. B **5**, 952 (1972).
[4] L. B. Coleman, M. J. Cohen, D. J. Sandman, F. G. Yamagishi, A. F. Garito, and A. J. Heeger, Solid State Commun. **12**, 1125 (1973).
[5] J. P. Farges, A. Brau, and F. Gutman, J. Phys. Chem. Solids **33**, 1723 (1972).
[6] A. Brau and J. P. Farges, Phys. Letters A **41**, 179 (1972).
[7] A. Brau, and J. P. Farges phys. stat. sol. (b) **61**, 257 (1974).
[8] R. G. Kepler, J. chem. Phys. **39**, 3528 (1963).
[9] D. B. Chesnut and W. D. Phillips, J. chem. Phys. **35**, 1002 (1961).
[10] Y. Suezaki, J. Phys. Soc. Japan **34**, 89 (1973).
[11] T. Kondow and T. Sakata, phys. stat. sol. (a) **6**, 551 (1971).
[12] Y. Iida, Bull. Chem. Soc. Japan **42**, 71 (1969); **42**, 637 (1969).
[13] E. Menefee and Yoh-Han Pao, J. chem. Phys. **36**, 3472 (1962).
[14] H. E. Benett and W. F. Koehler, J. Opt. Soc. Amer. **50**, 1 (1960).
[15] R. M. Vlasova and A. I. Gutman, Fiz. tverd. Tela **13**, 2808 (1971); Soviet Phys. — Solid State **13**, 2356 (1972).
[16] T. Kondow, K. Siratori, and H. Inokuchi, J. Phys. Soc. Japan **23**, 98 (1967).
[17] B. Lunelli and C. Pecile, J. chem. Phys. **52**, 2375 (1970).
[18] M. G. Kaplunov, T. P. Panova, and Yu. G. Borodko, phys. stat. sol. (a) **13**, K67 (1972).
[19] J. Bernasconi, P. Brüesch, D. Kuse, and H. R. Zeller, J. Phys. Chem. Solids, to be published.
[20] J. P. Farges and A. Brau, phys. stat. sol. (b) **61**, 669 (1974).

(Received December 10, 1973)

phys. stat. sol. (b) 62, 625 (1974)

Subject classification: 13.1 and 13.4; 22.2.1; 22.2.2

Istituto di Fisica Teorica dell'Università, Napoli
and Gruppo Nazionale di Struttura della Materia del C.N.R.

Resonant States in III–V Semiconductor Compounds

By

G. IADONISI and G. P. ZUCCHELLI

A realistic model to discuss resonant states in semiconductors as GaAs and InP is developed. Analytic expressions for the energies and the widths of the resonant states are found and compared with recent experimental optical results in InP.

Es wird ein realistisches Modell zur Behandlung resonanter Zustände in Halbleitern wie GaAs und InP entwickelt. Analytische Ausdrücke der Energien und Resonanzbreiten der Zustände werden berechnet und mit neueren Ergebnissen von optischen Versuchen an InP verglichen.

1. Introduction

In this paper a model [1, 2], which was already considered for the study of the impurity bound and resonant levels, is applied to some semiconductors like GaAs and InP and analytic expressions for the energies and the widths of the resonant levels are obtained.

Recent optical experiments, which have given evidence for the existence of resonant states degenerate with the continuum of the conduction band of cubic crystals, increase the interest in these calculations. Electronic transitions from the ground bound state to resonant states associated to the X secondary minima have been observed in donor doped GaP and InP. In GaP [3], the optical transition occurs from the shallow ground bound state associated to the X minima to resonant states associated to higher secondary minima at X; in InP [4] the absolute minimum is at the Γ point so that the transition is from the Γ impurity bound state to the resonant states associated with the X minima.

The experiments give also the values of the resonant energies and widths in InP with Te and Si impurities. We have resonant energies of about 100 and 170 meV below the X minima; the width is about 10 meV.

From the theoretical point of view, these resonant states have the same properties as the resonant states of the scattering theory, but in crystals they originate from the band structure rather than from the form of the impurity potential [5 to 8].

Indirect experimental evidence for the existence of resonant states has been given in n-doped semiconductors by transport measurements [9 to 13] and in alkali-halides with F-centres by optical absorption (L-bands) [14, 15, 2]. In this latter case, the width of the resonance is also known, but the phonon coupling is very strong so that the experimental values of the width should be very different from the intrinsic width of the resonances.

2. Model

We report briefly, for clarity, the method already discussed in previous papers [8, 2] to calculate impurity bound or resonant levels when the band structure is complicated. We consider the general impurity equation [8]

$$[E(\boldsymbol{k}) - E]\,\varphi(\boldsymbol{k}) + \int \mathrm{d}\boldsymbol{k}'\, \mathsf{U}(\boldsymbol{k}, \boldsymbol{k}')\, \varphi(\boldsymbol{k}') = 0\,, \tag{1}$$

where $E(\boldsymbol{k})$, $\mathsf{U}(\boldsymbol{k}, \boldsymbol{k}')$, and $\varphi(\boldsymbol{k})$ are respectively the band energy of the crystal, the matrix element of the impurity potential in the Bloch basis and the Bloch transform of the impurity state.

When the impurity potential is attractive, we divide the $\boldsymbol{k}$-space in a number of zones Ω_i such that in each zone the band energy $E(\boldsymbol{k})$ has a minimum; the function $\varphi(\boldsymbol{k})$ is made up of component functions $\varphi_i(\boldsymbol{k})$ each defined in the corresponding zone Ω_i and finally, in each zone, the zero-order solutions $\varphi^0_{in}(\boldsymbol{k})$ are considered (i and n indicate the zone and the level, respectively), obtained with the neglect of the interzones matrix elements. The solution of (1) is found by writing

$$\varphi(\boldsymbol{k}) = \sum_{i,n} c_{in}\, \varphi^0_{in}(\boldsymbol{k}) \tag{2}$$

and the energies are given by

$$\det |(E^0_{in} - E)\, \delta_{in,i'n'} + (1 - \delta_{i,i'})\, U_{in,i'n'}| = 0\,, \tag{3}$$

where E^0_{in} is the zero order eigenvalue corresponding to the n level in the Ω_i zone and $U_{in,i'n'}$ the quantity

$$\int \mathrm{d}\boldsymbol{k}\, \mathrm{d}\boldsymbol{k}'\, \varphi^{0*}_{in}(\boldsymbol{k})\, \mathsf{U}(\boldsymbol{k}, \boldsymbol{k}')\, \varphi^0_{i'n'}(\boldsymbol{k}')\,. \tag{4}$$

The method considered has the advantage that in each zone the zero order solution $\varphi^0_{in}(\boldsymbol{k})$ can be found by using suitable approximations (for example the effective mass approximation). Semiconductors like GaAs or InP have a conduction band with minima at the Γ and X points [18]. We consider then four zones in the $\boldsymbol{k}$-space with the following band energy [2]:

$$E(\boldsymbol{k}) = \begin{cases} \gamma k^2 & \text{in zone } \Omega_0 \equiv [-k_0 \leqq k_x \leqq k_0; \\ & -k_0 \leqq k_y \leqq k_0;\ -k_0 \leqq k_z \leqq k_0] \\ \alpha[(|k_x| - k_0)^2 + k_y^2 + k_z^2] + \Delta & \text{in zone } \Omega_1 \equiv [|k_x| \geqq k_0; \\ & -|k_x| \leqq k_y \leqq |k_x|;\ -|k_x| \leqq k_z \leqq |k_x|] \\ \alpha[(|k_y| - k_0)^2 + k_x^2 + k_z^2] + \Delta & \text{in zone } \Omega_2 \equiv [|k_y| \geqq k_0; \\ & -|k_y| \leqq k_x \leqq |k_y|;\ -|k_y| \leqq k_z \leqq |k_y|] \\ \alpha[(|k_z| - k_0)^2 + k_x^2 + k_y^2] + \Delta & \text{in zone } \Omega_3 \equiv [|k_z| \geqq k_0; \\ & -|k_z| \leqq k_x \leqq |k_z|;\ -|k_z| \leqq k_y \leqq |k_z|]\,, \end{cases} \tag{5}$$

where $k_0 = 2\pi/a$ (a lattice parameter), γ^{-1} and α^{-1} are proportional to the inverse of the effective mass at Γ and X points of the band structure, respectively, and Δ (>0) is the energy separation between the minima at X and at Γ point.

We are interested in studying the resonant states in the energy range $0 < E < \Delta$. Therefore we consider the following zero order solutions in expansion (2): a) the bound and continuum states of zone Ω_0; b) the bound states of zones Ω_1, Ω_2, and Ω_3. Moreover, we want to study resonant states with Γ_{15} symmetry of cubic group, because the experimental optical transitions refer to transitions from the ground state with Γ_1 symmetry. In [2] it is shown that, by considering in the zone Ω_1, Ω_2, and Ω_3 the states 1s, we do not find states with symmetry such that the dipole matrix element is not zero; only by considering the 2p-states we can find solutions of (1) with symmetry Γ_{15}. In fact it is found that the nine 2p-states of the Ω_1, Ω_2, and Ω_3 zones give two resonant triplets of symmetry Γ_{15} and a resonant triplet of symmetry Γ_{25}.

The energies and the widths of the Γ_{15} resonant states are the solutions in the complex energy plane of the following equations:

$$\left.\begin{aligned} E_1 &= \tfrac{1}{2}\,\{a(E_1) + b(E_1) + c(E_1) + \sqrt{[b(E_1) - c(E_1) + a(E_1)]^2 + 8d^2(E_1)}\}\,,\\ E_2 &= \tfrac{1}{2}\,\{a(E_2) + b(E_2) + c(E_2) - \sqrt{[b(E_2) - c(E_2) + a(E_2)]^2 + 8d^2(E_2)}\}\,,\end{aligned}\right\} \quad (6\text{a})$$

where

$$\left.\begin{aligned} a(E) &= E^0_{1,1} - \sum_i \frac{U^2_{0i,11}}{E^0_{0i} - E} - u_{11,11}\,,\\ b(E) &= U_{11,21} - \sum_i \frac{U^2_{0i,11}}{E^0_{0i} - E} - u_{11,11}\,,\\ c(E) &= E^0_{1,1} - \sum_i \frac{U^2_{0i,31}}{E^0_{0i} - E} - u_{31,31}\,,\\ d(E) &= U_{11,31} - \sum_i \frac{U_{11,0i}U_{0i,31}}{E^0_{0i} - E} - u_{11,31}\,,\end{aligned}\right\} \quad (6\text{b})$$

and

$$u_{in,i'n'} = \int\limits_{\boldsymbol{k}' \in \Omega_0} \mathrm{d}\boldsymbol{k}'\, \frac{U_{in,0\boldsymbol{k}'}U_{0\boldsymbol{k}',i'n'}}{\gamma k'^2 - E - i\varepsilon}. \quad (6\text{c})$$

In the above expressions we have indicated the discrete and the continuum levels of the Ω_0 zone by the indices i and k, respectively, and with the index 1 the 2p-levels (with azimutal number $m = 0$) of the zone Ω_1. In InP the quantity Δ, $m_1^* = \hbar^2/2\gamma$ and a have, respectively, the values 0.9 eV [16], $0.07m_e$ (m_e electronic mass) [17], 6 Å [18], while the value of the $m_2^* = \hbar^2/2\alpha$ mass is not well known. The condition $\gamma k_0^2 \gg \Delta$ holds and in Fig. 1 we show qualitatively the relative position of the quantity γk_0^2 and Δ in the direction (1, 0, 0) of k-space.

In these conditions we can limit the continuum spectrum of the zone Ω_0 to values of k less than $\sqrt{\Delta\gamma^{-1}}$, by assuming that the width Γ and the energy E of the resonant states satisfy the condition $\Delta - E \gg \Gamma$.

To calculate the above matrix elements we solve, in the zero order approximation the impurity equation in each zone in the effective mass approximation with the Coulomb impurity potential and neglecting central cell corrections. We have that, for example, the wave function $\varphi^0_{1n}(\boldsymbol{k})$ is connected to the Fourier

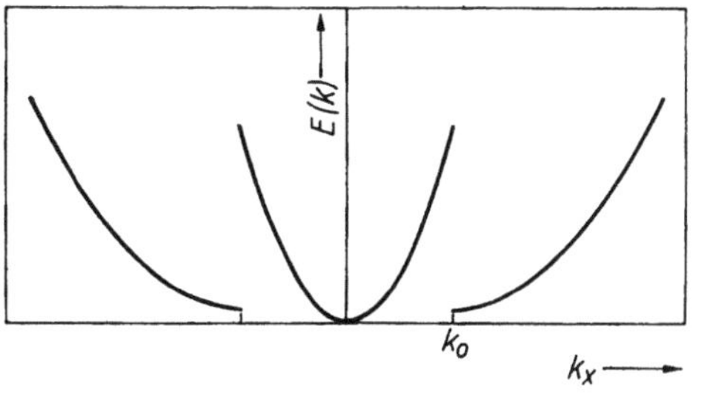

Fig. 1. Qualitative band structure $E(\boldsymbol{k})$ in the (1, 0, 0) direction

transform $f_{1n}(\boldsymbol{k})$ of the n-th state of an hydrogen atom by the relation

$$f_{1n}(\boldsymbol{k}) = \begin{cases} \varphi^0_{1n}(k_x + k_0, k_y, k_z)\,, & k_x > 0\,, \\ \varphi^0_{1n}(k_x - k_0, k_y, k_z)\,, & k_x < 0\,. \end{cases}$$

This can be easily understood, because the zone Ω_1 contains the two regions of the $\boldsymbol{k}$-space given by $[k_x \geqq k_0; -k_x \leqq k_y \leqq k_x; -k_x \leqq k_z \leqq k_x]$ and $[k_x \leqq -k_0; k_x \leqq k_y \leqq -k_x; k_x \leqq k_z \leqq -k_x]$ so that, when the zero order equation is considered, translations in the $\boldsymbol{k}$-space must be made to connect the two subzones [2]. We recall that $f_{1n}(\boldsymbol{k})$ depends on the effective mass $\hbar^2/2\alpha$ of the X point of the band structure and on $\varepsilon(0)$, which is the value at $q = 0$ of the q dependent static dielectric constant $\varepsilon(\boldsymbol{q})$.

Since $f_{1n}(\boldsymbol{k})$ is localized around $k = 0$, it follows that the function $\varphi^0_{1n}(\boldsymbol{k})$ is localized for $k_x > k_0$ around $(k_0, 0, 0)$ and for $k_x < -k_0$ around $(-k_0, 0, 0)$ and it is zero in the zone Ω_0.

In InP, assuming an isotropic effective electronic mass equal to the free electron mass at the X point, we find that $f_{1n}(\boldsymbol{k})$ is different from zero in a region, which is a small fraction of the Brillouin zone. For smaller effective masses, the localization in $\boldsymbol{k}$-space increases. This fact is very useful for an approximate evaluation of the matrix elements which appear in (6). In fact, using the odd parity of $f_{11}(\boldsymbol{k})$, we have

$$\begin{aligned} U_{11,21} &= \int\limits_{\substack{\boldsymbol{k}\in\Omega_1 \\ \boldsymbol{k}'\in\Omega_2}} \mathrm{d}\boldsymbol{k}\,\mathrm{d}\boldsymbol{k}'\,\varphi^{0*}_{11}(\boldsymbol{k})\,\mathsf{U}(\boldsymbol{k},\boldsymbol{k}')\,\varphi^0_{21}(\boldsymbol{k}') \\ &= \int\limits_{\substack{k_x>0 \\ k'_y>0}} \mathrm{d}\boldsymbol{k}\,\mathrm{d}\boldsymbol{k}'\,f^*_{11}(\boldsymbol{k})\,\mathsf{U}(\boldsymbol{k}+\boldsymbol{k}_{0x}, \boldsymbol{k}'+\boldsymbol{k}_{0y})\,f_{11}(\boldsymbol{k}') + \\ &\quad + \int\limits_{\substack{k_x>0 \\ k'_y<0}} \mathrm{d}\boldsymbol{k}\,\mathrm{d}\boldsymbol{k}'\,f^*_{11}(\boldsymbol{k})\,\mathsf{U}(\boldsymbol{k}+\boldsymbol{k}_{0x}, \boldsymbol{k}'-\boldsymbol{k}_{0y})\,f_{11}(\boldsymbol{k}') + \\ &\quad + \int\limits_{\substack{k_x<0 \\ k'_y>0}} \mathrm{d}\boldsymbol{k}\,\mathrm{d}\boldsymbol{k}'\,f^*_{11}(\boldsymbol{k})\,\mathsf{U}(\boldsymbol{k}-\boldsymbol{k}_{0x}, \boldsymbol{k}'+\boldsymbol{k}_{0y})\,f_{11}(\boldsymbol{k}') + \\ &\quad + \int\limits_{\substack{k_x<0 \\ k'_y<0}} \mathrm{d}\boldsymbol{k}\,\mathrm{d}\boldsymbol{k}'\,f^*_{11}(\boldsymbol{k})\,\mathsf{U}(\boldsymbol{k}-\boldsymbol{k}_{0x}, \boldsymbol{k}'-\boldsymbol{k}_{0y})\,f_{11}(\boldsymbol{k}') \\ &= 2\int\limits_{\substack{k_x>0 \\ k'_y>0}} \mathrm{d}\boldsymbol{k}\,\mathrm{d}\boldsymbol{k}'\,f^*_{11}(\boldsymbol{k})\,\mathsf{U}(\boldsymbol{k}+\boldsymbol{k}_{0x}, \boldsymbol{k}'+\boldsymbol{k}_{0y})\,f_{11}(\boldsymbol{k}') - \\ &\quad - 2\int\limits_{\substack{k_x>0 \\ k'_y>0}} \mathrm{d}\boldsymbol{k}\,\mathrm{d}\boldsymbol{k}'\,f^*_{11}(\boldsymbol{k})\,\mathsf{U}(\boldsymbol{k}+\boldsymbol{k}_{0x}, -\boldsymbol{k}'-\boldsymbol{k}_{0y})\,f_{11}(\boldsymbol{k}')\,. \end{aligned} \tag{7}$$

In these equations

$$\mathsf{U}(\boldsymbol{k},\boldsymbol{k}') = -\frac{4\pi e^2}{\varepsilon(|\boldsymbol{k}-\boldsymbol{k}'|)}\,\frac{1}{|\boldsymbol{k}-\boldsymbol{k}'|^2} \tag{8}$$

is the screened impurity potential when central cell corrections and anisotropy if the q-dependent dielectric constant are neglected [19, 20]. Since the function $f_{11}(\boldsymbol{k})$ is localized around the point $k = 0$, we put

$$\mathsf{U}(\boldsymbol{k}+\boldsymbol{k}_{0x}, \boldsymbol{k}'+\boldsymbol{k}_{0y}) = -\frac{4\pi e^2}{\varepsilon(|\boldsymbol{k}_{0x}-\boldsymbol{k}_{0y}|)}\,\frac{1}{|\boldsymbol{k}+\boldsymbol{k}_{0x}-\boldsymbol{k}'-\boldsymbol{k}_{0y}|^2}$$

and similarly for $\mathsf{U}(\boldsymbol{k}+\boldsymbol{k}_{0x}, -\boldsymbol{k}'-\boldsymbol{k}_{0y})$ so that we write

$$\mathsf{U}(\boldsymbol{k}+\boldsymbol{k}_{0x}, \boldsymbol{k}'+\boldsymbol{k}_{0y}) - \mathsf{U}(\boldsymbol{k}+\boldsymbol{k}_{0x}, -\boldsymbol{k}'-\boldsymbol{k}_{0y}) =$$

$$= -\frac{4\pi e^2}{\varepsilon\left(\sqrt{2}k_0\right)} \frac{4\,(\boldsymbol{k}+\boldsymbol{k}_{0x})\cdot(\boldsymbol{k}'+\boldsymbol{k}_{0y})}{[k^2+k'^2+2k_0^2+2k_xk_0+2k'_yk_0]^2 - 4\,[(\boldsymbol{k}+\boldsymbol{k}_{0x})\cdot(\boldsymbol{k}'+\boldsymbol{k}_{0y})]^2} \approx$$

$$\approx -\frac{4\pi e^2}{\varepsilon\left(\sqrt{2}k_0\right)} \frac{k'_x(k_x+k_0)+k_y(k'_y+k_0)+k_zk'_z}{k_0^4}. \tag{9}$$

Finally we have

$$U_{11,21} = -\frac{8\pi e^2}{\varepsilon\left(\sqrt{2}\,k_0\right)k_0^4}\left|\int_{k_x>0} k_z f_{11}(\boldsymbol{k})\,\mathrm{d}\boldsymbol{k}\right|^2 =$$

$$= -\frac{2\pi e^2}{\varepsilon\left(\sqrt{2}\,k_0\right)k_0^4}\left|i\left(\frac{\partial F_{11}}{\partial z}\right)_{r=0}\right|^2, \tag{10}$$

where F_{11} is the first hydrogenic p-state (with $m = 0$). For $U_{11,31}$ we have

$$\mathsf{U}(\boldsymbol{k}+\boldsymbol{k}_{0x}, \boldsymbol{k}'+\boldsymbol{k}_{0z}) - \mathsf{U}(\boldsymbol{k}+\boldsymbol{k}_{0x}, -\boldsymbol{k}'-\boldsymbol{k}_{0z}) \approx$$

$$\approx -\frac{4\pi e^2}{\varepsilon\left(\sqrt{2}\,k_0\right)} \frac{k'_x(k_x+k_0)+k_z(k'_z+k_0)+k_yk'_y}{k_0^4} \tag{9a}$$

so that

$$U_{11,31} = -\frac{2\pi e^2}{\varepsilon\left(\sqrt{2}\,k_0\right)k_0^4}\left|\left(\frac{\partial F_{11}}{\partial z}\right)_{r=0}\right|^2 -$$

$$-\frac{4\pi e^2}{\varepsilon\left(\sqrt{2}\,k_0\right)k_0^4}\left(i\,\frac{\partial F_{11}}{\partial z}\right)^*_{r=0}\left(-i\,\frac{2a_0}{\pi}\,\frac{\partial F_{11}}{\partial z}\right)_{r=0}$$

$$= -\frac{2\pi e^2}{\varepsilon\left(\sqrt{2}\,k_0\right)k_0^4}\left|\left(\frac{\partial F_{11}}{\partial z}\right)_{r=0}\right|^2\left(1+\frac{4k_0a_0}{\pi}\right), \tag{11}$$

where a_0 is the effective Bohr radius.

We note that the quantities $U_{11,21}$ and $U_{11,31}$ are not equal, since the function $f_{11}(\boldsymbol{k})$ has odd parity with respect to the k_z variable and the $\boldsymbol{k}$-integration is on the regions $k_x > 0$, $k'_y > 0$ and $k_x > 0$, $k'_z > 0$, respectively.

All the above quantities can be easily computed; their order of magnitude depends on the values of the effective mass at the X point. For the values of m_2^* considered in Section 3 the matrix elements $U_{11,21}$ and $U_{11,31}$ are at most of the order of some tens of meV and therefore smaller than Δ.

A similar procedure can be used to calculate the matrix elements $U_{0i,11}$ and $U_{0i,31}$; by using the same approximation of (9), we find that only the p-states of the zone Ω_0 give non-vanishing matrix elements of $U_{0i,11}$ and $U_{0i,31}$. Their form is similar to the one given by (10) or (11); the difference is that there appears a derivative of the p-wave function of the zone Ω_1 times the derivative of the p-wave function of the zone Ω_0.

We can conclude again that, for the values of m_2^* considered in Section 3 these matrix elements are very small with respect to Δ. On the other hand, the series of the type

$$-\sum_i \frac{U^2_{0i,11}}{E^0_{0i}-E} = \sum_i \frac{U^2_{0i,11}}{E-E_{0i}} < \frac{1}{E}\sum_i U^2_{0i,11}$$

are strongly convergent so that the contribution of this term is negligible as E is of the order of Δ.

We calculate now the matrix element $U_{0\boldsymbol{k},11}$ assuming that the wave functions of the continuum spectrum of zone Ω_0 is given by the free wave functions; we have, in the same approximations, used previously,

$$\left.\begin{aligned}
U_{0\boldsymbol{k},11} &= \int \mathrm{d}\boldsymbol{k}''\, f_{11}(\boldsymbol{k})\,[U(\boldsymbol{k}', \boldsymbol{k}_{0x}+\boldsymbol{k}'') - U(\boldsymbol{k}', -\boldsymbol{k}_{0x}-\boldsymbol{k}'')]\\
&= -\frac{4\pi e^2}{\varepsilon(k_0)}\,[\int \mathrm{d}\boldsymbol{k}''\, f_{11}(\boldsymbol{k}'')\, k_z'']\,\frac{k_z}{(k^2+k_0^2)^2-4k_0^2k_x^2}\\
&= -\frac{8\pi e^2}{\varepsilon(k_0)}\left[\left(-i\,\frac{\partial F_{11}}{\partial z}\right)_{r=0}\right]\frac{k_z}{(k^2+k_0^2)^2-4k_0^2k_x^2},\\
U_{0\boldsymbol{k},31} &= -\frac{4\pi e^2}{\varepsilon(k_0)}\left[\int_{k_z'>0}\mathrm{d}\boldsymbol{k}'\, f_{11}(\boldsymbol{k}')\right]\frac{4k_0k_z}{(k^2+k_0^2)^2-4k_0^2k_z^2}-\\
&\quad-\frac{8\pi e^2}{\varepsilon(k_0)}\left(-i\,\frac{\partial F_{11}}{\partial z}\right)_{r=0}\frac{k_z}{(k^2+k_0^2)^2-4k_0^2k_z^2}\\
&= -\frac{8\pi e^2}{\varepsilon(k_0)}\left(-i\,\frac{\partial F_{11}}{\partial z}\right)_{r=0}\left(1+\frac{4k_0a_0}{\pi}\right)\frac{k_z}{(k^2+k_0^2)^2-4k_0^2k_z^2}.
\end{aligned}\right\}\qquad(12)$$

The quantities $u_{11,11}$, $u_{11,31}$, and $u_{31,31}$ are therefore given by

$$\left.\begin{aligned}
u_{11,11} &= \frac{64\pi^2e^4}{\varepsilon^2(k_0)}\left|\left(\frac{\partial F_{11}}{\partial z}\right)_{r=0}\right|^2\int_0^{\sqrt{\Delta/\gamma}}\mathrm{d}\boldsymbol{k}'\,\frac{k_z'^2}{(\gamma k'^2-E-i\varepsilon)\,[(k'^2+k_0^2)^2-4k_0^2k_x'^2]^2},\\
u_{11,31} &= \frac{64\pi^2e^4}{\varepsilon^2(k_0)}\left|\left(\frac{\partial F_{11}}{\partial z}\right)_{r=0}\right|^2\left(1+\frac{4k_0a_0}{\pi}\right)\times\\
&\quad\times\int_0^{\sqrt{\Delta/\gamma}}\mathrm{d}\boldsymbol{k}'\,\frac{k_z'^2}{(\gamma k'^2-E-i\varepsilon)\,[(k'^2+k_0^2)^2-4k_0^2k_x'^2]\,[(k'^2+k_0^2)^2-4k_0^2k_z'^2]},\\
u_{31,31} &= \frac{64\pi^2e^4}{\varepsilon^2(k_0)}\left|\left(\frac{\partial F_{11}}{\partial z}\right)_{r=0}\right|^2\left(1+\frac{4k_0a_0}{\pi}\right)^2\times\\
&\quad\times\int_0^{\sqrt{\Delta/\gamma}}\mathrm{d}\boldsymbol{k}'\,\frac{k_z'^2}{(\gamma k'^2-E-i\varepsilon)\,[(k'^2+k_0^2)^2-4k_0^2k_z'^2]^2}.
\end{aligned}\right\}\qquad(13)$$

Since $\gamma k_0^2 \ll \Delta$ all the denominators in the integrals can be approximated by their first order expansion around $k' = 0$. We obtain finally

$$\left.\begin{aligned}
u_{11,11} &= \frac{128\pi^3 e^4}{\varepsilon^2(k_0)} \left|\left(\frac{\partial F_{11}}{\partial z}\right)_{r=0}\right|^2 \frac{1}{k_0^8} \times \\
&\times \left\{\left[\frac{2}{3\gamma^2}\left(\frac{\Delta}{3}\sqrt{\frac{\Delta}{\gamma}} + E\sqrt{\frac{\Delta}{\gamma}} + \frac{E}{2}\sqrt{\frac{E}{\gamma}}\log\frac{\sqrt{\Delta}-\sqrt{E}}{\sqrt{\Delta}+\sqrt{E}}\right) - \right.\right. \\
&- \left.\frac{8}{5\gamma^2(\gamma k_0^2)}\left(\frac{\Delta^2}{9}\sqrt{\frac{\Delta}{\gamma}} + E^2\sqrt{\frac{\Delta}{\gamma}} + \frac{E\Delta}{3}\sqrt{\frac{\Delta}{\gamma}} + \frac{E^2}{2}\sqrt{\frac{E}{\gamma}}\log\frac{\sqrt{\Delta}-\sqrt{E}}{\sqrt{\Delta}+\sqrt{E}}\right)\right] + \\
&+ \left.\frac{i\pi}{2\gamma}\left(\frac{E}{\gamma}\right)^{3/2}\left(\frac{2}{3} - \frac{8}{5}\frac{E}{\gamma k_0^2}\right)\right\}, \\
u_{11,31} &= \frac{128\pi^3 e^4}{\varepsilon^2(k_0)} \left|\left(\frac{\partial F_{11}}{\partial z}\right)_{r=0}\right|^2 \left(1 + \frac{4k_0 a_0}{\pi}\right) \frac{1}{k_0^8} \times \\
&\times \left\{\left[\frac{2}{3\gamma^2}\left(\frac{\Delta}{3}\sqrt{\frac{\Delta}{\gamma}} + E\sqrt{\frac{\Delta}{\gamma}} + \frac{E}{2}\sqrt{\frac{E}{\gamma}}\log\frac{\sqrt{\Delta}-\sqrt{E}}{\sqrt{\Delta}+\sqrt{E}}\right) - \right.\right. \\
&- \left.\frac{8}{15\gamma^2(\gamma k_0^2)}\left(\frac{\Delta^2}{9}\sqrt{\frac{\Delta}{\gamma}} + E^2\sqrt{\frac{\Delta}{\gamma}} + \frac{E\Delta}{3}\sqrt{\frac{\Delta}{\gamma}} + \frac{E^2}{2}\sqrt{\frac{E}{\gamma}}\log\frac{\sqrt{\Delta}-\sqrt{E}}{\sqrt{\Delta}+\sqrt{E}}\right)\right] + \\
&+ \left.\frac{i\pi}{2\gamma}\left(\frac{E}{\gamma}\right)^{3/2}\left(\frac{2}{3} - \frac{8}{15}\frac{E}{\gamma k_0^2}\right)\right\}, \\
u_{31,31} &= \frac{128\pi^3 e^4}{\varepsilon^2(k_0)} \left|\left(\frac{\partial F_{11}}{\partial z}\right)_{r=0}\right|^2 \left(1 + \frac{4k_0 a_0}{\pi^2}\right)^2 \frac{1}{k_0^8} \times \\
&\times \left\{\left[\frac{2}{3\gamma^2}\left(\frac{\Delta}{3}\sqrt{\frac{\Delta}{\gamma}} + E\sqrt{\frac{\Delta}{\gamma}} + \frac{E}{2}\sqrt{\frac{E}{\gamma}}\log\frac{\sqrt{\Delta}-\sqrt{E}}{\sqrt{\Delta}+\sqrt{E}}\right) + \right.\right. \\
&+ \left.\frac{8}{15\gamma^2(\gamma k_0^2)}\left(\frac{\Delta^2}{9}\sqrt{\frac{\Delta}{\gamma}} + E^2\sqrt{\frac{\Delta}{\gamma}} + \frac{E\Delta}{3}\sqrt{\frac{\Delta}{\gamma}} + \frac{E^2}{2}\sqrt{\frac{E}{\gamma}}\log\frac{\sqrt{\Delta}-\sqrt{E}}{\sqrt{\Delta}+\sqrt{E}}\right)\right] + \\
&+ \left.\frac{i\pi}{2\gamma}\left(\frac{E}{\gamma}\right)^{3/2}\left(\frac{2}{3} + \frac{8}{15}\frac{E}{\gamma k_0^2}\right)\right\}.
\end{aligned}\right\} \tag{14}$$

3. Calculations and Results

When the condition $\Gamma \ll E$, which is typical for the resonant states, is used and the real and imaginary parts of (6a) are separated and expanded in the ration Γ/E, we have to the first order in Γ/E and to the lowest order in $\Delta/\gamma k_0^2$:

$$x = A(x) \pm \frac{1}{\sqrt{2}}\sqrt{\varrho(x) + |C(x)|}\,, \qquad y = -B(x) \mp \frac{1}{\sqrt{2}}\sqrt{\varrho(x) - |C(x)|}\,,$$

where

$$x = \frac{E}{\Delta}, \qquad y = \frac{\Gamma}{\Delta}, \qquad \text{and} \qquad \varrho(x) = \sqrt{C^2(x) + D^2(x)}\,; \tag{15a}$$

$$\left.\begin{aligned}
A(x) &= \frac{1}{2}\left\{2\,\frac{E_{11}^0}{\Delta} + \frac{U_{11,21}}{\Delta} - T\left[\left(1 + \frac{4k_0a_0}{\pi}\right)^2 + 2\right] f(x)\right\},\\
B(x) &= -\frac{1}{2}\,T\left[\left(1 + \frac{4k_0a_0}{\pi}\right)^2 + 2\right]\pi x^{3/2},\\
C(x) &= \frac{1}{4}\left\{\left[\frac{U_{11,21}}{\Delta} + T\left(\left(1 + \frac{4k_0a_0}{\pi}\right)^2 - 2\right) f(x)\right]^2 - \right.\\
&\quad - T^2\left[\left(1 + \frac{4k_0a_0}{\pi}\right)^2 - 2\right]\pi^2x^3 + 8\left[\frac{U_{11,31}}{\Delta} - T\left(1 + \frac{4k_0a_0}{\pi}\right) f(x)\right]^2 - \\
&\quad \left. - 8T^2\left(1 + \frac{4k_0a_0}{\pi}\right)^2 \pi^2x^3\right\},\\
D(x) &= \frac{1}{2}\left\{T\left[\left(1 + \frac{4k_0a_0}{\pi}\right)^2 - 2\right]\pi x^{3/2}\left[\frac{U_{11,21}}{\Delta} + T\left(\left(1 + \frac{4k_0a_0}{\pi}\right)^2 - 2\right) f(x)\right] - \right.\\
&\quad \left. - 8T\left(1 + \frac{4k_0a_0}{\pi}\right)\pi x^{3/2}\left[\frac{U_{11,31}}{\Delta} - T\left(1 + \frac{4k_0a_0}{\pi}\right) f(x)\right]\right\};
\end{aligned}\right\} \tag{15b}$$

$$\left.\begin{aligned}
f(x) &= \frac{2}{3} + 2x - x^{3/2}\log\frac{1-\sqrt{x}}{1+\sqrt{x}},\\
T &= \frac{128\pi^3e^4}{\varepsilon^2(k_0)}\left|\left(\frac{\partial F_{11}}{\partial z}\right)_{r=0}\right|^2 \frac{1}{k_0^8}\,\frac{1}{3\gamma\Delta}\left(\frac{\Delta}{\gamma}\right)^{3/2}.
\end{aligned}\right\} \tag{15c}$$

The substitution of the value of $(\partial F_{11}/\partial z)_{r=0}$ in (10) and (15a) gives

$$U_{11,21} = \frac{1}{16(k_0a_0)^4}\,\frac{e^2}{\varepsilon(\sqrt{2}k_0)\,a_0},$$

$$T = \frac{4\pi^2}{3}\,\frac{1}{(k_0a_0)^3}\left(\frac{e^2}{\varepsilon(k_0)\,a_0(\gamma k_0^2)}\right)^2\left(\frac{\Delta}{\gamma k_0^2}\right)^{1/2}.$$

Notice that the dependence of T on the effective mass at the point Γ is of the form $m_1^{*5/2}$ while the dependences of $U_{11,21}$ and on the effective mass m_2^* at the X point are of the form m_2^{*5} and m_1^{*6} and thus much stronger. We give in Fig. 2, for the InP case, the calculated values of the splitting ΔE and the width Γ of the resonant states as functions of the effective mass of the X point. The value of the energy E is fixed by the hydrogenic formula, because the corrections are very small. We notice that the width of the two resonant states are practically coincident; furthermore, for small values of the mass at X, the splitting and the width are of the same order. When the mass increases, the splitting becomes larger than the width.

In InP the mass at X is not isotropic; we must therefore consider a mean mass. If the mean mass is smaller than $0.1m_e$ the two resonant states are not separated,

Fig. 2. Energy splittings ΔE and widths Γ of the two resonant states associated to the X minima in InP as functions of the effective mass at X point

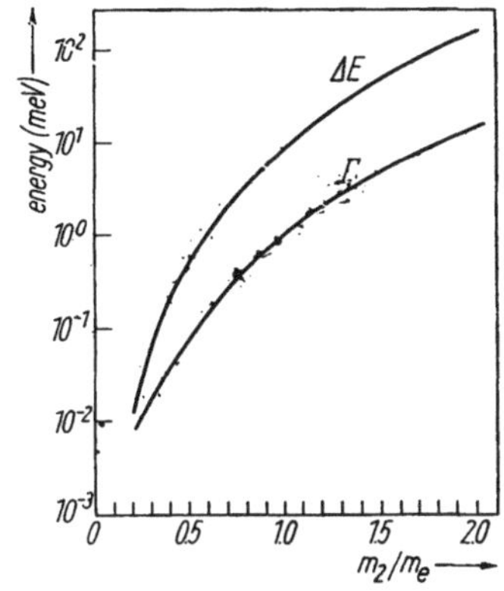

as the widths are larger than the splitting. For larger values of the mean mass the two resonant states appear separated: a splitting of 10 meV is obtained for the value of the mass is of the order of $1m_e$.

An effective mean mass of $0.1m_e$ gives a value of the resonant energy of 5 meV, which is much smaller than the observed value of $\approx$100 meV; an effective mass of $1m_e$ gives good values for the resonant energies, but the states should be split by $\approx$10 meV with a width of the order of 1 meV.

The longitudinal and transverse effective masses at X point have been calculated [21]; their values are 3.71 and $0.24m_e$, respectively. The eigenvalues of the hydrogenic Schrödinger equation with non isotropic effective mass may be found in [22]: the value of the isotropic mass which gives the same eigenvalue for the ground state is $m_2^* \approx 0.7m_e$. Consequently one obtains a splitting between the two Γ_{15} states larger than their widths.

In any case, in this model central cell corrections are not considered; the experimental values of the resonant energies are strongly dependent on the type of impurity. This indicates that the binding energies of the p-states are so large that central cell corrections must be considered, while they can be neglected in the calculations of the p-state energies in Si and Ge p-states [22]; this also implies a large mass at the X point. There is also a dependence of the interzone matrix elements on the impurity, but it is small in Si [23] and it should be the same also in InP.

In conclusion the model indicates that the two resonant states should appear resolved contrary to the experiments [4].

The widths of resonant states have been calculated in the zero gap semiconductors for acceptor impurities [24, 25]. For the ground state, a dependence of the width on the masses of the valence (m_1^*) and conduction (m_2^*) bands of the form $(m_2^*/m_1^*)^{3/2}$ is obtained; it must be compared with the $m^{5/2}$ dependence of the present model. The difference is due to the fact that we are considering p-states instead of s-states. To conclude we note that the strong dependence of the splitting on the effective mass strongly supports the hypothesis [2] that in F center in alkali halides the great separation between the L_1 and L_2 peaks is due to the strong interzone mixing.

References

[1] M. Altarelli and G. Iadonisi, Nuovo Cimento B **5**, 21 (1971).
[2] M. Altarelli and G. Iadonisi, Nuovo Cimento B **5**, 36 (1971).

[3] A. ONTON, Phys. Rev. B **4**, 4449 (1971).
[4] A. ONTON, Y. YACOBY, and R. J. CHICOTKA, Phys. Rev. Letters **28**, 966 (1972).
[5] G. A. PETERSON, Proc. VII. Internat. Conf. Phys. Semicond., Dunod, Paris 1964 (p. 771).
[6] J. CALLAWAY, J. Math. Phys. **5**, 783 (1964).
[7] Y. A. IZYUMOV, Adv. Phys. **14**, 559 (1965).
[8] F. BASSANI, G. IADONISI, and B. PREZIOSI, Phys. Rev. **186**, 735 (1969); Rep. Progr. Phys. **37** (1974) (to be published).
[9] R. T. BATE, J. appl. Phys. **33**, 26 (1962).
[10] A. G. FOYT, R. E. HALSTED, and W. PAUL, Phys. Rev. Letters **16**, 55 (1966).
[11] B. B. KOSICKI and W. PAUL, Phys. Rev. Letters **17**, 246 (1966).
B. B. KOSICKI, W. PAUL, A. J. STRAUSS, and G. W. ISELER, Phys. Rev. Letters **17**, 1175 (1966).
[12] P. M. ADLER, Appl. Phys. Letters **4**, 3554 (1969).
[13] A. VUL', G. L. YA. BIR, and V. YA. SHMARTSEN, Fiz. Tekh. Poluprov. **4**, 2331 (1970).
[14] F. LÜTY, Z. Phys. **160**, 1 (1960).
[15] G. CHIAROTTI and U. M. GRASSANO, Nuovo Cimento B **46**, 78 (1966).
[16] W. P. DUMKE, M. R. LORENZ, and G. D. PETTIT, Phys. Rev. B **1**, 4668 (1970).
[17] H. EHRENREICH, J. Appl. Phys. **32**, 2155 (1961).
[18] M. L. COHEN and J. K. BERGSTRESSER, Phys. Rev. **141**, 789 (1966).
[19] H. NARA, J. Phys. Soc. Japan **20**, 778 (1965).
[20] D. BRUST, Phys. Rev. B **5**, 435 (1972).
[21] F. H. POLLAK, L. W. HIGGINBOTHAN, and M. CARDONA, Proc. Internat. Conf. Phys. Semicond., Kyoto 1966 (p. 20).
[22] R. A. FAULKNER, Phys. Rev. **184**, 713 (1969).
[23] T. G. CASTNER, Phys. Rev. B **2**, 4911 (1970).
[24] L. LIU and D. BRUST, Phys. Rev. **173**, 777 (1967).
[25] B. L. GELMONT and M. I. DYAKONOV, Zh. eksper. teor. Fiz. **62**, 713 (1972); Soviet Phys. — J. exper. theor. Phys. **35**, 377 (1972).

(Received January 31, 1974)

phys. stat. sol. (b) **62**, 635 (1974)

Subject classification: 13.4 and 20.3; 6; 22.4.3

P. N. Lebedev Physical Institute, Academy of Sciences of the USSR, Moscow (a) and Institute of Radiophysics and Electronics, Academy of Sciences of the Armenian SSR, Erevan (b)

Electron–Phonon Interaction in the Photoluminescence of CdTe

By

A. A. GIPPIUS (a) and J. R. PANOSSIAN (b)

The intensity of electron–phonon coupling in the optical transition (characterized by the value of parameter $\bar{N}$ in the formula of relative intensity of phonon replicas, $I_n = I_0\bar{N}^n/n!$; $n = 0, 1, 2, \ldots$) was studied as a function of the ionization energy E_i of the luminescence centres in CdTe. It was found that $\bar{N} = 76\ (\mathrm{eV}^{-2})\ E_i^2$. This dependence is in agreement with the results of the theory assuming bound carrier coupling with LO-phonons for deformation potential coupling.

Исследована зависимость интенсивности электрон–фононного взаимодействия при оптических переходах (характеризуемой величиной параметра $\bar{N}$ в формуле распределения интенсивности фононных повторений: $I_n = I_0\bar{N}^n/n!$, $n = 0, 1, 2 \ldots$) от энергии ионизации (E_i) центров люминесценции в теллуриде кадмия. Показано, что $\bar{N} = 76(\mathrm{eV}^{-2})\ E_i^2$. Эта зависимость согласуется с результатами теоретического расчета, основанного на предположении о том, что взаимодействие LO-фононов со связанными носителями осуществляется через деформационный потенциал.

1. Introduction

In the present paper the emphasis is placed on the dependence of the intensity of electron–phonon coupling in the optical transitions on the nature and the properties of luminecsence centres. The nature of the luminescence centres was discussed partly in the previous paper of the authors [1], some additional data will be published elsewhere.

It is the characteristic property of luminescence spectra of CdTe (as well as other II–VI compounds) that the majority of the observed lines can be combined in series of equidistant lines, arising from the optical transitions of a definite type with simultaneous emission of LO-phonons [1 to 3]. The distribution of intensity in a series is given by the expression

$$I_n = I_0\bar{N}^n/n! \tag{1}$$

(Poisson distribution), where I_n is the intensity of the line due to the simultaneous emission of a photon and n phonons and the value of the parameter $\bar{N}$ is a measure of the intensity of interaction of phonons with the particular type of electronic transition [4, 5]. The dependence of $\bar{N}$ on properties of luminescence centres and, specifically, the increase of $\bar{N}$ with the increase of ionization energy of the centres, was noted in earlier works [1, 2, 5, 6]. In the present paper the study of the dependence of $\bar{N}$ on properties of luminescence centres is continued and the comparison of experimental data with the results of recent theory is made.

2. Experimental

The photoluminescence spectra of CdTe single crystals were recorded at temperatures 4.2, 20.4, and 77 °K. We used a double grating monochromator with reversed linear dispersion $1/D = 10$ Å/mm. The spectral slit width was (2 to 5) × × 10^{-4} eV. The luminescence was excited by the light of a high-pressure xenon lamp [7]. The excitation density varied between 10^{18} and 10^{20} photons/cm² s. To obtain reproducible surface conditions, chemical polishing or cleaving was used[1]). In the analysis of our data we used the results of the electrical measured performed by Chapnin [8].

3. Results

3.1 Dependence of $\bar{N}$ on the ionization energy of luminescence centres

The increase of electron–phonon coupling strength $\bar{N}$ with the increase of ionization energy of an acceptor E_i was traced in CdTe on the example of the edge emission lines and the lines of the group 1.39 to 1.45 eV [1].

If the phonon replica whose relative intensities are given by (1) are well resolved, the value of $\bar{N}$ is determined from the ratio of the intensity of the transition with one phonon to that with no-phonon: $\bar{N} = I_1/I_0$. The ionization energy of a luminescence centre is calculated from the equation

$$E_i = E_g - h\nu_0 , \tag{2}$$

where E_g is the band gap and $h\nu_0$ is the energy of zero-phonon transition; the latter in the case of well-resolved phonon replica can be determined directly from the luminescence spectra.

Luminescence centres with deep levels usually give broad structureless luminescence bands. In this case the procedure of determination of $\bar{N}$ and E_i (described in [9]) turns out to be more labourious. Now it is necessary first to choose the parameter $\bar{N}$ giving the best fit of the envelope of Poisson distribution to the observed contour of the luminescence band and then determine the energy corresponding to the zero-phonon transition in this distribution.

Basing upon the values of $\bar{N}$ and E_i for five acceptor centres we tried to find the quantitative dependence $\bar{N}(E_i)$. The dependence turned out to be given by the equation

$$\bar{N} = 76(\text{eV}^{-2})\, E_i^2 , \tag{3}$$

where E_i is measured in eV.

Table 1

E_i (eV)	$\bar{N}$ (exper.)	$\bar{N} = 76\, E_i^2$
0.05 ± 0.005	0.19	0.19
0.067 ± 0.005	0.34	0.34
0.14 ± 0.008	1.5	1.49
0.32 ± 0.02	7.5	7.72
0.38 ± 0.02	11	10.97

[1]) The CdTe crystals were grown in the Semiconductor Physics Laboratory of the P. N. Lebedev Institute of Physics of the Academy of Science of the USSR by S. N. Maksimovskii and Yu. V. Klevkov.

In Table 1 the values of ionization energy E_i are given along with the coupling strength $\bar{N}$ found experimentally and calculated from the (3). The acceptors with $E_i = 0.05$ eV and $E_i = 0.067$ eV are responsible for III–IV and I–II series of the edge emission respectively [1]. The acceptor with $E_i = 0.14$ eV gives the group of lines 1.39 to 1.45 eV [10]. The acceptors with $E_i = 0.32$ and 0.38 eV produce broad luminescence bands with the maxima at 1.11 and 1.02 eV, respectively.

3.2 *Comparison with theory*

It is of interest to compare the experimentally found dependence $\bar{N}(E_i)$ with the prediction of theory. As far as we know there are two works in which the dependence of coupling strength $\bar{N}$ on the properties of luminescence centres is analysed. In the work of Hopfield [5] the coupling strength $\bar{N}$ was shown to increase with the increase of localization of a carrier at a centre. This result is in qualitative agreement with the experimental data. Hopfield used in his calculation a Gaussian charge distribution for the trapped carrier:

$$\varrho(r) \sim \exp(-r^2/a^2) \tag{4}$$

and has shown that in such an approximation $\bar{N} \sim 1/a$. It is difficult however to obtain from Hopfield's work the quantitative dependence $\bar{N}(E_i)$ which would be sensible to compare to our data. First of all the parameter a in (4) is not Bohr's radius for the ground state and cannot be reliably calculated from the known values of E_i even for shallow levels. Moreover three of the five levels listed in Table 1 are deep.

In the work of Bebb and Chapman [11] the phonon coupling strength is calculated for the deep centres with the use of wave functions obtained by the quantum defect method. By this method an approximate form of a wave function can be found for the region of large r where the effective mass equations hold. It has been shown that for many applications (calculations of optical transition probability, spectral dependence of photoionization cross section etc.) it is sufficient to determine the wave function in the region of large r; the behaviour of the wave function in the core region does not affect the final result. Bebb and Chapman write the effective mass equation in the form

$$\left\{-\frac{\hbar^2}{2m^*}\nabla^2 - \frac{e^2}{\varkappa r} - \varepsilon\right\} F(r) = 0$$

and assuming that it is valid only in the region of large r find the solutions for $F(r)$ corresponding to the observed binding energies ε. The wave functions determined in this way depend upon a parameter $\nu = [R^*/\varepsilon]^{1/2}$ and R^* is the "effective Rydberg", i.e. the binding energy of a shallow hydrogenic centre. Using these wave functions some characteristics of deep centres were calculated, specifically the spectral dependence of the photoionization cross section and the dipole matrix element for transitions $\nu s - 2p$. The results of these calculations were in good agreement with experiment [11].

In Fig. 1 our experimental data are compared with the results of calculations of Bebb and Chapman (see Fig. 5 in [11]). The ordinate shows the ratio $\bar{N}/\bar{N}_H$ ($\gamma_{\nu m}/\gamma_{1m}$ in the notation of [11]), where $\bar{N}_H$ is the coupling strength for a hydrogenic centre. Since for an acceptor in CdTe the value of $R^* = 13.6 \dfrac{m_h^*}{m_0} \dfrac{1}{\varkappa^2} =$

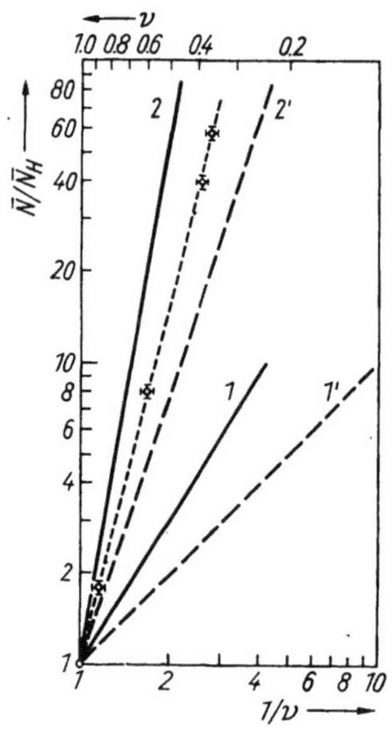

Fig. 1. Comparison between the dependence $\overline{N}/\overline{N}_H$ versus $1/\nu$ found experimentally and predicted by theory [11]. Curves 1 and 1′ correspond to a bound-carrier coupling with LO-phonons for polaron coupling, curves 2 and 2′ are for bound-carrier coupling with LO-phonons for deformation potential coupling. Curves 1 and 2 are the quantum defect result, curves 1′ and 2′ are the corresponding hydrogenic result scaled by varying the Bohr radius a^* to fit the observed binding energy. The dashed line passing through the experimental points is given by

$$\overline{N}/\overline{N}_H = (1/\nu)^4$$

$= 0.054$ eV (for $\varkappa = 10$ and $m_h^* = 0.4\, m_0$) coincides within the accuracy of our measurements with the ionization energy of the shallowest acceptor studied in the present work, we took $R^* = 0.05$ eV and $\overline{N}_H = 0.19$ (see Table 1). As illustrated in Fig. 1 the theory predicts a power law dependence of $\overline{N}/\overline{N}_H$ on $1/\nu$. The power index depends on the mechanism of electron–phonon coupling (curves 1 and 1′ are calculated for the case of polaron coupling of a bound carrier with optical phonons, curves 2 and 2′ for deformation potential coupling) and on the type of wave function used (curves 1 and 2 are obtained with the use of wave functions determined by the quantum defects method; curves 1′ and 2′ correspond to scaled hydrogenic wave functions where instead of Bohr radius a^* appears $b^* = \nu a^*$).

Referring to Fig. 1 our experimental data fall on a straight line which lies between curves 2 and 2′ (equation (3) gives $\overline{N}/\overline{N}_H = (1/\nu)^4$). Thus the experiment reveals a much stronger dependence $\overline{N}$ on the level depth than predicted by the theory assuming polaron coupling. It means that in optical transitions in CdTe electrons interact with LO-phonons mainly through the deformation potential. This result seems at first sight unexpected since it is customary to assume that it is due to macroscopic polarization created by LO-phonons that they couple strongly with electrons. Actually it should be born in mind that the ionicity in CdTe is less than in other II–VI compounds, therefore polaron coupling is not necessarily a major coupling mechanism. The experimentally determined dependence $\overline{N}(1/\nu)$ turns out to be somewhat weaker than shown by curve 2 (quantum defect wave functions, deformation potential coupling). This may be due to some admixture of polaron coupling.

It should be noted that the analysis of the experimentally determined dependence $\overline{N}(E_i)$ was performed also in [6]. The coupling strengths $\overline{N}$ were calculated on the basis of the theory of Hopfield [5] for a number of impurities in GaAs. In doing so it was assumed that the magnitude a in (4) was the Bohr radius.

The latter was calculated from the equation $E_i = Ze^2/2\varkappa a$ up to rather big values of E_i (0.11 to 0.15 eV) when this equation is scarcely valid. In the approximation used the coupling strength $\bar{N}$ varies linearly with E_i. The values of $\bar{N}$ calculated for Cu and Mn are close to those found experimentally. But as a whole the experimentally determined dependence $\bar{N}(E_i)$ proves to be non-linear. It is readily seen however that the experimental points on Fig. 7 in [6] (dependence $\bar{N}(E_i)$ plotted in double logarithmic scale) fall satisfactorily on a straight line with a slope 2, i.e. the dependence $\bar{N}(E_i)$ occurs to be the same as in the present work. In GaAs one can expect strong a dependence $\bar{N}(E_i)$ which is characteristic of deformation potential coupling since the ionicity in this material is very small. The close agreement between the dependence $\bar{N}(E_i)$ obtained in the present work and those which can be found in [6] is probably fortuitous since, as pointed out in [6], for shallow centres in GaAs $E_i \approx \hbar\omega_{LO}$; in this case the Born-Oppenheimer adiabatic approximation (which is the basis of the theories developed in [5, 11]) may break down [12]. In our work even for shallow acceptors $E_i \approx 3\hbar\omega_{LO}$, therefore the Born-Oppenheimer approximation holds.

3.3 Electron–phonon interaction in the edge emission

The analysis of edge emission spectra of CdTe performed in our previous work [1] led to the suggestion that it is the interaction of LO-phonons with the carrier belonging to the acceptor that played the leading role in forming the phonon structure of spectra of inter-impurities optical transitions. Let us try to estimate the difference in the intensity of the interaction with LO-phonons between acceptors and donors responsible for the edge emission.

The data concerning the intensity of LO-phonon interaction with donors are scarce. In CdTe crystals annealed in vacuum a group of lines is observed which is probably due to the radiative transition of a hole to the level of a deep donor E_c −0.15 eV. For this transition $\bar{N} \approx 0.65$. Assuming that the dependence $\bar{N}(E_i)$ for donors in CdTe is the same as that for acceptors (i.e. is given by the (3)) one can calculate that for shallow hydrogenic donor ($E_i = 0.015$ eV) the coupling strength $\bar{N} \approx 0.0065$. Thus the coupling strength for hydrogenic acceptor ($\bar{N} = 0.19$) is approximately 30 times higher than that for hydrogenic donor. Bearing in mind this estimate and taking into account the fast increase of $\bar{N}$ with the increase of the acceptor ionization energy we can conclude that in CdTe the phonon structure of spectra of inter-impurities optical transitions involving shallow donors (I and III series of edge emission, 1.39 to 1.45 eV) is indeed determined by the interaction of LO-phonons with acceptors.

4. Conclusion

The agreement found between the dependence $\bar{N}(E_i)$ obtained in the present work and that predicted by the theory of Bebb and Chapman confirms the view that the LO-phonon coupling strength depends primarily on the wave function outside the core region and is insensitive to the structure of the centres in the central cell (the latter is probably different for various centres studied since at least some of them represent defect–impurity complexes). In view of the fact that the dependence $\bar{N}(E_i)$ obtained in the present work for CdTe corresponds to deformation potential coupling of electrons with LO-phonons it would be of interest to perform a similar study on a material with higher degree of ionicity than CdTe.

Acknowledgements

The authors are indebted to S. N. Maksimovskii and Yu. A. Klevkov for providing the CdTe crystals, to V. A. Chapnin for making available his electric measurement data, to V. V. Ushakov and V. D. Kulakovskii for help in performing measurements.

References

[1] J. R. PANOSSIAN, A. A. GIPPIUS, and V. S. VAVILOV, phys. stat. sol. **35**, 1069 (1969).
[2] R. K. HALSTED, in: Physics and Chemistry of II–VI Compounds, Chapter 8, Eds. M. AVEN and J. S. PRENER, North-Holland Publ. Co., Amsterdam 1967 (p. 385).
[3] D. C. REYNOLDS, C. W. LITTON, and T. C. COLLINS, phys. stat. sol. **12**, 3 (1965).
[4] M. A. KRIVOGLAZ and S. I. PEKAR, Trudy Inst. Fiz. Akad. Nauk Ukr. SSR **4**, 37 (1953).
[5] J. J. HOPFIELD, J. Phys. Chem. Solids, **10**, 110 (1959).
[6] E. W. WILLIAMS, Brit. J. appl. Phys. **18**, 253 (1967).
[7] A. A. GIPPIUS, J. R. PANOSSIAN, and G. A. IVANOV, Prib. i Tekh. Eksper. Nr. 2, 212 (1970).
[8] V. A. CHAPNIN, Thesis, Moscow 1970.
[9] A. A. GIPPIUS, J. R. PANOSSIAN, and V. A. CHAPNIN, phys. stat. sol. (a) **21**, 753 (1974).
[10] V. S. VAVILOV, A. A. GIPPIUS, and J. R. PANOSSIAN, in: II–VI Semiconducting Compounds, Ed. D. G. THOMAS, New York/Amsterdam 1967 (p. 743).
[11] H. B. BEBB and R. A. CHAPMAN, Proc. 3. Internat. Conf. Photoconductivity, Pergamon Press, Oxford/New York/Toronto/Sydney/Braunschweig 1969 (p. 245).
[12] A. S. DAVYDOV, Kvantovaya Mekhanika, Fizmatgiz, Moscow 1963 (§ 116).

(Received January 22, 1974)

phys. stat. sol. (b) 62, 641 (1974)

Subject classification: 13.5

Department of Physics, University of Southern California, Los Angeles

Effect of a Brownian Sublattice on Polariton Dispersion

By

L. MERTEN[1]) and P. DA R. ANDRADE[2])

A set of coupled equations describing the effect of a disorder mechanism on polariton dispersion is given. The treatment differs from that one finds in the literature because the disorder mechanism is not taken as a second-order perturbation. A particular solution of these coupled equations shows that all originally undamped polaritons become damped waves. They all shift in frequency. Physical properties of different wave types in an order–disorder crystal are discussed. For the complex frequencies of the coupled waves a modified Kurosawa relation and a modified LST relation is valid. Numerical evaluations of this model shall be given in a forthcoming paper

Es wird ein Satz gekoppelter Grundgleichungen zur Beschreibung des Einflusses eines Unordnungsmechanismus auf die Polaritondispersion angegeben. Im Unterschied zu der Behandlung des Problems in der bisherigen Literatur wird der Unordnungsmechanismus hier nicht als Störung von nur zweiter Ordnung aufgefaßt. Die Lösung dieser gekoppelten Gleichungen zeigt, daß alle ursprünglich ungedämpften Polaritonwellen zu gedämpften Wellen werden. Außerdem verschieben sich ihre Frequenzen. Die physikalischen Eigenschaften der verschiedenen in einem Ordnungs-Unordnungs-Kristall möglichen Wellentypen werden diskutiert. Für die komplexen Frequenzen der gekoppelten Wellen gilt eine modifizierte Kurosawa- und eine modifizierte LST-Beziehung. In einer zweiten, in Vorbereitung befindlichen Veröffentlichung soll dieses Modell numerisch ausgewertet werden.

1. Introduction

In this paper we are interested to establish a theoretically fundamental structure for the coupling of long optical phonons and polaritons with a dielectric disorder mechanism which is itself intrinsic to the crystal structure. We have in mind crystals of order–disorder type. In other words crystals that have at least one sublattice of particles having one or more equivalent positions in a lattice site. These particles have an extra degree of freedom compared to that one for the particles which form the ordered lattice. In this sense, an order–disorder crystal is a crystal which has at least two intermixed sublattices. The ordered sublattice is a lattice formed by particles which are under a displacement motion, which means they oscillate around its equilibrium position. The disordered sublattice is a lattice formed by particles which have the possibility to assume at least two positions along a direction overcoming a potential barrier. These particles are random or "Brownian" particles. They have a time dependent statistical behaviour and its fluctuation is due to random "thermal" forces. One expects that a dipole system of random particles shows a dielectric relaxation behaviour. This relaxation of dipoles can be described by the theory of

[1]) Permanent address: Lehrstuhl für Physik, Westfälische Wilhelms-Universität, Münster.

[2]) On leave of absence from Univ. Fed. do R.G.S. (Brazil) as a Senior Research Fellow of Brazilian National Research Council.

Brownian motion [1, 2]. This means that the changes in orientation of dipoles in a relaxation process obey the same probability laws as the changes in position of colloidal particles. In other words it is assumed that the random force averages to zero. In solids besides the internal field acting on dipoles one has also the effect of crystalline field upon them. Crystalline field generates potential barriers between positions of low energy for possible orientations of the dipoles. So, the dipole system relaxes due to the flow of particles over potential barriers. The inverse of the probability of a "disorder" particle in one potential well jumping to the other well, for unit time, is given by

$$\tau_c = \tau_0 \exp\left(\frac{\Delta U}{kT}\right),$$

where ΔU is the activation energy. This result was obtained by Kramers [2] by finding analytical solutions for the Brownian motion of a particle in a double well potential for the case of a high ($\Delta U \gg kT$) potential barrier. Naturally, the relaxation time of the dipole system is some function of the inverse of the probability for the jumping particle. The dielectric relaxation is essentially one of the Debye-type.

Recently, the relationship between phonon structure, disorder mechanism and their respective contributions to the dielectric behaviour of order–disorder materials were reanalyzed using a frame of work based on the properties of an Brownian sublattice [3 to 5]. It turns out that the Lyddane-Sachs-Teller [6] relationship and soft mode theory [7] have no general validity as is frequently assumed. The dielectric behaviour of some order–disorder crystals was explained using Mason's [9] mechanism for self-diffusion with the additional assumption that the behavior of the frequency of the phonon associated with a disorder mechanism follows the behaviour of the activation energy as a function of temperature. In general one expects, however, that the disorder mechanism and the mechanism for the formation of polaritons are not independent, but can be strongly coupled by their polarization fields and their effective electric fields. It is this coupling that we want to describe in this paper, using a simple lattice model but which contains already most of the fundamental points.

2. Coupling of Long Optical Phonons and Polaritons and a Disorder Mechanism

To solve this problem algebraically and to discuss its physical background we must first derive the basic equations of the coupled system. We assume that we have an AB-lattice and that one of the sublattices, let us say the B-lattice, is disordered in the sense previously described (a Brownian sublattice). The particles of each sublattice can either be atoms, ions or molecule complexes. For example, for $NaNO_2$, the A-sublattice consists of Na^+ ions, whereas the B-sublattice is formed by NO_2^- ions. We assume that within the B-ions one of the atoms has two equivalent positions, forming a double well potential. In $NaNO_2$ the nitrogen atoms can take two equivalent positions along the b-direction of the crystal. They are randomly jumping to either the "right" or the "left" of the ac crystallographic plane. We furthermore assume for our model that besides the Coulomb-forces, short range forces are only effective between first neighbour ions.

In the following we treat only coupled waves in the direction of the disordering, that means along the x-direction (or b-direction) because for this direction the coupling of long optical phonons and polaritons must be the strongest.

Hence only the components u_1 and u_2 of the displacement vectors in x-direction of the two sublattices appear in the equation of motion. If f is the force constant between two neighbouring A- and B-particles of the two sublattices, the equation of motion (Newton's law) for both reads (see e.g. [8])

$$m_1\ddot{u}_1 = -f(u_1 - u_2) + e^*F\,, \tag{1a}$$

$$m_2\ddot{u}_2 = -f(u_2 - u_1) - e^*F\,. \tag{1b}$$

In this set of equations subscript one is referring to the A-sublattice, and subscript two to the B-sublattice. m_1 and m_2 are the masses of the ions A and B, and e^* and $-e^*$ their effective charges, respectively. F denotes the effective field acting on the lattice sites of the ions. We still assume that the A-ions (ordered sublattice) are positively charged, and the B-ions (disordered sublattices) are negatively charged.

Equations (1a) and (1b) can be combined to one equation:

$$M(\ddot{u}_1 - \ddot{u}_2) = -f(u_1 - u_2) + e^*F\,,$$

where $M = m_1m_2/(m_1 + m_2)$ denotes the reduced mass. With $\Delta u = u_1 - u_2$ one finally gets

$$M\Delta\ddot{u} = -f\Delta u + e^*F\,. \tag{1}$$

Next step we need to consider the dipole polarization for the combined system. Let us start considering the dipole polarization due to the self-diffusion process occurring in the B-sublattice. We are going to use the Mason's description [9]. The dipole polarization rate equation given by Mason is[3])

$$\dot{P}_\mathrm{d} = \left[\frac{N_-}{\tau_{\mathrm{c}+}} - \frac{N_+}{\tau_{\mathrm{c}-}}\right] 2\,\mu\,, \tag{2a}$$

where

$$\frac{1}{\tau_{\mathrm{c}\pm}} = \frac{1}{\tau_0} \exp\left[-\frac{\Delta U}{kT} \pm \frac{1}{2}\, F\bar{e}\delta(kT)^{-1}\right]$$

are the probabilities of a particle with charge $\bar{e}$ jumping in the positive x-direction per unit time (sign $(+)$) and in the negative direction (sign $(-)$), respectively. N_+ and N_- are the populations in the wells along the positive and negative direction. δ denotes the distance between the minima of the two wells and $\mu = \frac{1}{2}\,\bar{e}\delta$ the dipol moment. τ_0 is frequently assumed to obey Eyring's relationship $\tau_0 = h/(kT)$. In some cases, however, the exponential factor is so predominant that τ_0 can be taken as a constant. Regarding $P_\mathrm{d} = (N_+ - N_-)\,\mu$ it is more convenient for the following to write (2a) in the form

$$\dot{P}_\mathrm{d} = N\mu\left(\frac{1}{\tau_{\mathrm{c}+}} - \frac{1}{\tau_{\mathrm{c}-}}\right) - P_\mathrm{d}\left(\frac{1}{\tau_{\mathrm{c}+}} + \frac{1}{\tau_{\mathrm{c}-}}\right), \tag{2b}$$

where $N = N_+ + N_-$ is the total number of the jumping atoms per unit volume.

[3]) Compared with Mason's paper [9] the factor 2 must be added on the right-hand side of Mason's equation (4), p. 857, namely for the following reason: If a particle jumps from the "left" to the "right" potential well, a dipol $-\mu = -\frac{1}{2}\bar{e}\delta$ is annihilated on the left and a dipol $\mu = \frac{1}{2}\bar{e}\delta$ created on the right. Therefore, the change of dipols is 2μ, not μ, per jumping particle. Accordingly the change of dipols is -2μ, if a particle jumps from the right to the left. For the same reason the factor $\frac{1}{2}$ must be omitted in Mason's equation (8), p. 857, compared with our equation (2b). For more details of the "dipolar polarization" by double well potentials see e.g. [13] (p. 19 to 27).

It is clear now that we have three physically different contributions to the total polarization P of our AB-system given by

$$P = \frac{e^*}{v_a}(u_1 - u_2) + \frac{\alpha_1 + \alpha_2}{v_a} F + P_d =$$

$$= \frac{e^*}{v_a} \Delta u + \gamma' F + P_d \tag{3}$$

with $\gamma' = (\alpha_1 + \alpha_2)/v_a$, where v_a is the volume of the unit cell.

Let us see the physical meaning of each polarization term in (3). The first term $\frac{e^*}{v_a}(u_1 - u_2) = \frac{e^*}{v_a} \Delta u$ describes the polarization caused by the displacement u_1 and u_2 of the two ions. The second term is due to the deformation of the electronic clouds of the ions by the effective electric field (α_1 and α_2 are the polarizabilities of the A- and B-ions, respectively). The third contribution comes from the polarization P_d due to the self-diffusion mechanism described by (2).

In addition to our fundamental set of equations (1) to (3), we must now to take into account the Maxwell's equations. For plane waves, Maxwell's equations reduce to

$$P = \frac{n^2 - 1}{4\pi} E = \frac{\varepsilon_x - 1}{4\pi} E \tag{4a}$$

for transverse waves and

$$P = -\frac{1}{4\pi} E \tag{4b}$$

for longitudinal waves. $n = ck/\omega$ is the refractive index, and $\varepsilon_x(\omega)$ the dielectric function in x-direction.

In contrast to (1) to (3) the electric field in (4) is not the effective field, but the macroscopic field E. The relation between both fields is given by the Lorentz-relation:

$$F = E + \beta P\,. \tag{5}$$

β is called the Lorentz-factor. For isotropic crystals $\beta = 4\pi/3$, but in every case β can be calculated by lattice theory, so that for the following the Lorentz-factor can be regarded as known. In this way we have established the fundamental set of equations (from (1) to (5)) which is in principle able to determine the coupling between the polaritons of this simplified lattice model with a disorder-mechanism of Mason-Debye type. In other words, it is possible to determine through this set of equations the effect of a Brownian sublattice on the polariton dispersion or vice versa.

The dispersion relation $\omega = \omega(k)$ and the five variables (eigenvectors) Δu, F, E, P_d, P are completely determined by (1) to (5). Solving this system of basic equations is only complicated by the fact that one of the equations, namely (2), is in general a nonlinear equation, so that the total system becomes nonlinear.

A quantitative solution for the general case can be obtained only by numerical methods. For sake of simplicity, however, and in order to foresee how the dispersion will be affected by this coupling, we now will treat the special case of linearized theory.

3. Special Case of Linearized Theory

The main difficulty for the general case is the nonlinearity which is present in (2). But solving the problem for the linearized system, we can already see most of the interesting features of coupling between long optical phonons, polaritons and a disorder-mechanism. Physically linearization means that the effective electric field in (2) must be sufficiently small, i.e.

$$\frac{1}{2} F\bar{e}\delta(kT)^{-1} < 1 .$$

Regarding that

$$\exp\left[\pm \frac{1}{2} F\bar{e}\delta(kT)^{-1}\right] \approx 1 \pm \frac{1}{2} F\bar{e}\delta(kT)^{-1}$$

$$\approx 1 \pm lF$$

with

$$l = \frac{1}{2} \bar{e}\delta(kT)^{-1} = \mu(kT)^{-1} ,$$

equation (2) in this approximation reads

$$\dot{P}_{\mathrm{d}} = \frac{2}{\tau_{\mathrm{c}}} (N\mu lF - P_{\mathrm{d}})$$

or

$$\tau_{\mathrm{d}}\dot{P}_{\mathrm{d}} = BF - P_{\mathrm{d}} , \tag{2c}$$

where $B = N\mu^2(kT)^{-1}$ and $\tau_{\mathrm{d}} = \tau_{\mathrm{c}}/2 = (\tau_0/2) \exp(\Delta U/kT)$ according to our assumption of a Brownian sublattice (see e.g. Kramer's paper [2]).

Making the ansatz of plane waves and considering first transverse waves the set of equations from (1) to (5) can be rewritten as follows[4]):

$$-\omega^2 M\Delta u = -f\Delta u + e^*F , \tag{1'}$$

$$i\omega\tau_{\mathrm{d}}P_{\mathrm{d}} = BF - P_{\mathrm{d}} , \tag{2'}$$

$$P = \frac{e^*}{v_{\mathrm{a}}} \Delta u + \gamma'F + P_{\mathrm{d}} , \tag{3'}$$

$$P = \frac{n^2 - 1}{4\pi} E = \frac{\varepsilon_x - 1}{4\pi} E , \tag{4'}$$

$$F = E + \beta P . \tag{5'}$$

Our first aim is to get the dispersion relation $\omega = \omega(k) = ck/\sqrt{\varepsilon_x(\omega)}$ from this system of coupled equations. Mathematically this dispersion relation is obtained from the condition that the determinant of the system should be zero. But to get a more physical insight we prefer to use an equivalent method by constructing the dielectric function $\varepsilon_x(\omega)$ by eliminating all the variables given in the system.

---

[4]) The system for transverse and longitudinal waves differs only by equation (4'). In the case of longitudinal waves, one has $P = -\frac{1}{4\pi} E$ for (4').

Substituting[5]) Δu and P_d in (3') by (1') and (2') respectively, one obtains

$$P = \left[\frac{(e^*)^2}{v_a(f - M\omega^2)} + \gamma' + \frac{B}{1 + i\omega\tau_d}\right] F$$

$$= \left[\gamma' + \frac{S}{\omega_0^2 - \omega^2} + \frac{B}{1 + i\,\omega\tau_d}\right] F \tag{6}$$

with

$$\omega_0 = \sqrt{\frac{f}{M}} \quad \text{and} \quad S = \frac{(e^*)^2}{v_a M}. \tag{6a}$$

Finally instead of the effective field F one can introduce the macroscopic electric field E by using (5'), so one has

$$P = GF = G[E + \beta P] \tag{7a}$$

or

$$P = \frac{G}{1 - \beta G} E, \tag{7b}$$

if G denotes the bracket in (6).

Now comparing this equation with (4') one obtains

$$n^2 = \frac{c^2 k^2}{\omega^2} = \varepsilon_x(\omega) = \frac{4\pi G}{1 - \beta G} + 1 \tag{8a}$$

$$= \frac{(4\pi - \beta) G + 1}{1 - \beta G}, \tag{8b}$$

or introducing the bracket G explicitly from (6), (8) can be written as

$$\frac{c^2 k^2}{\omega^2} = \varepsilon_x(\omega) = \frac{a + \dfrac{b}{\omega_0^2 - \omega^2} + \dfrac{c}{1 + i\omega\tau_d}}{a' + \dfrac{b'}{\omega_0^2 - \omega^2} + \dfrac{c'}{1 + i\omega\tau_d}} \tag{9}$$

with $a \equiv (4\pi - \beta)\,\gamma' + 1$, (9a)

$$b \equiv (4\pi - \beta)\, S = \frac{(4\pi - \beta)\,(e^*)^2}{v_a M}, \tag{9b}$$

$$c \equiv (4\pi - \beta)\, B = (4\pi - \beta)\, N\mu^2 (kT)^{-1} \tag{9c}$$

$$a' \equiv -\beta\gamma' + 1, \tag{9d}$$

$$b' \equiv -\beta S = -\beta \frac{(e^*)^2}{v_a M}, \tag{9e}$$

$$c' \equiv -\beta B = -\beta N\mu^2 (kT)^{-1}. \tag{9f}$$

---

[5]) $\Delta u = \dfrac{e^*}{f - M\omega^2} F, \quad P_d = \dfrac{B}{1 + i\omega\tau_d} F.$

Since we are treating a coupled system of polaritons and a dielectric relaxation mechanism it is interesting to see how our equations can reproduce Mason's [9, 5] expression valid for an uncoupled system of dielectric relaxation in the limiting cases of $\omega \ll \omega_0$ and $\omega \gg \omega_0$.

In the limiting case of $\omega \ll \omega_0$ we can re-write the second term — the phonon term — in a contracted form by taking both the numerator and denominator in (9) in a modified shape by defining

$$\tilde{a} = a + \frac{b}{\omega_0^2} = (4\pi - \beta)\,\gamma + 1 \tag{9a'}$$

and

$$\tilde{a}' = a' + \frac{b'}{\omega_0^2} = 1 - \gamma\beta\,, \tag{9d'}$$

where γ is given by

$$\gamma = \gamma' + \frac{S}{\omega_0^2}. \tag{9h}$$

In this case γ contains the electronic part γ' of the polarization and the lattice part (atomic part) S/ω_0^2 as it is assumed in Mason's theory.

In the case of $\omega \gg \omega_0$ the phonon term in the numerator and in the denominator can be neglected against the other terms, so that γ contains only the electronic part γ'. Therefore, all the relevant equations for this limiting case are given by the following equation, if γ is substituted by γ'. Comparing the coefficients given in (9f) and (9d') with the coefficient A (see equations (10) and (29) in [9]), which appears in Mason's dielectric function, the following relationship results:

$$A = \frac{\bar{e}\delta\beta N\mu}{2(1 - \beta\gamma)\,kT} = \frac{\beta N\mu^2}{(1 - \beta\gamma)\,kT} = -\frac{c'}{\tilde{a}'} = \frac{\beta B}{\tilde{a}'}, \tag{10}$$

where $\bar{e}$ is the electronic charge of the jumping particle (Mason has treated the case for a proton jumping in H–O-bond).

Now in the limiting case of $\omega \ll \omega_0$ equation (9) reads as

$$\varepsilon_x(\omega) = \frac{\tilde{a} + \dfrac{c}{1 + i\,\omega\tau_\mathrm{d}}}{\tilde{a}' + \dfrac{c'}{1 + i\omega\tau_\mathrm{d}}},$$

which can be conveniently algebraically rearranged as

$$\varepsilon_x(\omega) = \frac{\tilde{a}}{\tilde{a}'}\left[1 + \frac{\left(\dfrac{c}{\tilde{a}} - \dfrac{c'}{\tilde{a}'}\right)}{1 - \left(-\dfrac{c'}{\tilde{a}'}\right) + i\omega\tau_\mathrm{d}}\right]$$

or

$$\varepsilon_x(\omega) = \frac{\tilde{a}}{\tilde{a}'} + \frac{m}{1 - \left(-\dfrac{c'}{\tilde{a}'}\right) + i\omega\tau_\mathrm{d}} \tag{11}$$

with

$$m = \frac{\tilde{a}}{\tilde{a}'}\left(\frac{c}{\tilde{a}} - \frac{c'}{\tilde{a}'}\right) = \frac{1}{\tilde{a}'}\,\frac{c\tilde{a}' - c'\tilde{a}}{\tilde{a}'}.$$

Obviously it follows from (9c), (9f), (9a′), and (9d′) that

$$\frac{c\tilde{a}' - c'\tilde{a}}{\tilde{a}'} = \frac{4\pi B}{1 - \gamma\beta} = \frac{4\pi}{\beta} A$$

and

$$m = \frac{1}{\tilde{a}'}\,\frac{4\pi}{\beta} A = \frac{4\pi}{\beta(1 - \beta\gamma)} A,$$

if one introduces the Mason-coefficient from (10).

With the notation

$$\varepsilon_x' = \frac{\tilde{a}}{\tilde{a}'} = \frac{4\pi\gamma}{1 - \gamma\beta} + 1$$

and regarding again (10) in the denominator, (11) finally reads

$$\varepsilon_x(\omega) = \varepsilon_x' + \frac{\dfrac{4\pi}{\beta(1 - \beta\gamma)} A}{1 - A + i\omega\tau_d} \tag{12}$$

which is essentially the Mason equation valid for the case of a zero spontaneous polarization.[6]) The relaxation term in (12) is basically a Debye-type relaxation term, with, however, two additional microscopic parameters: the A-coefficient and the activation energy ΔU through $\tau_d = (\tau_0/2) \exp(\Delta U/kT)$.

4. Physical Properties of Different Wave Types of an Order-Disorder Crystal

From the algebraical structure of the dispersion relation given by (9) it is possible to make some essential predictions concerning the physical properties of the wave types of an order–disorder crystal. (Further details from numerical calculations of the k-dependence of the complex eigenfrequencies as the roots of (9) will be given in a forthcoming paper [10].)

From a purely mathematical viewpoint it is possible to extend the Kurosawa [11] and Lyddane-Sachs-Teller relationship in order to include the dispersion at low frequencies. However, it is essential to keep in mind that two (at least) basic different physical phenomena are involved: one a resonant process and another a relaxation process. Both processes generate different wave-type as we can see in the following. In fact, (9) contains the generalization of Kurosawa equation, including the contribution of the relaxation process. In order to see this, let us write (9) in the following form:

$$\varepsilon_x(\omega) = \frac{c^2k^2}{\omega^2} = \frac{a(\omega_0^2 - \omega^2)(1 + i\omega\tau_d) + b(1 + i\omega\tau_d) + c(\omega_0^2 - \omega^2)}{a'(\omega_0^2 - \omega^2)(1 + i\omega\tau_d) + b'(1 + i\omega\tau_d) + c'(\omega_0^2 - \omega^2)}. \tag{13}$$

The numerator and the denominator are obviously polynominals of third degree. If we denote the three roots of the numerator as ω_1^1, ω_2^1, ω_3^1 and the three roots of

[6]) Remark: There remains a slight difference: The additional factor $1/(1 - \beta\gamma)$ in the numerator is not appearing in Mason's equation. But our result seems to be the more accurate one. The source of this difference is explained in the Appendix on page 652.

the denominator as ω_1^t, ω_2^t, ω_3^t we therefore can rewrite (13) in a factorized form (generalized Kurosawa equation [11])

$$\varepsilon_x(\omega) = \frac{c^2k^2}{\omega^2} = \varepsilon_x^\infty \frac{(\omega_1^l - \omega)(\omega_2^l - \omega)(\omega_3^l - \omega)}{(\omega_1^t - \omega)(\omega_2^t - \omega)(\omega_3^t - \omega)}. \tag{14}$$

The constant factor ε_x^∞ obviously is the dielectric function for $\omega \to \infty$, because in this limiting case the quotient has the limit one.

To show the physical properties of the waves corresponding to the roots ω_1^l, ω_2^l, ω_3^l; ω_1^t, ω_2^t, ω_3^t we first prove that there are some relations between the roots. Obviously (13) can be transformed into a polynomial of the fifth degree in ω. Now, if one takes the complex conjugate of both sides of (13), one gets

$$\frac{c^2k^2}{(-\omega^*)^2} = \frac{a[\omega_0^2 - (-\omega^*)^2][1 + i(-\omega^*)\tau_d] + b[1 + i(-\omega^*)\tau_d] + c[\omega_0^2 - (-\omega^*)^2]}{a'[\omega_0^2 - (-\omega^*)^2][1 + i(-\omega^*)\tau_d] + b'[1 + i(-\omega^*)\tau_d] + c'[\omega_0^2 - (-\omega^*)^2]}. \tag{15}$$

In other words: ω can be substituted by $-\omega^*$ without changing the polynomial. Therefore, if ω_n is a root of the polynomial, $-\omega_n^*$ must be a root too. In principle two different cases are possible:

1. If ω_n belongs to one wave, $-\omega_n^*$ belongs to another wave, i.e., the waves are occurring in pairs; one wave has the time dependence

$$e^{i\omega_n t} = e^{-(\gamma_n/2)t} e^{i\bar{\omega}_n t}, \tag{16a}$$

the other must have the time dependence

$$e^{i(-\omega_n^*)t} = e^{-(\gamma_n/2)t} e^{-i\bar{\omega}_n t}, \tag{16b}$$

where $\bar{\omega}_n = \mathrm{Re}(\omega_n)$ denotes the real part of ω_n and $\gamma_n/2 = \mathrm{Im}(\omega_n)$ its imaginary part. Physically we have two damped waves, which differ only by the opposite direction of propagation.

2. ω_n and $-\omega_n^*$ are belonging to the same "wave". Then the root itself must fulfil the condition

$$\omega_n = -\omega_n^*. \tag{17}$$

This is the condition that the frequency is a purely imaginary one with $\mathrm{Im}(\omega_n) = \gamma_n/2$, i.e. the wave must be of the type

$$e^{i(i \mathrm{Im}\, \omega_n)} = e^{-(\gamma_n/2)t} = e^{-t/\tau_n} \tag{18}$$

with $\tau_n = 2/\gamma_n$. In other words: We do not have a wave in the common sense, but a Debye-Mason-like relaxation. This can be seen somewhat more clearly, if we write the corresponding factor $\omega_n - \omega$ in the product form of the polynomial as

$$\omega_n - \omega = i\frac{\gamma_n}{2} - \omega = \frac{i}{\tau_n} - \omega = \frac{i}{\tau_n}(1 + i\omega\tau_n). \tag{19}$$

We call "waves" of this type in the following as relaxation waves.

In (14) at least one of the roots in the numerator and the denominator must correspond to a relaxation wave (index 1), because the three roots cannot be combined in pairs. But we assume from physical reasons[7]) that the remaining two roots (index 0) are of the first kind, corresponding to a pair of damped phonons with opposite direction of propagation. (14) then can be written in the following form:

$$\varepsilon_x(\omega) = \frac{c^2k^2}{\omega^2} = \varepsilon_x^\infty \frac{(\omega_0^1 - \omega)\,(-(\omega_0^1)^* - \omega)\left(i\frac{\gamma_1^1}{2} - \omega\right)}{(\omega_0^t - \omega)\,(-(\omega_0^t)^* - \omega)\left(i\frac{\gamma_1^t}{2} - \omega\right)} =$$

$$= \varepsilon_x^\infty \frac{\tau_1^t}{\tau_1^1} \frac{(|\omega_0^1|^2 - \omega^2 - i\gamma_0^1\omega)\,(1 + i\omega\tau_1^1)}{(|\omega_0^t|^2 - \omega^2 - i\gamma_0^t\omega)\,(1 + i\omega\tau_1^t)}. \tag{20}$$

Therein $|\omega_0^1|$ and $|\omega_0^t|$ are the absolute values of ω_0^1 and ω_0^t, respectively.

Equation (20) can be regarded as an extended Kurosawa relation in which relaxation waves are included. For $\omega \to 0$ we get from (20) an extended Lyddane-Sachs-Teller relation (LST relation):

$$\varepsilon_x^0 = \varepsilon_x^\infty \frac{\tau_1^t}{\tau_1^1} \frac{|\omega_0^1|^2}{|\omega_0^t|^2}$$

or

$$\frac{\varepsilon_x^0}{\varepsilon_x^\infty} = \frac{\tau_1^t}{\tau_1^1} \frac{|\omega_0^1|^2}{|\omega_0^t|^2}. \tag{21}$$

The interpretation of the frequencies occurring in the extended Kurosawa formula is analogous to the case of bare optical phonons and polaritons. The two zeros $\omega_1 = \omega_0^1, \omega_2 = -(\omega_0^1)^*$ of (20) correspond to a pair of damped optical transverse phonons for $k = 0\left(\varepsilon_x = \frac{c^2k^2}{\omega^2} = 0!\right)$, the zero $\omega_3 = i\gamma_1^1/2 = i\,(1/\tau_1^1)$to a relaxation wave for $k = 0$. Because ε_x must be zero as well for longitudinal waves in x-direction for each k (cf. e.g. footnote 1 on page 36 in [12]), the three zeros also give the frequencies of the longitudinal waves for each wave vector k. Therefore one has the result: One of the longitudinal waves is also a relaxation wave with τ_1^1.

The zeros of the denominator, i.e. the poles of (14) and (20), are the frequencies of the transverse waves for $k \to \infty$ because $n^2 = \varepsilon_x = c^2k^2/\omega^2 \to \infty$ (if ω is finite).

The two zeros $\omega_1' = \omega_0^t, \omega_2' = -(\omega_0^t)^*$ are the frequencies of the damped transverse optical and $\omega_3' = i\gamma_1^t/2 = i(1/\tau_1^t)$ corresponds to the transverse relaxation wave for large values of k.

We now treat the waves for finite wave vector k. If we re-write the dispersion relation (14) in form

$$c^2k^2\,(\omega_1^t - \omega)\,(\omega_2^t - \omega)\,(\omega_3^t - \omega) - \omega^2\,(\omega_1^1 - \omega)\,(\omega_2^1 - \omega)\,(\omega_3^1 - \omega) = 0 \tag{22}$$

it is seen immediately that the polynomial is of the fifth degree. Therefore five waves for general k exist corresponding to the five zeroes of the polynomial (22).

[7]) The only alternative possibility permitted from a pure algebraical and combinatorial point of view, namely that all three roots are purely imaginary corresponding to relaxation waves, is improbable, because in the limiting case of uncoupled waves these two frequencies correspond to undamped phonons. It seems that in real crystals the coupling mechanism cannot be so "strong", that pure polaritons and phonons drastically are converted into relaxation waves.

Fig. 1. Dispersion branches of polaritons for a diatomic lattice in x-direction

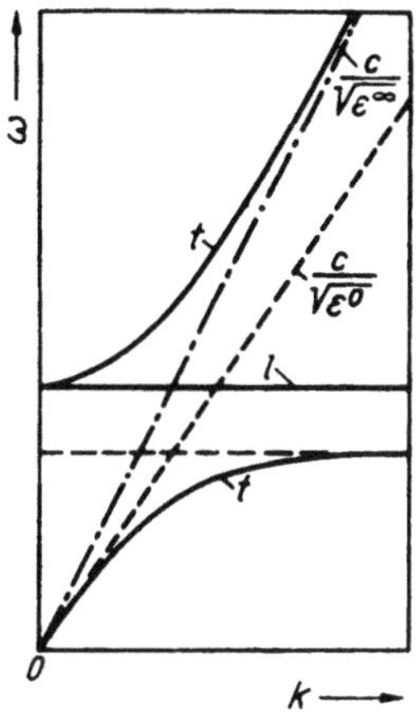

At least one of these waves again must be a relaxation wave

$$e^{i\omega_n(k)t} = e^{-(\gamma_n(k)/2)t} = e^{-t/\tau_n(k)} . \tag{23}$$

Regarding the remaining four solutions we again assume that they can be combined in two pairs of damped waves, the damped polaritons.[8])

These polariton waves correspond to the two transverse branches of undamped polaritons in the uncoupled state. Besides the two branches of transverse damped polaritons there is one branch of k-independent damped longitudinal polaritons, corresponding to the horizontal branch in Fig. 1. In addition there are two branches of relaxation waves, one transverse wave with a k-dependent relaxation time and one longitudinal wave with a k-independent relaxation time.

Summarizing we can say that by coupling of polaritons with a Debye-Mason-relaxation-mechanism all originally undamped polaritons are becoming more or less damped waves and they shift in frequency (real part).

As further comments we can say that whenever a crystal has an intrinsic sublattice of Brownian particles, one expects that the dielectric response at low frequency follows a Debye-Mason relaxation shape. And vice-versa, whenever a crystal shows a dielectric dispersion at low frequency, this crystal has a sublattice of Brownian particles. This relaxation process will be coupled to the resonant process due to the vibrating particles in different strengths which can be described by our fundamental set of equations given by (1) to (5). In fact many crystals show a dispersion at low frequencies as $BaTiO_3$, $LiTaO_3$, and many others.

However, all theories for the polariton dispersion made so far do not take the disorder-mechanism in the same base and degree that we are taking in this theory. It is clear, however, from our results that a disorder-mechanism generally cannot be neglected in polariton theory. Particularly some of the well-known discrepancies for the behaviour of the dielectric function of $BaTiO_3$ should be explained by our theory.

[8]) The two other possibilities from a purely algebraically point of view, namely that three or all five waves are relaxation waves, we exclude again for reasons already mentioned in footnote seven.

Appendix

In this Appendix we prove that in the limiting case $\omega \ll \omega_0$ and for vanishing spontaneous polarization P_s Mason's equation (29) in [9], p. 863, becomes identical with our equation (12), if the step from equation (27) to equation (29) in Mason's paper is treated somewhat more carefully and if in addition a factor 1/2 is added in the last term of the denominator of equation (27) and (29).

To prove this, we first show that Mason's equation (27), p. 863, is in complete agreement with the corresponding equation derived from our equation (1′) to equation (5′), p. 645, apart from the above mentioned factor 1/2 in the denominator.

For $\omega \ll \omega_0$ (3′) obviously reduces to

$$P = \left[\gamma' + \frac{S}{\omega_0^2}\right] F + P_d,$$

$$= \gamma F + P_d . \tag{A.1}$$

Introducing P into (5′), we get for the effective (or local) field F

$$F = E + \beta[\gamma F + P_d] \quad \text{or} \quad F = \frac{1}{1 - \beta\gamma}(E + \beta P_d), \tag{A.2}$$

in agreement with Mason's equation (7), p. 857. Finally, combining equation (A.2) with our equation (2′) on p. 645, it follows

$$i\omega\tau_d P_d = \frac{B}{1 - \beta\gamma}(E + \beta P_d) - P_d$$

or

$$\left(1 + i\omega\tau_d - \frac{\beta B}{1 - \beta\gamma}\right) P_d = \frac{B}{1 - \beta\gamma} E .$$

Regarding $A = (\beta B)/(1 - \beta\gamma)$ according to (10) and (9d′), p. 647, we finally get

$$P_d = \frac{A/\beta}{1 - A + i\omega\tau_d} E \tag{A.3}$$

with

$$\tau_d = \frac{\tau_0}{2} e^{\Delta U/kT} \quad (\tau_0 \text{ is } 1/\Gamma \text{ in Mason's notation}) .$$

This equation is exactly Mason's equation (27), p. 863, for $P_s = 0$, if the last term in the denominator of Mason's equation is corrected by a factor 1/2. Obviously this correction results from the missing factor 2 in equation (4), p. 857, in Mason's paper (see footnote 3 on p. 643). Therefore, if in the following we refer to Mason's equation (27), we always understand this equation with the corrected term $i\omega\tau_d = \left((i\omega)/2\right) e^{\Delta U/kT}/\Gamma$ in the denominator.

When going from equation (27) (our equation (A.3)) to equation (29), Mason regarded the dielectric susceptibility $\varepsilon_0 - 1$, connecting the electronic and atomic polarization term $4\pi P_0$ and the macroscopic field E, as constant, i.e. P_E as independent of P_0 ($= P_d$ in our notation). Actually, however, only the polarization P_E and the effective field F are connected by a constant factor, according to

$$P_E = \gamma F \tag{A.4}$$

(cf. Mason's unnumbered equation above equation (7), p. 857, or our equation (A.1)). But together with Mason's equation (7) or our equation (A.2) it follows from (A.4)

$$P_E = \frac{\gamma}{1-\beta\gamma}(E+\beta P_0) = \frac{\gamma}{1-\beta\gamma}E + \frac{\beta\gamma}{1-\beta\gamma}P_0 , \tag{A.5}$$

i.e. P_E also depends on P_0. It is just this second term of (A.5) which was disregarded by Mason, in going from (28) to (29). If this term is taken into account, equation (29) in Mason's paper reads explicitly

$$D = \left(4\pi\frac{\gamma}{1-\beta\gamma}+1\right)E + 4\pi\left(\frac{\beta\gamma}{1-\beta\gamma}+1\right)P_0 \tag{A.6}$$

or regarding equation (27) (or our equation (A.3)) and $\frac{\beta\gamma}{1-\beta\gamma}+1=\frac{1}{1-\beta\gamma}$:

$$D = \varepsilon E = \left(4\pi\frac{\gamma}{1-\beta\gamma}+1+\frac{4\pi}{1-\beta\gamma}\,\frac{A/\beta}{1-A+i\omega\tau_d}\right)E ,$$

i.e.

$$\varepsilon = 4\pi\frac{\gamma}{1-\beta\gamma}+1+\frac{\frac{4\pi}{\beta(1-\beta\gamma)}A}{1-A+i\omega\tau_d} . \tag{A.7}$$

This equation is in complete agreement with our equation (12).

Acknowledgements

Both authors want to express their gratitude to Prof. S. P. S. Porto for his kind hospitality at the University of Southern California (USA). One of the authors (L. M.) also wishes to thank the Deutsche Forschungsgemeinschaft for financial support.

References

[1] A. Einstein, Ann. Phys. (Germany) 17, 549 (1905).
[2] A. H. Kramers, Physica 7, 284 (1940).
[3] P. da R. Andrade, A. D. Prasad Rao, R. S. Katiyar, and S. P. S. Porto, Solid State Commun. 12, 847 (1973).
[4] P. da R. Andrade and S. P. S. Porto, Solid State Commun., in the press.
[5] P. da R. Andrade and S. P. S. Porto, Brazil Rev. Phys., in the press.
P. da R. Andrade, R. S. Katiyar, and S. P. S. Porto, to be pusblished.
[6] R. M. Lyddane, R. G. Sachs, and E. Teller, Phys. Rev. 59, 673 (1951).
[7] W. Cochran, Phys. Rev. Letters 3, 412 (1959).
[8] M. Born and K. Huang, Dynamical Theory of Crystal Lattices, Oxford Press, 1954.
[9] W. P. Mason, Phys. Rev. 72, 854 (1947).
[10] P. da R. Andrade, G. Borstel, and L. Merten, to be published.
[11] T. Kurosawa, J. Phys. Soc. Japan 16, 1298 (1961).
[12] G. Lamprecht and L. Merten, phys. stat. sol. (b) 55, 33 (1973).
[13] V. V. Daniel, Dielectric Relaxation, Academic Press, London/New York 1967.

(Received February 6, 1974)

phys. stat. sol. (b) **62**, 655 (1974)

Subject classification: 14.1; 15; 21.1; 21.1.1

The Research Institute for Iron, Steel and Other Metals, Tôhoku University, Sendai

Kondo Effect in the Transport Properties of the CsCl-Type Compounds $Fe_{1-x}Ti_{1+x}$

I. Their Anomalous Behaviors in the Titanium-Rich Compositions

By

K. IKEDA

A resistance minimum at 30 to 60 °K and a negative and giant valley of the thermoelectric power near 47 °K were observed in the CsCl-type compounds, $Fe_{1-x}Ti_{1+x}$, with titanium-rich compositions. This anomalous behavior of the transport phenomena, which are distinct from those of the stoichiometric compound showing an ordinary temperature dependence of the resistivity and a positive and monotonically increasing thermoelectric power, have features similar to the Kondo effect in dilute magnetic alloys. It is suggested that such anomalous phenomena may be caused by the existence of the iron atoms occupying the wrong atomic sites in a slightly disordered CsCl-type lattice.

Ein Widerstandsminimum bei 30 bis 60 °K und ein negatives Riesen-Tal der Thermokraft bei 47 °K wurde in den CsCl-Typ-Verbindungen, $Fe_{1-x}Ti_{1+x}$, bei Titan-reicher Zusammensetzung beobachtet. Dieses anomale Verhalten der Transportphänomene, das von der gewöhnlichen Temperaturabhängigkeit der Widerstandsänderung und der positiven und monoton steigenden Thermokraftänderung in der stöchiometrischen Probe verschieden ist, zeigt dem Kondo-Effekt in verunreinigten magnetischen Metallen ähnliche Eigentümlichkeiten. Es wird vorgeschlagen, daß der Ursprung dieser anomalen Phänomene die Existenz von Eisen-Atomen auf falschen Atomplätzen in einem leicht fehlgeordneten CsCl-Typ-Gitter ist.

1. Introduction

The equiatomic compounds in many binary alloy systems (A–B) containing the transition metal elements (B) have the CsCl-type structure [1, 2], especially when it has the mean electron concentration of six per atom. Most of them, with a few exceptions [3, 4], have a tendency to be non-magnetic in the perfect atomic ordering, even if B is iron. Electrical, magnetic, and thermal properties may give some valuable information on the band structure of these CsCl-type compounds [5], contributing to such a tendency. FeTi, which has some peculiar magnetic properties, is aimed at such a study. Nevitt [6] has measured the magnetization of FeTi and found that it is paramagnetic at room temperature, but the magnetization curve at temperatures lower than 80 °K shows a tendency towards saturation at fields beyond 10 kOe. He has interpreted that this behavior is due to a small concentration of uncompensated spins produced by the upset of atomic order in the CsCl-type lattice. Schröder and Cheng [7] have confirmed again this superparamagnetism by the occurrence of the anomalous behavior of the specific heat at low temperatures.

A certain atomic disordering of FeTi, causing such a superparamagnetism, may have some unusual effect on the transport phenomena. Therefore, we attempted to measure the electrical resistivity and the thermoelectric power of

Table 1

Purities and impurity elements of raw materials of high purity (in wt%)

material	purity (%)	impurity (p.p.m.)
Fe	99.999	Si 5, Cu 3, Mg 2, Mn 2, Al < 1, Ag < 1, O 47, H 0.66, N 198, C 10
Ti	99.9	Fe 100, Ag < 1, Ca 2, Cu 1, Mg 1, Mn 2, Na 5, Si 10, Zn 50

this compound for the first time in its single phase range [8], expecting the effect due to the variation of the degree of the atomic order and the electron concentration with composition. In the present work, the experimental results of the electrical resistivity and the thermoelectric power in $Fe_{1-x}Ti_{1+x}$ will be described in detail and some discussion will be given on the anomalous transport behavior in these compounds with titanium-rich compositions [9, 10].

2. Experimental Procedure

Sixteen $Fe_{1-x}Ti_{1+x}$ specimens were prepared in the composition ranging from 46 to 56 at% Ti by arc melting mixtures of 99.9% electrolytic iron and 99.5% sponge titanium in a way almost similar to that reported previously [11], except the final heat-treatment at 1000 °C for 72 h in a vacuum around 10^{-7} Torr. Furthermore, the specimens of high purity were prepared from 99.999% pure iron and 99.9% pure sponge titanium, the impurities of which are tabulated in Table 1. The initial balanced and final chemically analyzed compositions of the ingots are given in Table 2.

Table 2

Initial balanced and final chemically analyzed compositions of the $Fe_{1-x}Ti_{1+x}$ specimens. Two specimens of high purity are marked with HP

specimen no.	balanced compositions (at% Ti)	chemically analyzed compositions (at% Ti)
46	46.0	46.3
47	47.0	47.1
48	48.0	48.0
49	49.0	48.8
50 (a)	50.0	49.7
50 (b)	50.0	50.3
50 (c)	50.0	50.5
51 (a)	51.0	51.6
51 (b)	51.0	51.9
52 (a)	52.0	52.5
52 (b)	52.0	52.8
53	53.0	53.0
54	54.0	54.0
55	55.0	54.4
56	56.0	55.9
50 (HP)	50.0	49.8
52 (HP)	52.0	52.9

X-ray powder photographs were taken with FeKα radiation and the density was determined at 20 °C by a conventional picnometric method using toluene. The electrical resistivity was measured by a conventional potentiometric method in the temperature range from 1.2 to 300 °K by means of the same apparatus as reported previously [11]. A calibrated germanium thermometer and a 2% Co–Au and Cu thermocouple were used for temperature measurements. The measurements of the thermoelectric power were made in the temperature range from 1.4 to 100 °K by means of the cryostat illustrated in [12].

3. Experimental Results

3.1 Crystal structure and phase boundaries

Room-temperature X-ray powder patterns in the specimens containing from 49.7 to 52.5 at% Ti were definitely indexed on the basis of the CsCl-type structure, though only a faint (100) Kα superlattice line could be observed. For the specimens having a titanium composition less than 49.7 at% Ti, diffraction patterns showed several extra lines characteristic of the $MgZn_2$-type Laves phase in addition to lines for the CsCl-type FeTi phase. For the specimens with the composition more than 53.0 at% Ti, one faint additional line, belonging to the intermetallic compound $FeTi_2$ [13], was also observed between (110) Kβ and (110) Kα lines of the FeTi phase. From X-ray diffraction patterns and the lattice parameter vs. composition curve shown as a solid line in Fig. 1, we can see that the single FeTi phase ranges approximately from 49.5 to 52.5 at% Ti, where the lattice parameter increases linearly from 2.9735 to 2.9780 Å with increasing titanium content.

Fig. 1 also shows the results of the density measured at room temperature for three specimens within the single FeTi phase region. The (a) and (b) lines in this figure are the calculated density curves from the composition dependence of the lattice parameter on the assumption of the substitutional and vacancy types, respectively. According to these experimental results of the crystal structure in $Fe_{1-x}Ti_{1+x}$, the excess titanium atoms seem to exist substitutionally in the single-phase region in the CsCl-type ordered structure.

3.2 Electrical resistivity

Fig. 2 shows the results of the electrical resistivity ϱ measured in the temperature range from 4.2 °K to room temperature for fifteen specimens having the titanium composition from 46.3 to 55.9 at% Ti. The temperature variation of the electrical resistivity of the specimens near the stoichiometric composition and in the two-phase region with Fe_2Ti shows an ordinary metallic temperature dependence. Two specimens near the stoichiometry (49.7 and 50.3 at% Ti) have the residual resistivity of about 2 μΩcm. On the other hand, in the non-

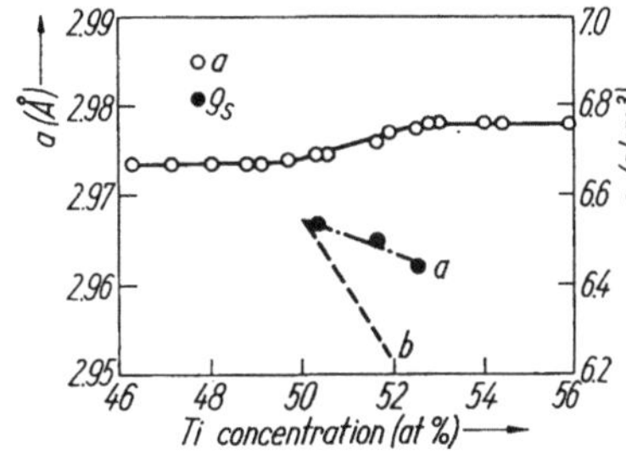

Fig. 1. The composition dependence of the lattice parameter (○) and the density (●) at 20 °C in the $Fe_{1-x}Ti_{1+x}$ compounds with the CsCl-type structure

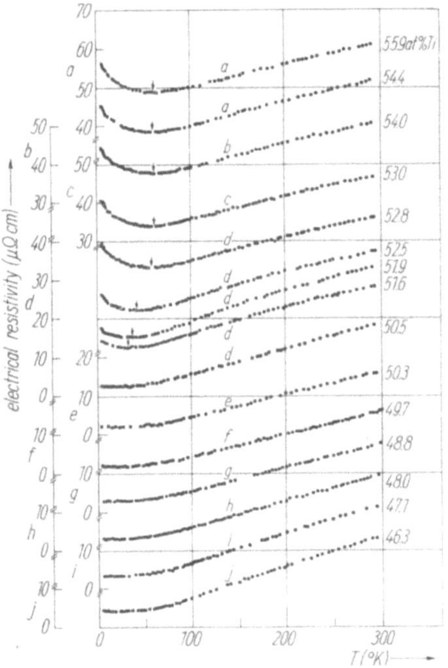

Fig. 2. The temperature variation of the electrical resistivity in $Fe_{1-x}Ti_{1+x}$

stoichiometric specimens with the titanium-rich compositions, one finds not only a considerably large residual resistivity but also a resistance minimum. This phenomenon becomes more pronounced in higher titanium concentrations, varying both the depth of the minimum and the temperature $T^{\min}$, at which the resistance minimum occurs, while, in the two-phase region with $FeTi_2$, we see that both quantities are hardly variable though the residual resistivity increases. At temperatures higher than about 100 °K, the electrical resistivity increases linearly with increasing temperature and shows similar slopes in all specimens. According to further measurements down to 1.2 °K, some superconductive behaviors were found in the specimens with 52.5 to 55.9 at% Ti, showing that the electrical resistivity makes the stepped variation or abruptly drops towards zero near 2.6 °K and indicating again the existence of the second phase of superconductive $FeTi_2$ [14] in accordance with the X-ray diffraction.

Fig. 3 shows the composition dependence of the electrical resistivity at 4.2 °K and that at $T_{\min}$, which are designated as $\varrho_{4.2\,°K}$ and $\varrho_{\min}$, respectively. The composition dependence of the residual resistivity shows a minimum near the stoichiometric composition and reveals a fast increase in the single-phase region and a slow increase in the two-phase region, with increasing departure of the composition from the stoichiometry.

3.3 Thermoelectric power

The thermoelectric powers were measured in the temperature range from 1.4 to 100 °K for five specimens within the single-phase region. Fig. 4 shows the temperature variation of the thermoelectric power of the $Fe_{1-x}Ti_{1+x}$ specimens near the stoichiometric composition (50.3 and 50.5 at% Ti) and in the titanium-rich ones (51.6, 51.9, and 52.5 at% Ti). The thermoelectric power of the former, in which the superparamagnetic behavior [6, 7] has been observed, is positive and shows a fairly large temperature variation and a tendency towards saturation above 50 °K. It is very similar to the thermoelectric power observed in chromium [15] having the same mean valence electron concentration $N_{vc} = 6$ as that of FeTi.

On the other hand, the thermoelectric power of the latter, in which the phenomenon of the resistance minimum has been observed, changes its sign near 10 °K, then increases its magnitude with negative sign and shows a deep minimum at about 47 °K. This anomaly of the thermoelectric power in the titanium-rich

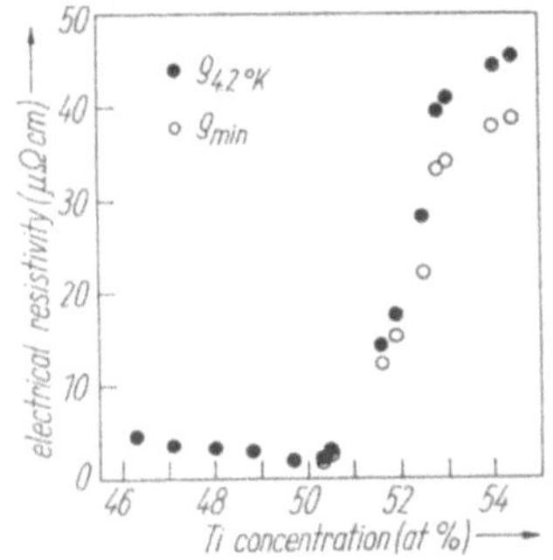

Fig. 3. The composition dependence of the electrical resistivity at 4.2 K (●) and T_{min}(○)

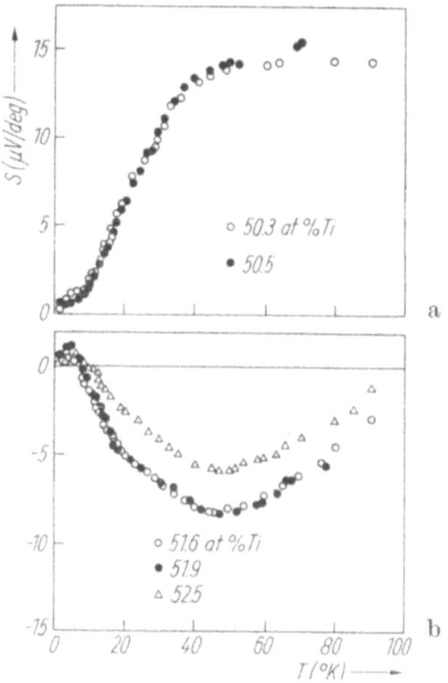

Fig. 4. The temperature variation of the thermoelectric power of the $Fe_{1-x}Ti_{1+x}$ specimens

specimens is similar, both in sign and absolute value, to those which have been observed in dilute magnetic alloys containing iron as an impurity [16]. Moreover the temperature dependence of the thermoelectric power of the specimen containing 52.5 at% Ti, in which the phenomenon of the resistance minimum is more pronounced, shows a more gentle slope than that of the specimens containing 51.6 and 51.9 at% Ti. Similar behavior has been observed in dilute magnetic alloys with slightly more impurity concentrations [16].

3.4 Experimental check by the high purity specimens

In two $Fe_{1-x}Ti_{1+x}$ specimens of high purity near the stoichiometric composition and in the titanium-rich one, the electrical resistivity and the thermoelectric power were measured. Those results are in good agreement with the experimental results mentioned in 3.2 and 3.3, which clarifies that the cause of the resistance minimum and the anomalous thermoelectric power in titanium-rich $Fe_{1-x}Ti_{1+x}$ specimens is not impurities contained in raw materials.

4. Discussion

4.1 Normal transport properties in $Fe_{1-x}Ti_{1+x}$ with $x \approx 0$

As described in the previous section, the temperature variation of the resistivity and the thermoelectric power in the stoichiometric $Fe_{1-x}Ti_{1+x}$ compound is normal in comparison with that in the titanium-rich ones. Such normal transport properties seem to be understood immediately in the same way as those of usual transition metals. Actually, the similarities of the resistivity and the thermoelectric power between the stoichiometric compound and chromium [15, 17] seem to reflect the extraordinary similarity between both state density curves, which has been reported in the theoretical calculations [5] and the experimental results of the specific heat at low temperatures [18, 19].

In transition metals, the existence of the d-band has a great influence [20] on the resistivity due to the phonon scattering, ϱ_T, which is distinct from the case of alkali and noble metals. Namely, ϱ_T consists of the T^3 term due to the inter-band s–d scattering and the T^5 term due to the intraband s–s scattering at temperatures far lower than the Debye temperature θ, i.e.

$$\varrho_T = C\left(\frac{T}{\theta}\right)^3 + C'\left(\frac{T}{\theta}\right)^5. \tag{1}$$

The constants C and C' contain the electron–phonon interaction energies of the s–d scattering and the s–s scattering, respectively. In usual transition metals, the parameter of the T^3 term is far larger [21] than that of the T^5 term, so that the T^3 dependence of ϱ_T is expected to dominate at low temperatures. On the other hand, assuming that the electrical conductivity is dominated by the s–d scattering, one may express the thermoelectric power S by the following equation [22]:

$$S = \frac{\pi^2 k^2 T}{3\,|e|}\left(\frac{1}{D_\mathrm{d}(\varepsilon)}\,\frac{\partial D_\mathrm{d}(\varepsilon)}{\partial \varepsilon} - \frac{1}{F}\,\frac{\partial F}{\partial \varepsilon}\right)_{\varepsilon=\zeta}, \tag{2}$$

where $D_\mathrm{d}(\varepsilon)$ is the state density of the d band, F is the area of the Fermi surface, and ζ is the Fermi energy.

The electrical resistivity of the stoichiometric FeTi specimen, ϱ_{st}, is considered to consist of ϱ_T and the residual resistivity ϱ_0 due to some crystal defects. ϱ_{st} is assumed here to have the same value as the resistivity of the specimen containing 49.7 at% Ti, as it has the lowest residual resistivity. Logarithmic plots of $(\varrho_{st} - \varrho_0)$ against $\log T$ agree with our expectations and show the T^3 dependence below 100 °K ($C/\theta^3 = 2.5\times10^{-6}\ \mu\Omega\mathrm{cm/deg^3}$), where ϱ_0 is estimated at $2\ \mu\Omega$cm. On the other hand, the thermoelectric power of the stoichiometric compound shows a linear temperature variation with its gradient of 0.2 to 0.3 (μV/deg^2) below 40 °K, as can be seen from Fig. 4. By neglecting the second term in (2) $[(1/D_\mathrm{d}(\varepsilon))\,(\partial D_\mathrm{d}(\varepsilon)/\partial\varepsilon)]_{\varepsilon=\zeta}$ is obtained as 200 to 300 (Ry^{-1}) from this value. This fairly large value of $[(1/D_\mathrm{d}(\varepsilon))\,(\partial D_\mathrm{d}(\varepsilon)/\partial\varepsilon)]_{\varepsilon=\zeta}$ is in substantial agreement with that estimated from the calculated state density curves, which has been reported in [5]. Thus, the resistivity and the thermoelectric power in the stoichiometric $Fe_{1-x}Ti_{1+x}$ compound seem to be explained in terms of its band structure. Therefore, the superparamagnetic behavior, which has been observed in the measurements of the magnetization and the specific heat at low temperatures, seems to have no influence on these transport properties.

4.2 Anomalous transport properties in $Fe_{1-x}Ti_{1+x}$ with $x > 0$

In the $Fe_{1-x}Ti_{1+x}$ specimens with titanium-rich compositions, the anomalous transport phenomena, namely, the resistance minimum and a negative and giant valley of the thermoelectric power near 47 °K, have been found. One may be able to analyze these anomalous transport properties, assuming that they consist of the contributions from the matrix and some magnetic scattering centers. In order to explain the phenomenon of the resistance minimum, we assume that the total resistivity can be expressed as the sum of ϱ_{st} and an additional term due to an anomalous scattering, causing the resistance minimum. Fig. 5 shows this excess resistivity of titanium-rich $Fe_{1-x}Ti_{1+x}$ specimens, $\Delta\varrho$, divided by the concentration deviation x. As can be seen from this figure, $\Delta\varrho/x$ decreases linearly with increasing log T in a certain temperature range, but does not fall in the same curve. Such a temperature dependence of the excess resistivity were previously found in dilute magnetic alloys and were explained theoretically in terms of the s–d interaction model by Kondo [23], who gave it by the following equation:

$$\Delta\varrho = \Delta\varrho_0\left(1 + 4JD \log \frac{kT}{W}\right), \tag{3}$$

where $\Delta\varrho_0$ is the residual resistivity due to the magnetic scattering centers and is proportional to their concentration c. J is the s–d exchange energy constant, D is the state density, and W is the band width. Clearly, (3) seems to be applied to our case in respect of the temperature dependence of $\Delta\varrho$. However, if it could be so, we must discuss the true character of the magnetic scattering center in this case, which seems to be neither the substitutional excess titanium atoms, nor the foreign magnetic impurity atoms besides constituent ones as described in 3.4, since c does not coincide with x as shown in Fig. 5. To infer the composition dependence of the concentration c of this magnetic scattering center, both $T_{\min}$ and $(\varrho_{4.2\,°K} - \varrho_{\min})$ are plotted logarithmically against log x in Fig. 6. According to Kondo [23], the depth of the resistance minimum should be proportional to c and $T_{\min}$ to $c^{1/3}$, if the temperature dependent term of the electrical resistivity due to the phonon scattering has a T^3 dependence. Analyzing the phenomenon of the resistance minimum in $Fe_{1-x}Ti_{1+x}$ as the Kondo effect, we find from Fig. 6 that c is proportional to $x^{1.4\text{ to }1.6}$. The true character of the magnetic scattering center will be suggested in the next section.

The anomalous thermoelectric power of the titanium-rich $Fe_{1-x}Ti_{1+x}$ specimens is also discussed in such a way as the phenomenon of the resistance minimum as described above. Namely, it consists of the contributions from the matrix and the magnetic scattering centers, so that the thermoelectric power S is expressed by the following equation:

$$S = \frac{\varrho_a S_a}{\varrho_a + \varrho_b} + \frac{\varrho_b S_b}{\varrho_a + \varrho_b}, \tag{4}$$

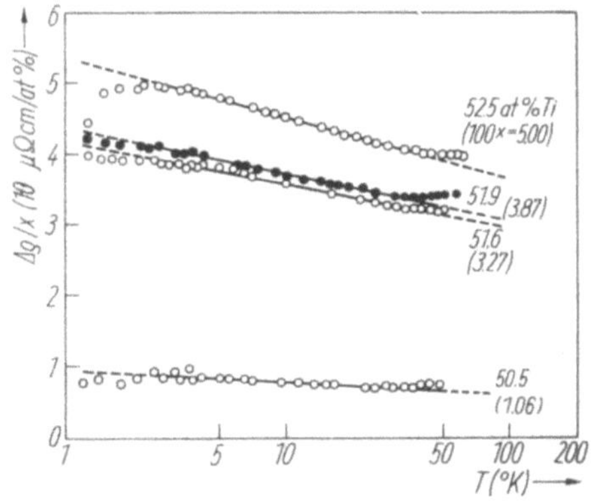

Fig. 5. The excess resistivity $\Delta\varrho(T)$ divided by x vs. log T curves in $Fe_{1-x}Ti_{1+x}$. (Read 100 μΩ cm/at% instead of 10 μΩcm/at%)

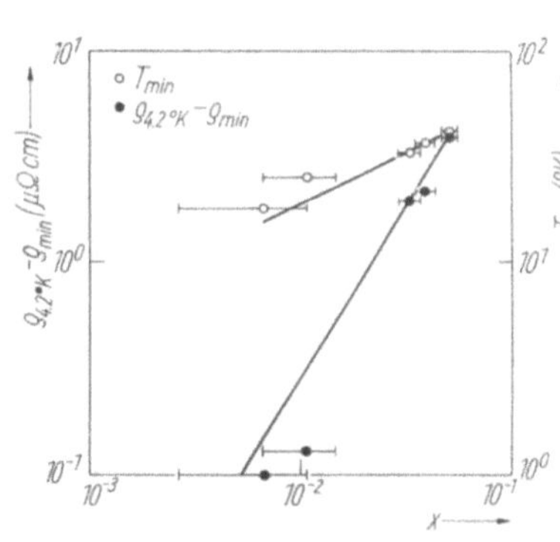

Fig. 6. Logarithmic plots of $T_{\min}$ and $(\varrho_{4.2K} - \varrho_{\min})$ against log x in the $Fe_{1-x}Ti_{1+x}$ specimens

where the resistivity ϱ_a and ϱ_b, and the thermoelectric power S_a and S_b are the parts due to the matrix (a) and to the magnetic scattering centers (b) themselves, respectively. The former contribution seems to correspond to that of the stoichiometric compound, because the thermoelectric power of all $Fe_{1-x}Ti_{1+x}$ specimens is positive below 10 °K. Kondo [24] gave a theoretical explanation of the giant thermoelectric power of dilute magnetic alloys as well as that of the phenomenon of the resistance minimum in terms of the s–d interaction. Moreover, the existence of a peak of thermoelectric power near the Kondo temperature T_K was predicted by Suhl and Wong [25], and by Maki [26] in the following expression:

$$S = -\frac{\pi}{2e} \frac{\sin(2\eta)}{1 - \cos(2\eta)\,\{\tau/(\tau^2 + \pi^2 s(s+1))^{1/2}\}} \frac{\pi^2 s(s+1)}{(\tau^2 + \pi^2 s(s+1))^{3/2}}, \qquad (5)$$

where η is the phase shift due to ordinary potential V only, s the spin quantum number of the magnetic scattering center, and $\tau = \ln(T/T_K)$. This formula gives the extremity near $\tau = 0$, as easily verified. As shown in Fig. 4, the experimental results of the thermoelectric power of three specimens with titanium-rich compositions show the expected extremity at the same temperature, giving the Kondo temperature of about 47 °K.

4.3 Origin of the anomalous transport phenomena

In Section 4.2, the anomalous transport phenomena of the titanium-rich $Fe_{1-x}Ti_{1+x}$ compounds, which are distinct from the normal behavior of the stoichiometric one, have been analyzed by assuming the existence of certain magnetic scattering centers from the similarity to those of dilute magnetic alloys. However, there are several efforts for the explanation of the resistance minimum in some transition metal alloys from the point of view of their special band structure. Rowland et al. [27] have reported that the anomalous resistivity at low temperatures in dilute Pd–(Rh) alloys is not due to the usual Kondo mechanism [28] but to temperature-dependent modification of the interband s to d scattering caused by the large value of $\partial D_d(\varepsilon)/\partial\varepsilon$. In this mechanism, however, one cannot explain such a large value of the depth of the resistance minimum in $Fe_{1-x}Ti_{1+x}$ with $x > 0$ ($\varrho_{4.2°K} - \varrho_{min} \leqq 6\ \mu\Omega cm$) and also different behavior between $x \approx 0$ and $x > 0$. Furthermore, Chakrabarti and Beck [29] have reported that the resistivity anomaly in the Cr–Al solid solution is understood from its semiconducting state caused by an energy gap in the d-band. Such a band structure could not be expected in the $Fe_{1-x}Ti_{1+x}$ compounds [5].

On the other hand, it may be corroborated by the following experimental evidences besides many features of the transport phenomena themselves in the present work that the anomalous transport phenomena in $Fe_{1-x}Ti_{1+x}$ ($x > 0$) are due to the Kondo effect. First, as described in our previous report [30], the stoichiometric specimen follows the Curie law and the titanium-rich ones follow the Curie-Weiss law with the negative paramagnetic Curie temperature. The second reason is that the coefficient γ of the term of the observed specific heat linear in temperature becomes fairly larger in the nonstoichiometric compositions [30] than the γ value expected from the state density curve [5] in the rigid band model. These facts support our situation, as can be seen from [28].

The most important problem seems to be the origin of the magnetic scattering center assumed in the present paper. Similar phenomena were also found in other CsCl-type compounds such as CoAl [31] and NiAl [32]. Sellmyer et al.

[31] have found that the phenomenon of the resistance minimum occurs in the non-stoichiometric CoAl compounds with cobalt-rich compositions. They have interpreted this phenomenon as the Kondo effect caused by the excess cobalt atoms in the non-magnetic CoAl matrix. However, in the $Fe_{1-x}Ti_{1+x}$ compounds, the only possibility for the origin of the magnetic scattering centers should be ascribed to the disordering of the CsCl-type structure. Now, we want to propose the following model for this purpose. In $Fe_{1-x}Ti_{1+x}$ as studied at present (1000 °C, 72 h, annealed and then furnace cooled), the atomic ordering of the specimen may be slightly upset, which makes some iron atoms occupying the wrong atomic sites. Such an iron atom is surrounded by eight neighbouring iron atoms, forming the same local environment as that in pure iron metal. The substitutional iron atom, therefore, is assumed to have a magnetic moment and behaves actually as a kind of magnetic scattering center like an isolated magnetic impurity atom in dilute magnetic alloys causing the Kondo effect.

Thus, many experimental results in $Fe_{1-x}Ti_{1+x}$ ($x > 0$), which have features similar to the Kondo effect in dilute magnetic alloys, seem to be explained by assuming a slight upset of the perfect CsCl-type ordering. The c–x relation shown in Fig. 6 suggests an important fact to us, if the magnetic scattering center has the origin proposed by us. Namely, such a relation can be confirmed from the composition dependence of the degree of atomic order in the CsCl-type lattice. Concerning this, the results of the neutron diffraction will be reported in a companion paper. Furthermore, the analyzed results of the transport phenomena in the stoichiometric and titanium-rich $Fe_{1-x}Ti_{1+x}$ compounds and a suggestion of their origin described above differ entirely from the usual viewpoint [6, 7] about the real character of their magnetic properties, because the degree of the atomic ordering in the former with strong magnetism is higher than that in the latter with weak magnetism [30]. According to the present point of view, the superparamagnetic behavior [6, 7] in the stoichiometric compound must be understood as being due to another origin besides atomic disordering. Concerning this, we will also present another interpretation on the results of measurements of the magnetization and the specific heat at low temperatures of $Fe_{1-x}Ti_{1+x}$ compounds in the near future.

Acknowledgements

The author is much indebted to Dr. T. Nakamichi for much valuable advice and assistance in preparing the manuscript, and to Dr. K. Noto for help with the experiments. Thanks are due to Prof. M. Yamamoto and Prof. Y. Muto, who have supported this research work. The author also wishes to thank Prof. T. Kasuya and Prof. T. Sugawara for helpful discussions, and Mr. T. Yamada, Mr. T. Sato, and Mr. T. Ono for the preparation of the specimens.

References

[1] P. A. Beck, Electronic Structure and Alloy Chemistry of the Transition Elements, John Wiley & Sons, New York 1963 (p. 138).

[2] B. H. Kear, C. T. Sims, N. S. Stoloff, and J. H. Westbrook, Ordered Alloys: Structural Applications and Physical Metallurgy, Claitor's Publishing Division, Baton Rouge 1970 (p. 49).

[3] M. V. Nevitt and A. T. Aldred, J. appl. Phys. **34**, 463 (1963).

[4] R. M. Bozorth, Ferromagnetism, D. van Nostrand Co., New York 1951 (p. 193).

[5] J. Yamashita and S. Asano, Prog. theor. Phys. **48**, 2119 (1972).

[6] M. V. Nevitt, J. appl. Phys. **31**, 155 (1960).
[7] K. Schröder and C. H. Cheng, J. appl. Phys. **31**, 2154 (1960).
[8] H. P. Stüwe and Y. Shimomura, Z. Metallk. **51**, 180 (1960).
[9] K. Ikeda, T. Nakamichi, and M. Yamamoto, J. Phys. Soc. Japan **32**, 280 (1972).
[10] K. Ikeda, J. Phys. Soc. Japan **34**, 272 (1973).
[11] K. Ikeda, T. Nakamichi, and M. Yamamoto, phys. stat. sol. (a) **12**, 595 (1972).
[12] K. Ikeda, T. Nakamichi, and M. Yamamoto, Sci. Rep. Res. Inst. Tohoku Univ., to be published.
[13] W. B. Pearson, A Handbook of Lattice Spacing and Structures of Metals and Alloys, Pergamon Press, London/New York 1958 (p. 662).
[14] B. Matthias, J. appl. Phys. **31** (Supplement to No. 5), 23S (1960).
[15] T. F. de Young, S. Arajs, and E. E. Anderson, J. less-common Metals **29**, 93 (1972).
[16] M. D. Daybell and W. A. Steyert, Phys. Rev. **167**, 536 (1968).
D. K. C. MacDonald, W. B. Peason, and I. M. Templeton, Proc. Roy. Soc. **A266**, 161 (1962).
[17] S. Arajs and G. R. Dunmyre, J. appl. Phys. **36**, 3555 (1965).
[18] E. A. Starke, Jr., C. H. Cheng, and P. A. Beck, Phys. Rev. **126**, 1746 (1962).
[19] C. H. Cheng, C. H. Wei, and P. A. Beck, Phys. Rev. **120**, 426 (1960).
[20] A. H. Wilson, Proc. Roy. Soc. (London) **A167**, 580 (1938).
[21] G. W. Webb, Phys. Rev. **181**, 1127 (1969).
R. V. Colvin and S. Arajs, J. appl. Phys. **34**, 286 (1963).
[22] V. E. Panin, V. P. Fadin, and V. M. Zhukova, Izv. vuzov Fiz. Moscow **11**, 94 (1968).
[23] J. Kondo, Prog. theor. Phys. **32**, 37 (1964).
[24] J. Kondo, Prog. theor. Phys. **34**, 372 (1965).
[25] H. Suhl and D. Wong, Physics **3**, 17 (1967).
[26] K. Maki, Progr. theor. Phys. **41**, 586 (1969).
[27] J. A. Rowlands, D. Greig, and P. Blood, J. Phys. F **1**, L29 (1971).
[28] J. Kondo, Solid State Phys. **23**, 183 (1969).
A. J. Heeger, Solid State Phys. **23**, 283 (1969).
J. E. van Dam and G. J. van den Berg, phys. stat. sol. (a) **3**, 11 (1970).
[29] D. J. Chakrabarti and P. A. Beck, J. Phys. Chem. Solids **32**, 1609 (1971).
[30] K. Ikeda, T. Nakamichi, K. Noto, Y. Muto, and M. Yamamoto, phys. stat. sol. (b) **51**, K39 (1972).
[31] D. J. Sellmyer, G. R. Caskey, and J. Franz, J. Phys. Chem. Solids **33**, 561 (1972).
[32] Y. Yamaguchi, T. Aoki, and J. O. Brittain, J. Phys. Chem. Solids **31**, 1325 (1970).

(Received February 4, 1974)

phys. stat. sol. (b) **62**, 665 (1974)

Subject classification: 13.1; 18; 20.1; 22.1.3

I. Physikalisches Institut der Rheinisch-Westfälischen Technischen Hochschule, Aachen

The Masses of Free Holes and Electrons in Tellurium

By

M. LUTZ, H. STOLZE, and P. GROSSE

The transverse effective hole masses of tellurium samples with $p = 10^{16}$ to 10^{18} cm^{-3} have been determined from Faraday rotation measurements at 10, 90, and 300 K. The comparison of the experimental values with the calculated Faraday masses enables to prove the validity of various band models for large k. The model presented by Bangert gives the best fit. Thus the strong temperature dependence of the hole mass determined from transport measurements by many authors is shown to be due to the nonparabolicity of the valence band. At higher temperatures (300 to 400 K) also the free electrons contribute to the Faraday effect. This rotation could be explained by Bangert's model, too.

Die transversale Löchermasse wurde für Tellurproben mit Löcherkonzentrationen zwischen 10^{16} und 10^{18} cm^{-3} aus Faradayeffekt-Messungen bei 10, 90 und 300 K bestimmt. Die experimentellen Faradaymassen wurden mit den nach den bekannten Bandmodellen berechneten verglichen. Dabei ergab das von Bangert angegebene Modell besonders für große k-Werte die beste Übereinstimmung. Außerdem erklärt sich so die starke Temperaturabhängigkeit der Löchermassen, die von vielen Autoren bei Transportmessungen gefunden wurde. Sie wird durch die Nichtparabolizität des Valenzbandes verursacht. Bei hohen Temperaturen (300 bis 400 K) tragen auch die freien Elektronen zum Faradayeffekt bei. Diese Drehung konnte ebenfalls mit dem Modell von Bangert beschrieben werden.

1. Introduction

Many experiments have been carried out by various authors to determine the hole masses of tellurium. It always appeared that the masses obtained from measurements of transport properties (IR-absorption [1, 3], reflectivity near the plasma-edge [2 to 4], Faraday effect [5 to 10]) are larger than those from magnetoresonance experiments (cyclotron resonance [11 to 16], Shubnikov-de Haas effect [17, 18]).

A possible explanation for this may be the fact that the transport mass is an average value over a large range in k-space, depending on doping and temperature. In order to study the dependence of the effective mass on temperature and carrier concentration we have performed Faraday experiments at 10, 90 and from 300 to 400 K, on samples with carrier concentrations which vary from 10^{16} to 10^{18} cm^{-3} and magnetic fields up to 6 T. In the spectral region from 5 to 25 μm the condition $\omega\tau \gg 1$ is fulfilled even at higher temperatures and concentrations, so that the Faraday effect is a very convenient method to determine the effective mass without knowledge of the relaxation time τ. In tellurium, however, Faraday effect measurements are possible only, if the magnetic field is parallel to the c-axis. For all other directions the linear birefringence caused by the optical anisotropy ($n_{\parallel} = 6$, $n_{\perp} = 4.8$) is much stronger than the circular birefringence caused by the Faraday effect. Therefore only the transverse masses can be studied.

Our samples were prepared from Sb-doped Czochralski single crystals by a combined mechanical and chemical method [19] for the sawing and polishing procedure. Much care had to be taken for orienting and mounting the samples into the magnet so that the angle between the direction of light propagation and the c-axis is less than 0.5° [20].

2. Experimental Method

For our experiments in the spectral region from 5 to 20 μm we used an IR-spectrophotometer with a globar-source, a KCl-prism-double monochromator (Leiss) and a thermocouple.

Magnetic fields up to 6 T are generated with a superconducting magnet, which allows to perform the measurements at sample temperatures of 10, 90, 300 to 500 K.

The sample was located between a polarizer rotating with 6.25 Hz and a fixed analyzer. This arrangement yields a 12.5 Hz ac signal at the detectors output. A change of the direction of polarization by the sample produces a phase shift in the ac signal. In this way the measurement of Faraday effect is related to a measurement of a phase shift of the ac signal in comparison to a phase fixed reference signal. For this procedure we used the phasemeter 406 L from Ad-YU-Electronics. The minimum detectable angle of rotation of our experimental setup was 0.1°.

Later the use of a KBr double monochromator (van Cittert/Kipp and Zonen) and a helium-cooled germanium bolometer (Infrared Laboratories, Inc.) allowed us to extend the experiments to 25 μm. In spite of these improvements the free carrier rotation is weak at room temperature and therefore the uncertainty is large especially for samples with low carrier densities. However, it was possible to improve further the experimental setup by using magnetic fields up to 15 T produced by Bitter magnets of the Hochmagnetfeldanlage, Technische Universität Braunschweig.

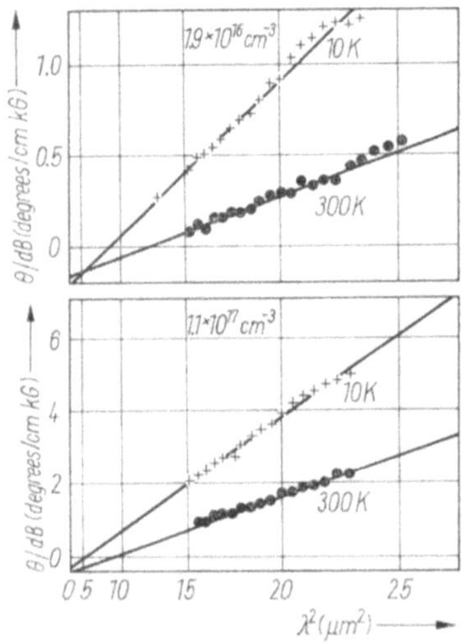

Fig. 1. Faraday rotation in tellurium, $B = 5.74$ T

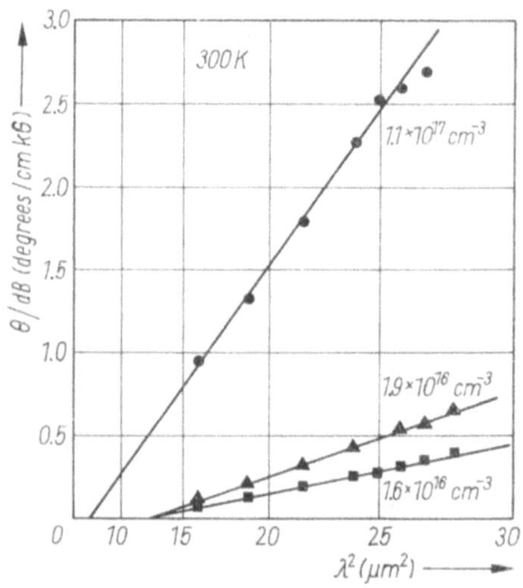

Fig. 2. Faraday rotation in tellurium, $B = 14.7$ T

3. Experimental Results

Typical spectra are shown in Fig. 1 and 2. The measurements were always taken in the configuration $\boldsymbol{B} \parallel \boldsymbol{c}$, $\boldsymbol{E} \perp \boldsymbol{c}$.

In Fig. 1 the Verdet constant (angle of rotation per magnetic field and sample thickness) is plotted versus λ^2, so the free carrier rotation is expected to be a straight line. The slope of this line is determined by the effective mass and the carrier density.

Since the carrier density does not depend on the temperature for these samples the increase of the slope between 300 and 10 K may be mainly caused by a decrease of the effective hole mass. For samples with a concentration $<5\times \times 10^{16}$ cm^{-3} the rotation at 300 K and 6 T is very small. Thus experiments performed in the Bitter magnet at 14.7 T give larger accuracy (Fig. 2).

In all cases the rotation varies linearly with the magnetic field. The rotation could not be extrapolated to zero for $\lambda = 0$ [7] for all samples. This is caused by a concentration-dependent contribution to the interband Faraday effect, which approaches a constant value for long wavelengths. This subject will be discussed in a further publication.

4. Theory

Usually for spectral regions with weak absorption ($\omega \gg 1/\tau$) the free carrier rotation θ is written in terms of the magneto-conductivity tensor $j_i = \sigma_{ij}(B)\, E_j$, with $\boldsymbol{B} \parallel z$ and $E_j = E_{0j} \exp i\omega t$,

$$\theta = -\frac{d}{2nc\varepsilon_0} \operatorname{Re} \sigma_{xy} \tag{1}$$

(d sample thickness and n refractive index).

Following the Drude-Zener theory and with the conditions $\omega \gg 1/\tau$ and $\omega \gg \omega_c$ the following expression for this magneto-conductivity coefficient is obtained:

$$\operatorname{Re} \sigma_{xy} = -\varepsilon_0 \frac{\omega_p^2 \omega_c}{\omega^2} = -\frac{e^3 B}{\omega^2} \frac{p}{m^2} \tag{2}$$

(p hole concentration, $\omega_p^2 = e^2 p/\varepsilon_0 m$, $\omega_c = eB/m$).

In the case of nonparabolic bands we calculated a more general expression deduced from Boltzmann equation which relates the conductivity to the energy band structure $E(k)$ and the Fermi distribution function $f_0(E)$. With $\omega \gg \omega_c$ we obtain

$$\operatorname{Re} \sigma_{xy} = \frac{z e^3 B}{(2\pi)^3 \hbar^4} \int \mathrm{d}\boldsymbol{k}\, \tau' \frac{\partial f_0}{\partial E} \frac{\partial E}{\partial k_x} \left[\frac{\partial E}{\partial k_y} \frac{\partial}{\partial k_x} - \frac{\partial E}{\partial k_x} \frac{\partial}{\partial k_y}\right] \tau' \frac{\partial E}{\partial k_y} \tag{3}$$

(z number of equivalent valleys including spin degeneracy).

This expression is the Hall effect term following from dc transport theory. However, in the ac problem the relaxation time τ' means the complex quantity

$$\tau' = \frac{\tau}{1 + i\omega\tau}. \tag{4}$$

Since $\omega \gg 1/\tau$, τ' in (3) becomes $-i/\omega$,

$$\operatorname{Re} \sigma_{xy} = -\frac{ze^3B}{(2\pi)^3 \hbar^4\omega^2} \int \frac{\partial f_0}{\partial E} \frac{\partial E}{\partial k_x} \left[\frac{\partial E}{\partial k_y} \frac{\partial^2 E}{\partial k_x \partial k_y} - \frac{\partial E}{\partial k_x} \frac{\partial^2 E}{\partial k_y^2}\right] \mathrm{d}\boldsymbol{k} . \tag{5}$$

Substituting (5) into (1) yields the expression for the Faraday rotation, which was first determined by Stephen and Lidiard [21] for cubic crystals. Comparing (2) and (5) one can see that the quantity p/m^2 in (2) is the following average:

$$\left\langle\frac{p}{m^2}\right\rangle = \frac{z}{(2\pi)^3 \hbar^4} \int \frac{\partial f_0}{\partial E} \frac{\partial E}{\partial k_x} \left[\frac{\partial E}{\partial k_y} \frac{\partial^2 E}{\partial k_x \partial k_y} - \frac{\partial E}{\partial k_x} \frac{\partial^2 E}{\partial k_y^2}\right] \mathrm{d}\boldsymbol{k} . \tag{6}$$

The range in the k-space, where the average has to be taken depends very strongly upon the experimental conditions as given by carrier concentration and temperature.

In tellurium the energy bands depend only on k_z and $k_\perp^2 = k_x^2 + k_y^2$. The valence band is not spin-degenerate and the number of equivalent valleys in the Brillouin zone is 2.

Therefore, we transform (6) to cylindric coordinates and the integration over the azimuth gives

$$\left\langle\frac{p}{m^2}\right\rangle_\perp = \frac{1}{2\pi^2\hbar^4} \int_0^\infty \int_0^\infty \frac{\partial f_0}{\partial E} \left(\frac{\partial E}{\partial k_\perp}\right)^3 \mathrm{d}k_\perp \, \mathrm{d}k_z . \tag{7}$$

For the hole concentration we have

$$p = \frac{1}{\pi^2} \int_0^\infty \int_0^\infty f_0 k_\perp \, \mathrm{d}k_\perp \, \mathrm{d}k_z . \tag{8}$$

Table 1

Valence band parameters and derived quantities for the models K1, K2, W, B

	Kamimura et al. [22, 23] K1	K2	Weiler [24] W	Braun [25] B	
A	-2.57	-2.67	-2.12	-3.26	10^{-15} eVcm^2
B	-4.09	-3.94	-4.20	-3.63	10^{-15} eVcm^2
β	1.1	6.0	3.0	6.0	10^{-29} eVcm^4
β'	0	0	2	4	10^{-29} eVcm^4
$2\Delta_1$	6.315	6.315	6.315	6.315	10^{-2} eV
η	-1.18	-1.31	-2.12	0	10^{-15} eVcm^2
ξ	4	7	0	0	10^{-30} eVcm^4
ξ'	0	0	0	0	eVcm^4
S	2.62	2.57	2.49	2.48	10^{-8} eVcm
k_m	2.11	2.15	1.54	2.77	10^6 cm^{-1}
ΔE	2.57	2.55	0.77	2.8	10^{-3} eV
$m_\perp/m_0$	0.102	0.096	0.090	0.117	$k_\perp = 0$, $k_z = 0$
$m_\perp/m_0$	0.110	0.104	0.098	0.125	$k_\perp = 0$, $k_z = k_\mathrm{m}$

If the band structure is known an "effective Faraday mass" can now be calculated:

$$m_\perp^2 = \frac{p}{\left\langle \dfrac{p}{m^2} \right\rangle_\perp} = 2\hbar^4 \frac{\displaystyle\int_0^\infty\int_0^\infty f_0 k_\perp \, \mathrm{d}k_\perp \, \mathrm{d}k_z}{\displaystyle\int_0^\infty\int_0^\infty \frac{\partial f_0}{\partial E}\left(\frac{\partial E}{\partial k_\perp}\right)^3 \mathrm{d}k_\perp \, \mathrm{d}k_z} . \tag{9}$$

A detailed structure of the valence bands near the point H for tellurium was derived by Doi et al. [22] using the $\boldsymbol{k}\cdot\boldsymbol{p}$-perturbation method.

For the upper valence band (H_4), which is the only one of interest for our problem, they obtained the following dispersion relation, containing a camelback shape for $E_4(k_z)$:

$$E_4 = Ak_\perp^2 + Bk_z^2 + \beta k_\perp^4 + \beta' k_\perp^2 k_z^2 + 2\varDelta_1 + \Delta E + \\ + \sqrt{(2\varDelta_1 + \eta k_\perp^2 + \xi k_\perp^4 + \xi' k_\perp^2 k_z^2)^2 + S^2 k_z^2} . \tag{10}$$

Terms describing the trigonal warping of the energy surface are neglected because a nonlinear B-field dependence of the rotation was not observed in our samples ($p < 2 \times 10^{18}$ cm^{-3}).

Successively various attempts have been made to deduce band parameters by a fit of the experimental results, particularly of cyclotron resonance experiments, intervalence band absorption as done by Kamimura et al. (here cited as model K1 [22], K2 [23]) by Weiler (model W [24]) and additional Shubnikov-de Haas effect experiments by Braun et al. (model B [25]) and by Anzin et al. [26], whose fit parameters agree with those of model B. The band parameters are listed in Table 1.

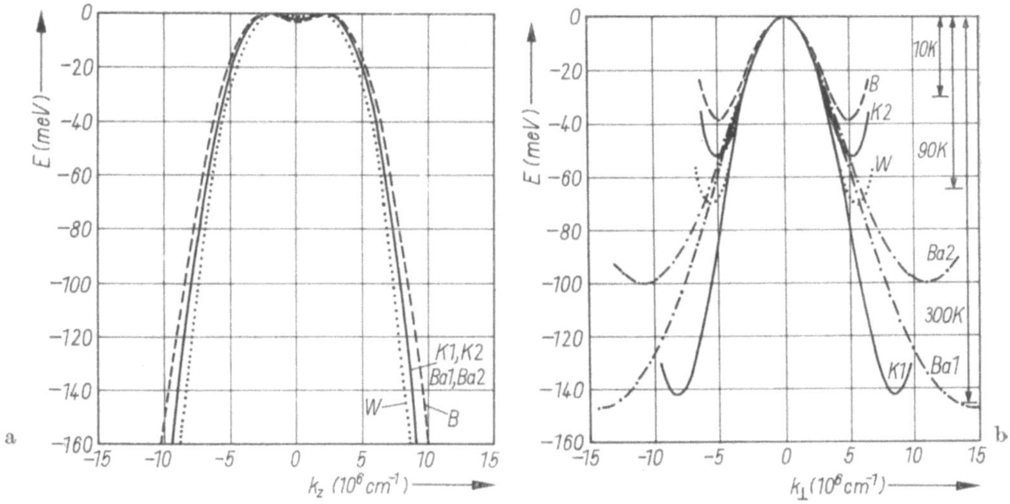

Fig. 3. a) k_z-dependence of the upper valence band near point H ($k_\perp = 0$). (For the notation of the different models see Tables 1, 2.); b) $k_\perp$-dependence of the upper valence band near point H ($k_z = k_m$). The arrows (↕) show the integration limits of energy, necessary to occupy the valence band states according to the hole concentration with an inaccuracy less than 1%

Furthermore also the following characteristic quantities are given: k_m position of the "camel-back" on the k_z-axis; ΔE energy difference between $k_z = 0$ and $k_z = k_m$; $m_{\perp p}$ transverse hole mass for $k_\perp = 0$, $k_z = 0$, and $k_\perp = 0$, $k_z = k_m$.

For the calculation of the Faraday mass from (9) not only the valence band structure in a small range near the point H but also a larger region of the k-space must be taken into account, where the models may differ from each other.

In order to show the difference among these models and to prove their usefulness for the interpretation of our experiments we have plotted $E_4(k_\perp, k_z)$ versus k_z (Fig. 3a) and $k_\perp$ (Fig. 3b).

The difference in the k_z-dependence is not very strong. The main difference among these models lies undoubtlessly in the $k_\perp$-dependence. All models show a minimum and a following increase of the energy for large $k_\perp$. The position of the minimum depends upon the magnitude of the positive $k_\perp^4$ terms in relation to the negative $k_\perp^2$-terms.

This re-increase of the energy curves gives rise to some problems in computing the integrals in (9).

In order to calculate the carrier density and the effective mass with a deviation from the exact value less than 1% the interval of integration must be taken down to an energy E_i lower than $E_F - 6k_BT$, if E_F lies in the valence band and down to $6k_BT$ if E_F lies in the gap. For the sample with highest carrier density ($p = 1.4 \times 10^{18}$ cm^{-3}) the E_i have the following values

$$E_i = -30 \text{ meV}, \quad T = 10 \text{ K},$$
$$E_i = -65 \text{ meV}, \quad T = 90 \text{ K},$$
$$E_i = -145 \text{ meV}, \quad T = 300 \text{ K}.$$

Table 2

Valence band parameters and derived quantities for the models Ba1, Ba2

	Bangert [27] Ba1	Ba2	
A	−0.65	−1.21	10^{-15} eVcm^2
B	3.77	3.77	10^{-15} eVcm^2
C	2.67	2.67	10^{-15} eVcm^2
D	7.6	7.6	10^{-15} eVcm^2
$2\Delta_1$	6.315	6.315	10^{-2} eV
E_G	0.335	0.335	eV
S	2.44	2.44	10^{-8} eVcm
Q	3.8	4.0	10^{-8} eVcm
k_m	1.95	1.95	10^6 cm^{-1}
ΔE	1.57	1.57	10^{-3} eV
$m_{\perp p}/m_0$	0.104	0.107	$k_\perp = 0$, $k_z = 0$
$m_{\perp p}/m_0$	0.114	0.119	$k_\perp = 0$, $k_z = k_m$
$m_{\perp n}/m_0$ H61	0.055	0.05	$k_\perp = 0$, $k_z = 0$
$m_{\perp n}/m_0$ H62	0.065	0.06	$k_\perp = 0$, $k_z = 0$

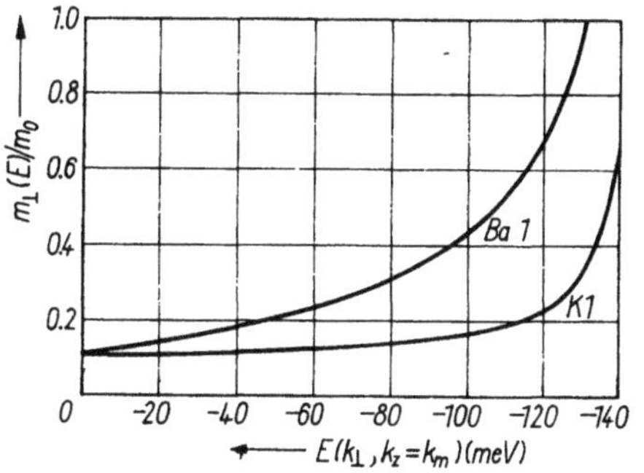

Fig. 4. Calculated energy dependence of the transverse mass

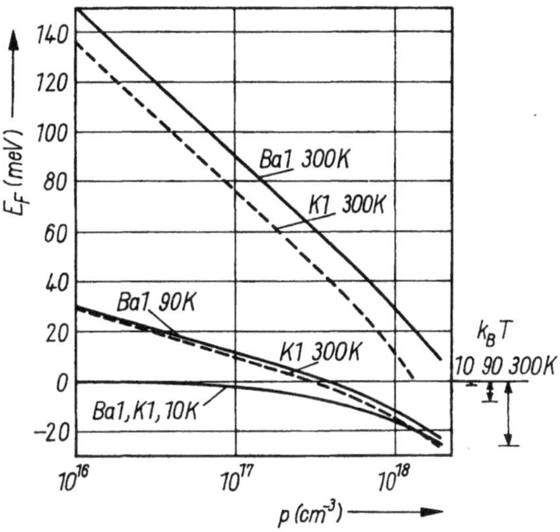

Fig. 5. Calculated concentration and temperature dependence of the Fermi energy

These integration limits are indicated in Fig. 3b by arrows. It now can easily be seen that from the four models only K1 allows the calculation of the effective mass in the whole temperature range. For the other models (K2, W, B) the integrals do not converge at 300 K or even at 90 K. As mentioned above the change of the sign of the curvature in these models for the energy bands is caused by the nonparabolic $k_\perp^4$-terms.

These nonparabolic terms are explained in Kamimura's dispersion relation by a coupling of the H_4, H_5 bands to the 2 H_6-valence bands only. A coupling to other H_6-bands e.g. the conduction bands is neglected. In the dispersion relation given by Bangert [27] however a strong coupling between the two H_6-conduction bands and the H_4, H_5 valence bands is considered. For the valence bands Bangert obtained the following expression:

$$E_{4/5} = \frac{1}{2}\left[E_G + \varepsilon_\pm + (C - A)\,k_\perp^2 + (D - B)\,k_z^2\right] - \Delta E - \sqrt{\frac{1}{4}\left[E_G - \varepsilon_\pm + (C + A)\,k_\perp^2 + (D + B)\,k_z^2\right]^2 + Q^2 k_\perp^2}\,, \tag{11}$$

$$\varepsilon_\pm = -2\Delta_1 \pm \sqrt{4\Delta_1^2 + S^2 k_z^2} \tag{12}$$

and for the conduction bands the expression with the opposite sign:

$$E_{61/62} = \frac{1}{2}\,[\ldots] + \sqrt{\ldots\ldots} \tag{13}$$

The band parameters are determined by a fit to cyclotron resonance, intervalence band and interband absorption (Table 2).

E_G is the energy gap, Q the matrix element for the transitions from the H_4 valence band to the H_6 conduction band. Q is estimated only very roughly by a comparison with other experiments to be 25 times greater than the matrix element for the transitions from the H_6 valence band to the H_4 valence band. Thus we expect a value of the order of 4×10^{-8} eVcm [28]. In order to study the

Bangert model in comparison to the other models, we also have plotted the H_4-band in Fig. 3a, b for two different values of Q (Table 2). An optimum value for Q will result from the fit to our measurements (see Section 5.1).

The k_z dependence of Ba1 and Ba2 agrees with that of K1 and K2. The same holds for the $k_\perp$-dependence, but only for small $k_\perp$.

For large $k_\perp$ the difference becomes more significant. Thus K1 and Ba1 show the minimum nearly at the same energy (−150 meV) but the nonparabolicity is much more stronger for Ba1, which can also be seen from Fig. 4, where the transverse mass is displayed, which is defined by

$$\frac{m_\perp(E)}{m_0} = \frac{\hbar^2 k_\perp}{\partial E/\partial k_\perp}, \tag{14}$$

$E = E(k_\perp, k_z = k_{\mathrm{m}})$, $k_\perp$ varying in the same range as in Fig. 3b. Fig. 5 shows the Fermi energy E_{F} versus carrier density.

5. Discussion

5.1 Valence band

From the experimental values of the free carrier rotation we calculated the quantity

$$\frac{\Delta\theta}{\Delta\lambda^2\, Bd} = \mathrm{const}\, \frac{p}{m_\perp^2} \tag{15}$$

by the least squares method. The carrier density was taken from Hall effect measurements at 77 K, by $p = 1/R_{\mathrm{H}}e$.

Since even for the sample with the lowest doping ($p = 1.6\times10^{16}$ cm^{-3}) the number of the thermally generated electrons is negligible in comparison to the number of the holes, $m_\perp$ obtained from (9) represents the hole mass. The results are shown in Fig. 6 in dependence on carrier density and temperature.

While the increase in the mass for higher concentration is only of the order of 10 to 20%, the mass changes from 10 to 300 K nearly by a factor 2. Such a strong temperature dependence was reported in recent papers by various authors [6 to 8].

In comparison with the experimental values the masses computed from model K1 and Ba1 using (9) are also plotted in Fig. 6.

Model Ba1 with the value $Q = 3.8\times10^{-8}$ eVcm agrees with the experimental data for the three temperatures within the accuracy of our measurements. We conclude therefore that the strong temperature dependence of the mass is mainly caused by the disproportion of the holes in the band when the temperature increases.

Beside this influence of the temperature one expects also a variation of the band parameters with temperature because of the strong thermal expansion coefficients of the lattice. Experiments, however, as cyclotron resonance and p-band absorption depending on a change of the band structure parameters much stronger than on the disproportion of the carriers, do not show any shift of the absorption line up to 60 and 100 K, respectively [29, 30]. We believe, therefore, that the parameters, especially those which determine the transverse hole mass, change only very weakly in the temperature range from 4 to 100 K and even to 300 K. This may be roughly proved by the following estimation.

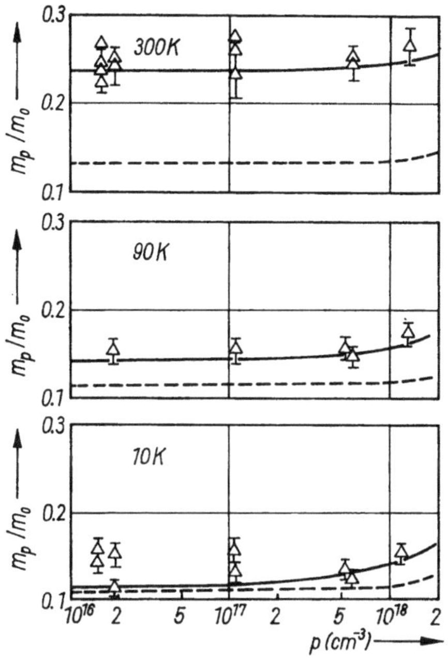

Fig. 6. Comparison of the experimental and theoretical values of the transverse mass

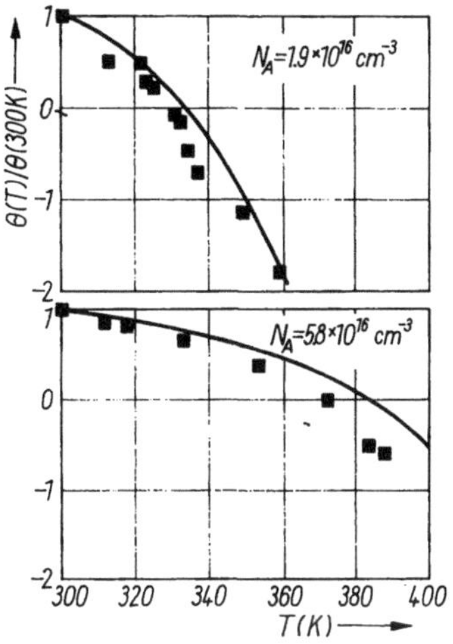

Fig. 7. Variation of free carrier rotation with temperature, $\lambda = 17.4$ μm, $B = 5.74$ T. ■ experimental values, (———) theory, model Ba1

From Bangert's model the transverse hole mass at the top of the H_4-band is given by

$$\frac{\hbar^2}{2m_\perp} = A + \frac{Q^2}{E_G}. \tag{16}$$

A is more than 5 times smaller than Q^2/E_G and so it can be neglected. Assuming that the transition probability Q depends weaker on the temperature than E_G, we obtain

$$\frac{\Delta m_\perp}{m_\perp} = \frac{\Delta E_G}{E_G}. \tag{17}$$

The energy gap is affected by the thermal lattice expansion and the thermal excitation of the phonons. The first effect gives positive, the second effect a negative temperature coefficient for the gap [3]. At 4.2 K the thermal expansion term is the leading one. Using $dE_G/dT = 1.6 \times 10^{-4}$ eV K^{-1} we calculate an increase in the mass between 4 and 60 K of 3%.

In the range 100 to 300 K the resulting temperature coefficient is $dE_G/dT = -0.4 \times 10^{-4}$ eV K^{-1}, which yields a decrease in the mass of 2%. Summarizing we can say that the thermal increase in the effective mass is produced by the nonparabolicity of the band.

5.2 Conduction band

In samples with hole concentrations $<10^{17}$ cm^{-3} the density of the thermally excited electrons is no longer negligible for temperatures greater than 300 K.

The free carrier rotation is now determined by an electron contribution, too:

$$\frac{\theta}{dB} \sim \left(\frac{p}{m_{\perp p}^2} - \frac{n}{m_{\perp n}^2}\right), \tag{18}$$

where p and n are connected by

$$p = N_A + n; \quad pn = n_i^2, \tag{19}$$

N_A density of acceptors, assuming total ionization.

Thus the Faraday effect gives some information on the free electron mass, too. Fig. 7 shows the variation of the free carrier rotation with temperature in relation to the value at 300 K for a fixed wavelength. The rotation decreases with increasing temperature as expected from equation (18). At $\theta = 0$ the hole contribution is just compensated by the electron contribution and for $\theta < 0$ the electron contribution predominates.

For a comparison with the experiment we have computed the rotation from model Ba1, where in the case of the electrons the integration had to be taken over the two H_6-conduction bands (Fig. 7).

The agreement between theory and experiment is very satisfactory. The deviations of the experimental points from the theoretical curve are of the order of the experimental determination of the rotation and the carrier density (maximum 10%). Summarizing we can say that the electron mass within an uncertainty of 10 to 20%, is equal to that given by model Ba1, that yields for $k_\perp = k_z = 0$ the transverse masses $m_{\perp n} = 0.055\, m_0$ (H_{61}) and $m_{\perp n} = 0.065\, m_0$ (H_{62}) and the longitudinal mass $m_{\parallel n} = 0.050\, m_0$.

The agreement between our experimental results and the description of the conduction band by the Bangert model should be a hint that the electrons of interest in our experiment are located near the point H in the Brillouin zone and not in point Z as assumed by Radoff and Dexter [12].

Acknowledgements

We thank Dr. E. Bangert, Würzburg, for correspondence and helpful discussions. We are grateful to Mr. R. Drope, Aachen, for the preparation of the tellurium single crystals. Further we are indebted to the Deutsche Forschungsgemeinschaft for financial support and the Hochmagnetfeldanlage, Technische Universität Braunschweig, for the possibility to perform some additional measurements there.

References

[1] P. GROSSE, Z. Phys. **193**, 318 (1966).
[2] P. GROSSE and M. LUTZ, Verhandl. DPG (VI) **2**, 30 (1967).
[3] P. GROSSE, Springer Tracts in Mod. Phys. **48** (1969).
[4] R. S. CALDWELL and H. Y. FAN, Phys. Rev. **114**, 664 (1959).
[5] J. L. CALLIES and C. RIGAUX, J. Phys. Chem. Solids **25**, 1363 (1964).
[6] M. LUTZ and P. GROSSE, Verhandl. DPG (VI) **5**, 283 (1970); (VII) **6**, 727 (1971).
[7] M. LUTZ and P. GROSSE, Europhysics Conference of the Physics on Selenium and Tellurium, Pont à Mousson 1971.
[8] H. O. HALLER and H. KAHLERT, Acta phys. Austriaca **33**, 33 (1971).
[9] H. KAHLERT and H. O. HALLER, Europhysics Conference on the Physics of Selenium and Tellurium, Pont à Mousson 1971.
[10] H. D. BAUMGART, Dissertation, Universität zu Köln 1972.
[11] J. H. MENDUM and R. N. DEXTER, Bull. Amer. Phys. Soc. **9**, 632 (1964).

[12] P. L. RADOFF and R. N. DEXTER, phys. stat. sol. **35**, 261 (1969).
[13] J. L. PICARD and D. L. CARTER, J. Phys. Soc. Japan **21** (Suppl.), 202 (1966).
[14] Y. COUDER, Phys. Rev. Letters **22**, 890 (1969).
[15] K. J. BUTTON, G. LANDWEHR, C. C. BRADLEY, P. GROSSE, and B. LAX, Phys. Rev. Letters **23**, 14 (1969).
[16] K. F. RENK, and P. GROSSE, Solid State Commun. **9**, 1175 (1971).
[17] E. BRAUN and G. LANDWEHR, J. Phys. Soc. Japan **21** (Suppl.), 380 (1966).
[18] C. GUTHMANN and J. M. THUILLIER, phys. stat. sol. **38**, 635 (1970).
[19] D. FISCHER and P. GROSSE, Z. angew. Phys. **30**, 154 (1970).
[20] M. LUTZ and H. STOLZE, Verhandl. DPG (VI) **8**, 353 (1973).
[21] M. J. STEPHEN and A. B. LIDIARD, J. Phys. Chem. Solids **9**, 43 (1958).
[22] T. DOI, K. NAKAO, and H. KAMIMURA, J. Phys. Soc. Japan **30**, 1400 (1971).
[23] H. KAMIMURA, K. NAKAO, and T. DOI, Proc. Internat Conf. Phys. Semicond., Boston 1970 (p. 347).
[24] M. H. WEILER, Solid State Commun. **8**, 1017 (1970).
[25] E. BRAUN, L. J. NEURINGER, and G. LANDWEHR, phys. stat. sol. (b) **53**, 635 (1972).
[26] V. B. ANZIN, M. S. BRESLER, I. I. FARBSTEIN, E. S. ITSKEVICH, YU. V. KOSICHKIN, V. A. SUKHOPAROV, A. S. TELEPNEV, and V. G. VESELAGO, phys. stat. sol. (b) **48**, 531 (1971).
[27] E. BANGERT. Verhandl. DPG (VI) **8**, 294 (1973).
[28] E. BANGERT, D. FISCHER, and P. GROSSE, phys. stat. sol. (b) **59**, 419 (1973).
[29] M. v. ORTENBERG, K. J. BUTTON, G. LANDWEHR, and D. FISCHER, Phys. Rev. B **6**, 2100 (1972).
[30] D. FISCHER, E. BANGERT, and P. GROSSE, phys. stat. sol. (b) **55**, 527 (1973).

(Received January 29, 1974)

phys. stat. sol. (b) **62**, 677 (1974)

Subject classification: **13.4**

Institute of Physics, Leningrad State University

On the Theory of an Impurity Molecule

II. The Born-Oppenheimer Method

By

A. V. LIAPZEV and A. A. KISELEV

A nonlinear impurity molecule is considered. The molecule is assumed to show almost free rotations about one axis which can librate. The Born-Oppenheimer method is applied to solve the Schrödinger equation transformed from cartesian coordinates to coordinates defined by the Eckart conditions and describing the vibrations and rotations. The Hamiltonian, the eigenvalue, and eigenfunction are expanded in power series in the Born-Oppenheimer parameter and these expansions are inserted in the Schrödinger equation. The eigenfunction and eigenvalue expansions are calculated up to the first and the fifth terms, respectively. Some kinetic corrections caused by the libration are shown to appear in the first order term of the eigenfunction expansion and in the fourth order term of the eigenvalue expansion.

Рассматривается нелинейная примесная молекула в предположении что она может свободно вращаться около оси способной либрировать. Для решения уравнения Шредингера, преобразованного от декартовых координат к координатам, определенным условиями Эккарта, применяется метод Борна-Оппенгеймера. Именно, оператор Гамильтона, собственное значение и собственная функция разложены по степеням борн-оппенгеймеровского параметра и эти разложения подставлены в уравнение Шредингера. Разложения для собственной функции и собственного значения вычислены до членов первого и пятого порядка соответственно. Показано, что в члене первого порядка в разложении собственной функции и в члене четвертого порядка в разложении собственного значения появляются кинетические поправки, обусловленные либрацией.

1. Introduction

In the preceding paper [1] we considered the kinetic energy operator for an impurity molecule. Having assumed the molecule to be normal we introduced the coordinates describing translations, rotations, and vibrations of the molecule and the crystal by means of the Eckart conditions. Then we transformed the kinetic energy operator $\mathcal{T}$ to the new coordinates and obtained the following expression (see the notations in [1]):

$$\left.\begin{aligned} \mathcal{T} &= b^{1/2}\, T\, b^{-1/2}\,, \\ T &= \frac{1}{2m}\left[\sum_{tja} p_{ja}^{(t)^2} + \sum_{t\lambda} \frac{m}{M_t} p_\lambda^{(t)^2} + \boldsymbol{M}\sigma\boldsymbol{M} + \frac{m}{M}\sum_\lambda \left(p_\lambda - l_\lambda^{(e)}\right)^2 + \mathcal{F}\left(\xi_\lambda^{(1)}\right)\right], \\ \boldsymbol{M} &= \boldsymbol{M}^{(e)} - \boldsymbol{M}^{(n)} + \frac{m}{M_1}\boldsymbol{M}^{(v)}\,. \end{aligned}\right\} \tag{1}$$

In this paper we try to solve the transformed Schrödinger equation with the help of the Born-Oppenheimer method [2], i.e. we expand the Hamiltonian, the eigenvalue, and eigenfunction in power series in the Born-Oppenheimer parameter $\varkappa = (m/M)^{1/4}$, insert these series in the Schrödinger equation and equate the coefficient of each power of $\varkappa$ to zero. Thus we obtain equations of motion for our system. Considering the case when a normal impurity molecule shows almost free rotations on only one axis and when this axis may only librate we obtain the equations of the electronic, vibrational, and rotational motions in the zeroth, second, and fourth orders of the method, respectively. Anharmonic and some kinetic corrections appear in the fourth order term of the eigenvalue expansion and in the first order term of the eigenfunction expansion. When carrying out these derivations we assume that the temperature of the crystal is equal to 0 °K and that the crystal is in its ground state.

2. On the Choice of the Parameter $\varkappa$

According to Born and Oppenheimer [2] one gets the true orders of the electronic, vibrational, and rotational energy levels choosing $\varkappa = (m/M)^{1/4}$ to be the decomposition parameter and making the following substitution for the vibrational coordinates ξ:

$$\xi = \varkappa\zeta\,. \tag{2}$$

It is not difficult to understand the formal reasons for this choice and for making the scale transformation (2).

Using the adiabatic approximation we can derive from the Schrödinger equation the equation of harmonic vibrations of the nuclei. Separating the equation for a mode ξ we obtain

$$\left[-\frac{\hbar^2}{2m}\left(\frac{m}{M}\right)\frac{d^2}{d\xi^2} + \frac{m\omega^2}{2}\,\xi^2\right] a_\nu(\xi) = \varepsilon_\nu a_\nu(\xi)\,.$$

The solution is given by

$$a_\nu = \alpha^{1/2}(\pi^{1/2}\,2^\nu\nu!)^{-1/2}\,H_\nu(\alpha\xi)\,e^{-(1/2)\alpha^2\xi}\,,$$

where $\alpha = (m\omega/\hbar)^{1/2}\,(m/M)^{-1/4}$. Hence we find that $|a_\nu(\xi)|^2$ tends to $\delta(\xi)$ when (m/M) tends to zero (see for instance [3]). This means that the nuclei stop in their equilibrium positions when their masses tend to infinity. Having done the scale transformation (2) we see that the functions $a_\nu(\zeta)$ do not depend on $\varkappa$. Therefore we can hope that choosing $\varkappa$ to be the decomposition parameter and making the substitution (2) we obtain the eigenfunction expansion without singularities in the zeroth term (see also [4, 5]).

3. Expansion of the Hamiltonian

Let us consider the matrices $G^{(n)}_{31,\,31}$ and $G^{(n)}_{31,\,41}$ and let us write explicitly the parameter $\varkappa$ in the expressions for the matrix elements (see the formulae (A1) and (A2) in [1]). We find that

$$G^{(n)}_{31,\,31} = \varkappa^{-4}I\,, \qquad G^{(n)}_{31,\,41} = \varkappa^{-4}L\,.$$

Hence

$$\sigma = \varkappa^4\sigma'\,, \qquad \sigma' = (I - LL^{\mathrm{T}})^{-1}\,. \tag{3}$$

Making the substitutions

$$\xi_\lambda^{(t)} = \varkappa(\mu_t)^{-1/2}\,\zeta_\lambda^{(t)}\,, \qquad \xi_\lambda = \varkappa\zeta_\lambda\,, \tag{4}$$

$$\left(\mu_t = \frac{M_t}{M}\right)$$

and introducing the operator $\boldsymbol{M}^{(\mathrm{v})'}$ given by

$$\boldsymbol{M}^{(\mathrm{v})} = \mu_1 \varkappa^{-4} \boldsymbol{M}^{(\mathrm{v})'}$$

we obtain

$$T = \frac{1}{2m}\left[\sum_{tja} p_{ja}^{(t)^2} + \varkappa^2\left(\sum_{t,\lambda} p_\lambda^{(t)^2} + \sum_\lambda p_\lambda^2\right) - \right.$$

$$\left. -2\varkappa^3 \sum_\lambda p_\lambda l_\lambda^{(e)} + \varkappa^4\left(\boldsymbol{M}\sigma\boldsymbol{M} + \sum_\lambda l_\lambda^{(t)^2} + \mathcal{F}\right)\right],$$

where we omit primes when writing σ' and $\boldsymbol{M}^{(\mathrm{v})'}$ and where now σ and $\mathcal{F}$ depend on $\varkappa$ owing to the substitution (4). Expanding σ and $\mathcal{F}$ in power series in $\varkappa$ (see Appendix I) we get the expansion for the operator T:

$$T = \sum_k \varkappa^k T^{(k)'}\,. \tag{5}$$

After the substitution (4) the potential energy U becomes a function of $\varkappa$ and can also be expanded in power series in $\varkappa$:

$$U = \sum_k \varkappa^k U^{(k)'}\,. \tag{6}$$

Combining the expansions (5) and (6) we obtain the expansion for the Hamiltonian. Evidently, this expansion is convenient to consider an impurity molecule showing free (or almost free) rotations over all the three Eulerian angles. However, we are interested here in the case when the rotation on only one axis is possible.

4. Libration

Let us assume now that only small vibrations (librations) take place over the nutation angle $\vartheta_3 \equiv \vartheta$ and that the corresponding equilibrium position is given by $\vartheta = 0$. Instead of the two other Eulerian angles ϑ_1 and ϑ_2 we introduce new variables

$$\Phi = \tfrac{1}{2}(\vartheta_2 - \vartheta_1)\,, \qquad \Psi = \vartheta_1 + \vartheta_2$$

and consider ϑ and Φ to be polar coordinates of the libration.[1]) Then the following librational coordinates can be introduced:

$$\eta_1' = \vartheta \cos\Phi\,, \qquad \eta_2' = \vartheta \sin\Phi\,.$$

The coordinates η_λ' ($\lambda = 1, 2$) describe a vibrational motion, therefore the scale transformation similar to (4) must be done,

$$\eta_\lambda' = \varkappa\eta_\lambda\,. \tag{7}$$

[1]) As we consider the motion over ϑ to be a small vibration we replace the upper limit π in the corresponding normalization integral by ∞.

After the substitution (7) the operators $M_j^{(n)}(j=1,2)$ become functions of $\varkappa$ and can be written in the following way:

$$M_j^{(n)} = \varkappa^{-1}\mathcal{M}_j^{(n)} + \sum_{k=0}^{\infty} \varkappa^{2k+1} M_j^{(nk)}, \tag{8}$$

where

$$\left.\begin{aligned} \mathcal{M}_j^{(n)} &= \frac{\hbar}{i}\sum_k R_{jk}\frac{\partial}{\partial\eta_k}, \\ R = \{R_{jk}\} &= \begin{pmatrix} \cos\frac{\Psi}{2} & \sin\frac{\Psi}{2} \\ -\sin\frac{\Psi}{2} & \cos\frac{\Psi}{2} \end{pmatrix}, \\ M_j^{(nk)} &= f_{j1}^{\prime(k)}\mathcal{N}_1 + f_{j2}^{\prime(k)}\mathcal{N}_2, \\ \mathcal{N}_1 = \frac{\hbar}{i}\frac{\partial}{\partial\Phi}, &\qquad \mathcal{N}_2 = \frac{\hbar}{i}\frac{\partial}{\partial\Psi}, \end{aligned}\right\} \tag{9}$$

and where $f_{jp}^{\prime(k)}$ are functions of η_1, η_2.

The operators $l_\lambda^{(e)}$ are given by (see [1] equation (16)),

$$l_\lambda^{(e)} = \sum_{jk} C_{jk} Z_{j\lambda}^{(1)} P^{(e1)},$$

where the matrix C depends on the Eulerian angles and thus after the substitution (7) becomes a function of $\varkappa$. Expanding the matrix elements C_{jk} in power series in $\varkappa$ we obtain

$$l_\lambda^{(e)} = \sum_k \varkappa^k l_\lambda^{(ek)}, \tag{10}$$

where

$$l_\lambda^{(em)} = \sum_{jk} C_{jk}^{(m)} Z_{j\lambda}^{(1)} P^{(e1)}$$

(the matrices $C^{(m)}$ $(m = 0, 1)$ are given in Appendix II).

Inserting the expansions (8) and (10) into the expansion (5) we get another expansion for the kinetic energy operator

$$T = \sum_k \varkappa^k T^{(k)}. \tag{11}$$

The expansions (5) and (11) differ beginning with the second order terms. We have in particular

$$T^{(2)} = \frac{1}{2m}\left(\sum_{t\lambda} p_\lambda^{(t)^2} + \sum_\lambda p_\lambda^2 + \sum_i I_i^{(0)^{-1}} \mathcal{M}_i^{(n)^2}\right), \tag{12}$$

where $I_i^{(0)}$ is the moment of inertia about i-th coordinate axis for the equilibrium internuclear distances. Now it is convenient to introduce new librational coordinates q_k:

$$\eta_i = \sum_k R_{ik}^{-1} I_k^{(0)^{-1/2}} q_k,$$

where the matrix R is given by (9). Then the third term on the right hand side of (12) takes the form

$$\frac{1}{2m}\sum_k \mathcal{P}_k^2, \qquad \mathcal{P}_k = \frac{\hbar}{i}\frac{\partial}{\partial q_k}.$$

It should be noted that having introduced the coordinates q_k the following relation must be taken into account:

$$\left.\frac{\partial}{\partial \Psi}\right|_{\eta_i = \text{const}} = \left.\frac{\partial}{\partial \Psi}\right|_{q_i = \text{const}} - \frac{1}{2}\frac{\partial}{\partial \Phi}.$$

Then the expression for the operators $M_j^{(\text{n}k)}$ changes,

$$M_j^{(\text{n}k)} = f_{j1}^{(k)}\mathcal{N}_1 + f_{j2}^{(k)}\mathcal{N}_2,$$

where $f_{jp}^{(k)}$ are functions of q_1, q_2 (see also Appendix II).

We are going to develop the Born-Oppenheimer method up to the fourth order, therefore we write now the operators $T^{(k)}$ up to $k = 4$:

$$T^{(0)} = \frac{1}{2m}\sum_{tja} p_{ja}^{(t)^2}, \qquad T^{(1)} = 0,$$

$$T^{(2)} = \frac{1}{2m}\left(\sum_{\lambda t} p_\lambda^{(t)^2} + \sum_\lambda p_\lambda^2 + \sum_k \mathcal{P}_k^2\right),$$

$$T^{(3)} = \frac{1}{2m}\Bigg[-2\sum_{k=1,2} I_k^{(0)-1/2}\,\mathcal{P}_k(M_k^{(\text{e})} + M_k^{(\text{v})}) + \sum_{i,k=1,2}\sigma_{ik}^{(1)} I_i^{(0)1/2} I_k^{(0)1/2}\mathcal{P}_i\mathcal{P}_k - 2\sum_\lambda p_\lambda l_\lambda^{(\text{e}0)}\Bigg],$$

$$T^{(4)} = \frac{1}{2m}\Bigg[\sum_{k=1,2} I_k^{(0)-1/2}\,(\mathcal{P}_k M_k^{(\text{n}1)} + M_k^{(\text{n}1)}\mathcal{P}_k) + \sum_{k=1,2} I_k^{(0)^{-1}}\,(M_k^{(\text{e})} + M_k^{(\text{v})})^2 + I_3^{(0)^{-1}}\,(M_3^{(\text{e})} + M_3^{(\text{v})} + \mathcal{N}_1 - \mathcal{N}_2)^2 + \sum_{k=1,2} I_k^{(0)^{-1}}\Big[\mathcal{P}_k\sigma_{k3}^{(1)}\,(M_3^{(\text{e})} + M_3^{(\text{v})} + \mathcal{N}_1 - \mathcal{N}_2) + (M_3^{(\text{e})} + M_3^{(\text{v})} + \mathcal{N}_1 - \mathcal{N}_2)\,\sigma_{3k}^{(1)}\mathcal{P}_k + \sum_{i,k=1,2}\sigma_{ik}^{(2)} I_i^{(0)1/2} I_k^{(0)1/2}\mathcal{P}_i\mathcal{P}_k + \sum_\lambda l_\lambda^{(\text{e}0)^2} - 2\sum_\lambda p_\lambda l_\lambda^{(\text{e}1)} + \mathcal{F}^{(0)}\Bigg].$$

Evidently, after the substitution (7) we have also to write instead of the expansion (6) another expansion for the potential energy U:

$$U = \sum_k \varkappa^k U^{(k)}.$$

Combining this expansion with the expansion (11) we obtain the expansion for the Hamiltonian H:

$$H = \sum_k \varkappa^k H^{(k)}. \tag{13}$$

5. Equations of Motion

Now instead of the wave function φ (see [1], equation (8)) we introduce the wave function $\psi = b^{-1/2}\varphi$. We have for the function ψ

$$H\psi = W\psi\,, \tag{14}$$

$$H = T + U\,, \qquad \int |\psi|^2\, b\, d\tau_u = 1\,.$$

Let us expand the eigenfunction ψ and the eigenvalue in power series in $\varkappa$:

$$\psi = \sum_k \varkappa^k \psi^{(k)}\,, \tag{15}$$

$$W = \sum_k \varkappa^k W^{(k)}\,, \tag{16}$$

where the function $\psi^{(0)}$ is supposed to be normalized in the same way as the function ψ. Inserting the expansions (13), (15), and (16) in (14) we get the following equation:

$$\sum_m \varkappa^m \sum_{k=0}^{m} (H^{(k)} - W^{(k)})\, \psi^{(m-k)} = 0\,. \tag{17}$$

According to the Born-Oppenheimer method we should now equate the coefficient of each power of $\varkappa$ to zero. For a free molecule we would then obtain the equations of the electronic, vibrational, and rotational motions from the equations of the zeroth, second, and fourth orders, respectively [2] (see also [6]). For the case of a rotating impurity molecule we have to modify this scheme.

The difficulty arises in the zeroth order. The operator $H^{(0)} = T^{(0)} + U^{(0)}$ describes the electronic motion when the nuclei are fixed in their equilibrium positions and we suppose that the eigenfunctions φ_k and eigenvalues V_k of the operator are known. Evidently in our case φ_k and V_k depend on the angle Ψ. If now we equated the zeroth order coefficient in (17) to zero then for the eigenvalue V_n (in which we are interested and which we assume to be nondegenerate) we would get $W^{(0)} = V_n$; however this is impossible as V_n is a function of Ψ and $W^{(0)}$ must be a number. As the kinetic energy operator for the rotation appears in the fourth order term $T^{(4)}$ we cannot uncouple the equations coming from (17) and thus have to solve them together (at any rate up to the fourth order).

To get a possibility to consider these equations separately we write the eigenvalue V_n in the following way:

$$V_n(\Psi) = \mathsf{V}_n + \mathsf{v}_n'(\Psi)\,,$$

where V_n is constant and we assume that

$$\mathsf{v}_n'(\Psi) = \varkappa^4 \mathsf{v}_n(\Psi)\,. \tag{18}$$

Formally this means that the rotation becomes more free when $\varkappa$ tends to zero or that the molecule stops with an arbitrary orientation relative to the crystal when the average mass of the nuclei tends to infinity. Though there are no physical reasons for such a behaviour of impurity molecules still for realistic values of $\varkappa$ and for low barriers (as, for example, in the case of the molecular ion NO_2^- in alkali halides [7]) we get a reasonable description of our system.

Now we change the vectors $\boldsymbol{y}_b^{(t)0}$ in the operator $H^{(0)}$ by small quantities $\Delta \boldsymbol{y}_b^{(t)}$ and express $\Delta \boldsymbol{y}_b^{(t)}$ in terms of $\xi_\lambda^{(t)}$; besides we shift the origin 0_1 (see [1]), turn the axis of rotation by a small angle ϑ, and express these displacements in terms of ξ_λ and η_i. Then the eigenvalue V_n (adiabatic potential) becomes a function of the coordinates $\xi_\lambda^{(t)}$, ξ_λ, and η_i'. If we make the scale transformations (4) and (7), express η_i in terms of q_i and expand the operator $H^{(0)}$, the eigenfunction φ_n and the eigenvalue V_n in power series in $\varkappa$ we can express the derivatives of the adiabatic potential with respect to $\zeta_\lambda^{(t)}$, ζ_λ and q_i in terms of matrix elements involving electronic functions and the corresponding derivatives of the potential energy U. We omit these calculations and only give in Appendix III the expression for the m-th term $V_n^{(m)}$ of the adiabatic potential expansion (see also [2] and [6] Appendix 5). In accordance with (18) we assume that

$$V_n^{(m)} = \mathsf{V}_n^{(m)} + \varkappa^4 \mathsf{v}_n^{(m)} , \tag{19}$$

where $\mathsf{V}_n^{(m)}$ does not depend on Ψ.

Let us expand the functions $\psi^{(m)}$ in the eigenfunctions φ_l:

$$\psi^{(m)} = \sum_l a_l^{(m)} \varphi_l$$

and insert these series in (17):

$$\sum_m \varkappa^m \sum_{k=0}^{m} \sum_l (H^{(k)} - W^{(k)}) \, a_l^{(m-k)} \varphi_l = 0 . \tag{20}$$

The term with $m = 0$ has the form

$$\sum_l (V_l^{(0)} - W^{(0)}) \, a_l^{(0)} \, \varphi_l . \tag{21}$$

According to (19) it contains the term $\varkappa^4 \mathsf{v}_n^{(0)}$ which must be joined to the fourth order term. Having done this we may equate the rest of (21) to zero. We get

$$\left. \begin{aligned} a_l^{(0)} = 0 \, (l \neq n), \qquad W^{(0)} = \mathsf{V}_n^{(0)} , \\ \psi^{(0)} = a_n^{(0)} \varphi_n , \end{aligned} \right\} \tag{22}$$

the function $a_n^{(0)}$ remains undefined.

Let us now multiply equation (20) by $\overline{\varphi}_r(z_a, \Psi)$ and integrate over the electronic coordinates z_a. Taking into account (22) we obtain

$$\sum_{m=1}^{\infty} \varkappa^m \left[a_r^{(m)} (V_r^{(0)} - \mathsf{V}_n^{(0)}) + \sum_{k=1}^{m} \sum_l (H_{rl}^{(k)} - W^{(k)} \delta_{kl}) \, a_l^{(m-k)} \right] + \\ + \varkappa^4 \mathsf{v}_n^{(0)} a_n^{(0)} = 0 . \tag{23}$$

The term with $m = 1$ has the form

$$a_r^{(1)} (V_r^{(0)} - \mathsf{V}_n^{(0)}) + (H_{rn}^{(1)} - W^{(1)} \delta_{rn}) \, a_n^{(0)} . \tag{24}$$

Assuming that $r = n$ we obtain

$$\varkappa^4 a_n^{(1)} \mathsf{v}_n^{(0)} + (H_{nn}^{(1)} - W^{(1)}) \, a_n^{(0)} . \tag{25}$$

According to the formulae given in Appendix III $H_{nn}^{(1)} = U_{nn}^{(1)} = \mathsf{V}_n^{(1)} + \varkappa^4 \mathsf{v}_n^{(1)}$. Moving the terms containing $\mathsf{v}_n^{(k)}$ $(k = 1, 2)$ to the fifth order and equating the rest of (25) to zero we obtain $W^{(1)} = \mathsf{V}_n^{(1)}$. Hence $W^{(1)} = 0$ because $\mathsf{V}_n^{(1)}$ is a linear function of the vibrational and librational coordinates and $W^{(1)}$ must be a number. The equations[2])

$$\frac{\partial V_n}{\partial q_i} = 0 , \qquad \frac{\partial V_n}{\partial \zeta_\lambda} = 0 , \qquad \frac{\partial V_n}{\partial \zeta_\lambda^{(t)}} = 0$$

define the equilibrium positions for the rotatory and translatory movements and for vibrations of the molecule and the crystal.

Assuming in (24) that $r \neq n$ we obtain

$$a_r^{(1)} = \frac{U_{nn}^{(1)}}{V_n^{(0)} - V_r^{(0)}} a_n^{(0)} \qquad (r \neq n) .$$

The term in (23) with $m = 2$ takes now the form

$$a_r^{(2)}(V_r^{(0)} - V_n^{(0)}) + \sum_{l \neq n} \frac{U_{rl}^{(1)} U_{ln}^{(1)}}{V_n^{(0)} - V_l^{(0)}} a_n^{(0)} + (H_{rn}^{(2)} - W^{(2)} \delta_{rn}) a_n^{(0)} , \tag{26}$$

where $H_{rn}^{(2)} = T^{(2)} \delta_{rn} + U_{rn}^{(2)}$. Assuming that $r = n$ we obtain

$$\varkappa^4 a_n^{(2)} v_n^{(0)} + \left(T^{(2)} + U_{nn}^{(2)} + \sum_{l \neq n} \frac{U_{nl}^{(1)} U_{ln}^{(1)}}{V_n^{(0)} - V_l^{(0)}} - W_n^{(2)} \right) a_n^{(0)} + O(\varkappa^4), \tag{27}$$

where the terms $O(\varkappa^4)$ arise because we have substituted $V_n^{(0)}$ for $\mathsf{V}_n^{(0)}$ in the denominator. According to Appendix III

$$U_{nn}^{(2)} + \sum_{l \neq n} \frac{U_{nl}^{(1)} U_{ln}^{(1)}}{V_n^{(0)} - V_l^{(0)}} = V_n^{(2)} = \mathsf{V}_n^{(2)} + \varkappa^4 v_n^{(2)} .$$

Removing from (27) the higher order terms and equating the rest of (27) to zero we obtain

$$(\mathcal{H}_n^{(v)} - W^{(2)}) a_n^{(0)} = 0 , \tag{28}$$

where

$$\mathcal{H}_n^{(v)} = T^{(2)} + \mathsf{V}_n^{(2)} .$$

This equation describes harmonic vibrations of our system. If we could consider the vibrations of the molecule separately of the lattice vibrations and if the quadratic form $\mathsf{V}_n^{(2)}$ were diagonalized then the equation (28) would describe the normal vibrations of the molecule and of the lattice. In the limit of large $\mathcal{N}_2$ we could then consider the vibrational spectrum of the crystal to be continuous and replace in the usual way the sum over the values of the wave vector $\boldsymbol{k}$ by the corresponding integration. Introducing the normal coordinates in the general case we mix our vibrational coordinates q_i, ξ_λ, and $\xi_\lambda^{(t)}$ but still quasi-continuous spectra will appear. In the following derivation we keep the summing over the vibrational states, but we suppose that the wave vector $\boldsymbol{k}$ can be approximately introduced to classify the lattice vibrations and that the corresponding sums can be replaced by integrals.

---

[2]) The derivatives are taken at the point where all the displacements are equal to zero.

Let σ_{ns} and ε_{ns} be the eigenfunctions and the eigenvalues of the operator $\mathcal{H}_n^{(v)}$, respectively. We expand the functions $a_n^{(m)}$ in the eigenfunctions σ_{ns}:

$$a_n^{(m)} = \sum_{s'} R_{ns'}^{(m)} \sigma_{ns'} , \tag{29}$$

where the coefficients $R_{ns'}^{(m)}$ depend on Ψ. Let ε_{ns} be the eigenvalue in which we are interested. Assuming for the sake of simplicity that this eigenvalue is non-degenerate we obtain from equation (28)

$$a_n^{(0)} = R_{ns}^{(0)} \sigma_{ns} , \qquad W^{(2)} = \varepsilon_{ns} .$$

The derivations concerning the expression (26) for $r \neq n$ and also the higher order terms of (23) are simple and similar to those already made. Therefore in the following we omit them and write only basic equations and results.

Considering the term with $m = 3$ in (23) we obtain the following equation:

$$\left(\mathcal{H}_n^{(v)} - \varepsilon_{ns}\right) a_n^{(1)} = \left(W^{(3)} - V_n^{(3)} - T_{nn}^{(3)}\right) a_n^{(0)} . \tag{30}$$

Inserting the expansion (29) ($m = 1$) into this equation, multiplying it by a function $\sigma_{ns''}$ and integrating over the vibrational coordinates we obtain the following results:

$$R_{ns'}^{(1)} = \frac{(V_n^{(3)} + T_{nn}^{(3)})_{s's}}{\varepsilon_{ns} - \varepsilon_{ns'}} R_{ns}^{(0)} ,$$

$$W^{(3)} = (V_n^{(3)} + T_{nn}^{(3)})_{ss} .$$

Hence $W^{(3)} = 0$ because $V_n^{(3)} + T_{nn}^{(3)}$ is an odd function of the vibrational coordinates and momenta.

We can obtain the equation for the function $R_{ns}^{(0)}$ considering the term with $m=4$ in (23). First we derive the equation for the function $a_n^{(2)}$ which is similar to (30) and can be solved in the same way; then assuming $s' \neq s$ we obtain

$$\left(\mathcal{H}_n^{(r)} - W_{nr}\right) R_{ns}^{(0)} = 0 ,$$

where

$$\mathcal{H}_n^{(r)} = - \frac{\hbar^2}{2m} I_3^{(0)^{-1}} \frac{\partial^2}{\partial \Psi^2} + E_n(\Psi) ,$$

$$E_n(\Psi) = V_n^{(0)} + \sum_{l \neq n} \frac{[U_{nl}^{(1)} (\mathcal{H}_n^{(v)} - \varepsilon_{ns}) U_{ln}^{(1)}]_{ss}}{V_n^{(0)} - V_l^{(0)}} +$$

$$+ \frac{1}{2m} \left[\sum_{k=1}^{3} I_k^{(0)^{-1}} (M_k^{(e)^2})_{nn} + \sum_{\lambda} (l_\lambda^{(e0)^2})_{nn} \right], \tag{31}$$

$$W_{nr} = W^{(4)} - w_{ns}^{(4)},$$

$$w_{ns}^{(4)} = (V_n^{(4)} + T_{nn}^{(4)'})_{ss} + \sum_{s' \neq s} \frac{(V_n^{(3)} + T_{nn}^{(3)})_{ss'} (V_n^{(3)} + T_{nn}^{(3)})_{s's}}{\varepsilon_{ns} - \varepsilon_{ns'}}, \tag{32}$$

where $(T_{nn}^{(4)'})_{ss}$ is the part of $(T_{nn}^{(4)})_{ss}$ which does not depend on Ψ:

$$(T_{nn}^{(4)})_{ss} = \frac{1}{2m}\Big\{\mathcal{F}^{(0)} + \sum_{k=1,2} I_k^{(0)-1/2}\,(\mathcal{P}_k f_k^{(11)}\mathcal{N}_1 + f_k^{(11)}\mathcal{N}_1\mathcal{P}_k)_{ss} + \\ + \sum_{k=1}^{3} I_k^{(0)^{-1}}\,(M_k^{(v)^2})_{ss} + I_3^{(0)^{-1}}\,[\mathcal{N}_1(\mathcal{N}_1 + 2\,M_3^{(v)})]_{ss} + \\ + \sum_{k=1,2} I_k^{(0)^{-1}}\,[\mathcal{P}_k\sigma_{k3}(M_3^{(v)} + \mathcal{N}_1) + (M_3^{(v)} + \mathcal{N}_1)\,\sigma_{3k}^{(1)}\mathcal{P}_k]_{ss} + \\ + \sum_{k=1,2} (I_i^{(0)}I_k^{(0)})^{1/2}\,(\sigma_{ik}^{(2)}\mathcal{P}_i\mathcal{P}_k)_{ss}\Big\}$$

(one can easily prove that the second term on the right hand side of (31) does not depend really on the vibrational quantum number s). Denoting an eigenfunction of the operator $\mathcal{H}_n^{(r)}$ by ϱ_{nr} we have $R_{ns}^{(0)} = \varrho_{nr}$ and

$$\psi^{(0)} = \varphi_n\sigma_{ns}\varrho_{nr}\,.$$

To complete calculation of the function $\psi^{(1)}$ we must find the function $R_{ns}^{(1)}$. We derive the equation for this function considering the term with $m = 5$ in (23) in the same way as was done for the lower orders:

$$(\mathcal{H}_n^{(r)} - W_{nr})\,R_{ns}^{(1)} = -\,W^{(5)}R_{ns}^{(0)}\,.$$

Hence using the normalization condition for the function $\psi^{(0)}$ we find that $W^{(5)} = 0$ and $R_{ns}^{(1)} = 0$ and thus we have finally

$$\psi^{(1)} = \left[\varphi_n \sum_{s' \neq s} \frac{(V_n^{(3)} + T_{nn}^{(3)})_{s's}}{\varepsilon_{ns} - \varepsilon_{ns'}}\,\sigma_{ns'} + \sigma_{ns} \sum_{l \neq n} \frac{U_{ln}^{(1)}}{V_n^{(0)} - V_l^{(0)}}\,\varphi_l\right]\varrho_{nr}\,. \tag{33}$$

As we have already noted we suppose that the lattice vibrations can be approximately classified with the help of the wave vector $\boldsymbol{k}$ and that the sum over the values of $\boldsymbol{k}$ can be replaced by the corresponding integral. Separating the terms caused by the lattice vibrations in the formulae (32) and (33) and introducing the integration we can obtain the anharmonic corrections of the same form as those obtained by Van Hove et al. [8] for the eigenvalue and the eigenfunction in the second and in the first orders of perturbation theory, respectively.

Comparing the formulae (32) and (33) with the corresponding results obtained for a free normal molecule [6] we see that some additional terms caused by kinetic corrections appear in the case of an impurity molecule. We come to this result because we have to move 'a part' of rotation (the libration) from the fourth to the second order of the Born-Oppenheimer method. Evidently we would obtain a similar result in other similar cases, for example, if we investigated an impurity molecule with completely hindered rotation or a molecule adsorbed on a surface.

Appendix I

Expansion of the matrix σ

This expansion can be obtained in the same way as in the paper [6]. After the scale transformation of the vibrational coordinates the matrix $I' \equiv I - LL^{\mathrm{T}}$

takes the form

$$I' = I_0 + \varkappa I_1 + \varkappa^2 I_2 = I_0(E + \varkappa a + \varkappa^2 b) \equiv I_0 A ,$$

where I_0 is the diagonal matrix of the momenta of inertia for the equilibrium internuclear distances. Expanding in the formula $\sigma = A^{-1} I_0^{-1}$ the matrix A^{-1} in power series in $\varkappa$ we get

$$\sigma = \sum_{l=0}^{\infty} \varkappa^l \sum_{1/2l \leqq n \leqq l} (-1)^n \sum_{\substack{j_1, \ldots, j_n = 0 \\ (j_1 + \cdots + j_n = l - n)}}^{1} a^{1-j_1} b^{j_1} \ldots a^{1-j_n} b^{j_n} I_0^{-1} .$$

We have in particular $\sigma_{ik}^{(0)} = \delta_{ik} I_k^{(0)^{-1}}$.

Appendix II

The matrices $C^{(m)}$ ($m = 0, 1$) and the functions $f_{jp}^{(k)}$

$$C^{(0)} = \begin{pmatrix} \cos \Psi & -\sin \Psi & 0 \\ \sin \Psi & \cos \Psi & 0 \\ 0 & 0 & 1 \end{pmatrix},$$

$$C^{(1)} = \begin{pmatrix} 0 & 0 & I_1^{(0)-1/2} q_1 \sin \Psi + I_2^{(0)-1/2} q_2 \cos \Psi \\ 0 & 0 & I_1^{(0)-1/2} q_1 \cos \Psi - I_2^{(0)-1/2} q_2 \sin \Psi \\ -I_2^{(0)-1/2} q_2 & I_1^{(0)-1/2} q_1 & 0 \end{pmatrix},$$

$$f_{jp}^{(k)} = \frac{|B_{2k}|}{(2k)!} \left(I_1^{(0)^{-1}} q_1^2 + I_2^{(0)^{-1}} q_2^2 \right)^{k-1} \tilde{f}_{jp}^{(k)} ,$$

$$\tilde{f}_{11}^{(k)} = 2^{2k} I_2^{(0)-1/2} q_2 , \qquad \tilde{f}_{12}^{(k)} = -2(2^{2k} - 1) I_2^{(0)-1/2} q_2 ,$$

$$\tilde{f}_{21}^{(k)} = -2^{2k} I_1^{(0)-1/2} q_1 , \qquad \tilde{f}_{22}^{(k)} = 2(2^{2k} - 1) I_1^{(0)-1/2} q_1 ,$$

where B_n is the Bernoulli number.

Appendix III

Expansion for the adiabatic potential

Assuming that $V_n^{(0)}(\Psi) - V_p^{(0)}(\Psi) \neq 0$ one can prove by induction that the following formulae hold (see also [6]):

$$V_n = \sum_m \varkappa^m V_n^{(m)} ,$$

$$V_n^{(m)} = (U^{(m)})_{nn} + \sum_{l=1}^{m-1} \sum_{k \neq n} (U^{(l)})_{nk} C_{nk}^{(m-l)} ,$$

$$C_{np}^{(m)} = \left(V_n^{(0)} - V_p^{(0)} \right)^{-1} \left[(U^{(m)})_{pn} + \sum_{l=1}^{m-1} \left\{ \sum_{k \neq n} (U^{(l)}) C_{pknk}^{(m-l)} - V_n^{(l)} C_{np}^{(m-l)} \right\} \right].$$

References

[1] A. A. Kiselev and A. V. Liapzev, phys. stat. sol. (b) **62**, 271 (1974).
[2] M. Born and R. Oppenheimer, Ann. Phys. (Germany) **4**, 457 (1927).
[3] L. Schwartz, Méthodes mathématiques pour les sciences physiques, Chap. 2, Section 4, Hermann, Paris 1961.
[4] A. A. Kiselev, Optika i Spektroskopiya **22**, 195 (1967).
[5] V. F. Bratsev, Vestnik Leningradsk. Univ. **16**, 20 (1970).
[6] A. A. Kiselev, J. Phys. **B3**, 904 (1970).
[7] L. A. Rebane, Proc. Internat. Seminar Selected Problems in the Theory of Impurity Centers in Crystals, Tallinn, September 1970, Tallinn 1972 (p. 353).
[8] L. Van Hove, N. M. Hugenholtz, and L. P. Howland, Problems in Quantum Theory of Many-Particle Systems, W. A. Benjamin, Inc., New York 1961 (pp. 1 to 103).

(Received October 15, 1973)

phys. stat. sol. (b) 62, 689 (1974)

Subject classification: 19; 13.4; 22.8

Le Centre de Recherches sur les Atomes et les Molécules, Université Laval, Quebec (a), and Eaton Electronics Research Laboratory, Physics Department, McGill University, Montreal (b)

Electron Spin–Lattice Relaxation Measurements of the Jahn-Teller Complex $Cu(6H_2O)^{++}$

By

K. P. LEE[1]) (a) and D. WALSH (b)

Spin–lattice relaxation measurements at microwave frequencies in the liquid helium temperature range yield relaxation rates which are four orders of magnitude faster than the corresponding rate in a system without Jahn-Teller complications. The results also emphasize that Jahn-Teller parameters for this system cannot be determined to-date with any certainty.

Les mesures de relaxation du spin–réseau à la fréquence de microonde pour des tempér-atures de l'hélium liquide donnent des taux de relaxation de quatre ordres de grandeur plus élevé que le taux dans un système sans complications de Jahn-Teller. Les résultats soulignent aussi qu'on ne peut pas déterminer aujourd'hui avec certitude les paramètres Jahn-Teller.

1. Introduction

The ground state of $Cu^{++}(3d^9)$ is ^2D. Under the perturbation of a cubic crystalline electric field of octahedral symmetry, the fivefold degenerate orbital states are split into a lower doublet, E, and an excited triplet, T_2, the latter having an energy Δ above the former. In lanthanum magnesium nitrate (LMN) X-ray analysis shows that the $Mg(6\ H_2O)$ complexes are trigonally distorted [1]. Since a trigonal field does not split an E-state and the spin–orbit splitting to first order is zero, the ground state of Cu^{++} as a substitutional impurity at the Mg site is ^2E within this approximation. According to the Jahn-Teller theorem a non-linear molecular configuration is unstable against distortions which lower its symmetry when the electronic state is degenerate [2]. Consequently, we expect that the $Cu(6\ H_2O)$ complexes will be similarly affected and their magnetic properties modified accordingly (Fig. 1). In fact, using the results of Opik and Pryce [3] and assuming the stabilizational energy to be much greater than the zero-point vibrational energy, there are three equivalent distortions corresponding to elongation or compression along the principal axes. This prediction is supported by electron paramagnetic resonance results in the liquid helium temperature range at microwave frequencies by Bleaney et al. [4] and more recently by similar EPR results [5, 6]. At higher temperatures only one electronic transition is observed, the corresponding g-value being approximately the average of the g-values for the low-temperature spectra. This was first interpreted by Abragam and Pryce [7] to be the "motional narrowed" absorption line corresponding to the "dynamic" situation (in contrast with the "static" distorted one), when the tunneling rate between equivalent distorted configura-

[1]) Now at the Department of Physics, University of New Brunswick, Fredericton, N.B., Canada.

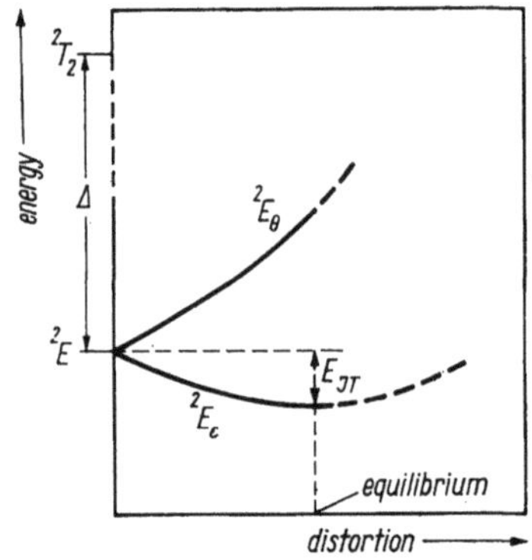

Fig. 1. Energy splitting of an orbital doublet, E, due to linear Jahn-Teller distortions. E_{JT} is the Jahn-Teller stabilization energy

tions becomes comparable with the measuring frequency. However, to-date a satisfactory explanation has yet to be found which is consistent with recent estimates of this tunneling rate based on spin–lattice relaxation measurements [8].

Electron spin–lattice relaxation measurements are usually performed at microwave frequencies by the pulse saturation method, the recovery of the absorption line to equilibrium being observed by standard EPR techniques using a low-power monitor signal. Consequently, even at liquid helium temperatures the "dynamic" situation can manifest itself in relaxation phenomena. In the presence of a dc magnetic field the three possible distortions of the complex are not equivalent energetically; however, tunneling can still proceed with the absorption or emission of one or more phonons. If this reorientation is accompanied by a spin-flip and if the corresponding reorientation rate is faster than the usual spin–lattice relaxation rate between a Kramers doublet, the observed phenomena is tunneling-assisted relaxation. Some preliminary spin–lattice relaxation results for Cu^{++} in LMN by Lee and Walsh [9] gave a relaxation rate which was four orders of magnitude faster than that expected for a system free of Jahn-Teller complications and a theoretical calculation based on tunneling-assisted relaxation was in general agreement with this enhancement of the recovery. Independently Breen et al. [6] using both spin–lattice relaxation and spin-echo techniques reported a similar enhancement of their recovery signal, while the theory of Williams et al. [8] similarly involved reorientation with spin-flip.

In addition to the very short relaxation times the recovery signal after pulse saturation was non-exponential. Breen et al. [6] decided that the initial portion of the recovery was due to spectral diffusion and consequently, this part of the recovery was omitted from their analysis. On the other hand, our experiments indicated that between 1.6 and 4.2 °K, the recovery signal was non-exponential for pulse lengths of up to 30 ms. The power level of the saturating pulse used was much less than that employed by Breen et al. [6]. As shown in Section 3, two time constants were deduced from the initial portion and the tail portion of the recovery signal. Since the initial portion of the recovery shows a linear temperature dependence, it cannot be due to spectral diffusion. However, it is indicative of a direct spin–lattice relaxation process. The tail portion of the recovery has considerable scatter and could be fitted to a T- or T^2-dependence. A T^2-dependence implies the presence of a phonon bottleneck [10], but we have failed to confirm its existence due to the absence of size and concentration dependence.

Unfortunately, the g-values of Cu^{++} are insensitive to level shifts of tens of wave numbers. Consequently, the effect of random crystal strain cannot be measured directly using EPR techniques. Using the absence of phonon bottleneck as a clue, Williams et al. [8] estimated strain splittings to be of the order of 1 or 40 cm^{-1} depending on their theoretical model used. From a study of linewidths of Ni^{++} in LMN and MgO, we have deduced a random strain of the order of 10^{-5} in LMN leading to a strain splitting of about 0.005 cm^{-1} [11]. Accordingly, our theoretical calculations assumed some strain to be present, but, nevertheless, considered it to be much less than the Zeeman splitting. In our preliminary report [9] we associated the initial recovery with the spin–lattice relaxation and the tail portion with the reorientation of the complex without spin–flip, the rate for the latter being reduced due to the bottleneck effect. In view of the absence of a definite bottleneck we now associate both portions of the recovery with spin–lattice relaxation which implies that the local strain is larger than the normal random strain expected in the crystal. In either case, however, since Breen et al. [6] discarded the initial portion of the recovery, their results do not agree with ours.

Section 2 deals with the experimental details. The summary of the results will be given in Section 3, while discussion and conclusion follow in Sections 4 and 5, respectively.

2. The Experiment

2.1 Sample preparation

Lanthanum magnesium nitrate hydrate can be written as

$$2\,\{La(NO_3)_6\}\cdot 3\,\{Mg(H_2O)_6\}\cdot 6\,H_2O$$

to emphasize the coordination of the divalent and trivalent ions. A detailed X-ray diffraction study has been made by Zalkin et al. [1] on the isomorphous cerium crystal. Two trigonal sites for the Mg atoms were distinguished, one being more cubic than the other. In our experiment we will only be concerned with the copper complex in the former site [6].

The copper-doped crystals were prepared by mixing in stoichiometric quantity three nitrates of lanthanum, magnesium, and copper. The saturated solution at room temperature was seeded and allowed to grow at 0 °C. The crystals usually grew in hexagonal plates and only those visually free of imperfection were used in the experiment. At the beginning seeds may be obtained by slowly adding concentrated nitric acid to the double-nitrate solution while stirring.

The lanthanum nitrate was bought from American Potash and Chemical Corporation (99.997% purity) and from Johnson, Matthey, and Co., the latter's nitrate having been prepared from spectrographically standardised La_2O_3.

Natural copper consists of two isotopes Cu^{63} and Cu^{65} with relative abundance 69.1 and 30.9%, respectively. Their nuclear quadrupole moments are comparable resulting in partial overlap of hyperfine lines although the hyperfine structure belonging to each isotope is well resolved. We have therefore grown crystals from isotopically pure copper nitrate prepared by Oak Ridge National Laboratory (USA) from 99.62% enriched Cu^{63}.

2.2 The spectrometer

The pulse saturation experiments were performed using a conventional superheterodyne spectrometer operating at 9.37 GHz [12]. A schematic diagram is shown in Fig. 2. A Varian V-58 klystron was stabilised by a Dymec oscillator synchronizer (Model Dy-2650A) to 1 ppm per week. A microwave amplifier (Varian VA-824B) increased the power to 1.2 W. It was estimated that the power in the cavity was about 120 mW. The mixing of the main frequency at

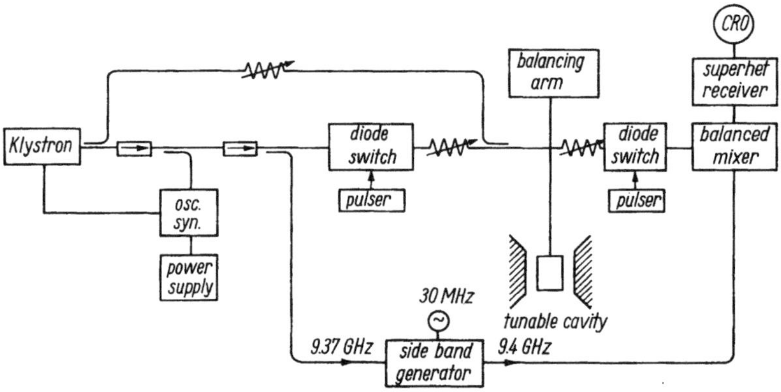

Fig. 2. Block diagram of an X-band spectrometer for pulse saturation experiment

9.37 GHz with a sideband at 9.4 GHz produces a 30 MHz IF which was amplified and demodulated before presenting on an oscilloscope.

The sideband frequency was generated by allowing a portion of the main klystron power to fall on a microwave diode which was being modulated at 30 MHz creating two sidebands separated by 60 MHz. The center frequency component was mostly balanced out by a magic tee arrangement and one of the sidebands (here 9.4 GHz) was selected by a high-Q transmission cavity before feeding it into one arm of the balanced mixer.

Three Tektronix pulse generators and one waveform generator (160 series) were used to produce pulses of varying widths, voltages, time delays, and recurrent rates. One of the pulse units drives the switching diode (Somerset Model X416) which controls the saturating power pulse. The second drives another diode that switches on the IF signal and also triggers the oscilloscope display of the EPR recovery signal. The third delivered a second trigger to the oscilloscope to establish a base line for reference. The overall response of the system was about 1 μs being limited only by the rise times of the pulse generators.

2.3 Procedure

All relaxation measurements were made on the imaginary part of the magnetic susceptibility by setting the external magnetic field to the center of the resonance line. The results presented here are for fields applied along the z-axis of the complex and for the lowest field component of the hyperfine lines since the theoretical derivation of the relaxation rate was done for $\boldsymbol{H} \parallel z$ for simplicity.

To insure working in the linear region of the detector, the recovery signal was adjusted so as to obtain a maximum IF output when the spin system returns to thermal equilibrium. Further a portion of the initial recovery was omitted from the plot as it was found that the detection system behaved fairly linearly for IF output above 0.2 V. For reasons of accuracy, it was often necessary to photograph the initial portion of the recovery in an expanded time scale. A quick check showed that the relaxation rate was independent of the hyperfine components within 30%. An angular study of the relaxation rate in $(10\bar{1}0)$ plane revealed no change in the behaviour of the recovery signal up to $\theta = 80°$. The rate increased steadily away from the z-axis, but the maximum increase (at $\theta = 45°$) was about a factor of two only. The microwave power pulse was repeated every 130 ms to ensure the return to thermal equilibrium condition of the system, i.e. lattice + spin.

As the bubbling of liquid helium was quite severe above the λ-point, it was necessary to exclude liquid helium from the tunable brass cavity. This was achieved by enclosing the cavity in a brass cylinder and sealed just above the cavity coupling flange so that the stainless steel waveguide remained in contact with liquid helium. Helium exchange gas entered the cavity via a hole in the waveguide above the helium liquid. To ensure efficient thermal conduction, the sample was always mounted in such a way as to allow maximum contact area between the crystal and the cavity wall. A comparative study at 4.2 °K with and without liquid helium contacting sample showed no observable change in the measured relaxation times within experimental error. Below the λ-point liquid helium was allowed into the cavity. Temperatures were then obtained from vapour pressure readings. Above the λ-point they were given by a carbon resistor attached to the plunger, though it was found that the increase in temperature of the cavity was usually small during pulsing being less than 0.1 °K with 10 ms pulse at normal power level.

3. Results

The recovery of the absorption signal from saturation was recorded on Polaroid film. A trace of a typical recovery signal is shown in Fig. 3. By plotting the signal amplitude against time on semi-logarithmic paper, the time constant can be determined from the slope. During the experiment it was found that with the maximum power available and using pulse lengths of up to 30 ms, the recovery signal was non-exponential over the full range of temperature reported.

In the presence of a dc magnetic field and some local strain, the three possible configurations of the complex are not equivalent energetically. Our interpretation of the direct spin-lattice rate is essentially a relaxation process via non-Kramers conjugate pairs as a result of phonon-assisted tunneling. There are two independent paths for this process and consequently we treat the recovery signal as the sum of two exponential terms. Since the initial portion of the recovery has an approximate linear temperature dependence, this rate gives a more reliable estimate of the spin-lattice relaxation.

Fig. 4 shows the rate constants for the initial portion of the recovery. Each point in the figure (including those in Fig. 5) represents an average of the results of at least five measurements. Some are the averages of some twenty to thirty measurements. Although there is still some scatter, it may be noted that the relaxation rate is independent of concentration, and for a 1% sample it is also independent of size. A study was made to determine the character of the variation of relaxation rate with pulse length. It was found that the rate constant for the initial recovery increases with decreasing pulse length. For example, at a pulse length of 0.03 ms the deduced rate was about 8000 T. However, the variation becomes smaller at longer pulses and quite small above pulse length of 1 ms.

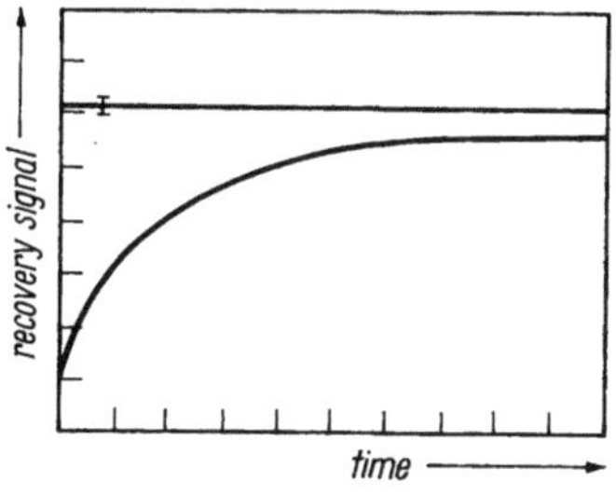

Fig. 3. Trace of recovery signal from photograph for a 1% thick sample at 4.2 K using 10 ms saturating pulse. The horizontal reference line was established by a trigger pulse delayed by about 15 ms to ensure return of a thermal equilibrium of the spin system. The error bar indicates the amount of noise present in the signal

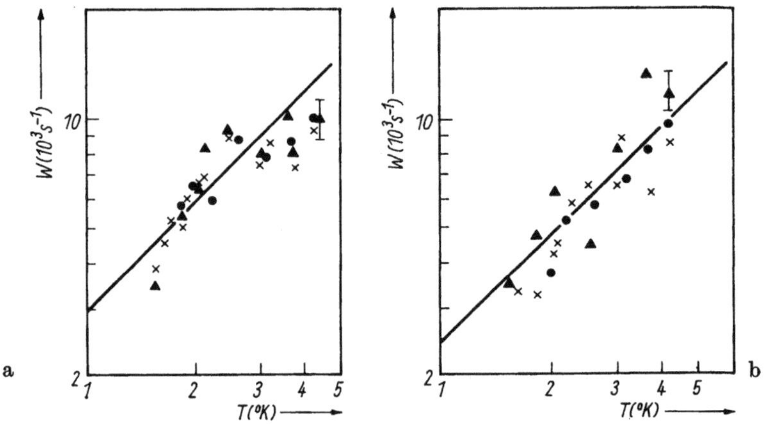

Fig. 4. Rate constants obtained for a microwave saturating pulse of a) 1 ms and b) 10 ms. The symbols are ▲ for 0.1%, thickness $t = 3.0$ mm; ● 1%, $t = 1.5$ mm, and × 1%, $t = 3.4$ mm. The vertical bars (in Fig. 4 and 5) signify the amount of scatter which is somewhat reduced for stronger samples

The rate constants for the tail portion of the recovery signal is plotted in Fig. 5. The deduced rate varies between a T- and a T^2-dependence. One observes that the temperature dependence is more T^2 for longer saturating pulses, and for samples of about the same size it is also more T^2 for 1% than for 0.1% samples. Thus there is some evidence of a phonon bottleneck. Nevertheless, it must be concluded that the experiments fail to confirm the normal bottleneck condition. For example, there is practically no size dependence whatsoever.

It may be noted that the experimental points shown in Fig. 4 and 5 have considerable scatter. This is due not only to the difficulties associated with weak signals and short relaxation times, but also to the mere fact of extracting two time constants from a single non-exponential curve, since this procedure greatly reduces the portion of the curve used for linear approximation in a semilogarithmic plot and thus increases the experimental error.

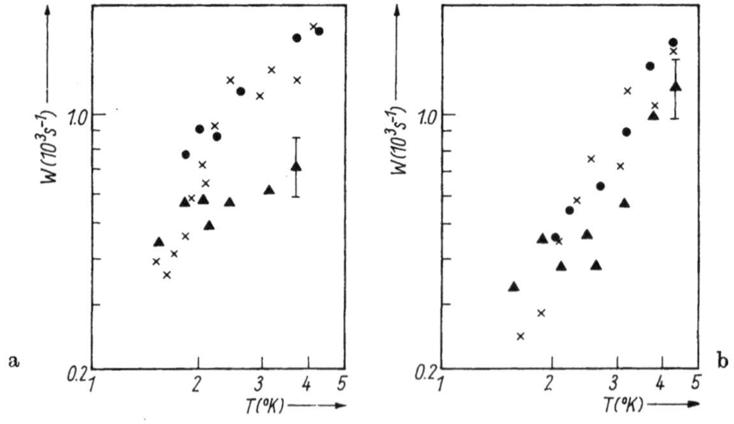

Fig. 5. Rate constants for the tail portion of the recovery signal. For symbols see Fig. 4

In some copper samples, Ni^{++} impurities were detected. Since Ni^{++} has a faster rate of relaxation than copper, it might be thought that cross relaxation may increase the relaxation rate of copper [13]. This is contrary to our observations: Along the z-axis one nickel line was only 40 Oe away. As the magnetic field is rotated away from the z-axis in one direction, the nickel line moved progressively further away from the copper spectrum. However, the deduced relaxation rate of Cu^{++} was slowest at $\theta = 0°$. As the nickel line moved away, the rate increases instead of decreasing as would be the case if cross relaxation had been taking place.

4. Discussion

Our results are at variance with the spin–lattice relaxation results of Breen et al. [6]. We do not think that the initial recovery was due to spectral diffusion. Firstly, if it were spectral diffusion, one should be able to observe an exponential recovery at sufficiently long saturating pulse. This was not observed by us and neither by Breen et al. [6] below 8 °K. Secondly, as stated by them, the non-exponentiality of some recoveries only is consistent with a model of spectral diffusion and careful measurement was performed on only one sample which happened to be the most concentrated. Lastly, the diffusion theory of Klauder and Anderson [16] and that of Mims et al. [17] require the same diffusion constant to be derived from primary and stimulated echo measurements. The disagreement by an order of magnitude between these two measurements of Breen et al. [6] has placed some doubt on their interpretation of the origin of the non-exponentiality. Thus the evidence for spectral diffusion is not strong. Since we have observed a linear temperature dependence for the initial recovery, we have accordingly interpreted it as a direct process.

Table 1

nominal concentration (at%)	actual (at%)	concentration (ions/cm³)
0.1	0.003	0.9×10^{17}
1	0.03	9.0×10^{17}

Regarding the tail portion of the recovery, one would expect agreement between the two experiments. The disagreement is disturbing at first sight. Looking more closely one finds that our rates are not very different from theirs; for example, at 4 °K our rate is about five times slower than their protonated rate. This is not serious considering that their sample is at least 2.5 times more concentrated than ours and the rates are concentration-dependent at high concentrations as shown by them. Table 1 shows the actual concentration of the crystals as determined by iodometric analysis [14]. In view of the greater scatter of our data, the most obvious viewpoint to adopt is that the tail portion of the recovery also has a linear T-dependence. In that case the non-exponential decay is simply the sum of the two possible spin–lattice relaxation processes. We assume that the reorientation rate without spin-flip is very fast and not observed. This rate corresponds to that measured by Breen et al. [6] using spin-echo techniques.

Since the initial recovery has less scatter we shall quote its rate for the direct process. The deduced rate is 2300 T. The corresponding Jahn-Teller parameters such as E_{JT} and the potential barrier V are shown in Table 2.

Table 2

Some J-T parameters deduced by various authors

E_{JT} (cm^{-1})	V (cm^{-1})	reference
3000	1200	Opik and Pryce [3]
1346	407	Liehr and Ballhausen [18]
1250	1620	Williams et al. [8] (Model D)
1100	900	Williams et al. [8] (Model T)
960	240	Lee and Walsh [11]

5. Conclusion

In spite of the contribution from spin–lattice relaxation measurements to our understanding of Cu^{++} in LMN, it is quite apparent that the ground state properties of Cu^{++} are not completely understood. The most significant result is the enhancement of the spin–lattice relaxation rate by four orders of magnitude due to phonon-assisted tunneling, and this observation has revived great interest in Jahn-Teller complexes [15]. The less exciting result is shown in Table 2, where it is apparent that reliable Jahn-Teller parameters are difficult to obtain for these systems. A similar observation was made recently by Ham [15] concerning Cu^{++} in MgO: "... From presently available data one cannot yet determine 3Γ or other Jahn-Teller parameters of the system with any certainty."

Acknowledgement

The authors wish to thank NRC for financial assistance during the course of this experimental work, which was performed in partial fulfilment of the Ph.D. requirements by one of the authors (K. P. Lee).

References

[1] A. Zalkin, J. D. Forrester, and D. H. Templeton, J. chem. Phys. **39**, 2881 (1963).
[2] H. A. Jahn and E. Teller, Proc. Roy. Soc. **A161**, 220 (1937).
[3] U. Opik and M. H. L. Pryce, Proc. Roy. Soc. **A238**, 425 (1957).
[4] B. Bleaney, K. D. Bowers, and R. S. Trenam, Proc. Roy. Soc. **A228**, 157 (1955).
[5] K. P. Lee, Ph. D. Thesis, McGill University, Montreal 1969.
[6] D. P. Breen, D. C. Krupka, and F. I. B. Williams, Phys. Rev. **179**, 241 (1969).
[7] A. Abragam and M. H. L. Pryce, Proc. Roy. Soc. **A205**, 135 (1951).
[8] F. I. B. Williams, D. C. Krupka, and D. P. Breen, Phys. Rev. **179**, 255 (1969).
[9] K. P. Lee and D. Walsh, Phys. Letters A **27**, 17 (1968).
[10] P. L. Scott and C. D. Jeffries, Phys. Rev. **127**, 32 (1962).
[11] K. P. Lee and D. Walsh, Canad. J. Phys. **49**, 1620 (1971).
[12] G. Feher, Bell Syst. tech. J. **36**, 449 (1957).
[13] K. P. Lee and D. Walsh, Canad. J. Phys. **47**, 1753 (1969).
[14] I. M. Kolthoff and E. B. Sandell, Textbook of Quantitative Inorganic Analysis, 3rd edition, Macmillan Co., New York 1952 p. (585).
[15] F. S. Ham, in: Electron Paramagnetic Resonance, Ed. S. Geshwind, Plenum Press, New York 1972.
[16] J. R. Klauder and P. W. Anderson, Phys. Rev. **125**, 912 (1962).
[17] W. B. Mims, K. Nassau, and J. D. McGee, Phys. Rev. **123**, 2059 (1961).
[18] A. D. Liehr and C. J. Ballhausen, Ann. Phys. (U.S.A.) **3**, 304 (1958).

(Received August 9, 1973)

phys. stat. sol. (b) 62, 697 (1974)

Subject classification: 13.1 and 14.3; 22.7

Institute of Semiconductor Physics, Academy of Sciences of the Lithuanian SSR, Vilnius (a), and Institute of Physics, Academy of Sciences of the Azerbaidzhan SSR, Baku (b)

Investigation of Magnetoplasma Wave Dispersion in Pure and Doped $Bi_{92}Sb_8$ at Liquid Nitrogen Temperature

By

R. Brazis (a), J. Požela (a), B. Tairov (b), and M. Shakhtakhtinskii (b)

The bulk magnetoplasma wave dispersion and its relation to peculiarities of the energy band structure in $Bi_{92}Sb_8$ are investigated. The electron mobility tensor components and the carrier cyclotron mobilities are estimated in pure single crystals and those doped with tin and tellurium.

На основе исследования связи дисперсии магнитоплазменных волн (0,1 до 20 MHz) с особенностями зонной структуры $Bi_{92}Sb_8$ найдены компоненты тензора подвижности электронов и циклотронные подвижности электронов и дырок в нелегированных, а также легированных оловом и теллуром кристаллах $Bi_{92}Sb_8$ при $T = 77$ K.

1. Introduction

In recent years, as a result of a number of theoretical [1, 2] and experimental [3, 6] investigations, the main characteristics of energy band structure transformation in Bi–Sb alloys with the change of composition in the range of $x <$ < 30 at% Sb have been revealed. Semiconducting properties are shown to be inherent in Bi–Sb alloys in a limited range of composition. At $T = 4.2$ K this range was found to extend from approximately $x \approx 6$ to 25 at% [4, 5] while at 77 K it amounted to $7 < x < 20$ at% [6]; with the increase of temperature the range was found to narrow, disappearing completely at $T >$ > 180 K [6].[1]) The alterations in the band structure with composition are so pronounced that alloys, the composition of which differs by 1%, should be considered as completely separate semiconductors with their appropriate parameters such as effective mass, momentum and energy relaxation times, occupation of energy valleys by carriers, etc. Some of the parameters were measured only for several values of composition x, most of the measurements being made at liquid helium temperature. Different parameters of Bi–Sb at liquid nitrogen temperature were reported in [6 to 13]; the most complete measurements of current carrier mobility parameters in semiconducting Bi–Sb have been carried out by Yazaki and Abe [13]. The interpretation of their data, obtained from galvanomagnetic measurements, was based on an a priori model of three tilted ellipsoidal valleys both for electrons and for holes. However, as it was pointed out by Goldsmid [7], the above model for holes does not seem to have much justification. On the other hand the pecularities of helicon wave dispersion are closely related to the band structure of semiconductor [14] (see Appendix). Investigations reported in [15 to 17] demonstrated the existence of such correla-

[1]) General progress in the investigation of Bi–Sb was reviewed by Goldsmid [7].

tion in the Bi–Sb crystals. Recent investigations [18] of n-type $Bi_{92}Sb_8$ have shown the magnetoplasma wave dispersion to be sensitive to the change of the band structure induced by hydrostatic pressure. Thus, the results obtained from investigation of helicon and other magnetoplasma wave dispersion investigations may serve as a basis not only for determining the parameters of a semiconductor at a given energy band model, but also for the choise of the band model itself.

The aim of the present work is to determine the carrier mobility parameters of $Bi_{92}Sb_8$ at 77 K by means of magnetoplasma waves on the basis of qualitative and quantitative comparison between theory and experiment.

2. Experimental Technique and Results

$Bi_{92}Sb_8$ single crystals have been grown following the Czochralski technique under continuous feed of melt with due regard for segregation. The samples in the form of square plates with thicknesses of approximately 1 mm and other

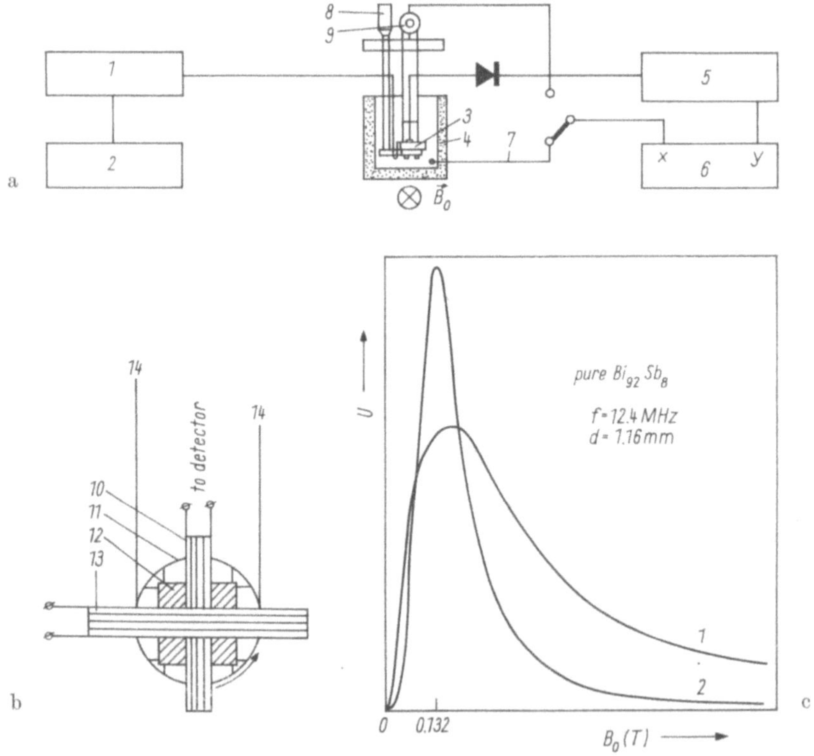

Fig. 1. a) Experimental set-up: rf generator (1), amplitude modulator (2), sample and coil plexiglass holder (3), styroform dewar (4), low-frequency amplifier (5), *xy* recorder (6), Hall magnetic field indicator (7), micrometer screw (8) for the correction of the angle between coils, precision potentiometer (9) for the indication of the cassette rotation angle. b) The cassette (11) with Bi–Sb sample (12) between crossed receiving (10) and exciting (13) coils; driving filament (14) is connected with the angle indicator (9). c) Typical directly measured dependences of signal in the receiving coil versus magnetic field. (1) $\boldsymbol{B}_0 \parallel$ binary axis, (2) $\boldsymbol{B}_0 \parallel$ bisectrix axis

dimensions of 5 to 8 mm were cut by the electroerosion technique with subsequent polishing and etching. The samples had square faces directed perpendicularly either to trigonal (x_3), or binary (x_1), or bisectrix (x_2) axes. The accuracy of the crystallographic orientation was checked by X-ray method. The measurement technique was similar to that in [15, 17] (Fig. 1). The cassette with a sample was mounted in a holder and could be rotated between the two crossed coils. Uniformity and linear polarization of the exciting rf magnetic field was obtained because of large dimensions of the exciting coil as compared to the sample dimensions. For better coupling with the field of magnetoplasma wave in the sample the receiving coil was wound close to the sample. Firstly, rotating the cassette the sample was oriented so that the exciting rf magnetic field was made parallel to one of the main crystallographic axes, and zero rf signal in the receiving coil was achieved by correction of the angle between the coils. After that the dc magnetic field, directed normal to the square faces of the sample, was turned on. The dependences between the rf oscillation amplitude U in the receiving coil and the dc magnetic field induction B_0 at 0.1 MHz $\leqq f \leqq$ 20 MHz were measured. For each frequency the magnetic induction B_{0m} corresponding to the maximum of the function $U = U(B_0)$ was noted. Moreover, the dependences $U = U(\varphi)$ at various magnetic field inductions were measured (φ is the angle between the definite crystallographic axis and the coil axis in the plane normal to the magnetic field).

Experimental results at $T = 77$ K for undoped $Bi_{92}Sb_8$ crystals are shown in Fig. 2 to 4, 6 and for Te-doped $Bi_{92}Sb_8$ in Fig. 5.

3. Discussion and Comparison with the Theory

The rf signal amplitude in the receiving coil is related to the resistivity tensor components [19] as follows:

$$U(B_0) = \left| GI \frac{\varrho_{xy}}{\varrho_{xx} + \varrho_{yy}} \sum_m \frac{1}{m^2} \left[1 + iQ\left(\frac{\omega}{\omega_m} - \frac{\omega_m}{\omega}\right)\right]^{-1} \right|, \quad (1)$$

where G is the factor depending on the frequency and the coil geometry, I is the ac amplitude in the exciting coil, $Q = (\varrho_{xx}\varrho_{yy} - \varrho_{xy}\varrho_{yx})^{1/2} (\varrho_{xx} + \varrho_{yy})^{-1}$. In the low-frequency limit, $\omega \ll \omega_m$ where $\omega_m = \pi^2\mu_0^{-1}d^{-2}m^2(\varrho_{xx}\varrho_{yy} - \varrho_{xy}\varrho_{yx})^{1/2}$, the dependence $U(B_0)$ is given by the formula [17]

$$U(B) = \left| GI \frac{\varrho_{xy}}{\varrho_{xx}\varrho_{yy} - \varrho_{xy}\varrho_{yx}} \right|. \quad (2)$$

In (1) and (2) $\varrho_{ij} = \varrho_{ij}(B_0, \varphi)$ are the resistivity tensor components, d is the sample thickness, μ_0 the magnetic permittivity of the vacuum and $m = 1, 3, 5, \ldots$ The coordinate axes x,y,z are parallel to the exciting coil axis, receiving coil axis, and dc magnetic induction $\boldsymbol{B}_0$, respectively.

As it follows from (2), the function $U = U(B_0)$ has a maximum at certain magnetic induction B_{0m} corresponding to the maximum of the non-diagonal conductivity tensor component σ_{xy}, provided $\sigma_{xz} = \sigma_{yz} = 0$; the position of this maximum does not depend on the frequency. If only one group of carriers takes part in the electric transport, this maximum defines the carrier cyclotron mobility $\mu_c = B_{0m}^{-1}$. The vertical parts of the dependences $fd^2B_{0m}^{-1}$ versus B_{0m} in Fig. 2 show that pure $Bi_{92}Sb_8$ has different cyclotron mobilities for different crystallographic directions. At $\boldsymbol{B}_0 \parallel \boldsymbol{x}_3$ the cyclotron mobility is lowest indicat-

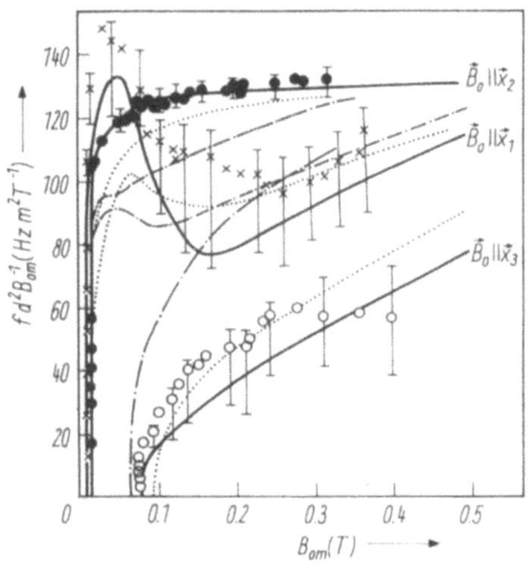

Fig. 2. Dependences between the resonance values of frequency f and magnetic induction B_{0m} for pure $Bi_{92}Sb_8$; d is the sample thickness; x_1, x_2, x_3 are binary, bisectrix, and trigonal axes, respectively. Points and vertical bars showing a scatter are experimental data. Solid lines are adjusted by computer selected proper parameters γ_i and n; dotted and dotted-dashed curves illustrate the sensitivity of the theory to the choice of these parameters; the electron concentration is $n = 6 \times 10^{22}$ cm^{-3} for all the curves. Calculated: dotted line: γ_i from 4th row; dashed-dotted line: γ_i from 2nd row solid line: this work, 3rd row, Table 1;

ing the largest cyclotron mass. At $\boldsymbol{B}_0 \| \boldsymbol{x}_1$ or $\boldsymbol{B}_0 \| \boldsymbol{x}_2$ the cyclotron mobilities are significantly higher as compared with the case $\boldsymbol{B}_0 \| \boldsymbol{x}_3$. It is worthwhile to note that similar relations between the cyclotron mobilities exist for electrons in bismuth [20]. As the frequency rises, the transition from the dependence (2) to helicon dimensional resonance is observed. It appears in Fig. 2 as a transition from the vertical to the horizontal part of the curves. The magnitude of $fd^2B_{0m}^{-1}$ is inversely proportional to the concentration of the carriers participating in the helicon wave propagation process. Therefore the negative slope of the curve when $\boldsymbol{B}_0 \| \boldsymbol{x}_1$ presents an evidence for the existence of at least two carrier groups with different cyclotron mobilities. One group gives a maximum $U(B_0)$ at $\omega \ll \omega_m$ (vertical part of the curve, Fig. 2) and the other group, owing to its lower cyclotron mobility, joins in the wave propagation process at higher magnetic fields, viz. 0.1 to 0.2 T. When $\boldsymbol{B}_0 \| \boldsymbol{x}_2$, the horizontal dependence of $fd^2B_{0m}^{-1}$ versus B_{0m} typical for helicon dimensional resonance is observed. Therefore at $\boldsymbol{B}_0 \| \boldsymbol{x}_2$ the cyclotron mobilities for all the carrier groups should be very close in magnitude and the ellipticity of the carrier cyclotron orbit should not be very high. These facts completely agree with the three-ellipsoidal-valley model for electrons in the L-point of the Brillouin zone.

We shall now consider the angular dependence $U = U(\varphi)$. According to [15] this dependence coincides with the angular dependence of the tensor component $\varrho_{xy}(\varphi)$:

$$U(\varphi) \sim |\varrho_{xy}(\varphi)| = |\varrho^{a}_{xy}(B_0, 0) + \{[\varrho^{s}_{xy}(B_0, 0)]^2 + \tfrac{1}{4}[\varrho_{yy}(B_0, 0) - \varrho_{xx}(B_0, 0)]^2\}^{1/2}| \times \times \sin(2\varphi + \theta)\,, \tag{3}$$

where

$$\theta = \frac{1}{2} \operatorname{arctg} \frac{2\varrho^{s}_{xy}(B_0, 0)}{\varrho_{yy}(B_0, 0) - \varrho_{xx}(B_0, 0)}\,,$$

the component $\varrho_{xy}(\varphi)$ is given in the coordinate system tied with the coils, and $\varrho_{ij}(B_0, 0)$ are the resistivity tensor components in the coordinate system fixed to the crystallographic axes.

The dependence $U(\varphi)$ for $\boldsymbol{B}_0 \| \boldsymbol{x}_1$ is shown in Fig. 3. It is of interest to note that even for $B_0 = 0$ the angular dependence $U(\varphi)$ is observed in the form of

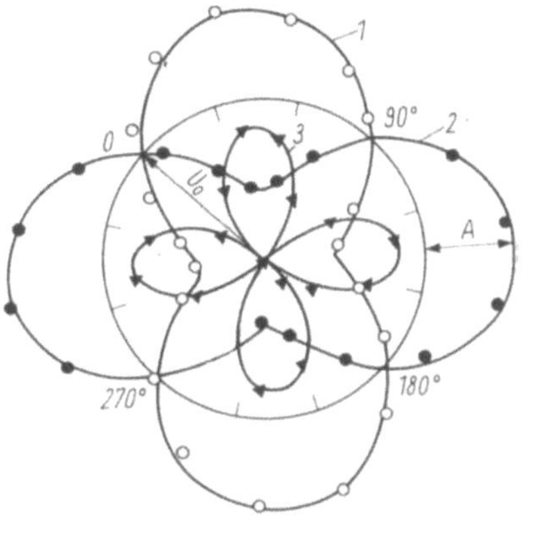

Fig. 3. Experimental dependences $U = U(\varphi)$ for pure $Bi_{92}Sb_8$ in polar coordinates (dots) and sine curves $U = |U_0 + A \sin 2\varphi|$ fitted to the experimental points (solid lines). $f = 0.8$ MHz. (1) $B_0 = 0.03$ T, $B_0 > 0$; (2) $B_0 = 0.03$ T, $B_0 < 0$; (3) $B_0 = 0$ (the scale of curve 3 is 50 times enlarged as compared with that of curves 1 and 2)

a four-petal curve centered at $U = 0$ (Fig. 3) It is evident that the conductivity anisotropy of $Bi_{92}Sb_8$ even in the absence of magnetic field is quite high. As ϱ_{ij} depends on the magnetic field, the quantity $\eta = 2A/U_0$ depends on the magnetic field as well (Fig. 4). The quantities A and U_0, as it follows from (3), are proportional to the symmetrical and antisymmetrical parts of ϱ_{xy}. One can see in Fig. 4 that the quantity $\eta = 2A/U_0$ at $\boldsymbol{B}_0 \parallel \boldsymbol{x}_2$ is in general much lower as compared to that at $\boldsymbol{B}_0 \parallel \boldsymbol{x}_1$. This suggests that the cross-section of the effective mass ellipsoids by the plane perpendicular to the magnetic field is more elliptical at $\boldsymbol{B}_0 \parallel \boldsymbol{x}_1$ as compared to $\boldsymbol{B}_0 \parallel \boldsymbol{x}_2$.

The conventional galvanomagnetic measurements showed that the sign of the Hall e.m.f. was due to electrons.

On the basis of the above data one can state that the essential features of the electrical conductivity in pure $Bi_{92}Sb_8$ are governed by electrons distributed in a similar way as in pure bismuth, i.e. in the valleys at L-points of the Brillouin zone. The holes do not seem to participate. Therefore at the first step we interpret all the experimental data by the model where holes are not involved. According to this model we have carried out the calculations of the dependences $fd^2B_{0m}^{-1}$ versus B_{0m} using (1) and the dependences $\eta = \eta(B_0)$ from the following equation:

$$\eta = \frac{2A}{U_0} = \left| \frac{2\,\{[\varrho_{xy}(B_0, 0) + \varrho_{yx}(B_0, 0)]^2 + [\varrho_{yy}(B_0, 0) - \varrho_{xx}(B_0, 0)]^2\}^{1/2}}{\varrho_{xy}(B_0, 0) - \varrho_{yx}(B_0, 0)} \right|. \quad (4)$$

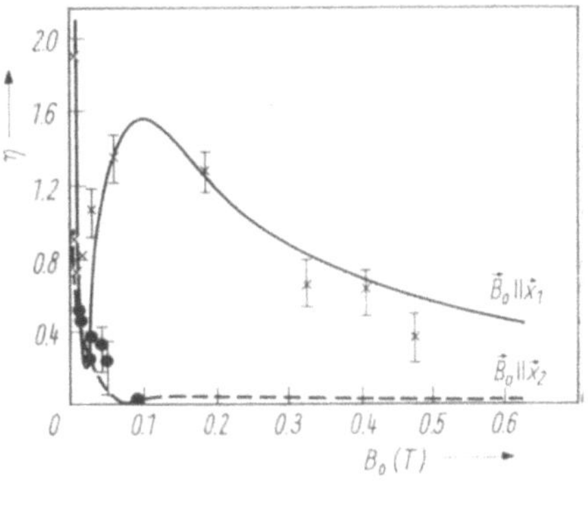

Fig. 4. Dependences $\eta = \eta(B_0)$ for pure $Bi_{92}Sb_8$. Experimental points are obtained at $f = 0.8$ MHz, ×, ● experiment; solid and dashed lines: theory

Table 1

Electronic parameters of the Bi–Sb at $T = 77$ K

	μ_{c1b} ($m^2 V^{-1} s^{-1}$)	μ_{c3} ($m^2 V^{-1} s^{-1}$)	γ_i (m^{-2}Vs)				n ($10^{23} m^{-3}$)	K_{21}	ref.
			γ_1	γ_2	γ_3	γ_4			
$Bi_{85}Sb_{15}$	35.8	8.0	0.01734	2.89	0.108	0.464	1.09	167	[13]
$Bi_{90}Sb_{10}$	50.7	10.85	0.01339	0.874	0.0301	0.0851	0.9	65	[13]
$Bi_{92}Sb_8$	100	13.15	0.00725	1.2	0.0155	0.0787	0.6	166	this work
$Bi_{95}Sb_5$	76.9	16.65	0.01418	0.83	0.02933	0.1299	1.39	59	[7]
$Bi_{97}Sb_3$	52.1	14.25	0.01255	0.933	0.0527	0.169	1.61	74	[13]
Bi	31	4.08	0.02083	8.3	0.1086	0.77		400	[21]

We also took into consideration the following inequalities satisfied in the experiment: $B_0 \gg B_\sim$, $\nu \gg k v_F$, $\omega \ll \omega_c$, $\omega \ll \nu$, where $B_\sim$ is the amplitude of the excited magnetoplasma wave, ν the collision frequency, ω_c the cyclotron frequency, ω the angular frequency, $\boldsymbol{k}$ the wave vector, and v_F the Fermi velocity. The displacement current could be neglected as compared to the conductivity current in the frequency range of interest (0.1 to 20 MHz). It was also assumed that (i) the constant energy surfaces for electrons at L-points are ellipsoidal, (ii) both the collision frequency tensor and the effective mass tensor could be diagonalized by the same transformation of the coordinate system. Under such assumptions it is convenient to introduce the reciprocal mobility tensor

$$\hat{\gamma} = \begin{pmatrix} \gamma_1 & 0 & 0 \\ 0 & \gamma_2 & \gamma_4 \\ 0 & \gamma_4 & \gamma_3 \end{pmatrix}, \tag{5}$$

where $\gamma_1 = (\hat{m}\hat{\nu}q^{-1})_{11}$, $\gamma_2 = (\hat{m}\hat{\nu}q^{-1})_{22}$, $\gamma_3 = (\hat{m}\hat{\nu}q^{-1})_{33}$, $\gamma_4 = (\hat{m}\hat{\nu}q^{-1})_{23}$ and the subscripts 1, 2, 3 correspond to the coordinate axes x_1, x_2, x_3. Moreover, the collision frequency was assumed to be independent of the carrier energy. This could not lead to significant errors since in the extremely degenerate semiconductors, like $Bi_{92}Sb_8$, only the carriers with energy in the narrow range near the Fermi level take part in the collisions. The calculations were made using the expressions for the conductivity tensor components given in [16] taking into account all the above-mentioned inequalities and assumptions. Computer calculations were made to find the set of parameters γ_i and the carrier concentration n for the best fit of theoretical and experimental dependences. Such parameters are given in Table 1[2]) and the corresponding theoretical curves are shown in Fig. 2, 4 by solid lines. The error in the parameters γ_i is about 25% and in the electron concentration less than 5%.

[2]) Some authors [7, 21] give the reciprocal parameters $\mu_i \equiv \gamma_i^{-1}$. These parameters are not the mobility tensor components. They can be used only for convenient writing of the formulas.

Using the parameters γ_i and n obtained by means of the magnetoplasma waves we have calculated the galvanomagnetic coefficients R_{123}, R_{231}, σ_{11}, and σ_{33}. According to [16] and taking into account the electrons in L-valleys only, one can obtain the following expressions for the dc galvanomagnetic coefficients in low magnetic fields:

$$\sigma_{11} = \sigma_{22} = \frac{1}{2} ne \left(\mu_1 + \frac{\mu_{c3}^2}{\mu_1}\right), \qquad \sigma_{33} = ne \frac{\mu_{c1a}^2}{\mu_2}, \tag{6}$$

$$R_{231} = \frac{1 + K_{21}}{ne\left[\left(K_{21} + \frac{\mu_{c3}^2}{\mu_1 \mu_2}\right) + \frac{1}{2} \mu_{c1a}^2 B_0^2 (1 + K_{21})^2\right]}, \tag{7}$$

$$R_{123} = \frac{1}{ne\left[\frac{1}{4}\left(\frac{\mu_1}{\mu_{c3}} + \frac{\mu_{c3}}{\mu_1}\right)^2 + \mu_{c3}^2 B_0^2\right]}, \tag{8}$$

where n is the electron concentration, $\mu_i = \gamma_i^{-1}$, $K_{21} = \gamma_2/\gamma_1$. Cyclotron mobilities are

$$\mu_{c1a} = \left(\frac{\gamma_1}{\Delta}\right)^{1/2}, \qquad \mu_{c3} = \left(\frac{\gamma_3}{\Delta}\right)^{1/2}, \tag{9}$$

where $\Delta = \gamma_1(\gamma_2\gamma_3 - \gamma_4^2)$ [16]. The terms with B_0^2 in (7) and (8) are included for a more precise definition of the magnetic field to be considered as sufficiently low.

Measurements of these coefficients have been made at low magnetic fields. When R_{231} was measured the current $\boldsymbol{I}$ was directed along the binary axis x_1, the magnetic field $\boldsymbol{B}_0$ was parallel to the bisectrix axis x_2, and R_{123} was measured with $\boldsymbol{I} \parallel \boldsymbol{x}_1$ and $\boldsymbol{B}_0 \parallel \boldsymbol{x}_3$. The results of dc galvanomagnetic measurements are, in general, in satisfactory agreement with the calculated values (see Table 2). It is worthwhile to note that additional calculations of the dependences $fd^2 B_{0m}^{-1}$ versus B_{0m} when some amount of holes has been included showed essential discrepancy with the experimental data. All these results show the electrical conductivity of an undoped $Bi_{92}Sb_8$ to be essentially governed by the electrons. On the other hand, it is known [7] that Bi–Sb alloys could have equal densities of electrons and holes since the valencies of bismuth and antimony atoms are equal. The discrepancy between our results and this statement could be removed if the effective mass of holes in $Bi_{92}Sb_8$ were much higher as compared to that of electrons. This is possible when there are valleys with high density of states

Table 2
Galvanomagnetic coefficients of pure $Bi_{92}Sb_8$ at $T = 77$ K

σ_{11} ($10^3\ \Omega^{-1}\ cm^{-1}$)	σ_{33} ($10^3\ \Omega^{-1}\ cm^{-1}$)	R_{231} ($cm^3\ C^{-1}$)	R_{123} ($cm^3\ C^{-1}$)
7.1	7.78	106	5*)
6.68	6.43	103	3.7**)

*) Galvanomagnetic measurements.
**) Calculated (using γ_i obtained from magnetoplasma waves (Table 1)).

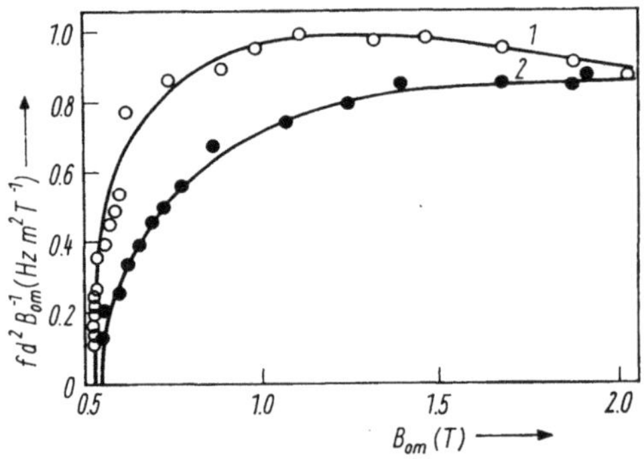

Fig. 5. Experimental dependences $fd^2 B_{0m}^{-1}$ vs. B_{0m} for n-type $Bi_{92}Sb_8$ (doped with 0.05 at% Te). (1) $\boldsymbol{B}_0 \parallel \boldsymbol{x}_1$; (2) $\boldsymbol{B}_0 \parallel \boldsymbol{x}_2$

in the valence band of $Bi_{92}Sb_8$ and therefore only a negligible number of holes is present in the L-valleys. This question will be discussed later on in connection with the investigation of p-type $Bi_{92}Sb_8$.

Now we pass to consideration of the properties of doped $Bi_{92}Sb_8$. The samples of n-type were obtained from tellurium-doped crystals. The experimental curves (Fig. 5) qualitatively agree with the model of electron valleys situated at L-points. When $\boldsymbol{B}_0 \parallel \boldsymbol{x}_1$ the cyclotron mobility $\mu_{c1b} = 1.87\ m^2\ V^{-1}\ s^{-1}$, and $\mu_{c3} \approx 0.6\ m^2\ V^{-1}\ s^{-1}$ when $\boldsymbol{B}_0 \parallel \boldsymbol{x}_3$. The electron concentration measured from the helicon dimensional resonance was $(0.8\ to\ 0.9) \times 10^{25}\ m^{-3}$. Since the tellurium ion concentration was $N_{Te} = 1.45 \times 10^{25}\ m^{-3}$ the donor efficiency of tellurium could be obtained: $n/N_{Te} = 0.55$ to 0.6 is in satisfactory agreement with galvanomagnetic data [22].

The sample of p-type $Bi_{92}Sb_8$ containing 0.01% Sn showed the following parameters: $\mu_c = 7.69\ m^2\ V^{-1}\ s^{-1}$ when $\boldsymbol{B}_0 \parallel \boldsymbol{x}_1$, $\mu_c = 8.0\ m^2\ V^{-1}\ s^{-1}$ when $\boldsymbol{B}_0 \parallel \boldsymbol{x}_2$ and $\mu_c = 0.56\ m^2\ V^{-1}\ s^{-1}$ when $\boldsymbol{B}_0 \parallel \boldsymbol{x}_3$; hole concentration $p = 3.25 \times \times 10^{23}\ m^{-3}$. The smallest cyclotron mobility of holes when $\boldsymbol{B}_0 \parallel \boldsymbol{x}_3$ shows that in contrast to pure bismuth the holes at T-points did not dominate. The explicit experimental data [17] suggest a model for holes in which large axes of effective mass ellipsoids are directed along the twofold axes. The number of such ellipsoids could be three or odd multiple of three. The hole valleys with such a symmetry in bismuth were investigated at helium temperature in [23]; some indications for the existence of such valleys in Bi–Sb at 4.2 K are given in [24]. These new valleys are localized apparently at the TW lines in the Brillouin zone. The energy separation between T, L, and TW valleys changes with the rise of temperature or the content of antimony in Bi–Sb and therefore in $Bi_{92}Sb_8$ at 77 K the new hole valleys can dominate the electrical conductivity. However, there are no data available on the energy separation between the valleys and the other parameters of $Bi_{92}Sb_8$ for a quantitative comparison of the experiment with the existing valence band structure models.

In p-$Bi_{92}Sb_8$ the damped helicon wave was also observed at room temperature but the phase of the measured signal $U(B_0)$ was opposite to that at $T = 77$ K. This shows that at room temperature the majority carriers (holes) make up only a low-mobility damping background, and the helicon wave dispersion is governed by electrons.

4. Anomalies in the Magnetoplasma Wave Propagation

There were observed some pecularities of the magnetoplasma wave propagation in $Bi_{92}Sb_8$ that overstep the limits of the conventional magnetoplasma wave theory.

(i) The angular dependence $U(\varphi) \approx U_0 + A \sin 3\varphi$ was observed when $\boldsymbol{B}_0 \parallel \boldsymbol{x}_3$ (Fig. 6). Such a dependence was observed earlier in $Bi_{95}Sb_5$ [25]. This is associated with the threefold symmetry of the crystal with respect to rotation about the axis x_3. However, the theory predicts no angular dependence when $\boldsymbol{B}_0 \parallel \boldsymbol{x}_3$ because all the electron ellipsoids for this direction are spaced symmetrically and $\varrho_{xx} = \varrho_{yy}$, $\varrho_{xy} = -\varrho_{yx}$. The dependence $U \sim \sin 3\varphi$ was observed in the magnetic field range where $1 \lesseqgtr \mu_{c3}B_0 < 10$ disappearing quickly with the increase of the magnetic induction B_0.

(ii) Non-reciprocal effects $U(\boldsymbol{B}_0) \neq U(-\boldsymbol{B}_0)$ were observed when $\boldsymbol{B}_0 \parallel \boldsymbol{x}_3$ under condition $\mu_{c3}B_0 \lesseqgtr 1$. These effects were especially clearly observed in Te- or Sn-doped crystals.

On the other hand the theory predicts the equality $U(\boldsymbol{B}_0) = U(-\boldsymbol{B}_0)$ because, when $\boldsymbol{B}_0 \parallel \boldsymbol{x}_3$, $\varrho_{xy}(\boldsymbol{B}_0) = -\varrho_{xy}(-\boldsymbol{B}_0)$ and the experimentally measured quantity $U(\boldsymbol{B}_0)$ is proportional to the modulus of the ϱ_{xy}, as it follows from (1). Such non-reciprocal effects could be caused by waves of any other nature; the phase of the signal induced by these waves would be insensitive to the magnetic field inversion. The signal induced by the magnetoplasma wave adds or subtracts with the signal of the other nature depending on the direction of $\boldsymbol{B}_0$.

We can assume that these anomalies result from the coupling between magnetoplasma and acoustic waves [26]. In our experiment the conditions $\omega \ll \omega_c$, $\omega \ll \nu$, and $\omega_c \approx \nu$ were satisfied, therefore direct electromagnetic generation of acoustic waves [27] was possible. If the velocity of the transverse acoustic wave propagating in $Bi_{92}Sb_8$ in the trigonal direction is the same as in bismuth, $v_s \approx 10^5$ cm s^{-1} [28], then for the frequency $f = 0.8$ MHz we have the acoustic wavelength $\lambda_s \approx 2d$, where d is the sample thickness. Thus the excitation of standing acoustic waves was possible and the amplitude of electromagnetic field coupled with the acoustic wave could not be small as compared to the amplitude of the magnetoplasma wave. The latter are highly damped when $\boldsymbol{k} \parallel \boldsymbol{B}_0 \parallel \boldsymbol{x}_3$ due to strong ellipticity of the electron cyclotron orbits [15]. In the cases $\boldsymbol{B}_0 \parallel \boldsymbol{x}_2$ and $\boldsymbol{B}_0 \parallel \boldsymbol{x}_1$ the damping of the helicon wave is smaller as compared to that of $\boldsymbol{B}_0 \parallel \boldsymbol{x}_3$ and probably therefore the perturbation connected with the acoustic waves for these directions was not observed. We do not present here a comparison of the experimental data with theory because the existing theory [27] does not include the anisotropy both of the mobility and

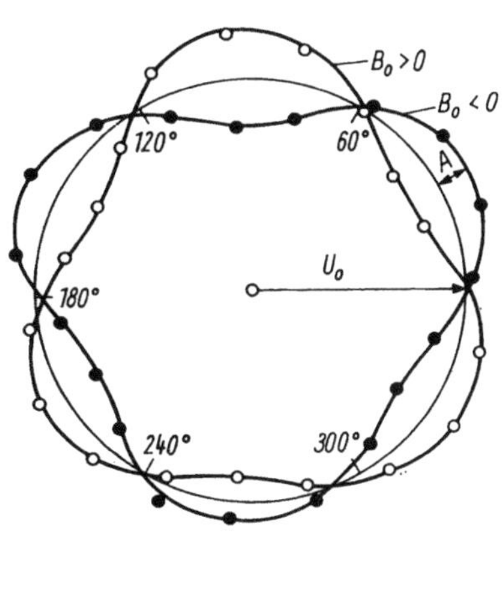

Fig. 6. Experimental dependences $U = U(\varphi)$ for pure $Bi_{92}Sb_8$ when $\boldsymbol{B}_0 \parallel \boldsymbol{x}_3$. $B_0 = 0.3$ T; $f = 0.8$ MHz

the elastic properties on the coupled acoustic and magnetoplasma wave dispersion and polarization.

5. Concluding Remarks

The magnetoplasma wave dispersion in Bi–Sb is found to be strongly affected by the mobility ansiotropy and the multi-valley structure of the energy spectrum. On its basis it is possible to verify the band structure model and to estimate the carrier mobility and conductivity parameters. In the investigated pure $Bi_{92}Sb_8$ alloys electrons are shown to dominate the electrical conductivity. Experimental results are in satisfactory agreement with the model of three ellipsoidal valleys at L-points of the Brillouin zone. The strong decrease of the electron cyclotron mobility observed in n-type $Bi_{92}Sb_8$, as compared to that of undoped $Bi_{92}Sb_8$, agrees with the non-parabolic dispersion law for electrons in the L-valleys. The existence of other filled valleys remains an open question but if they existed then their mobility would be much lower than that of L-electrons because their influence does not manifest itself markedly in the used magnetic fields. The experiment on p-type $Bi_{92}Sb_8$ shows essential difference between the valence band structure of this alloy and that of pure bismuth, namely, the T- and L-holes in $Bi_{92}Sb_8$ do not dominate the electrical conductivity.

Some anomalies observed in the magnetoplasma wave propagation are probably related to the electromagnetic generation of acoustic waves. They appear in polarization of the wave when $\boldsymbol{B}_0 \parallel \boldsymbol{x}_3$ and do not change the general features in magnetoplasma wave dispersion.

It should be noted that information on the peculiarities of band structure enters mainly the imaginary part of free carrier effective dielectrical permittivity. This imposes natural restrictions on the resonance linewidth and, consequently, on the precision of the method. The application of the above method for the investigation of fine effects as deviation from ellipticity etc., evidently, would be difficult. But the main peculiarities of band structure are truly reflected by this method. The cut-off of the resonance frequency dependence on magnetic induction at $\mu_c B_0 = 1$ allows to determine the cyclotron mobility for different crystallographic directions with great accuracy. This allows to investigate the band structure even in that temperature range where the cyclotron resonance method yielding similar information is already unsuitable due to collision damping.

Appendix

In what sense is the helicon dispersion equation related to the peculiarities of the semiconductor band structure?

It is a widely spread opinion that the helicon dispersion law within the limit $\mu B_0 \gg 1$ depends only on the value of the non-compensated charge $(n-p)e$ and it is not sensitive to the effective mass and to the peculiarities of the band structure (μ mobility, n, p electron and hole concentration, respectively, e elementary charge). This oversimplified model is suitable, strictly speaking, only for one-valley semiconductors with isotropic effective mass.

Let us consider a one-valley semiconductor with ellipsoidal isoenergetic surface. It is not an abstract model. For example, in n-type germanium, uniaxially compressed in [111] direction, under sufficiently high pressure all the electrons transfer into one [111] valley [29].

In agreement with [19], the equation of electromagnetic wave dispersion in a local approach is

$$k_\pm^2 = \frac{\omega\mu_0}{\varrho_{xx}\varrho_{yy} - \varrho_{xy}\varrho_{yx}} \left\{ -i\,\frac{\varrho_{xx} + \varrho_{yy}}{2} \pm \left[-\frac{(\varrho_{yy} - \varrho_{xx})^2}{4} - \varrho_{xy}\varrho_{yx} \right]^{1/2} \right\}, \tag{A1}$$

where displacement currents are neglected and the wave is supposed to be propagating in z-direction. In our experiment $\omega \ll \nu$, therefore the small-signal electrical resistivity tensor ϱ_{ij} depends only on the external magnetic field B_0. In the accepted one-valley model the mobility tensor for $B_0 = 0$ possesses only two components: μ_T and μ_L. When $\boldsymbol{B}_0$ is perpendicular to the great axis of mass ellipsoid, then we receive from (A1):

$$k_\pm^2 = \frac{\omega\mu_0 ne}{1 + \mu_c^2 B_0^2} \left\{ -i\,\frac{\mu_T + \mu_L}{2} \pm \left[-\frac{(\mu_L - \mu_T)^2}{4} + \mu_c^4 B_0^2 \right]^{1/2} \right\}, \tag{A2}$$

where $\mu_c = \sqrt{\mu_L \mu_T}$ is the cyclotron mobility. It is evident that even under the condition of $\mu_c^2 B_0^2 \gg 1$ the dispersion law remains sensitive to the mobility anisotropy. Helicon or, to be more exact, elliptically polarized waves appear under the condition

$$\mu_c^2 B_0^2 \gg \frac{(\mu_L - \mu_T)^2}{4\mu_c^2}, \tag{A3}$$

but even in this case the dispersion law

$$k_\pm^2 = \frac{\omega\mu_0 ne}{B_0} \left(\pm 1 - i\,\frac{\mu_T + \mu_L}{2\mu_L\mu_T B_0} \right) \tag{A4}$$

preserves data on mobility anisotropy. The condition of a weak helicon damping is

$$\mu_c B_0 \gg \frac{1}{2}\,\frac{\mu_T + \mu_L}{\mu_c}. \tag{A5}$$

If, e.g., $\mu_T \ll \mu_L$ then for a weak helicon damping the magnetic field should be $\mu_T/2\mu_c$ times higher than in the case of isotropic mobility. Thus, for L-electrons in bismuth (see Table 1) $\mu_1/\mu_{c3} \approx 12$ and the condition of a weak helicon damping (A5) at $\boldsymbol{B}_0 \parallel \boldsymbol{x}_3$ is

$$\mu_{c3} B_0 \gg 6\,.$$

The magnetic field necessary for satisfying (A5) can give rise to a shift of conductivity and valence band edges and a redistribution of current carriers among the valleys, quantum effects, etc. that will reflect themselves on the wave dispersion again. If a semiconductor possesses several groups of carriers with different cyclotron mobilities then (A4) cannot be satisfied for all groups of carriers simultaneously and the helicon dispersion law will contain information on the parameters of different carriers groups. This information can be got e.g. from the dependence of the helicon dimensional resonance frequency on the magnetic field. The function $k_\pm^2 = k_\pm^2(B_0)$ has a maximum at $\mu_c^2 B_0^2 = 1$

that allows to determine the cyclotron mobility. Thus the "helicon method" is of the same importance for determining of mobility anisotropy as the cyclotron resonance method for determining of effective mass anisotropy.

References

[1] S. Mase, J. Phys. Soc. Japan **13**, **434** (1958); **14**, 584 (1959).
[2] S. Golin, Phys. Rev. **176**, 830 (1968).
[3] L. Esaki, J. Phys. Soc. Japan Suppl. **21**, 89 (1966).
[4] N. B. Brandt, Ch. Dittmann, J. G. Ponomarev, and S. M. Chudinov, Zh. eksper. teor. Fiz., Pisma **11**, 250 (1970).
[5] N. B. Brandt, Ch. Dittmann, and J. G. Ponomarev, Fiz. tverd. Tela **13**, 2860 (1971).
[6] V. M. Grabov, G. A. Ivanov, V. L. Naliotov, V. S. Ponaryadov, and T. J. Yakovleva, Fiz. tverd. Tela **11**, 3653 (1969).
[7] H. J. Goldsmid, phys. stat. sol. (a) **1**, 7 (1970).
[8] M. G. Shakhtakhtinskii, D. S. Tomtiev, and A. A. Kuliev, Izv. Akad. Nauk Azerbaidzhan SSR, Ser. fiz.-tekh. mat. Nauk **4**, 27 (1968).
[9] M. I. Shakhtakhtinskaya and D. S. Tomtiev, phys. stat. sol. (b) **46**, 425 (1971).
[10] J. Neisecke and G. Schneider, Z. Naturf. **26a**, 1309 (1971).
[11] M. Dugué, phys. stat. sol. **11**, 149 (1965).
[12] A. A. Kuliev, M. G. Shakhtakhtinskii, and D. S. Tomtiev, Izv. Akad. Nauk Azerbaidzhan SSR, Ser. fiz.-tekh. mat. Nauk **3/4**, 17 (1967).
[13] T. Yazaki and Y. Abe, J. Phys. Soc. Japan **24**, 290 (1968); **25**, 633 (1968).
[14] J. K. Požela and R. B. Tolutis, Fiz. Tekh. Poluprov. **3**, 828 (1969).
[15] R. S. Brazis and J. K. Požela, Fiz. Tekh. Poluprov. **5**, 1009 (1971).
[16] R. S. Brazis and J. K. Požela, Liet. fiz. rink. **12**, 105 (1972); **12**, 510 (1972).
[17] R. S. Brazis, J. K. Požela, and B. A. Tairov, Fiz. Tekh. Poluprov. **6**, 2278 (1972).
[18] R. Tolutis, V. Riauka, J. Požela, and B. Tairov, Liet. fiz. rink. **13**, 534 (1973).
[19] P. A. Penz, J. appl. Phys. **38**, 4047 (1967); **39**, 1922 (1968).
[20] W. S. Boyle and G. E. Smith, Progr. Semicond. **7**, 1 (1960).
[21] T. C. Harman, J. M. Honig, and B. M. Tarmy, Proc. 7th Internat. Conf. Phys. Semicond., Paris 1964, Dunod, Paris 1964 (p. 407).
[22] D. V. Gitsu, G. A. Ivanov, V. I. Veraksa, B. P. Korolevski, and A. S. Fedorko, Izv. Akad. Nauk SSSR, Ser. neorg. Mater. **7**, 1063 (1971).
[23] M. Giura and R. Marcon, Phys. Rev. B **1**, 1528 (1970).
[24] N. B. Brandt and J. G. Ponomarev, Zh. eksper. teor. Fiz. **55**, 1215 (1968).
[25] R. S. Brazis and J. K. Požela, Liet. fiz. rink. **9**, 1111 (1969).
[26] E. A. Kaner and V. G. Skobov, Adv. Phys. **17**, 605 (1968).
[27] J. Mertsching, phys. stat. sol. **37**, 465 (1970).
[28] Y. Eckstein, A. W. Lawson, and D. H. Reneker, J. appl. Phys. **31**, 1534 (1960).
[29] A. K. Walton and S. F. Metcalfe, Infrared Phys. **13**, 183 (1973).

(Received December 10, 1973)

phys. stat. sol. (b) 62, 709 (1974)

Subject classification: 13.5.1 and 20.1; 22

Institute for Theoretical Physics, Academy of Sciences of the Ukrainian SSR, Kiev

A Theory of Two-Exciton Light Absorption in Antiferromagnetic Dielectrics

By

YU. B. GAIDIDEI and V. M. LOKTEV

The light absorption in antiferromagnetic dielectrics by the creation of two Frenkel excitons in the crystal is considered. It is shown that the absorption lines accompanied by the formation of bound states split into three perpendicularly polarized components. The form of the absorption spectrum both in the presence and absence of bound states for different polarizations of incident light, as well as the ratio of the intensities of the components are found. The object of the investigation is a crystal with a body-centered orthorhombic lattice.

Es wird die Lichtabsorption in antiferromagnetischen Dielektrika durch Bildung von zwei Frenkelexzitonen im Kristall betrachtet. Es wird gezeigt, daß die von der Bildung von gebundenen Zuständen begleiteten Absorptionslinien in drei zueinander senkrecht polarisierte Komponenten aufspalten. Die Form des Absorptionsspektrums bei Anwesenheit und Abwesenheit von gebundenen Zuständen für unterschiedliche Polarisation des einfallenden Lichts und das Verhältnis der Intensitäten der Komponenten wurden gefunden. Objekt der Untersuchung ist ein Kristall mit raumzentriertem rhombischem Gitter.

1. Introduction

The study of the optical properties of antiferrodielectrics (AFD) has resulted in the identification of a great number of new lines unobserved in nonmagnetic crystals and corresponding to a simultaneous creation of various quasi-particles, namely — two magnons or excitons, an exciton and a magnon [1]. It is essential that the pair absorption in AFD is connected with the excitation of two paramagnetic ions from different magnetic sublattices (the removal of spin forbiddenness in a pair [2, 3]). The double absorption in AFD (as well as in other crystals [4, 5]) is interesting, above all, in that it enables one to directly investigate the quasi-particle interaction which, as shown in a great number of papers [4 to 10], can lead to the appearence of new bound states of a pair. These states whose existence in certain cases may be considered to be experimentally proved [11, 12] manifest themselves in the form of narrow spectral lines of absorption [4, 5, 9].

While the two-magnon and exciton–magnon light absorption is investigated in detail (see, for example, [1]), the study of the two-exciton absorption in AFD has just been begun [12 to 15]. But the latter is of special interest just in that unlike the magnon–magnon and exciton–magnon interactions which have chiefly an exchange character the exciton–exciton interaction may have a Coulomb part [9]. This circumstance results in the fact that the bound states of two excitons are more feasible than the bound states of other quasi-particle pairs in AFD. The present paper is devoted to a study of the optical properties of the bound two-exciton states in AFD. Besides that, we investigate the shape

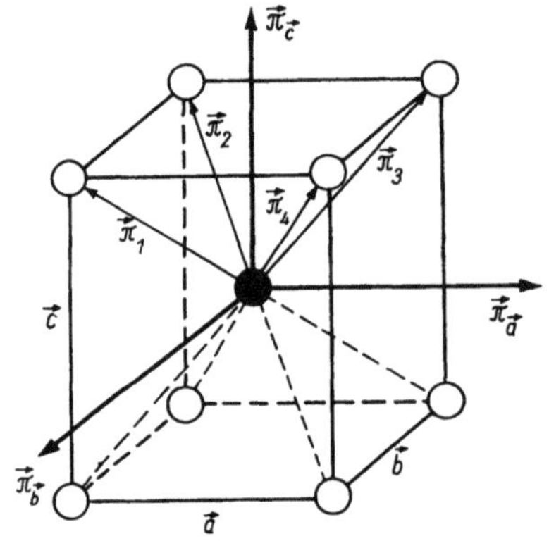

Fig. 1. Unit cell of AFD. The arrows indicate the directions of the dipole moments of transitions in ion pairs (π_1, π_2, π_3, π_4) and in the crystal (π_a, π_b, π_c). The light and dark circles are the ions of different sublattices

and polarization properties of wide (or continuous) absorption bands always accompanying two-particle excitations [4, 9]. Since a great number of AFD has a body-centered orthorombic lattice (Fig. 1), in what follows we dwell on the two-exciton absorption in crystals of just this symmetry.

2. The Hamiltonian and Dielectric Constant Tensor of AFD

Consider an AFD whose unit cell has two paramagnetic ions with oppositely directed spins. We assume that the ions are fixed in lattices points and have a strict spin orientation in the ground state. The lowest electron excited states of a paramagnetic ion correspond to the change in the value of its total spin by one, the energy of these states equalling $\approx 10^4\ \text{cm}^{-1}$. This allows one to restrict ourself to the processes, conserving the number of quasiparticles, in the Hamiltonian of a crystal. Therefore, the two-exciton states of an AFD are convenient to investigate using the creation $C_\nu^+(\boldsymbol{n};\boldsymbol{m})$ and annihilation $C_\nu(\boldsymbol{n};\boldsymbol{m})$ Bose-operators of double excitations [9]. The operator $C_\nu^+(\boldsymbol{n};\boldsymbol{m})$ corresponds to the creation of two excitations one of which is on the ion of the first magnetic sublattice in the $\boldsymbol{n}$-th cell, and the other is on the second magnetic sublattice in the $\boldsymbol{m}$-th cell. The index ν shows the parity of such state with respect to the permutation of excitations between the excited ions [9]. Using these operators the Hamiltonian of a crystal with two excited ions is represented as [9, 16]

$$\left.\begin{aligned}
H = {} & \sum_{\boldsymbol{n},\boldsymbol{m}} \sum_{\nu} [E_0 + V_{\boldsymbol{nm}}(\nu)]\, C_\nu^+(\boldsymbol{n};\boldsymbol{m})\, C_\nu(\boldsymbol{n};\boldsymbol{m}) + \\
& + \sum_{\boldsymbol{n},\boldsymbol{m},\boldsymbol{p}} \sum_{\nu,\nu'} C_\nu^+(\boldsymbol{n};\boldsymbol{m}) \{ M_{\boldsymbol{n}1,\boldsymbol{p}1}(\nu;\nu')\, C_{\nu'}(\boldsymbol{p};\boldsymbol{m}) + M_{\boldsymbol{m}2,\boldsymbol{p}2}(\nu;\nu')\, C_{\nu'}(\boldsymbol{n};\boldsymbol{p}) \} \;; \\
V_{\boldsymbol{nm}}(\nu) = {} & V_{1\boldsymbol{nm}}^{fg} - (-1)^\nu\, V_{2\boldsymbol{nm}}^{fg} \;; \\
M_{\boldsymbol{n}1,\boldsymbol{p}1}(\nu;\nu') = {} & M_{\boldsymbol{n}1,\boldsymbol{p}1}^{f} + (-1)^{\nu+\nu'}\, M_{\boldsymbol{n}1,\boldsymbol{p}1}^{g} \;; \\
M_{\boldsymbol{m}2,\boldsymbol{p}2}(\nu;\nu') = {} & M_{\boldsymbol{m}2,\boldsymbol{p}2}^{g} + (-1)^{\nu+\nu'}\, M_{\boldsymbol{m}2,\boldsymbol{p}2}^{f} \,.
\end{aligned}\right\} \tag{1}$$

Here the indices f and g denote the sets of quantum numbers of each of the excitations available in AFD, E_0 is the total excitation energy of a pair of free ions, $M_{\boldsymbol{n}\alpha,\boldsymbol{m}\alpha}^{f}$ and $M_{\boldsymbol{n}\alpha,\boldsymbol{m}\alpha}^{g}$ are the matrix elements of the resonance transfer of the excitations f and g in the α-th sublattice (in a collinear AFD the excitations do not go from one sublattice to another [16]); V_{1nm}^{fg} and V_{2nm}^{fg} are, respectively, the static and resonance interactions of the excited ions $\boldsymbol{n}1$ and $\boldsymbol{m}2$ [9].

The operator of interaction of AFD with light is characterized by the effective transition dipole moment $\boldsymbol{\Pi}_{\text{eff}}$ [3]. The value of $\boldsymbol{\Pi}_{\text{eff}}$ is obtained in the second order of pertubation theory by taking account the off-diagonal exchange interaction between paramagnetic ions from various magnetic sublattices of AFD. This interaction removes the spin forbiddenness, and makes a pair transition electrodipolar. The expression for the operator of an effective transition dipole moment is obtained in [9]. It has the form

$$\boldsymbol{\Pi}_{\text{eff}} = \sum_{\boldsymbol{n},\boldsymbol{m}} \boldsymbol{\pi}_{\boldsymbol{n},\boldsymbol{m}}\, C_2^+(\boldsymbol{n};\boldsymbol{m}) + \text{h.c.}\,, \tag{2}$$

where $\boldsymbol{\pi}_{\boldsymbol{n},\boldsymbol{m}}$ is the vector characterizing the dipole moment of transition of a pair. Because $\boldsymbol{\pi}_{\boldsymbol{n},\boldsymbol{m}}$ is defined by the exchange interaction between the ions, only the nearest-neighbour ions can be excited simultaneously by the light [2, 3]. In the AFD with a body-centered lattice every ion has eight neighbours of the other sublattice (Fig. 1). If the crystal has the inversion through origin, the symmetry properties of $\boldsymbol{\pi}_{\boldsymbol{n},\boldsymbol{m}}$ are such that

$$\boldsymbol{\pi}_{\boldsymbol{n},\boldsymbol{m}} = -\boldsymbol{\pi}_{\boldsymbol{m},\boldsymbol{n}}\,; \quad \boldsymbol{\pi}_{\boldsymbol{n},\boldsymbol{m}} = -\boldsymbol{\pi}_{-\boldsymbol{n},-\boldsymbol{m}}\,.$$

For definitnes, we assume that $\boldsymbol{\pi}_{\boldsymbol{n},\boldsymbol{m}}$ is directed along the line connecting the excited ions. Following (2), the operator H_{int} of interaction of the AFD with light is written as

$$H_{\text{int}} = \boldsymbol{\Pi}_{\text{eff}} \cdot \boldsymbol{E}(t) = \sum_{\boldsymbol{n},\boldsymbol{m}} (\boldsymbol{E}\boldsymbol{\pi}_{\boldsymbol{n},\boldsymbol{m}})\, C_2^+(\boldsymbol{n};\boldsymbol{m})\, \mathrm{e}^{i(\boldsymbol{Q}\boldsymbol{n}-\omega t)} + \text{h.c.}\,, \tag{3}$$

where $\boldsymbol{Q}$ and ω are the wave vector and the frequency of light, respectively. With the symmetry of $\boldsymbol{\pi}_{\boldsymbol{n},\boldsymbol{m}}$ chosen by us, only the operators of double excitations with $\nu = 2$ remain in the expressions (2) and (3), since the state corresponding to $\nu = 1$ is not excited by the light in the long-wave approximation. Going to an exciton representation using the transformation

$$C_\nu(\boldsymbol{n};\boldsymbol{m}) = \frac{1}{N} \sum_{\boldsymbol{K},\boldsymbol{q}} \mathrm{e}^{i\frac{\boldsymbol{K}}{2}(\boldsymbol{n}+\boldsymbol{m})}\, \mathrm{e}^{i\boldsymbol{q}(\boldsymbol{n}-\boldsymbol{m})}\, C_\nu(\boldsymbol{K};\boldsymbol{q})\,,$$

we rewrite (1) and (3) as

$$H = \sum_{\substack{\boldsymbol{K},\boldsymbol{q}\\ \nu}} E_{\boldsymbol{K}}(\boldsymbol{q})\, C_\nu^+(\boldsymbol{K};\boldsymbol{q})\, C_\nu(\boldsymbol{K};\boldsymbol{q}) + \frac{1}{N} \sum_{\substack{\boldsymbol{K},\boldsymbol{q},\boldsymbol{q}'\\ \nu}} V_\nu(\boldsymbol{q}-\boldsymbol{q}')\, C_\nu^+(\boldsymbol{K};\boldsymbol{q})\, C_\nu(\boldsymbol{K};\boldsymbol{q}')\,, \tag{4}$$

$$H_{\text{int}} = \sum_{\boldsymbol{\Delta},\boldsymbol{q}} \boldsymbol{\pi}_{\boldsymbol{\Delta},\boldsymbol{O}}\, \mathrm{e}^{i\boldsymbol{q}\boldsymbol{\Delta}} C_2^+(\boldsymbol{Q};\boldsymbol{q})\, \mathrm{e}^{-i\omega t} + \text{h.c.}\,; \tag{5}$$

$$E_{\boldsymbol{K}}(\boldsymbol{q}) = E_0 + \frac{1}{2}\left\{L_a \cos\frac{\boldsymbol{K}\boldsymbol{a}}{2} \cos \boldsymbol{q}\boldsymbol{a} + L_b \cos\frac{\boldsymbol{K}\boldsymbol{b}}{2} \cos \boldsymbol{q}\boldsymbol{b} + L_c \cos\frac{\boldsymbol{K}\boldsymbol{c}}{2} \cos \boldsymbol{q}\boldsymbol{c}\right\};$$

$$V_\nu(\boldsymbol{q}-\boldsymbol{q}') = \sum_{\boldsymbol{\Delta}} V_{\nu\boldsymbol{\Delta}\boldsymbol{O}}\, \mathrm{e}^{i(\boldsymbol{q}-\boldsymbol{q}')\boldsymbol{\Delta}}$$

$$L_a = 4(M^f_{\boldsymbol{n}\alpha,\boldsymbol{n}+\boldsymbol{a}\alpha} + M^g_{\boldsymbol{n}\alpha,\boldsymbol{n}+\boldsymbol{a}\alpha}) \text{ and so on}\,.$$

Here the vector $\boldsymbol{\Delta}$ connects the nearest-neighbours ions; L_a, L_b, and L_c are the double exciton band widths in the direction of the principal axes of a crystal. Since the values $M^f_{\boldsymbol{n}\alpha,\boldsymbol{m}\alpha}$ and $M^g_{\boldsymbol{n}\alpha,\boldsymbol{m}\alpha}$ of the resonance transfer of the excitation from one ion to another have an exchange character, and the interaction between excited ions is quadrupole–quadrupolar [9], we keep in (4) and (5) only the matrix elements relating to the nearest neighbours. Moreover, in the long-wave approximation the terms which do not conserve the index ν are proportional to $\boldsymbol{Qa}$, $\boldsymbol{Qb}$, and $\boldsymbol{Qc} \approx 10^{-3}$. Therefore, in what follows we shall keep in (4) only the terms with an optically active state $\nu = 2$, omitting the index ν.

The light absorption coefficient is expressed in terms of the imaginary part of the dielectric constant tensor, $\varepsilon''_{\mu\mu'}(\boldsymbol{Q}; \omega)$, which, in its turn, is related to the imaginary part of the Fourier transform of a retarded Green's function by

$$\varepsilon''_{\mu\mu'}(\boldsymbol{Q}; \omega) = -\frac{16\pi}{\Omega} \sum_{\boldsymbol{q},\boldsymbol{q}'} \boldsymbol{\pi}^{(\mu)}_{\Delta O} \boldsymbol{\pi}^{(\mu')}_{\Delta' O}\, \mathrm{e}^{i(\boldsymbol{q}\Delta - \boldsymbol{q}'\Delta')}\, \mathrm{Im}\, G(\boldsymbol{q}; \boldsymbol{q}'; \omega)\,, \tag{6}$$

$$G(\boldsymbol{q}; \boldsymbol{q}'; \omega) = \frac{1}{2\pi i} \int_0^\infty \mathrm{e}^{-i\omega t} \langle [C(\boldsymbol{Q}; \boldsymbol{q}; t), C^+(\boldsymbol{Q}; \boldsymbol{q}')] \rangle\, dt$$

(Ω is the crystal volume, and the operators $C(\boldsymbol{Q}; \boldsymbol{q})$ are taken in the Heisenberg representation with the Hamiltonian (4)). For our further investigation it is useful to introduce the dipole moments corresponding to the transitions of AFD into eigenstates, namely:

$$\left.\begin{aligned}
\boldsymbol{\pi}_a &= \boldsymbol{\pi}_1 - \boldsymbol{\pi}_2 - \boldsymbol{\pi}_3 + \boldsymbol{\pi}_4 \parallel \boldsymbol{a}\,; &\quad \boldsymbol{\pi}_1 &\equiv \boldsymbol{\pi}_{\Delta O} \quad \text{at} \quad \Delta = \frac{\boldsymbol{a}+\boldsymbol{b}+\boldsymbol{c}}{2}\,;\\
\boldsymbol{\pi}_b &= \boldsymbol{\pi}_1 + \boldsymbol{\pi}_2 - \boldsymbol{\pi}_3 - \boldsymbol{\pi}_4 \parallel \boldsymbol{b}\,; &\quad \boldsymbol{\pi}_2 &\equiv \boldsymbol{\pi}_{\Delta O} \quad \text{at} \quad \Delta = \frac{\boldsymbol{a}-\boldsymbol{b}+\boldsymbol{c}}{2}\,;\\
\boldsymbol{\pi}_c &= \boldsymbol{\pi}_1 + \boldsymbol{\pi}_2 + \boldsymbol{\pi}_3 + \boldsymbol{\pi}_4 \parallel \boldsymbol{c}\,; &\quad \boldsymbol{\pi}_3 &\equiv \boldsymbol{\pi}_{\Delta O} \quad \text{at} \quad \Delta = \frac{-\boldsymbol{a}+\boldsymbol{b}+\boldsymbol{c}}{2}\,;\\
\boldsymbol{\pi} &= \boldsymbol{\pi}_1 - \boldsymbol{\pi}_2 + \boldsymbol{\pi}_3 - \boldsymbol{\pi}_4 = 0\,; &\quad \boldsymbol{\pi}_4 &\equiv \boldsymbol{\pi}_{\Delta O} \quad \text{at} \quad \Delta = \frac{-\boldsymbol{a}-\boldsymbol{b}+\boldsymbol{c}}{2}\,.
\end{aligned}\right\} \tag{7}$$

On substituting (7) into (6), the tensor $\varepsilon''_{\mu\mu'}(\boldsymbol{Q}; \omega)$ becomes diagonalized, and its diagonal components are defined by:

$$\varepsilon''_{aa}(\boldsymbol{Q}; \omega) = -\frac{64\pi}{\Omega} |\boldsymbol{\pi}_a|^2 \sum_{\boldsymbol{q},\boldsymbol{q}'} \left\{\sin\frac{\boldsymbol{qa}}{2} \cos\frac{\boldsymbol{qb}}{2} \cos\frac{\boldsymbol{qc}}{2} \exp\left[i\,\frac{\boldsymbol{q}}{2}(\boldsymbol{a}+\boldsymbol{b}+\boldsymbol{c})\right]\right\} \times$$
$$\times \{\text{idem } (\boldsymbol{q} \to -\boldsymbol{q}')\}\, \mathrm{Im}\, G(\boldsymbol{q}; \boldsymbol{q}'; \omega) \equiv -\frac{8\pi}{v} |\boldsymbol{\pi}_a|^2\, \mathrm{Im}\, G_a(\omega)\,; \tag{8}$$

$$\varepsilon''_{bb}(\boldsymbol{Q}; \omega) = -\frac{64\pi}{\Omega} |\boldsymbol{\pi}_b|^2 \sum_{\boldsymbol{q},\boldsymbol{q}'} \left\{\cos\frac{\boldsymbol{qa}}{2} \sin\frac{\boldsymbol{qb}}{2} \cos\frac{\boldsymbol{qc}}{2} \exp\left[i\,\frac{\boldsymbol{q}}{2}(\boldsymbol{a}+\boldsymbol{b}+\boldsymbol{c})\right]\right\} \times$$
$$\times \{\text{idem } (\boldsymbol{q} \to -\boldsymbol{q}')\}\, \mathrm{Im}\, G(\boldsymbol{q}; \boldsymbol{q}'; \omega) \equiv -\frac{8\pi}{v} |\boldsymbol{\pi}_b|^2\, \mathrm{Im}\, G_b(\omega)\,; \tag{9}$$

$$\varepsilon''_{cc}(\boldsymbol{Q}; \omega) = -\frac{64\pi}{\Omega} |\boldsymbol{\pi}_c|^2 \sum_{\boldsymbol{q},\boldsymbol{q}'} \left\{\cos\frac{\boldsymbol{qa}}{2} \cos\frac{\boldsymbol{qb}}{2} \sin\frac{\boldsymbol{qc}}{2} \exp\left[i\,\frac{\boldsymbol{q}}{2}(\boldsymbol{a}+\boldsymbol{b}+\boldsymbol{c})\right]\right\} \times$$
$$\times \{\text{idem } (\boldsymbol{q} \to -\boldsymbol{q}')\}\, \mathrm{Im}\, G(\boldsymbol{q}; \boldsymbol{q}'; \omega) \equiv -\frac{8\pi}{v} |\boldsymbol{\pi}_c|^2\, \mathrm{Im}\, G_c(\omega) \tag{10}$$

($v \equiv \boldsymbol{a}[\boldsymbol{b}\times\boldsymbol{c}]$). The imaginary parts of the Green's functions $G_{\boldsymbol{a}}(\omega)$, $G_{\boldsymbol{b}}(\omega)$, and $G_{\boldsymbol{c}}(\omega)$ introduced into (8) to (10) define the absorption of the light polarized along the principal axes $\boldsymbol{a}$, $\boldsymbol{b}$, and $\boldsymbol{c}$, respectively.

To find $G_{\boldsymbol{a}}(\omega)$, $G_{\boldsymbol{b}}(\omega)$, and $G_{\boldsymbol{c}}(\omega)$ the equation of motion for $G(\boldsymbol{q};\boldsymbol{q}';\omega)$,

$$G(\boldsymbol{q};\boldsymbol{q}';\omega) = [\omega - E_Q(\boldsymbol{q})]^{-1}\left\{\delta_{\boldsymbol{q}\boldsymbol{q}'} + \frac{1}{N}\sum_{\boldsymbol{q}''} V(\boldsymbol{q}-\boldsymbol{q}'')\, G(\boldsymbol{q}'';\boldsymbol{q}';\omega)\right\},$$

must be multiplied at the left and right by the factors written out in (8) to (10) and summed over the wave vectors $\boldsymbol{q}$ and $\boldsymbol{q}'$ of the first Brillouin zone. On performing this procedure, we find that

$$\left.\begin{aligned}
G_{\boldsymbol{a}}(\omega) &= \frac{V^{-1}}{1 - V\{J_{000} - J_{111} + J_{001} - J_{110} - J_{100} + J_{001} - J_{101} + J_{010}\}} - V^{-1};\\
G_{\boldsymbol{b}}(\omega) &= \frac{V^{-1}}{1 - V\{J_{000} - J_{111} + J_{001} - J_{110} + J_{100} - J_{011} + J_{101} - J_{010}\}} - V^{-1};\\
G_{\boldsymbol{c}}(\omega) &= \frac{V^{-1}}{1 - V\{J_{000} - J_{111} - J_{001} + J_{110} + J_{100} - J_{011} - J_{101} + J_{010}\}} - V^{-1},
\end{aligned}\right\} \quad (11)$$

where

$$\left.\begin{aligned}
J_{nmp} \equiv J_{nmp}(\omega) &= \frac{1}{\pi^3}\int_0^{\pi}\!\!\int_0^{\pi}\!\!\int_0^{\pi} \frac{\cos nx \cos my \cos pz \,\mathrm{d}x\,\mathrm{d}y\,\mathrm{d}z}{\omega - E_0 - \frac{1}{2}(L_{\boldsymbol{a}}\cos x + L_{\boldsymbol{b}}\cos y + L_{\boldsymbol{c}}\cos z)},\\
&(n, m, p = 0, 1)
\end{aligned}\right\} \quad (12)$$

and V is the interaction of the nearest excited neighbours.

3. Bound Two-Exciton States

As mentioned above, the two-exciton absorption spectrum may consist of two parts — the narrow lines which correspond to the formation of bound ("one-particle") states of a pair in a crystal, and the wide band absorption whose width is defined by the quasi-particle band widths [4, 9]. And the stronger the bound state, the weaker the manifestation of a continuous absorption in the spectrum. In general, the bound states of two Frenkel excitons can occur, when the interaction between excitons exceeds the one-exciton band widths [9, 10]. Since the interaction between excitons has not only an exchange part, but also a Coulomb one, and the exciton band widths are defined by the interaction which has exclusively an exchange character [9] (moreover, the excited ions are situated closer than the ions of the same sublattice (Fig. 1)), it seems quite reasonable to consider a case when

$$|V| > |L_{\boldsymbol{a}}| + |L_{\boldsymbol{b}}| + |L_{\boldsymbol{c}}|\,. \quad (13)$$

Then in the frequency range $|\omega - E_0| > |L_{\boldsymbol{a}}| + |L_{\boldsymbol{b}}| + |L_{\boldsymbol{c}}|$ the functions $G_{\boldsymbol{a}}(\omega)$, $G_{\boldsymbol{b}}(\omega)$, and $G_{\boldsymbol{c}}(\omega)$ have their poles on the real axis. The derivation of analytic expression for the poles can be conviniently carried out in dimensionless quantities:

$$x \equiv \frac{\omega - E_0}{L};\quad a \equiv \frac{L_{\boldsymbol{a}}}{2L};\quad b \equiv \frac{L_{\boldsymbol{b}}}{2L};\quad c \equiv \frac{L_{\boldsymbol{c}}}{2L};\quad \alpha \equiv \frac{V}{L};\quad L \equiv \frac{1}{2}\sqrt{L_{\boldsymbol{a}}^2 + L_{\boldsymbol{b}}^2 + L_{\boldsymbol{c}}^2}\,. \quad (14)$$

Then in the range

$$J_{nmp}(x) \equiv LJ_{nmp}(\omega) = \int\limits_0^\infty \mathrm{d}t\ \mathrm{e}^{-xt}\ \mathrm{I}_n(at)\ \mathrm{I}_m(bt)\ \mathrm{I}_p(ct)\ ,$$

where $\mathrm{I}_n(at)$ . . . is the Bessel function of the imaginary argument. By representing $J_{nmp}(x)$ as an inverse power series in x and using the theorem on inversion of series [17], we obtain that the poles of the Green's functions $G_{a,b,c}(\omega)$ in the above frequency range are defined by

$$X_a = \alpha + \frac{b + c - a}{2} + \frac{1}{4\alpha} + \cdots ;$$

$$X_b = \alpha + \frac{a + c - b}{2} + \frac{1}{4\alpha} + \cdots ;$$

$$X_c = \alpha + \frac{a + b - c}{2} + \frac{1}{4\alpha} + \cdots .$$

To the poles of the Green's functions situated on the real axis there correspond δ-like peaks in the light absorption spectrum, and here the state with energy X_a is excited by the light polarized along the axis $\boldsymbol{a}$, and if $\boldsymbol{E} \parallel \boldsymbol{b}$ or $\boldsymbol{E} \parallel \boldsymbol{c}$ then X_b or X_c, respectively, is excited. These polarized absorption peaks correspond to the stationary states of a crystal which, because of the dependence of their energies on the wave vector $\boldsymbol{Q}$, may be attributed to the exciton type.

It is to be emphasized that the values of the splittings a–b, b–c and a–c are defined by the total one-exciton band widths. But the splitting itself can occur only for double excitations in which the excited centres are spatially separated (this splitting will not be observed, for example, in the case of two-phonon or vibron absorptions).

Note also that with our assumption on the direction of the transition dipole moment of a pair, the ratio of the intensities of the polarized two-exciton absorption lines is equal to the ratio of the squares of the lattice periods $|\boldsymbol{a}|^2$: $|\boldsymbol{b}|^2$:$|\boldsymbol{c}|^2$. This fact is quite interesting in that it relates the optical properties to the size of a crystal cell.

(11) and (12) yield the following particular cases. If the AFD has a body-centered orthorhombic lattice ($|\boldsymbol{a}| = |\boldsymbol{b}|$), then the bound states in the absorption spectrum will be represented by the two lines: 1. $X_a = X_b = \alpha + (c/2) + \cdots$ and the absorption is isotropic in the ab-plane ($\boldsymbol{E} \perp \boldsymbol{c}$); 2. $X_c = \alpha + a - (c/2) + \cdots$ for $\boldsymbol{E} \parallel \boldsymbol{c}$. For cubic crystals all the levels are degenerate $X_a = X_b = X_c = \alpha + a/2 + \cdots$ and the absorption is entirely isotropic.

The available possibility to define the exciton band widths using the two-exciton absorption spectra seems quite important. The calculations of magnon band widths [1] and triplets exciton bands [18] show that the splitting will be of 5 to 10 cm^{-1}.

4. Particular Cases Allowing an Exact Solution

Since the problem (the definition of the absorption spectrum with arbitrary parameters) cannot be completely solved, we consider particular cases A and B, labelling the physical quantities with relevant indices.

4.1 Case A

Suppose that the structure of ADF is such that the inequality $a \gg b, c$ holds. In this case one can obtain exact solutions both for the bound states and the wide band. In (11) and (12) we set $L_b = L_c = 0$ for the functions $G_a(\omega)$, $G_b(\omega)$, and $G_c(\omega)$ in the dimensionless units of (14), obtaining

$$G_a(\omega) = \frac{V^{-1}}{1 - V\,(J_0 - J_1)} - V^{-1} \equiv \frac{1}{L}\, G_1^{(A)}(x)\,; \tag{15}$$

$$G_b(\omega) = G_c(\omega) = \frac{V^{-1}}{1 - V\,(J_0 + J_1)} - V^{-1} \equiv \frac{1}{L}\, G_2^{(A)}(x)\,; \tag{16}$$

$$J_n \equiv \frac{1}{\pi} \int\limits_0^{\pi} \frac{\cos nx}{\omega - E_0 - L\cos x}\, dx\,; \qquad (n = 0, 1)\,.$$

For the function $G_1^{(A)}(x)$ (15), the solution in the range $|x| > 1$ exists subject to

$$\frac{\alpha - 1}{\alpha} > 0$$

and the root itself is equal to

$$X_1^{(A)} = \frac{2\alpha^2}{2\alpha - 1} - 1\,.$$

For the function $G_2^{(A)}(x)$ (16), the solution has the form

$$X_2^{(A)} = \frac{2\alpha^2}{2\alpha + 1} + 1$$

and exists with

$$\frac{\alpha + 1}{\alpha} > 1\,.$$

In both the cases the solutions exist at $|\alpha| > 1$, and there is always at least one bound state (in one of the polarizations) at any finite α. The splitting between the poles $X_2^{(A)} - X_1^{(A)} \approx 2$ is defined by the wide band width. The imaginary parts of the Green's functions (15) and (16) are equal to

$$\operatorname{Im} G_{1,2}^{(A)}(x) = -2\pi\, \frac{4\alpha(\alpha - 1)}{(2\alpha^2 \mp 1)^2}\, \delta(x - X_{1,2}^{(A)})\,.$$

The wide band absorption region $|x| \leqq 1$ is defined by the formulas

$$\operatorname{Im} G_{1,2}^{(A)}(x) = -\frac{\sqrt{1 - x^2}}{(1 \pm x)\,(1 \mp \alpha)^2 + \alpha^2(1 \mp x)}\,.$$

These formulas show that the band absorption vanishes at the edges of the band, and has its maxima at the points $x \approx 1/X_{1,2}^{(A)}$, respectively. It is also seen that the integral intensity of the absorption by the bound states tends to unity, as α increases, and the intensity of the absorption by the band states is $\approx \alpha^{-2}$. The total integral intensity in both the polarizations is conserved and equal to 1.

4.2. Case B

Let $a = b \gg c$. This case applies to a great number of AFD [1] which has a body-centered orthorhombic lattice. Since the values a, b, and c have an exchange character, and die fast with distance, one can assume that case B is close to the real situation. To obtain the result sought for, we set $L_c = 0$ in (12). Then we have for the Green's functions $G_a(\omega)$, $G_b(\omega)$, and $G_c(\omega)$:

$$G_a(\omega) = G_b(\omega) = \frac{V^{-1}}{1 - V(J_{00} - J_{11})} - V^{-1} \equiv \frac{1}{L} G_\perp^{(B)}(x),$$

$$G_c(\omega) = \frac{V^{-1}}{1 - V(J_{00} + J_{11} - 2J_{10})} - V^{-1} \equiv \frac{1}{L} G_{\parallel}^{(B)}(x),$$

$$J_{nm} \equiv J_{nm}(\omega) = \frac{1}{\pi^2} \int_0^\pi \int_0^\pi \frac{\cos nx \cos my \, dx \, dy}{\omega - E_0 - L(\cos x + \cos y)}; \qquad (17)$$

$$(n, m = 0, 1).$$

By going to the dimensionless quantities (14) and calculating the integral (17), we obtain that in the region $|x| > 2$ the poles of the functions $G_\perp^{(B)}(x)$ and $G_{\parallel}^{(B)}(x)$ are defined by

$$\frac{1}{2} x \mathrm{E}\left(\frac{2}{x}\right) - \left(\frac{x}{2} - \frac{2}{x}\right) \mathrm{K}\left(\frac{2}{x}\right) = \frac{\pi}{2\alpha} \qquad \text{for} \quad G_\perp^{(B)}(x); \qquad (18)$$

$$\mathrm{K}\left(\frac{2}{x}\right) + \frac{x}{2}\left[\mathrm{K}\left(\frac{2}{x}\right) - \mathrm{E}\left(\frac{2}{x}\right)\right] = \frac{\pi}{2\alpha}(\alpha + 1) \qquad \text{for} \quad G_{\parallel}^{(B)}(x), \qquad (19)$$

where K and E are the complete elliptic integrals of the first and second kinds, respectively.

The investigation of (18) shows that it has a single solution provided the condition $|\alpha| > (\pi/2)$ is met. The expression for the root of this equation has the form

$$X_\perp^{(B)} = \alpha + \frac{1}{2\alpha} + O\left(\frac{1}{\alpha^2}\right),$$

and the imaginary part of the function $G_\perp^{(B)}(x)$ in this case is defined by

$$\operatorname{Im} G_\perp^{(B)}(x) = -\pi\left(1 - \frac{1}{2\alpha^2}\right) \delta(x - X_\perp^{(B)}).$$

The condition of existence of the bound state when the light is polarized along **c**, or the solution of (19), depends upon the sign of α. If α is negative, then the solution exists at

$$\alpha < -\pi(\pi - 2)^{-1}.$$

But if α is positive, then the solution exists at any finite α. For $|\alpha| > 3$ the position of the root of (19) is defined by

$$X_{\parallel}^{(B)} = \alpha + 1 + \frac{1}{2\alpha} + O\left(\frac{1}{\alpha^2}\right)$$

and

$$\operatorname{Im} G_{\parallel}^{(B)}(x) = -\pi\left(1 - \frac{1}{2\alpha^2}\right) \delta(x - X_{\parallel}^{(B)}).$$

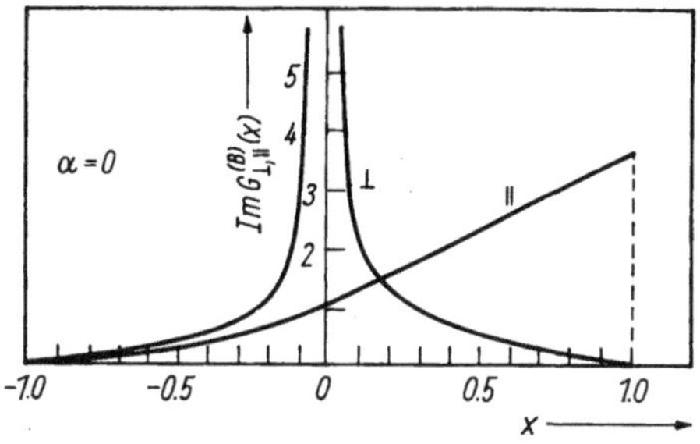

Fig. 2. The absorption spectrum for different polarizations of light at $\alpha = 0$

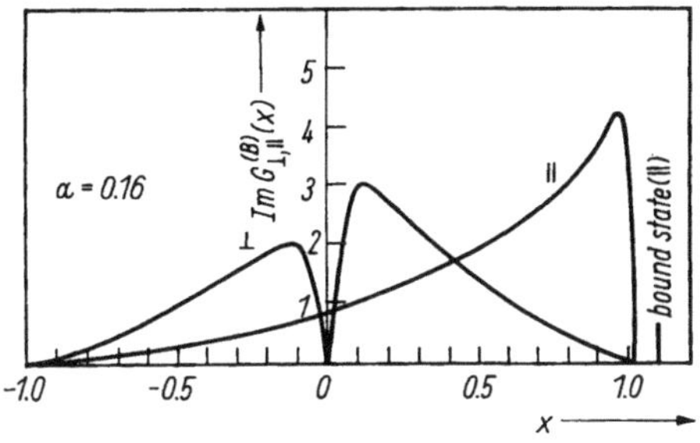

Fig. 3. The absorption spectrum for different polarizations of light at $\alpha = 0.16$

When α is a small positive value, the position of the pole is given by

$$x_{\parallel}^{(B)} \approx 1 + \exp\left(-\frac{\pi}{2\alpha}\right),$$

and the Green's function corresponding to it reads

$$\operatorname{Im} G_{\parallel}^{(B)}(x) \approx -\frac{\pi^2}{2\alpha} \exp\left(-\frac{\pi}{2\alpha}\right) \delta(x - X_{\parallel}^{(B)}) .$$

The latter expression shows that when the exciton–exciton interaction exceed little the threshold values, the intensity of the absorption accompanied by the creation of bound states is exponentially small, and practically all the absorption is defined by the wide band part (Fig. 3).

Consider the band absorption of light. Direct calculations show that in the region $|x| \leqq 2$

$$\operatorname{Im} G_{\perp}^{(B)}(x) = -\frac{2}{\pi} \frac{K'(x) - E'(x)}{\left[1 - \frac{2\alpha}{\pi} \frac{|x|}{x} E(x)\right]^2 + \left(\frac{2\alpha}{\pi}\right)^2 [K'(x) - E'(x)]^2}; \tag{20}$$

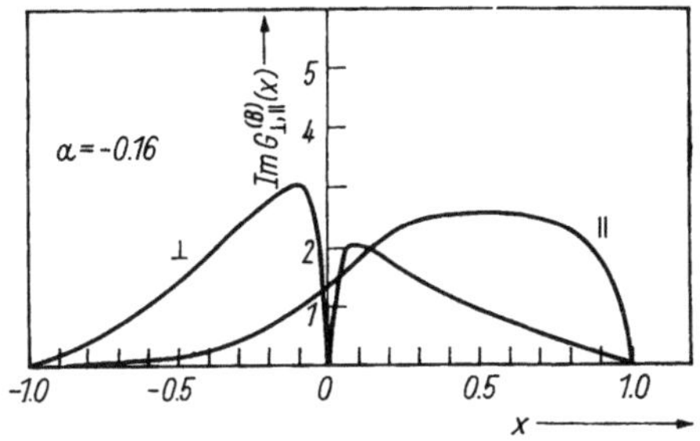

Fig. 4. The absorption spectrum for different polarizations of light at $\alpha = -0.16$

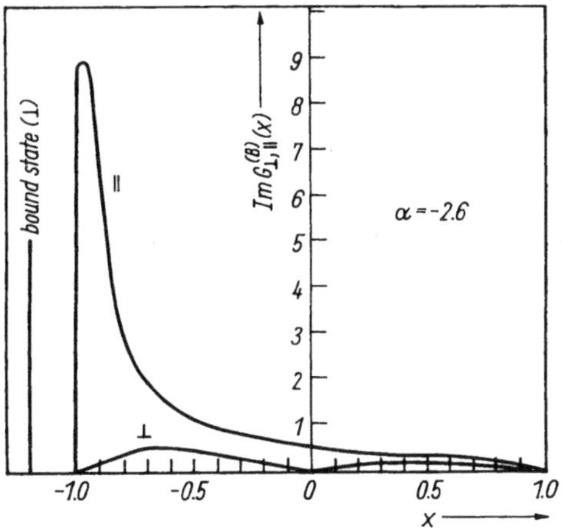

Fig. 5. The absorption spectrum for different polarizations of light at $\alpha = -2.6$

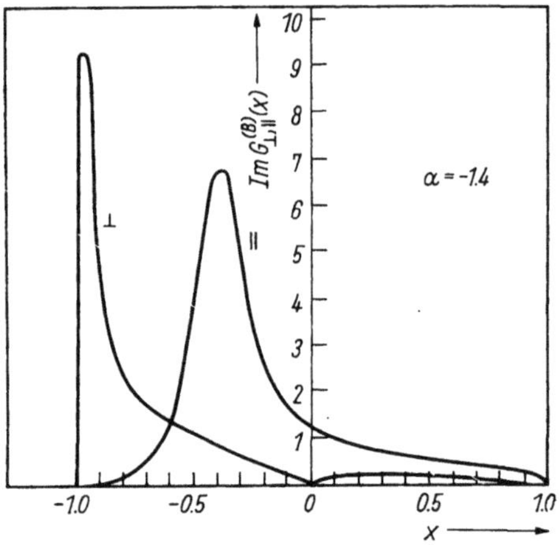

Fig. 6. The absorption spectrum for different polarizations of light at $\alpha = -1.4$

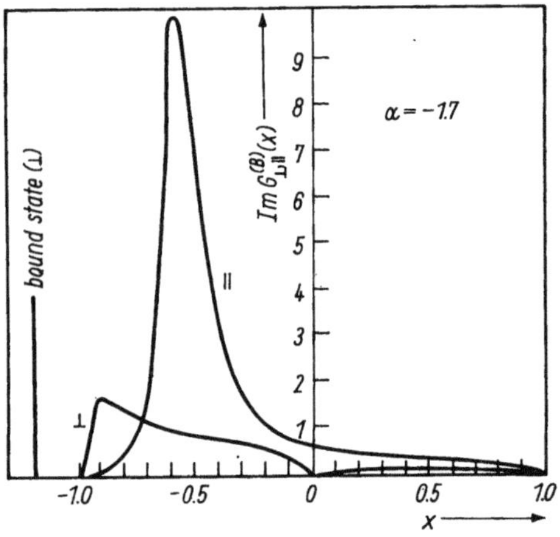

Fig. 7. The absorption spectrum for different polarizations of light at $\alpha = -1.7$

$$\operatorname{Im} G_{||}^{(B)}(x) = -\frac{2}{\pi} \frac{x\mathrm{K}'(x) + \mathrm{E}'(x)}{\left\{1 + \alpha - \frac{2\alpha}{\pi} \frac{|x|}{x} [\mathrm{E}(x) - (x+1)\mathrm{K}(x)]\right\}^2 + \left(\frac{2\alpha}{\pi}\right)^2 [x\mathrm{K}'(x) + \mathrm{E}'(x)]^2}, \tag{21}$$

where $\mathrm{K}'(x) \equiv \mathrm{K}(x')$, $\mathrm{E}'(x) \equiv \mathrm{E}(x')$, and $x' = \sqrt{1 - x^2}$. The shape of the absorption curves when the exciton–exciton interaction is absent is given in Fig. 2. It is seen that the curve (20) has a cut, and becomes infinite at the band centre, while the curve (21) remains finite. At any finite α both the curves vanish at the band edges $|x| = 1$. The forms of the absorption spectra for different values of α are represented by the plots showing that the shape of the absorption curve is essentially dependent on the polarization of the absorbed light. In a number of cases (Fig. 5 to 7) the continuous absorption curves have pronounced maxima whose position changes with changing polarization. Calculations show that if the exciton interaction is small and positive, then at $\boldsymbol{E} \perp \boldsymbol{c}$ the curve (20) has two maxima at the points $x \approx \pm\exp(-\pi/2\alpha)$ near the band centre, and the function (21) has its maximum near the right-hand edge of the band (Fig. 3). If $\alpha = -\pi/(\pi - 2)$, the function (21) has a maximum near the left-hand edge of the band absorption (Fig. 5).

Thus the band absorption has pronounced maxima when the interaction parameter α is close to the values which define the existence conditions in the system of bound states. At $|\alpha| \gg 1$ the wide band absorption is very low in intensity and practically structureless.

5. Conclusion

The results obtained indicate a sharp dependence of the two-exciton absorption spectrum of AFD on the light polarization. It is noteworthy that the band splitting is connected with the values defined by one-exciton processes — the excitation transfer from one ion to another. But the splitting manifests itself only in the two-exciton absorption. The integral intensity of the bands depending upon the polarization is defined by the ratio of the squares of the lattice periods, and this allows one to compare the two-exciton absorption in various AFD.

Furthermore, it is to be noted that if the AFD has a face-centered lattice, then the number of polarized bands may decrease. In particular, for plane structures ($NiCl_2$, K_2NiF_4, α-O_2) the spectrum can consist of the band doublets orthogonally polarized when the light falls perpendicularly to the layers.

References

[1] V. V. Eremenko and A. I. Belyaeva, Uspekhi fiz. Nauk **98**, 27 (1969).
[2] I. Halley and S. S. Ivera, Phys. Rev. Letters **15**, 565 (1965).
[3] Y. Tanabe, T. Moriya, and S. Sugano, Phys. Rev. Letters **15**, 1023 (1965).
[4] E. I. Rashba, Zh. eksper. teor. Fiz. **50**, 1064 (1966); **54**, 978 (1968).
[5] V. M. Agranovich, Fiz. tverd. Tela **12**, 562 (1970).
[6] S. A. Moskalenko, Optika i Spektroskopiya **5**, 146 (1958).
[7] H. van Kranendonk, Physica **25**, 1080 (1959).
[8] J. Freeman and P. Hopfield, Phys. Rev. Letters **21**, 910 (1968).
[9] E. G. Petrov, V. M. Loktev, and Yu. B. Gaididei, phys. stat. sol. **41**, 117 (1970).
[10] M. A. Koyushner, Zh. eksper. teor. Fiz. **60**, 220 (1971).
[11] V. L. Broude, E. I. Rashba, and E. F. Sheka, phys. stat. sol. **19**, 395 (1967).
[12] A. F. Prikhotko and O. S. Pakhomova, Ukr. fiz. Zh. **13**, 19 (1968).
[13] J. Ferguson, Austral. J. Chem. **21**, 307 (1968); **21**, 323 (1968).
[14] P. Van der Ziel and H. Guggenheim, Phys. Rev. **166**, 479 (1958).
[15] S. Stokowski and D. Seel, Phys. Rev. **B4**, 3541 (1971).
[16] E. G. Petrov and V. M. Loktev, Zh. eksper. teor. Fiz. **61**, 1570 (1971).
[17] M. A. Lavrentev and B. V. Shabat, The Methods of Theory of Complex Variables, Chapter 5, Nauka, Moscow 1965.
[18] S. Choi, J. Jortner, and S. Rice, J. chem. Phys. **41**, 3294 (1964).

(Received December 3, 1973)

phys. stat. sol. (b) 62, 721 (1974)

Subject classification: 13; 22.5

Institute of Physics, Polish Academy of Sciences, Warsaw

Low-Symmetry Crystal Fields in $CrCl_3$

By

M. MAJER and H. SZYMCZAK

A crystal field study has been performed for $CrCl_3$ crystals. The energy matrix has been diagonalized within the 120 states of the complete d^3 configuration. In the calculation the trigonal field, spin-orbit interaction as well as Tree's correction have been taken into account. The results of the calculations have been compared with the experimental data on the specific magnetic rotation, optical absorption, and EPR spectra.

Es wurde eine Kristallfelduntersuchung für $CrCl_3$-Kristalle durchgeführt. Die Energiematrix wurde innerhalb der 120 Zustände der vollständigen d^3-Konfiguration diagonalisiert. In der Berechnung wurde sowohl das trigonale Feld, die Spin–Bahn-Wechselwirkung als auch die Treesche Korrektur berücksichtigt. Die Rechenergebnisse wurden mit experimentellen Werten der spezifischen magnetischen Drehung, der optischen Absorption und mit EPR-Spektren verglichen.

1. Introduction

The crystals of chromium trihalides exhibit interesting magneto-optical properties. The specific magnetic rotation for these crystals is very large, e.g. it amounts to over 600,000°/cm at the band edge for $CrBr_3$. The figure of merit in chromium trihalides is also very large thus offering a possibility of technical and experimental applications. To account for the nature of the magneto-optical properties of these crystals, the crystal field acting on chromium ions should be determined. In general, the crystal field parameters are calculated from optical absorption spectra and electron paramagnetic resonance (EPR) spectra. The majority of experimental data on optical and EPR spectra refer to chromium trichloride ($CrCl_3$) crystals. Until now the crystal field spectrum of $CrCl_3$ has been interpreted only in the cubic [1] and trigonal [2] field approximation.

In this study, we present the results of a detailed crystal field calculation for chromium ions in $CrCl_3$ crystals, including the effects of spin–orbit coupling disregarded in the previous calculations [1, 2]. Therefore, it was also possible to calculate the parameters of the spin Hamiltonian determined by the EPR method.

The symmetry of $CrCl_3$ crystals (below 238 °K) corresponds to the space group C_{3i}^2. The point symmetry of Cr^{3+} ions is C_3. Chromium trichloride is metamagnetic, however, a magnetic field of the intensity of only few thousands oersteds is sufficient to induce a ferromagnetic ordering.

2. Calculations

The Hamiltonian for Cr^{3+} ions in $CrCl_3$ crystals can be presented in the form

$$\mathcal{H} = \mathcal{H}_1(Dq) + \mathcal{H}_2(B, C) + \mathcal{H}_3(v, v') + \mathcal{H}_4(\xi) , \qquad (1)$$

where B and C are the Racah parameters; Dq cubic crystal field parameter; v, v' trigonal crystal field parameters; ξ spin–orbit coupling parameter.

The diagonalization of the Hamiltonian (1) was performed within the 120 states of the complete d^3 configuration in the weak-field basis, using the method described by Macfarlane [3]. Unlike in the Macfarlane calculations [3], Tree's parameter α was inserted in the matrices.

The energy matrix was diagonalized for various values of the parameters B, C, Dq, v, v', ξ and α. The results of these calculations were used for interpretation of the absorption and specific magnetic rotation spectra measured by Limido et al. [2] and the EPR spectra obtained by Egorov and Yablokov [4]. The best fit with experimental results has been found for the following values of theoretical parameters (in cm^{-1}):

$$B = 520,\ C = 3520,\ Dq = 1435,\ v = -1800,\ v' = 400,\ \xi = 200,\ \alpha = 20.$$

In Table 1, the results of calculations are compared with the experimental data [2]. In this table, the energy levels are denoted by irreducible representations of the unitary double group C_3^* ($\overline{E}_+$, $\overline{E}_-$ and $2\overline{A}$). In the absence of magnetic field, the energy levels $\overline{E}_+$ and $\overline{E}_-$ are degenerate and will be denoted by $\overline{E}$. The obtained set of parameters accounts satisfactorily for not only the optical and specific magnetic rotation spectra but also for the spin Hamiltonian parameters. The spin Hamiltonian for Cr^{3+} ions in trigonal crystal field has the form

$$\mathcal{H}_s = \beta[g_{\parallel} H_z S_z + g_{\perp}(H_x S_x + H_y S_y)] + D(S_z^2 - \tfrac{5}{4}), \tag{2}$$

where $[H_x, H_y, H_z]$ is the vector of the external magnetic field. The parameters $g_{\parallel}$ and $g_{\perp}$ were measured previously [4] and the following values were obtained:

$$g_{\parallel} = 1.984 \pm 0.001\ ,$$
$$g_{\perp} = 1.989 \pm 0.001\ .$$

The parameter D was semi-empirically determined by Dillon et al. [5] from the magnetic anisotropy constant K_1 and estimated as $D = -0.1\ cm^{-1}$. Unfortunately, this result is very inaccurate. It follows from our calculations that $D = -0.2\ cm^{-1}$. Calculation of the parameters $g_{\parallel}$ and $g_{\perp}$ was carried out using the eigenvectors of the Hamiltonian (1). Knowing the exact wave functions $\overline{E}_+$ and $\overline{E}_-$ of the

Table 1

Calculated and observed energy levels (in cm^{-1}) of Cr^{3+} in $CrCl_3$

calculated	observed (from [2])	attribution
13715	13904	$2\overline{A}$
13746		$\overline{E}$
14252	14249	$\overline{E}$
14252		$2\overline{A}$
14470	14425	$\overline{E}$
14502	14513	$\overline{E}$
14600	14641	$2\overline{A}$
14729	14848	$\overline{E}$
14947		$\overline{E}$
15056	14936	$2\overline{A}$
15077	15267	$\overline{E}$
19237	19193	$\overline{E}$
19253		$2\overline{A}$
19818		$\overline{E}$
19856	19821	$2\overline{A}$
19887		$\overline{E}$
19917	20000	$\overline{E}$

ground state, the matrix elements of the magnetic moment $\bar{\mu}$ operator can be calculated:

$$\bar{\mu} = k\bar{L} + g_s\bar{S} , \tag{3}$$

where k is the orbital reduction factor; g_s is the free-spin g value ($g_s = 2.0023$).

In this way, the following values of the parameters $g_{||}$ and $g_{\perp}$ were obtained:

$$g_{||} = 1.985 , \quad g_{\perp} = 1.989 , \text{ and } k = 0.4 .$$

3. Discussion

An abnormally small value of k is to be compared with a low value of the Racah parameter B. In both cases, the decrease has been related to covalency. Explanation of the influence of covalency on the energy spectrum of chromium trihalides seems to be essential for understanding the mechanisms of magneto-optical phenomena in these crystals.

The results of previous calculations of crystal field parameters in $CrCl_3$ are summarized in Table 2. It is noticeable that the Racah parameters B and C as well as the parameter Dq are very similar in all cases. The most significant differences arise among the trigonal crystal field parameters v and v'. These differences are due to the fact that the spin–orbit interactions as well as the EPR data were not taken into account in the previously performed calculations.

Table 2

Crystal field parameters (in cm^{-1}) for Cr^{3+} ions in $CrCl_3$

	free ion	present paper	from [2]	from [2]	from [1]
B	920	520	550	550	550
C	3680	3520	3495	3530	3400
Dq		1435	1411	1409	1370
v		−1800	−137	−700	
v'		400	−536	−100	
ξ	273	200			170
α		20			
k		0.4			

References

[1] D. L. Wood, J. Ferguson, K. Knox, and J. F. Dillon, J. chem. Phys. **39**, 890 (1963).
[2] C. Limido, G. Pedroli, and G. Spinolo, Solid State Commun. **11**, 1385 (1972).
[3] R. M. Macfarlane, J. chem. Phys. **39**, 3118 (1963).
[4] G. A. Egorov and Yu. V. Yablokov, Zh. eksper. teor. Fiz. **39**, 265 (1960).
[5] J. F. Dillon, H. Kamimura, and J. P. Remeika, J. Phys. Chem. Solids **27**, 1531 (1966).

(Received January 16, 1974)

Erratum

To the paper by H. Buck and G. Alefeld

The Chemical Potential for Interstitials in Solid Solution

phys. stat. sol. (b) **47**, 193 (1971)

the authors request the following rectifications:

The calculation of the configurational entropy is not correct. Equations (1) and (2) on page 194 should read as

$$S = k \ln \left[\frac{N_0!\, 2^N}{(N_0 - N)!\, N!}\right]$$

and

$$S = k \ln \left[\frac{N_0!\, n^N}{(N_0 - N)!\, N!}\right].$$

The chemical potential is then given by

$$\mu = -uc + kT \ln \left[\frac{c}{n\,(r - c)}\right].$$

This chemical potential differs only by the additional term $-kT \ln n$ from Lacher's result (J. R. Lacher, Proc. Roy. Soc. **A161**, 525 (1937)), and therefore the model considered does not describe the observed asymmetry of the chemical potential as a function of concentration.

The authors wish to thank Prof. H. Wagner for pointing out the error in the expression for the entropy. This error has also been noticed by R. McLellan (private communication).

Erratum

To the paper by G. KOREN

Two-Photon Photoconductivity Measurements
in a Hexagonal ZnO Crystal

phys. stat. sol. (b) **61**, 277 (1974)

the author requests the following rectification:

The units of the ordinates of Fig. 2 to 4 should be Ω^{-1} cm^{-1} instead of s^{-1}.

See Short Note by I. A. Tsekhanovskii and S. A. Vorobev

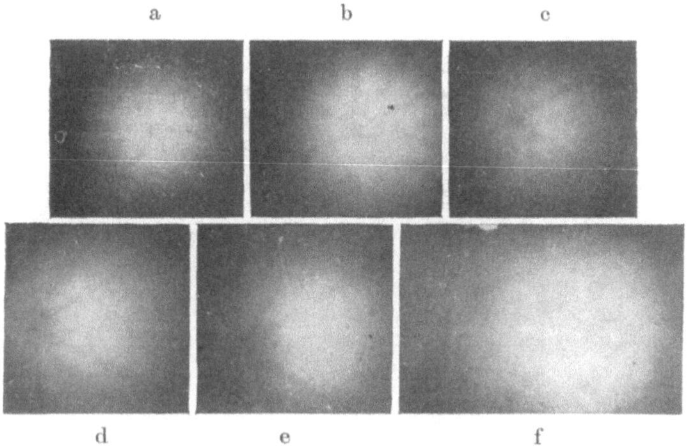

Fig. 1. Photographic records of the transmitted electron intensity behind a 8 μm thick silicon crystal for various incidence directions relative to a ⟨111⟩ axis; a) 0°, b) 0.33°, c) 1.0°, d) 1.5°, e) 2.0°, f) 3.0°

Short Notes

phys. stat. sol. (b) 62, K63 (1974)

Subject classification: 11; 22.1.2

Nuclear Physics Research Institute, Polytechnical Institute, Tomsk

Influence of Crystal Lattice on Multiple Scattering of Electrons

By

I.A. TSEKHANOVSKII and S.A. VOROBEV

The investigation of the angular distribution of electrons transmitted through a crystal may indicate the significant features of directional effects that have not been observed in wide-angle scattering measurements. In such experiments one can observe the anomalous transmission of electrons, the peculiarity of the transverse energy conservation in the collisions of the electrons with the atomic rows, and the effects of the crystal lattice on the multiple scattering of the swift electrons.

In the present experiment we have measured the intensity distributions of the swift electrons from a (Sr^{90} + Y^{90}) radioactive source behind the 8 μm Si single crystal. The angular divergence of the incident beam was about 0.1°. Fig. 1 a to f present the photographic records of the angular distributions of the transmitted electrons for different incidence directions relative to the ⟨111⟩ Si axis. The observed patterns are more complex in their intensity distribution than that expected from the ordinary multiple scattering theory (1). Obviously the less intense portions correspond to low index planar directions and the ⟨111⟩ crystallographic axis. The radial "deficit" streaks are surrounded on each side by an enhanced intensity that may be caused by anomalous transmission of the electrons close to the planar directions. By changing the crystal orientation relative to the incident beam direction the "star patterns" are moved at the same angle relative to the Gaussian distribution of the multiply scattered electrons. A similar behaviour of the transmitted intensity distribution can be observed for proton channelling or blocking phenomena (2, 3) and for Kikuchi patterns in electron diffraction (4). However, we observed a distinct expansion of strong intensity areas along the "deficit" streaks with the increase of the incidence angle of the electron beam. In this case the angular distribution of the transmitted electrons is found to be broader than that measured with the Si crystal in random direction. When the incident beam was aligned in the ⟨111⟩ direction of a Si crystal the angular distribution of the transmitted electrons was rather narrow that may appear from the conspicuous rise in the wide-angle scattering due to the close collisions of the electrons with the atomic rows. Indeed the measurement of the counting rate of electrons scattered into the angular range 3° to 5° with respect to the transmitted beam shows a pronounced dip in the ⟨111⟩ direction while tilting the crystal. Since the acceptance angle of the scintillation detector corresponded to the multiple scattering angle this observation confirms the narrowing of the angular distribution of the electron beam transmitted along the crystal axis. On the other hand this phenomenon may be caused by the effect of the anomalous transmission of the electron beam that is observed as six symmetric spots with the strong intensity at the cross points of low indexed planar streaks around the axial dip (Fig. 1 a to c). These spots of enhanced intensity are most pronounced for electron incidence parallel to a ⟨111⟩ axis of the crystal. In (5) from the standpoint of wave mechanics the phenomenon of anomalous transmission was related to the Borrman effect. According to the classical model the high transmittivity at small incidence angles supposes the capture of the swift electrons in the bound motion with regard to the atomic row (6, 7).

In the present experiments we have also measured the variation of the Rutherford scattering yield at 15° with the rotation angle of a Si single crystal relative to the incident beam. In the case of a beam aligned with the ⟨111⟩ axis the scattering yield displays an ordinary maximum caused by the close collisions of the electrons with the atomic rows. On the sides of this maximum there are any symmetric dips and humps at the angles, which correspond to the location of the spots of anomalous transmission in the photographic record of the angular distribution of electrons behind the crystal. Hence the observed dips in the wide-angle scattering yield may be caused by the electrons, which were captured on stable trajectories after multiple scattering in the crystal.

In Fig. 1 d to f for large angles of a crystal axis to the incident beam an arch-like intensity distribution centered around the axial direction is observed. Since the angular radius of the arch corresponds to the angle of incidence of the electron beam at the crystal this phenomenon is related to correlated scattering of electrons with conservation of the total transverse energy (8, 9). The diffusion of the particle distribution on the arch may be caused by multiple scattering of the electrons in the crystal. At the angle of incidence of about 3°, which is equal to several critical angles, there appear "excess" radial streaks with enhanced intensity of the electrons in the wide-angle scattering region (Fig. 1f). These "excess" bands are formed by electrons, which are originally scattered in close collisions with the atomic rows. In such a case the electrons are practically emitted from lattice atom positions and should be captured in bound motion with regard to the atomic planes.

References

(1) A.A. VOROBEV and B.A. KONONOV, Transmission of Electrons through Matter, Tomsk University, 1968.

(2) L.T. CHADDERTON, Phys. Letters (Netherlands) 23, 303 (1966).

(3) A.M. MARKUS, YA.E. GEGUSIN, and A.L. FAINSTEIN, Zh. eksper. teor. Fiz. 61, 332 (1971).

(4) P.B. HIRSCH, A. HOWIE, R.B. NICHOLSON, D.W. PASHLEY, and M.J. WHELAN, Electron Microscopy of Thin Crystal, Butterworths, London 1965.

(5) F. FUJIMOTO, S. TAKAGI, K. KOMAKI, H. KOIKE, and Y. UCHIDA, Radiation Effects 12, 153 (1972).

(6) H.C.H. NIP, M.J. HOLLIS, and J.C. KELLY, Phys. Letters A 28, 324 (1968).

(7) H.J. KREINER, F. BELL, R. SIZMANN, D. HARDER, and W. HÜTTL, Phys. Letters A 33, 135 (1970).

(8) H. KUMM, F. BELL, R. SIZMANN, H.J. KREINER, and D. HARDER, Radiation Effects 12, 53 (1972).

(9) V.V. KUDRIN and S.A. VOROBEV, to be published.

(Received January 2, 1974)

phys. stat. sol. (b) 62, K67 (1974)
Subject classification: 14.3 and 18.3; 22.8.2

Department of Electronics, University of Osaka Prefecture, Sakai, Osaka

Microwave Conductivity of Single Crystal Mn Ferrites

By
Y. WATANABE, K. UDA[1], S. SAITO, and S. TAKEMOTO

This paper concerns measurements of the microwave conductivity carried out on single crystals of $Mn_xFe_{3-x}O_4$ ($x \approx 0.9$ and ≈ 0.8) with an aim of throwing light on the origin of the ferromagnetic resonance (FMR) linewidth reported previously for the same specimens (1). Microwave dielectric properties were first used by Mizushima and Iida (2) for examining the relaxation process operating in excess-iron manganese ferrites. However, there have been few experiments in which both FMR and dielectric properties were measured, for this purpose, on the same specimen. Furthermore, the ordinary method which has been used for the microwave dielectric measurements of insulating materials is not appropriate to the present specimens since the effect of skin depth should be very significant on account of their high conductivities. We therefore measured the conductivity (σ_m) at 9.3 GHz by means of the Kohane and Sirvetz method (3), where σ_m was determined from the eddy current loss in spherical samples and not disturbed by the skin depth effect.

Measurements were made on polished spheres with diameter 0.6 to 2 mm under a large dc magnetic field (9000 G) applied parallel to the rf magnetic field, in order to reduce the effective permeability to unity. In Fig. 1 are shown $T\sigma_m$ vs. $1/T$ plots together with the corresponding dc ($T\sigma_{dc}$) and linewidth (ΔH) data (1). σ_{dc} observed is well described by the formula (reference (4))

$$\sigma_{dc} = (ne^2a^2/\tau_o kT)\exp(-E_g/kT), \quad (1)$$

indicating that the conduction is caused by electron hopping between Fe^{2+} and Fe^{3+} ions in the octahedral site. Here, n and a are the carrier concentration and jump distance, respectively, and the activation energy E_g is 30 to 40 meV. The value of x (or n) was estimated from equation (1) by using a = 3 Å (the distance between near-

1) Present address: Nippon Electric Co., Ltd., Kawasaki, Japan.

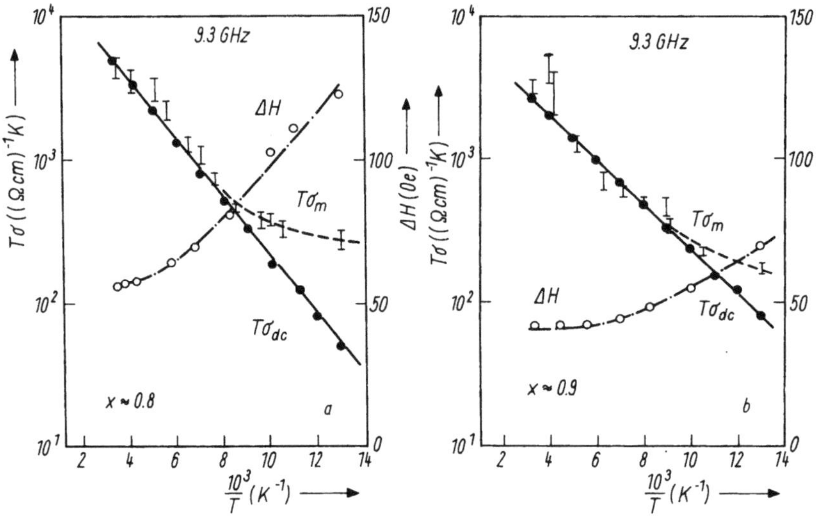

Fig. 1. Temperature dependence of σ_m, σ_{dc}, and ΔH; a) x ≈ 0.8, b) x ≈ 0.9

est neighbour cations) and $\tau_o = 3x10^{-13}$ s (4). σ_m exhibits a characteristic deviation from σ_{dc} below ca. 100 K, the magnitude of which is strongly dependent upon the concentration of Fe^{2+} ions, 1 - x. That is, the difference $\Delta\sigma = \sigma_m - \sigma_{dc}$ at 77 K is 2.9 and 1.3 $(\Omega cm)^{-1}$ for x ≈ 0.8 and ≈ 0.9, respectively. These behaviors in σ_m are quite similar to those which have been observed for compensated Si (5). The behaviors may thus be explained as due to the dielectric loss arising from hopping motion of carriers weakly bound around certain lattice defects, although details of the defects have not been clarified yet.

It is to be noted that there is a close correlation between $\Delta\sigma$ and ΔH, as clearly found by the dependences of these quantities upon the number of Fe^{2+} ions and also by the temperatures below which they show anomalous behaviors. This fact indicates that the increase of linewidth at low temperatures originates mainly from the valence-exchange mechanism. This mechanism has not been established in this system clearly but is well known in nickel-iron ferrites (6). The linewidth measurements in a lower temperature region are now in progress and will be published elsewhere.

The authors wish to express their sincere thanks to Prof. E. Uchida for supplying single crystals.

References

(1) Y. WATANABE, J. Phys. Soc. Japan 35, 716 (1973).

(2) K. MIZUSHIMA and S. IIDA, J. Phys. Soc. Japan 22, 1300 (1967).

(3) T. KOHANE and M.H. SIRVETZ, Rev. sci. Instrum. 30, 1059 (1959).

(4) N. MIYATA, J. Phys. Soc. Japan 16, 206 (1961).
F.K. LOTGERING, J. Phys. Chem. Solids 25, 95 (1964).

(5) M. POLLAK and T.H. GEBALLE, Phys. Rev. 122, 1742 (1961).
J. APPEL, Solid State Phys. 21 (1964).

(6) A.M. CLOGSTON, Bell Syst. tech. J. 34, 739 (1955).

(Received January 4, 1974)

phys. stat. sol. (b) 62, K71 (1974)

Subject classification: 6 and 13.4; 20.3; 22.4.1

III. Physikalisches Institut der Technischen Universität Berlin

The Dynamic Jahn-Teller Effect of Cubic ZnS:Cu^{2+} - Attempt of Coupling to T_2 Modes

By

U. SAUER, U. SCHERZ, and H. MAIER[1)]

In a foregoing paper (1) two of the authors discussed in detail the interpretation of the optical vibronic spectra of Cu^{2+} centres in cubic ZnS. The complicated emission spectrum with many narrow lines can be understood in terms of a dynamic Jahn-Teller effect acting on the $^2T_2$ ground state of the ion in a tetrahedral coordination. In an intermediate coupling scheme where the Jahn-Teller energy is of the same order of magnitude as the static lattice spin-orbit splitting a rigorous calculation was reported: The coupling of an E mode was performed by simultaneous diagonalization of the spin-orbit and Jahn-Teller Hamiltonians. The results could be fitted to the vibronic emission spectrum if two different E modes, which adequately represent the spectrum of the vibrational frequencies of the crystal, were considered. The strongly quenched gyromagnetic factor of the Γ_7 ground state (experimental value $g_{\Gamma_7} = 0.71$ (2)), however, could not be explained in this model. Recently Clerjaud and Gelineau (3) performed a similar calculation on this system including the excited electronic ^2E state. The low g_{Γ_7} factor was well reproduced, but some difficulties still remain in representing the higher vibronic states.

The introduction of an additional coupling to T_2 vibrational modes of comparable strength seems to be crucial as experimentally verified for the F^+ centre in CaO (4).

This is a very elaborate computational problem. In the present paper we therefore first treat the centre with more coupling to T_2 modes but, in contrast to (5), with simultaneous diagonalization of the Jahn-Teller and spin-orbit Hamiltonians.

We were able to take up to 5 phonons into account so that energy matrices with maximum dimension of 56 result (dimension of the matrix without group-theoretical reduction is 336). The Jahn-Teller Hamiltonian was taken linear in the lattice and

---

1) Present address: AEG-TFK-Forschungsinstitut Frankfurt/Main.

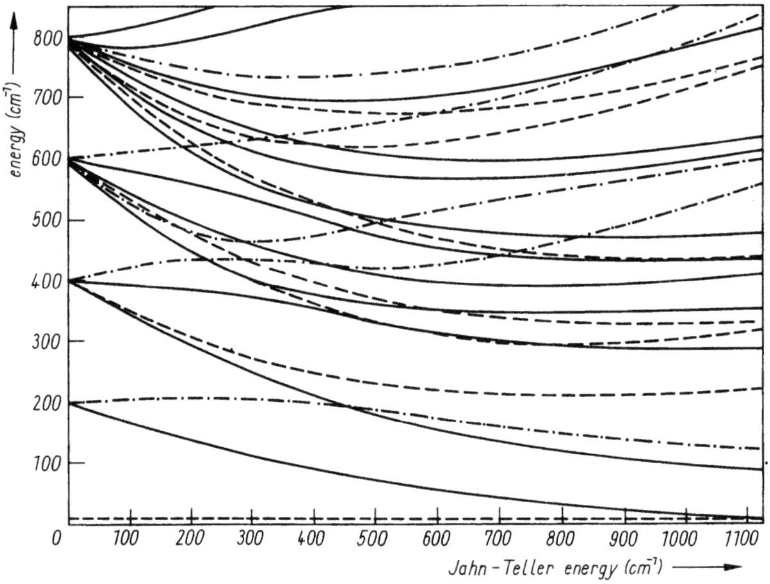

Fig. 1. Vibronic energy levels of a $^2T_2$ electronic multiplet coupled to a T_2 vibrational mode vs. Jahn-Teller energy:
$\hbar\omega = 200\ cm^{-1}$, $\lambda = -580\ cm^{-1}$, $k = 0.7$; —— Γ_8, -- -- Γ_7, —·— Γ_6

orbital operators:

$$H_{JT} = V_{T_2} (Q_\xi T_\xi + Q_\eta T_\eta + Q_\zeta T_\zeta),$$

where ξ, η, ζ indicate the basis functions of irreducible representation T_2. The static Jahn-Teller energy is then $E_{JT} = 2V_{T_2}^2 / 3\mu\omega^2$ with μ, ω oscillator mass and frequency, respectively.

For a T_2 mode of average energy $\hbar\omega = 200\ cm^{-1}$ and a spin-orbit coupling constant $\lambda = -580\ cm^{-1}$ (which is reduced from its free ion value by a factor $k = 0.7$ due to covalency and 4p mixing) the numerial results are shown in Fig. 1. Applying them to the experimental spectra reported in (1), we cannot account quantitatively for all the observed transitions. This has actually been expected, since at least two frequencies, i.e. the maxima at 100 and 300 cm^{-1} of the phonon density of states in ZnS should be regarded. But even the expansion of the calculation to two modes failed to match the total system of low energy vibronic states and there is no reason to assume that this should be possible by extending the cluster model and by taking a continuous

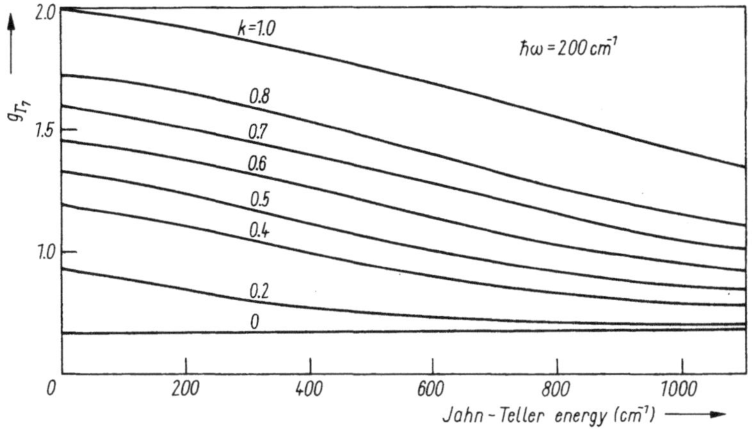

Fig. 2. g factor of the Γ_7 ground state of the ${}^2T_2$ multiplet coupled to a T_2 vibrational mode vs. Jahn-Teller energy for various orbital reduction factors k

spectrum of phonon energies into account. We therefore infer that the ground state is equally coupled to both T_2 and E modes. The same conclusion can be drawn considering the Zeeman splitting of the lowest Γ_7 state (Fig. 2). The orbital reduction factor is again assumed to be k = 0.7. Thus with reasonable Jahn-Teller energies we obtain values for the gyromagnetic factor g_{Γ_7} which considerably exceed its experimental magnitude of 0.71. It should be noted that the calculated g factors are nearly independent of the phonon energy.

These results support the conclusion made in (1) that only a simultaneous, comparably strong Jahn-Teller interaction with E and T_2 modes is capable of producing all the experimental results.

We thank Dr. M. Wöhlecke for many helpful discussions and the Recheninstitut of the Technische Universität Berlin for the support and provision of computing facilities. We are also grateful to Dr. B. Clerjaud and Dr. A. Gelineau for making available their results prior to publication.

References

(1) H. MAIER and U. SCHERZ, phys. stat. sol. (b) 62, 153 (1974).
(2) I. BROSER, U. SCHERZ, and M. WÖHLECKE, J. Luminescence 1, 39 (1970).
(3) B. CLERJAUD and A. GELINEAU, to be published.

(4) Y. MERLE D'AUBIGNÉ and A. ROUSSEL, Phys. Rev. B 3, 1421 (1971).

(5) R. ENGLMAN, M. CANER, and S. TOAFF, J. Phys. Soc. Japan 29, 306 (1970).

(Received January 21, 1974)

phys. stat. sol. (b) 62, K75 (1974)

Subject classification: 13.5.1

Central Research Institute for Chemistry,
Hungarian Academy of Sciences, Budapest

An Extension of Takeuti's Green's Function Formalism for Excitons to the Many-Band Case

By

M. KERTÉSZ

An analogue of the Koster-Slater (1) Green's function theory was developed by Takeuti (2) for the exciton problem in insulators and semiconductors. A different version of the formalism has been derived by Altarelli and Bassani (3) suggesting new approximations for the treatment of the integrals. Localized Wannier functions (WF) enter into the formalism, which are becoming now more and more easy to obtain from results of energy band calculations. It has been very recently proposed (4) to use this formalism without the neglect of the off-diagonal electron-hole interaction matrix with the aid of localized WF's (5), the latter having been calculated on the basis of an approximate (finite neighbour linear combination of atomic orbitals) Hartree-Fock crystal orbital scheme (6) directly. The two main restrictions of the formalism are: the neglect of other than one-particle excitations as well as the restriction to two bands only.

The aim of the present note is to point out that the formalism can be extended to the case of many bands. The exciton representation will be applied using the notation of Knox (7). The Green's function for any pair of bands (n stands for a conduction and m for a valence band) is

$$G_{n\ m}^{n'\ m'}(\vec{K}, E, \vec{\beta}-\vec{\beta}') = \delta_{nn'}\ \delta_{mm'} \int d^3\vec{k}\, \frac{e^{i\vec{k}(\vec{\beta}-\vec{\beta}')}}{E - \left[\varepsilon_n(\vec{k}) - \varepsilon_m(\vec{k}-\vec{K})\right]}\ , \qquad (1)$$

where $\vec{K}$ is the wavevector of the exciton, E its energy, $\varepsilon_i(\vec{k})$ the i-th energy band obtained from an approximate Hartree-Fock calculation. The interaction matrix involving integrals over WF's is

$$V_{n\ m}^{n'\ m'}(\vec{k},\vec{\beta},\vec{\beta}') = \sum_{\vec{R}} e^{i\vec{k}\vec{R}}\, 2\delta_M(n\vec{\beta}m'\,\vec{R}|1/r_{12}|m\vec{0}n'\vec{R}+\vec{\beta}') - \tag{2}$$

$$- (n\vec{\beta}m'\vec{R}|1/r_{12}|n'\vec{R}+\vec{\beta}'m\vec{0}) ,$$

where $n\vec{R}$ stands for a WF of band n localized around the site $\vec{R}$. Since the electron-hole pairs corresponding to different band pairs (n, m) and (n', m') are connected by equation (2), the two-band model is, of course, an approximation. However, it can be shown that it is possible to use a Green's function formalism similar to that of reference (2) in order to eliminate the large matrix elements in the neighbourhood of the diagonal in the eigenvalue problem of the exciton (reference (7)) and to reduce the dimension of the problem afterwards. The results is a linear homogeneous equation for the exciton wave function $U_{mn\vec{K}}(\vec{\beta})$:

$$\sum_{\vec{\beta}\vec{\beta}'}\sum_{n'm'} G_{mn}(\vec{K}, E, \vec{\beta}-\vec{R}) V_{m\ n}^{m'\ n'}(\vec{K},\vec{\beta},\vec{\beta}')U_{m'n'}(\vec{\beta}') = U_{mn}(\vec{R}). \tag{3}$$

As a consequence of the localized character of the WF's the electron-hole interaction matrix of equation (2) is localized around the site $\vec{R} = 0$. Therefore, the dimension of the problem of be treated is determined by the number of bands (this is limited in the LCAO approximation by the number of atomic orbitals used in an elementary cell) as well as by the degree of localization of the WF's. The solution of equation (3) is then straightforward (reference (4)).

It is worthwile to mention that in the limiting case of small intercellular interactions the exciton energy E tends to a value which can be obtained from a limited configuration interaction including only singly excited determinants for the elementary cell as a molecule. The present formalism is expected to give better results than the Frenkel model in systems with not very small intercellular interactions. It should be compared with the "supermolucule extension of the Frenkel model" (8), which represents an alternative treatment of the intermediate exciton problem, based on a Hartree product for the ground state in contrast to the present formalism, which is based on a SCF Hartree-Fock wave function for the ground state.

If we would have used the formalism of references (2) or (7) not based on Hartree-Fock wave function for the ground state there would not appear any additional difficulty except the fact that G would be no more diagonal in the band pair indices. This speciality of the Hartree-Fock method is based on the well-known possibility to choose the "off-diagonal Lagrangian multipliers" zero (9).

Work is in progress in order to apply this formalism to some organic semiconductors.

Stimulative discussions with Dr. G. Biczó are gratefully acknowledged.

References

(1) G. F. KOSTER and J.C. SLATER, Phys. Rev. 95, 1167 (1954).

(2) Y. TAKEUTI, Progr. theor. Phys. (Kyoto) 18, 421 (1957).

(3) J.M. ALTARELLI and B.F. BASSANI, J. Phys. C. 4, L328 (1971).

(4) M. KERTÉSZ, to be published in Chem. Phys. Letters.

(5) J. DES CLOIZEAUX, Phys. Rev. 135, A698 (1964).
E.I. BLOUNT, Solid Satte Phys. 13, 305 (1963).
M. KERTÉSZ and G. BICZÓ, phys. stat. sol. (b) 60, 249 (1973) and references therein.

(6) G. DEL RE, J. LADIK, and G. BICZÓ, Phys. Rev. 155, 997 (1967).

(7) R.S. KNOX, Solid State Phys. Suppl. 5, 1 (1963).

(8) S.A. RICE and J. JORTNER, in: Physics and Chemistry of the Organic Solid States, Vol.3, Eds. D. FOX, M. LABES, and A. WEISSBERGER, Interscience Publishers,1967.

(9) See e.g. F. SEITZ, Modern Theory of Solids, McGraw-Hill Publ.Co., 1940 (p.243).

(Received January 15, 1974)

phys. stat. sol. (b) 62, K79(1974)

Subject classification: 13

Sektion Mathematik und Physik der Technischen Hochschule
Otto von Guericke, Magdeburg

Density Functional Theory and Statistical Pseudopotentials

By

P. STREITENBERGER

Recently there appeared some papers concerned with the application of the Kohn-Sham self-consistent scheme to metallic properties (1). In the present paper an alternative treatment is given, which is suitable for a realistic perturbational calculation in the density functional formalism. Hohenberg and Kohn (2) have shown that the ground state energy of an interacting electron gas in a local static potential $V(\vec{r})$ is a unique functional of the electron density $n(\vec{r})$. If we want to calculate the total energy of a metal as a function of lattice structure we must determine the density of the valence electrons by minimization of the total electronic energy that can be written

$$E = \int V(\vec{r})n(\vec{r})d^3\vec{r} + \frac{1}{2}\iint \frac{n(\vec{r})n(\vec{r}\,')}{|\vec{r} - \vec{r}\,'|} d^3\vec{r}\, d^3\vec{r}\,' + T[n] + E_{xc}[n] , \qquad (1)$$

where T and E_{xc} are the kinetic energy and the exchange and correlation energy respectively. According to Kohn and Sham (3) this problem is aquivalent to a one-particle Schrödinger equation with the local self-consistent potential

$$W = \frac{\delta\Phi[n]}{\delta n} , \qquad (2)$$

where

$$\Phi[n] = E - T[n] . \qquad (3)$$

If we identify $n(\vec{r})$ in equation (1) with the valence electron density only, we must approximate the complicated nonlocal electron-ion interaction by a local one to obtain the external static potential $V(\vec{r})$. But there is no approximation if we include into $n(\vec{r})$ all electrons of the crystal inclusive of the electrons of the ion cores. In this case $V(\vec{r})$ is the simple local Coulomb potential of the atomic cores and all suppositions of the Hohenberg-Kohn-Sham theory are fulfilled. Writing $n(\vec{r}) = n_c(\vec{r}) + n_v(\vec{r})$

and consider the valence electron density $n_v(\vec{r})$ as the basic variable when the core electron density $n_c(\vec{r})$ is kept unchanged, we obtain from equations (1), (2), and (3) after a little algebra

$$W_v = V_{Hi}(\vec{r}) + \int \frac{n_v(\vec{r}\,')}{|\vec{r} - \vec{r}\,'|} d^3\vec{r}\,' + \frac{\delta}{\delta n_v} E_{xc}\left[n_c + n_v\right] + \frac{\delta}{\delta n_v}\left\{T\left[n_c + n_v\right] - T\left[n_v\right]\right\}. \tag{4}$$

The first two terms are the Hartree parts of the electron-ion and electron-electron interaction, the third term is the exchange and correlation potential, where there is no separation into electron-ion and electron-electron interaction. The last term is an additional repulsive potential arising due to the fact that the core and valence electron systems are not statistically independent. $W_v(\vec{r})$ is a statistitical pseudopotential acting on the non-interacting valence electrons and is exactly in the spirit of the density functional theory. For practical purposes W_v can be written formally in two parts which can be handled by different approximations:

$$W_{ei} = V_{Hi}(\vec{r}) + \frac{\delta}{\delta n_v}\left\{E_{xc}\left[n_c + n_v\right] - E_{xc}\left[n_v\right]\right\} + \frac{\delta}{\delta n_v}\left\{T\left[n_c + n_v\right] - T\left[n_v\right]\right\}, \tag{5}$$

$$W_{ee} = \int \frac{n_v(\vec{r}\,')}{|\vec{r} - \vec{r}\,'|} d^3\vec{r}\,' + \frac{\delta E_{xc}\left[n_v\right]}{\delta n} . \tag{6}$$

W_{ei} we also call "electron-ion" interaction, W_{ee} is the common electron-electron interaction. Taking the exchange part of equation (5) in the slowly varying density approximation (2) we have lowest order

$$W_{xi} = -4\left(\frac{3}{8\pi}\right)^{1/3}\left[(n_c + n_v)^{1/3} - n_v^{1/3}\right]. \tag{7}$$

Recently it was shown that an expression like (7) leads to essential improvements of the results for noble metals owing to the extended ion cores (4). For the repulsive potential we obtain in the same approximation

$$W_R = \frac{1}{2}(3\pi^2)^{2/3}\left[(n_c + n_v)^{2/3} - n_v^{2/3}\right] . \tag{8}$$

Long time ago a similar expression was derived by Gombás and Hellmann (5) starting from the Sommerfeld model of the free electron gas. For small ion cores we have approximately the density independent repulsive potential

$$W_R \approx \frac{1}{2}(3\pi^2)^{2/3}\left[n_c(\vec{r})\right]^{2/3} \tag{9}$$

which leads to a statistical pseudopotential for simple metals like the usual local pseudopotentials of band theory.

In the case of almost constant density we can make formally a series expansion in the density fluctuation n_s:

$$T\left[n_c + n_v\right] = T\left[n_c + n_o\right] + \int\left(\frac{\delta T}{\delta n_v}\right)_{n_o} n_s(\vec{r})\, d^3\vec{r} + \frac{1}{2}\iint\left(\frac{\delta^2 T}{\delta n_v^2}\right)_{n_o} n_s(\vec{r}) n_s(\vec{r}\,')\, d^3\vec{r}\, d^3\vec{r}\,' + \ldots , \tag{10}$$

where n_o is the mean value of $n_v(\vec{r}) = n_o + n_s(\vec{r})$. Contrary to the expansion about the uniform density situation (2) the linear term does not vanish because there is no translational invariance of the coefficient of n_s. Therefore we get a weaker potential and a justification for a low-order perturbational expansion of the total energy in the density functional theory.

References

(1) B. Y. TONG, Phys. Rev. B 6, 1189 (1972).

(2) P. HOHENBERG and W. KOHN, Phys. Rev. 136, B864 (1964).

(3) W. KOHN and L. J. SHAM, Phys. Rev. 140, A1133 (1965).

(4) J.A. MORIARTY, Phys. Rev. B 6, 1239 (1972).

(5) P. GOMBÁS, Handb. Physik, Vol 36, Ed. S. FLÜGGE, Springer-Verlag, 1956.

(Received January 7, 1974)

phys. stat. sol. (b) 62, K83 (1974)

Subject classification: 2 and 13.1; 20.1; 22.1; 22.2

Zentralinstitut für Elektronenphysik der Akademie
der Wissenschaften der DDR, Berlin

The Penn Gap in Ordered and Disordered Tetrahedrally Bonded Semiconductors

By

G. JUNGK

The Penn relation $\varepsilon_o - 1 = (\omega_p/\omega_g)^2$ is surprisingly well fulfilled for crystalline Ge, Si, and some III-V compounds taking ω_p as the plasma frequency of the valence electrons and ω_g as that frequency for which $\varepsilon_1(\omega) = 1$. The deviations from this relation for a-Ge and a-Si point to disorder-caused electronic transitions above ω_p.

The above relation was calculated (1) from an essentially one-oscillator model taking into account the wave-number k as a good quantum number (direct transition model). For comparison with experimental results the frequency for which the real part of the dielectric function $\varepsilon(\omega) = \varepsilon_1 + i\varepsilon_2$ equals unity (in the range of interband transitions) seems to be a good estimate for the effective gap frequency ω_g, because it is less influenced by broadening effects than the maximum of $\varepsilon_2(\omega)$. In spite of the inaccuracy of the published absolute values of optical constants of crystalline materials, the deviations from the predicted result $B = (\varepsilon_o - 1)(\omega_g/\omega_p)^2 = 1$ are relatively small (remembering the crudness of the above model). Table 1 summarizes experimental results from (2 to 5). The energy gaps given in braces were measured with an ellipsometric null-method (6) which enables an accurate determination of the absolute vaules of optical constants and therefore of ω_g. These gaps give indeed a better agreement with the Penn relation than other published values for these semiconductors. For the amorphous materials we took ω_g from the optical constants in (7, 8) obtained in the same manner, and found an essential deviation from B = 1. This result is beyond experimental errors and requires a more thorough explanation.

Van Vechten's (9) d-electron correction of the Penn formula cannot lower the discrepancy for a-Ge and is not applicable to silicon. Hübner's (10) consideration of the density effects on the above quantities cannot explain our results because a) this estimate would give a wrong prediction for the static dielectric constant ε_o and the plasma frequency of the amorphous state and b) we used the effective gaps from

Table 1

substance	ε_o	$\hbar\omega_p$(eV)	$\hbar\omega_g$(eV)	$B = \frac{\varepsilon_o - 1}{\omega_p^2}\omega_g^2$
GaP	8.6	16.5	5.5	0.85
GaAs	10.8	15.7	4.7	0.9
GaSb	14.5	13.3	(3.5)	0.95
InP	9.6	14.3	5.0	1.05
InAs	11.8	13.8	4.5	1.15
InSb	15.5	12.8	(3.2)	0.9
Si	11.8	16.45	4.5	0.8
Ge	16.4	16.0	(4.2)	1.05
a-Si	11.8	16.1	(3.4)	0.5
a-Ge	16.4	15.8	(3.0)	0.55

optical constants of nearly void-free material, with nearly the same density as their crystalline counterparts. Phillips' (11) incorporation of an oscillator strength $f \sim \omega_g^2$ into the Penn relation can be hardly proved, but an unchanged oscillator strength is more likely according to the similar nearest-neighbour order (12) in the crystalline and amorphous phase. Moreover, in all estimates the existence of a wave-number as quantum-number (and therefore lattice periodicity) is supposed.

There is a derivation of the Penn formula (13) without any assumption of periodicity, which eventually can give some insight. The Kramers-Kronig relation

$$\varepsilon_1(\omega) - 1 = \frac{2}{\pi}\int_0^\infty \frac{\omega'\varepsilon_2(\omega')}{\omega'^2 - \omega^2}\, d\omega'$$

is founded on causality only and gives in the low-frequency limit $\omega \rightarrow 0$

$$\varepsilon_o - 1 = \frac{2}{\pi}\int_0^\infty \frac{\varepsilon_2(\omega)}{\omega}\, d\omega \approx \frac{2}{\pi}\frac{1}{\omega_g}\int_0^{\omega_o} \varepsilon_2(\omega)d\omega \qquad (1)$$

if we assume one transition peaking at ω_g and no essential transitions above some frequency ω_o (or that the refractive index is determined by the valence electrons).

The sum rule

$$\frac{\pi}{2}\Omega_p^2 = \int_0^\infty \omega\,\varepsilon_2(\omega)\,d\omega \approx \omega_g \int_0^{\omega_0} \varepsilon_2(\omega)\,d\omega \qquad (2)$$

is derivable without taking into account periodicity (14, 15), where Ω_p is the plasma frequency for the total electron density. The above experimental agreement for the crystalline semiconductors favours the assumption that we can truncate the integral and take the plasma frequency ω_p of the valence electrons in formula (2). After that, the desired Penn relation follows immediately from (1) and (2). The experimental result $\varepsilon_0 - 1 < (\omega_p/\omega_g)^2$ implies therefore that the effective number of electrons contributing to optical transitions for energies $\leq \hbar\omega_p$ is reduced in the amorphous phases. According to the validity of formula (2) we must conclude that for a-Ge and a-Si there are electronic transitions above ω_p which are more pronounced than in the crystalline phases. This conclusion might be tested, for instance, by energy-loss experiments in a large energy range. The same deduction was taken by Bauer (16) for a-Ge from an experimental check of the sum rule. The assumed shift of the transition probabilities to d-electrons is, however, not sufficient here, because the same behaviour was observed for silicon, too.

The author is indebted to Dr. R.S. Bauer (Xerox, Palo Alto Research Center, USA) for a preprint of his Garmisch-Partenkirchen paper.

References

(1) D.R. PENN, Phys. Rev. 128, 2093 (1962).

(2) C. VON FESTENBERG, Z. Phys. 227, 453 (1969).

(3) N. DIDDEN, Z. Phys. 257, 310 (1972).

(4) J. STUKE and G. ZIMMERER, phys. stat. sol. (b) 49, 513 (1972).

(5) G.A.N. CONNELL and W. PAUL, J. non-cryst. Solids 8-10, 215 (1972).

(6) G. JUNGK, phys. stat. sol. (a) 3, 965 (1970).

(7) G. JUNGK, phys. stat. sol. (b) 44, 239 (1971).

(8) G. JUNGK, phys. stat. sol. (b) 55, 579 (1973).

(9) J.A. VAN VECHTEN, Phys. Rev. 182, 891 (1969).

(10) K. HÜBNER, phys. stat. sol. (b) 51, K125 (1972).

(11) J.C. PHILLIPS, phys. stat. sol. (b) 44, K1 (1971).

(12) S.H. WEMPLE, Phys. Rev. B 7, 3767 (1973).

(13) M. KASTNER, Phys. Rev. B 6, 2273 (1972).

(14) P. NOZIÈRES and D. PINES, Phys. Rev. 113, 1254 (1959).

(15) H. PUFF, Preprint, Zentralinst. Elektronenphys. Akad. Wiss. DDR, 1969.

(16) R.S. BAUER, Proc. 5th Conf. Phys. Amorphous and Liquid Semicond., Garmisch-Partenkirchen 1973.

(Received January 14, 1974)

phys. stat. sol. (b) 62, K87 (1974)

Subject classification: 18.2.1

Zentralinstitut für Festkörperphysik und Werkstofforschung der Akademie der Wissenschaften der DDR, Institutsteil für magnetische Werkstoffe, Jena

Nonlinear Mixing of Magnetostatic Volume Modes

By

H. PFEIFFER and W. HAUBENREISSER

Collinear interaction of elastic waves and magnetostatic waves of equal frequencies has been considered in (1 to 3). Different frequency and non-collinear mixing of elastic waves has been discussed in (4, 5).

The aim of this note is to examine the nonlinear mixing of two collinear magnetostatic volume modes with different frequencies in an infinite ferromagnetic slab. The sample geometry and coordinate system are chosen according to (6). The propagation direction, the external magnetic field $\vec{H}_o$, and the saturation magnetization $\vec{M}_o$ are assumed to be parallel to the z-axis.

The starting point is the equation of motion of the magnetic moment

$$\dot{\vec{M}} = \gamma\left[(\vec{M}_o + \vec{M}) \times (\vec{H}_o + \vec{H})\right] \tag{1}$$

and the Maxwell equations in the magnetostatic approximation

$$\text{rot}\ \vec{H} = 0\ ,\ \text{div}\ \vec{B} = 0,\ \vec{B} = \vec{H} + 4\pi\vec{M}\ . \tag{2}$$

The boundary conditions require the continuity of the tangential component of the field (H_t) and the normal component of the magnetic flux (B_n) on the surface of the slab. The nonlinear equation (1) is treated by perturbation theory in lowest order.

We put

$$\vec{M} = \vec{M}^{(1)} + \vec{M}^{(2)}\ ,\ \vec{H} = \vec{H}^{(1)} + \vec{H}^{(2)}\ .$$

In first order one obtains non-interacting magnetostatic volume modes, described by $\vec{M}^{(1)}$ and $\vec{H}^{(1)}$ (6). The second order equations read

$$\dot{\vec{M}}^{(2)} = \gamma\left[\vec{M}_o \times \vec{H}^{(2)} + \vec{M}^{(2)} \times \vec{H}_o + \vec{M}^{(1)} \times \vec{H}^{(1)}\right], \tag{3}$$

$$\text{rot}\,\vec{H}^{(2)} = 0\,,\ \text{div}\,(\vec{H}^{(2)} + 4\pi\vec{M}^{(2)}) = 0\,, \tag{4}$$

$$H_t^{(2)}\,,\ B_n^{(2)}\ \text{continuous}\,. \tag{5}$$

We consider the interaction of two free waves with the frequencies $\omega_1(k_1)$ and $\omega_2(k_2)$:

$$\vec{M}^{(1)} = \frac{1}{2}(\vec{m}_1\, e^{i\omega_1 t} + \vec{m}_2\, e^{i\omega_2 t} + \text{c.c.})\,,$$
$$\vec{H}^{(1)} = \frac{1}{2}(\vec{h}_1\, e^{i\omega_1 t} + \vec{h}_2\, e^{i\omega_2 t} + \text{c.c.})\,. \tag{6}$$

From equation (3) we see that the correction $\vec{M}^{(2)}$ contains constant-, double-, sum-, and difference frequency terms. In the following we confine ourselves to the sum frequency contribution; the extension of the calculation to the other terms is straightforward.

We obtain

$$\vec{M}_+^{(2)} = \vec{m}\, e^{i\omega_+ t} + \vec{m}^*\, e^{-i\omega_+ t}\,, \tag{7}$$

$$\vec{H}_+^{(2)} = \vec{h}\, e^{i\omega_+ t} + \vec{h}^*\, e^{-i\omega_+ t} \tag{8}$$

with $\omega_+ = \omega_1 + \omega_2$ and

$$\vec{m} = \chi(\omega_+)\,\vec{h} + \frac{1}{4}\, C(\omega_+)\,(\vec{m}_1 \times \vec{h}_2 + \vec{m}_2 \times \vec{h}_1)\,, \tag{9}$$

χ is the Polder tensor and the components of the tensor C are

$$C_{ij}(\omega_+) = \left(\frac{i\omega_+}{\gamma}\,\delta_{ijk} + \varepsilon_{ijk}\, H_{ok}\right)^{-1} \tag{10}$$

with

$$\vec{h} = \text{grad}\,\psi_+\,. \tag{11}$$

Equations (4) and (9) can be combined to an inhomogeneous differential equation inside the slab

$$(1+\varkappa_+)\left(\frac{\partial^2\psi_+}{\partial x^2}+\frac{\partial^2\psi_+}{\partial y^2}\right)+\frac{\partial^2\psi_+}{\partial z^2}=-\pi\,\mathrm{div}\left[C(\omega^+)\,(\vec{m}_1\times\vec{h}_2+\vec{m}_2\times\vec{h}_1)\right]. \tag{12}$$

Outside the slab

$$\Delta\psi_+ = 0 . \tag{13}$$

According to reference (6)

$$m_{ix}=\frac{\varkappa_i}{4\pi}h_{ix}\,,\quad m_{iy}=\frac{i\nu_i}{4\pi}h_{ix}\,,\quad m_{iz}=0 \tag{14}$$

with $\varkappa_i=\varkappa(\omega_i)$ and $\nu_i=\nu(\omega_i)$. The functions $\varkappa$ and ν are defined in reference (6). In the following we consider the case of antisymmetric modes with respect to the magnetic potential (in reference (6) denoted as a, c, ...):

$$\vec{h}_i=\mathrm{grad}\,\psi_i\,,\quad \psi_i=\frac{h_i^o}{\lambda_i}\sin\lambda_i x\, e^{-ik_i z} \tag{15}$$

with $\lambda_i=\dfrac{k_i}{[-(1+\varkappa_i)]^{1/2}}$.

The solution of the differential equations (12) and (13) with the boundary condition (5) yields the magnetic potential inside the slab corresponding to the generated sum frequency correction

$$\psi_+=\frac{h_1^o h_2^o}{D(\omega_+,k_+)}\left[L(\omega_+,k_+)\cos\lambda_+x-\frac{D(\omega_+,k_+)}{1+\varkappa_+}\left(I_-\cos(\lambda_1-\lambda_2)x+\right.\right.$$

$$\left.\left.+I_+\cos(\lambda_1+\lambda_2)x\right)\right]e^{-ik_+z} \tag{16}$$

with

$$D(\omega_+,k_+)=\cot\lambda_+s-\frac{1}{2}\left(\frac{1}{[-(1+\varkappa_+)]^{1/2}}-[-(1+\varkappa_+)]^{1/2}\right), \tag{17}$$

$$L(\omega_+,k_+)=A\left(\cot\lambda_+\frac{s}{2}-\frac{1}{[-(1+\varkappa_+)]^{1/2}}\right), \tag{18}$$

where $\lambda_+=\dfrac{k_+}{[-(1+\varkappa_+)]^{1/2}}$, $\quad k_+=k_1+k_2$.

$I_\pm$ and A are complicated functions of ω_i, ω_+, k_i, k_+, H_o, M_o, and the thickness of the slab s; they are not necessary for the further considerations and therefore their explicit expressions are not given here.

From (16) we see that the mixing of two antisymmetric (a) initial waves give a symmetric (s) contribution or for the general case: $a + a \rightarrow s$, $s + a \rightarrow a$, $s + s \rightarrow s$.

From (16) the magnetization distribution of the mixed magnetostatic volume mode can be obtained by (7), (9), (10), and (11). The calculations in reference (2) are included as a special case for $k_1 = -k_2$. A detailed evaluation of (16) and (9) shows that for $\Delta k = |k_1| - |k_2| \rightarrow 0$ the components $(H_+^{(2)})_x$, $(M_+^{(2)})_x$, $(M_+^{(2)})_y$ vanish linearly and $(H_+^{(2)})_z$ quadratically with Δk, but the component $(M_+^{(2)})_z$ remains finite in this limiting case. If ω_+ and k_+ fulfil the dispersion relation for free volume modes (equation (22) in reference (6)) the denominator $D(\omega_+, k_+)$ in (16) vanishes. A comparison of (17) and (18) shows that only resonance occurs (where ψ_+ becomes infinite), if the generated wave due to nonlinear mixing is symmetric. Thus, the phase matching condition $\omega(k_+) = \omega_+$ alone is not sufficient for resonance. It can be shown for the general case that the three modes $\omega_+(k_+)$, $\omega_1(k_1)$, $\omega_2(k_2)$ are in resonance, if they are all symmetric, or two of them are antisymmetric and one symmetric. Because the volume mode spectrum lies between the boundaries Ω_H and $(\Omega_H^2 + \Omega_H)^{1/2}$ (see reference (6)) the phase matching condition can only be fulfilled, if

$$\Omega_H = \frac{H_o}{4\pi M_o} < 1/3 .$$

A detailed treatment of the problem discussed here will be published elsewhere.

The authors wish to thank Prof. Dr. F. Voigt and Dr. U. Röpke for helpful discussions.

References

(1) R.B. THOMPSON and C.F. QUATE, J. appl. Phys. 42, 907 (1971).

(2) M. SCHULZ, J. appl. Phys. 43, 4752 (1972).

(3) J.D. ADAM, J.H. COLLINS, and J.M. OWENS, Electronics Letters 8, 229 (1972).

(4) G.J. JONES and D. KOBETT, J. Acoust. Soc. Amer. 35, 5 (1963).

(5) V.L. NEWHOUSE, C.L. CHEN, and K.L. DAVIS, J. appl. Phys. 43, 2603 (1972).

(6) R.W. DAMON and J.R. ESHBACH, J. Phys. Chem. Solids 19, 308 (1961).

(Received February 15, 1974)

phys. stat. sol. (b) 62, K91 (1974)

Subject classification: 8; 23

Physico-Technical Institute of Low Temperatures,
Academy of Sciences of the Ukrainian SSR, Kharkov

Hindered Rotational Effects in High-Temperature Phase N_2 and CO Crystals

By

V. A. SLUSAREV and YU. A. FREIMAN

As known there is no long-range orientational order in the β-phase of N_2 and CO crystals. The analysis of the temperature dependence of the rotational heat capacity C_{rot} in N_2 crystals (1) allows to conclude that the molecular rotation in the β-phase is hindered. This conclusion is based upon the fact that C_{rot} exceeds the heat capacity of free rotators (2 cal/mol deg). The decrease of C_{rot} with temperature clearly points to a liberation of rotational degrees of freedom, but purely experimental problems prevent this tendency to be traced up as far as to the triple point. Similar data for CO crystals are not available as yet. In the present paper we have estimated thermal properties for the high-temperature phase of N_2 and CO crystals.

To calculate the free energy and heat capacity of the system of rotators, we use the high-temperature expansion of the partition function

$$Z = Z_o \, \mathrm{Tr} \exp(-\beta H) \tag{1}$$

into a series in powers of $\beta = 1/T$:

$$Z = Z_o \left(1 - \frac{1}{1!} \frac{\langle H \rangle}{T} + \frac{1}{2!} \frac{\langle H^2 \rangle}{T^2} - \frac{1}{3!} \frac{\langle H^3 \rangle}{T^3} \right) \tag{2}$$

(Z_o is the partition function of free rotators). The trace in (1) implies averaging over angular variables.

The Hamiltonian of the system

$$H = \frac{1}{2} \sum_{i,k} H_{ik}$$

may be written in the form

$$H_{ik} = V_1 \Omega^i_{\alpha\beta} \Omega^k_{\gamma\delta} n_\alpha n_\beta n_\gamma n_\delta + V_2 \Omega^i_{\alpha\beta} \Omega^k_{\gamma\beta} n_\alpha n_\gamma + V_3 \Omega^i_{\alpha\beta} \Omega^k_{\alpha\beta} , \tag{3}$$

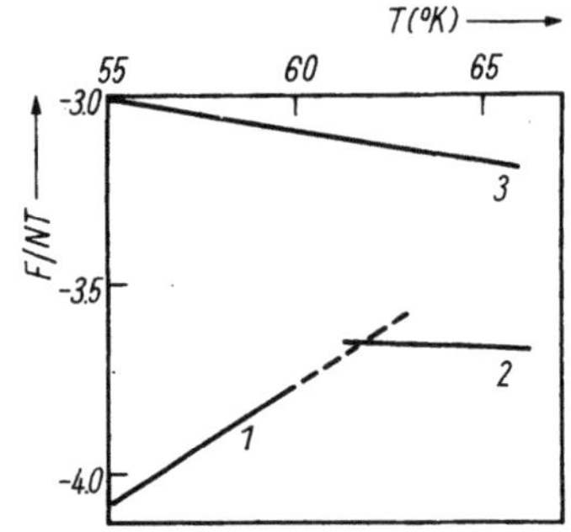

Fig. 1

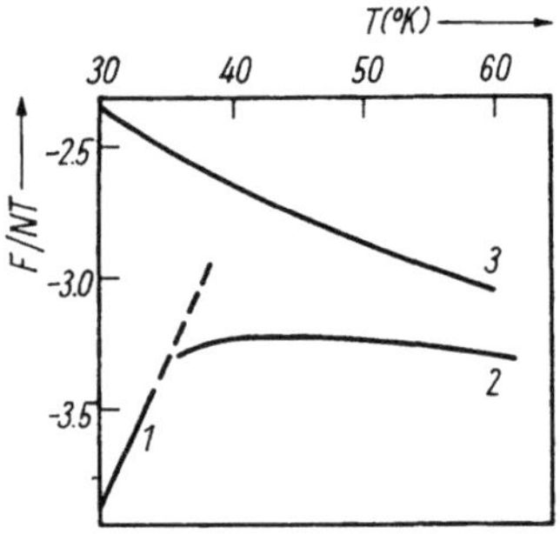

Fig. 2

Fig. 1. Temperature dependence of the free energy of N_2 crystal; (1) low-temperature phase,(2) high-temperature phase with account of correlational effects, (3) free rotators

Fig. 2. Temperature dependence of the free energy of CO crystal; (1) low temperature phase,(2) high-temperature phase with account of correlational effects. (3) free rotators

where

$$\Omega^i_{\alpha\beta} = \omega^i_\alpha \omega^i_\beta - \frac{1}{3}\delta_{\alpha\beta} \qquad (4)$$

and where $\vec{\omega}^i$ is the unit vector along the axis of the i-th molecule, $\vec{n}$ is the unit vector along the line connecting the centres of two molecules. The parameters of the Hamiltonian V_1, V_2, and V_3 are expressed in terms of molecular constants, such as the quadrupole moment Q, the Lennard-Jones molecular potential parameters ε and σ, the polarizability anisotropy factor k, and the repulsive force parameter B. The intermolecular interaction parameters used in the subsequent discussion are summarized in Table 1 of reference (2).

Taking into account that $\langle H \rangle = 0$ and omitting terms of order higher than $1/T^3$, we obtain for the free energy and the heat capacity

$$\frac{F_{rot} - F_o}{T} = -\frac{1}{2!}\frac{\langle H^2 \rangle}{T^2} + \frac{1}{3!}\frac{\langle H^3 \rangle}{T^3}, \qquad (5)$$

$$C_{rot} - C_o = \frac{\langle H^2 \rangle}{T^2} - \frac{\langle H^3 \rangle}{T^3}, \qquad (6)$$

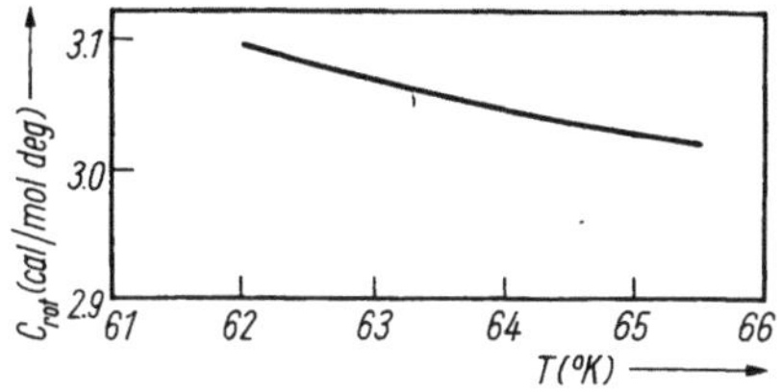

Fig. 3

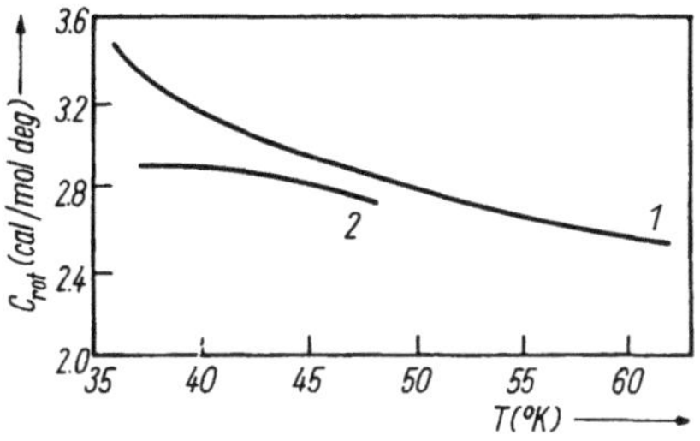

Fig. 4

Fig. 3. Temperature dependence of the high-temperature rotational heat capacity of N_2 crystal; (1) theoretical results, (2) results of paper (1)

Fig. 4. Temperature dependence of the high-temperature rotational heat capacity of CO crystal

where F_o and C_o are the free energy and the heat capacity of free rotators, respectively.

For purposes of numerical calculations we have used the Padé approximation for the truncated sums (5) and (6).

Fig. 1 and 2 show the results for the β-phase rotational free energy of N_2 and CO crystals, respectively. Extrapolating the low-temperature curves calculated in (2), we find the equilibrium transition temperature. The agreement between the orientational disordering temperatures thus evaluated with the measured phase transition temperatures is fair (35 and 35.61 ^oK for N_2 and 61.7 and 61.56 ^oK for CO, respectively). This suggests the contributions due to correlation effects to low--temperature-phase thermodynamic properties to be small.

Fig. 3 and 4 give the calculated β-phase heat capacities of N_2 and CO crystals, respectively. Data obtained in (1) are given in Fig. 3 as well. In the latter paper the rotational heat capacity at constant volume was obtained by substracting the translational part C_{tr} from the total one C_v. The former was calculated in the Debye approximation using $\Theta_D = 79\ ^oK$ determined in ultrasonic measurements (1). The β-phase translational heat capacity is close to its classical limit and insensitive to the specific model used in the calculations. In reference (1) the heat capacity C_v was evaluated using the data on heat capacity at constant pressure (3, 4), heat expansion (5), and isothermal compressibility (1).

The disagreement between the theoretical and experimental results, expect the nearest vicinity of the phase transition point, does not exceed the total experimental error.

As seen from Fig. 3 and 4, molecular rotation within the whole region of the β-phase in N_2 and CO crystals remains hindered.

Acknowledgement

The authors are indebted to V.G. Manzhelii for constructive discussion of the results of the present paper.

References

(1) P.A. BEZUGLY, L.M. TARASENKO, and YU.S. IVANOV, Fiz. tverd. Tela 10, 2119 (1968).

(2) V.A. SLUSAREV, YU.A. FREIMAN, I.N. KRUPSKII, and I.A. BURAKHOVICH, phys. stat. sol. (b) 54, 745 (1972).

(3) K. CLUSIUS, Z. phys. Chem. B3, 41 (1929).

(4) W.F. GIAQUE and J.O. CLAYTON, J. Amer. Chem. Soc. 55, 4875 (1933).

(5) V.G. MANZHELII, A.M. TOLKACHEV, and E.I. VOITOVICH, phys. stat. sol. 13, 351 (1966).

(Received February 15, 1974)

phys. stat. sol. (b) 62, K95 (1974)

Subject classification: 13.5.3 and 18

Department of Physics, Marathwada University, Aurangabad, Maharashtra

On the Theory of the Magnetic Polaron

By

S.S. SHAH

Recently much interest arose in the study of the magnetic polaron, i.e. an electron moving in the cloud of magnons. The s-d exchange coupling was used which was first suggested by Zener. The motion of conduction electrons in ferromagnetic non-metallic crystals is investigated by many authors (1, 2). The Green's function method or the variational techniques were used by these authors. This problem has been approached by the usual perturbation theory. In the available literature, no calculations of an interesting parameter, i.e. the average number of magnons associated with the conduction electrons moving in the magnetic insulator are found. This has been worked out for the simple polaron, i.e. the electron moving in the cloud of phonons (3). The calculation of this parameter will reveal some interesting properties.

We take the following Hamiltonian of the magnetic system:

$$H = H_e + H_m + H_{em} , \tag{1}$$

$$H_e = \sum E_\sigma(\vec{k}) C^+_{\vec{k}\sigma} C_{\vec{k}\sigma}, \quad H_m = \sum E^m_{\vec{\lambda}} a^+_{\vec{\lambda}} a_{\vec{\lambda}},$$
$$H_{em} = -\sum_{\vec{\lambda}} I_m (C^+_{\vec{k}\downarrow} C_{\vec{k}-\vec{\lambda}\uparrow} a_{\vec{\lambda}} + C^+_{\vec{k}-\vec{\lambda}\uparrow} C_{\vec{k}\downarrow} a^+_{\vec{\lambda}}) . \tag{2}$$

Here $(C^+_{\vec{k}\sigma}, C_{\vec{k}\sigma})$ and $(a^+_{\vec{\lambda}}, a_{\vec{\lambda}})$ are the electron and magnon (creation, annihilation) operators respectively in the corresponding states with $\vec{k}$ and $\vec{\lambda}$ representing wave-vectors, σ is the spin index. The single particle electron $E_\sigma(\vec{k})$ and magnon $E^m_{\vec{\lambda}}$ energies already incorporate the corrections arising from the interaction. The electron energies are $E_\sigma(\vec{k}) = \frac{\hbar^2 k^2}{2m^*} + \frac{\sigma}{2}$ IS, so that the exchange splitting is IS. σ takes $\pm(\uparrow, \downarrow)$ values and I is the exchange constant of the contact type interaction between conduction electron and localised spin of magnitude S. I_m is the electron-magnon coupling constant with

$$I_m = \frac{1}{2} \left(\frac{2S}{N}\right)^{1/2} I \ . \tag{3}$$

The unperturbed state of the system, i.e. the state in which no magnons are excited is taken as $|\vec{k}, 0\rangle$ and $|\vec{k}, \vec{\lambda}\rangle$ as the perturbed state of the system in first order.

$$|\vec{k}, \vec{\lambda}\rangle = |\vec{k}, 0\rangle + \frac{\sum_{\vec{\lambda}} |\vec{k} - \vec{\lambda}, 1_{\vec{\lambda}}\rangle\langle \vec{k} - \vec{\lambda}, 1_{\vec{\lambda}}| H_{em} |\vec{k}, 0\rangle}{E_{\vec{k}} - E_{\vec{k}-\vec{\lambda}} - W_{\vec{\lambda}}} , \tag{4}$$

where $W_{\vec{\lambda}}$ is the one-magnon energy in the mode. The total number of magnons $\langle N_{\vec{\lambda}}\rangle$ accompanying the electron is given by taking the expectation value of $\sum a^+_{\vec{\lambda}} a_{\vec{\lambda}}$ over the state $|\vec{k}, \vec{\lambda}\rangle$.

For the electron with spin $-\frac{1}{2}(\downarrow)$

$$\langle N_{\vec{\lambda}}\rangle = \frac{\sum_{\vec{\lambda}} |\langle \vec{k} - \vec{\lambda}, 1_{\vec{\lambda}}| H_{em} |\vec{k}, 0\rangle|^2}{\left[E(\vec{k})\downarrow - E(\vec{k} - \vec{\lambda})\uparrow - W_{\vec{\lambda}}\right]^2} \tag{5}$$

is the total number of magnons. For ($\sigma = 1/2$) electrons, this number will be modified. (But at absolute zero the formation of this polaron is ruled out.) In fact these results could be obtained by using the Green's function technique as well (4). When we substitute the wave functions and proper interaction in equation (5), we get

$$\langle N\rangle = \sum_{\vec{\lambda}} \frac{I_m^2}{\left[E(\vec{k})\downarrow - E(\vec{k} - \vec{\lambda})\uparrow - W_{\vec{\lambda}}\right]^2} . \tag{6}$$

We assume strong s-d exchange and take $E(\vec{k})\downarrow - E(\vec{k} - \vec{\lambda})\uparrow = IS$ and $W_{\vec{\lambda}} = D\lambda^2$, where $D = 2JSa^2$. We take $IS \gg W_{\vec{\lambda}}$ (strong s-d exchange), then equation (6) reduces after changing the summation into integration and substituting the value of I_m from equation (3)

$$\langle N\rangle = \frac{1}{12S} , \tag{7}$$

taking the value $S = 7/2$ as in the case of the ferromagnetic semiconductor EuO we get the number of magnons as 0.024.

From the magnitude of the above number we can conclude that even for strong s-d exchange interaction perturbation theory is valid. In case of simple polaron we come across the difficulties in case of strong binding (3). There will be an instability if the s-d exchange is weak and direct exchange is strong, i.e. $IS \leqq D\lambda^2$. The Green's function method and the formulation of the polaron problem in two Bose fields will appear elsewhere (4).

References

(1) P. RICHMAND, J. Phys. C 3, 2402 (1970).

(2) S. METHFESSEL and D.C. MATTIS, Hdb. Phys., Vol. 28/I, 1968.

(3) D. PINES, Polarons and Excitons, Ed. C.G. KUPER and WHITEFIELD, Oliver and Boyd, London 1963 (p. 35).

(4) S.S. SHAH and K.P. SINHA, to be published in Pramana.

(Received December 14, 1973)

phys. stat. sol. (b) 62, K99 (1974)

Subject classification: 13 and 20; 21

Institute of Metal Physics, Ural Scientific Centre, Academy of Sciences of the USSR, Sverdlovsk

The Peculiarities of the Electronic Structure of V_3Pd Exhibited in the X-Ray Emission Spectra of the Constituents

By

E.Z. KURMAEV and M.F. SOROKINA

According to (1, 2) the valence band of the V_3Pd alloy consists of two slightly overlapping subbands of quite certain origin: the bottom of the band arises from Pd4d-states, and the high-energy part is formed mainly of V3d-states. In this report we attempt to apply the theory (3, 4) to the constituent X-ray L-spectra of V_3Pd and to understand why the d-states of palladium are displayed in a more pronounced way in the vanadium L-spectrum than the d-states of vanadium in the palladium L-spectrum where only a slight increase of the high-energy intensity is found with respect to the pure Pd spectrum (Fig. 1).

According to (3) the intensity of the A-component X-ray spectrum from the alloy A-B is expressed in the following way:

$$I^{(A)}(\omega - \varepsilon^{(A)}) \sim \left|W^{(A)}\right|^2 \mathrm{Im}\left\{G_{oo}^{(A)}(\omega) + \sum_{\delta} S_{o\delta}^{(A)}\left[G_{o\delta}^{(A)}(\omega) + G_{\delta o}^{(A)}(\omega)\right] + \right.$$
$$\left. + \sum_{\delta} S_{o\delta}^{(A)} G_{\delta\delta}^{(A)}\right\},$$

where W is the matrix element of the corresponding X-ray transition and $S_{o\delta}$ is a magnitude which is determined by the overlap integral of the atomic wavefunctions for the valence states l and l_δ centred at the sites 0 and δ, respectively, $S_{o\delta} = - 1/2 \langle 0l \mid \delta l_\delta \rangle$, the site 0 being occupied by the emitting atom and the site δ by one of its nearest neighbours, $G_{oo}(\omega)$ (and $G_{\delta\delta}(\omega)$) is the "local density of states" of the component at the site 0 (or δ), $G_{o\delta}(\omega)$ the Green propagator in the site representation. All terms except of Green function are energy-independent.

In contrast to the case of disordered alloy spectra where an averaging over all possible configurations of the atomic environment is required, in the ordered alloy where the coordination of each atom is well defined such averaging may be omitted.

Therefore the expression for the A-component intensity is written as

$$I^{(A)}(\omega - \varepsilon^{(A)}) \sim |W^{(A)}|^2 \, \mathrm{Im} \left\{ G^{(A)}_{oo}(\omega) + 2 \left[Z^{(A)}_{A} S_{AA} G^{(A)}_{o\delta_A}(\omega) + Z^{(A)}_{B} S_{AB} G^{(A)}_{o\delta_B} \right] + S^2_{AA} Z^{(A)}_{A} G^{(A)}_{oo}(\omega) + S^2_{AB} Z^{(a)}_{B} G^{(B)}_{oo}(\omega) \right\} \equiv \tag{1}$$

$$\equiv I^{(A)}_{o}(G^{(A)}_{oo}) + I^{(A)}_{1}(G^{(A)}_{o\delta}) + I^{(A)}_{2}(G^{(A)}_{oo}) + I^{(A)}_{2'}(G^{(B)}_{oo}).$$

Here the factors of the type Z^{A}_{B} describe the number of B-kind atoms in the closest environment of an atom A and the terms of the type $I^{(A)}_{n}$ are the contributions to the intensity of the A-component spectrum from the terms of n-overlap order. An analogous expression may be written for the B-component intensity.

It is seen that the terms of zero order and of the second order are expressed through the local component density of states, the main contribution to intensity being supplied by the local density of states of the emitting component, while the local density of states of the second component contributes only in the second overlap-order. In the case when the component bands slightly overlap in energy, "own" and "foreign" contributions are in different spectral regions. The terms of the first order are characterized by a more complex energy dependence. Ignoring the contribution from these terms let us determine the relative contribution to the intensity

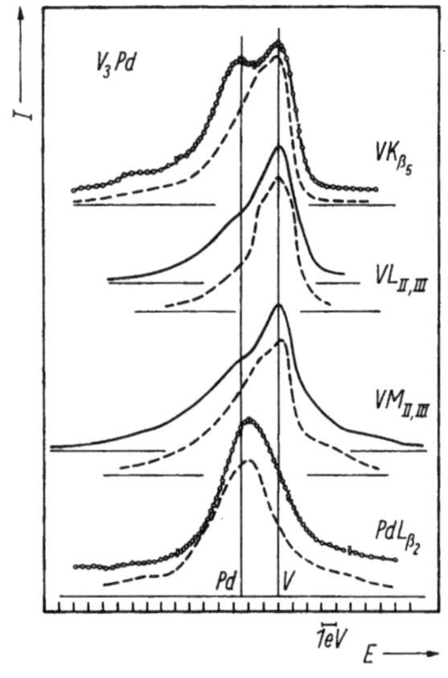

Fig. 1. X-ray spectra of constituents in V_3Pd alloy and of the corresponding pure metals (dashed lines) (1, 2)

from the "foreign" local density of states:

$$\frac{I_2^{(A)}(G_{oo}^{(B)})}{I_o^{(A)}(G_{oo}^{(A)}) + I_2^{(A)}(G_{oo}^{(A)})} = \frac{S_{AB}^2 Z_B^{(A)} \operatorname{Im} G_{oo}^{(B)}(\omega)}{(1+S_{AA}^2 Z_A^{(A)}) \operatorname{Im} G_{oo}^{(A)}(\omega)} . \tag{2}$$

This expression (and an analogous expression for the B-component intensity) refer to the model alloy with one valence electron per atom and we shall regard them to be valid also in the case when the components have more than one valence electron per atom (assuming the effects of correlation to be of little importance). To describe roughly the situation for the L-spectra of transition metal alloys we assume the valence states of A and B atoms to be 10-fold degenerate and filled with n_A and n_B electrons, respectively, so that neglecting the effects of electron redistribution in the alloy

$$\frac{\int_{-\infty}^{\omega_F} \operatorname{Im} G_{oo}^{(A)}(\omega)\, d\omega}{\int_{-\infty}^{\omega_F} \operatorname{Im} G_{oo}^{(B)}(\omega)\, d\omega} = \frac{n_A}{n_B}, \tag{3}$$

where w_F corresponds to the Fermi level. Then the relative integrated contribution from the "foreign" local density on the basis of equations (2) and (3) will be

$$i^{(A)} = \frac{\int_{-\infty}^{\omega_F} I_2^{(A)}(G_{oo}^{(B)})\, d\omega}{\int_{-\infty}^{\omega_F} \left[I_o^{(A)}(G_{oo}^{(A)}) + I_2^{(A)}(G_{oo}^{(A)})\right] d\omega} = \frac{S_{AB}^2 Z_B^{(A)} n_B}{(1+S_{AA}^2 Z_A^{(A)}) n_A} . \tag{4}$$

In the case of the ordered alloy V_3Pd, substituting for the number of d-electrons $n_A = 3$, $n_B = 10$ and, according to the atomic coordination for the A15-structure, $Z_A^{(A)} = 10$, $Z_B^{(A)} = 4$; $Z_B^{(B)} = 0$, $Z_A^{(B)} = 12$ (for the first coordination sphere of each component) and assuming as in (3) that $S_{AA} = S_{BB} = S_{AB} = -0.1$, we get for the relative intensity of the "palladium" peak in the vanadium spectrum $i^{(V)} = 0.12$ and for the relative intensity of the "vanadium" peak in the palladium spectrum a value three times smaller ($i^{(Pd)} = 0.04$) despite the fact that the nearest environment of a palladium atom consists of only vanadnium atoms and vanadium atoms have but four palladium atoms among 12 neighbours. This results from the larger number of d-electrons in Pd.

A transfer of electrons from V to Pd (which in fact occurs in V_3Pd according to our investigation of X-ray absorption spectra) can only serve to increase the effect, i.e. to increase still more $i^{(V)}$ and decrease $i^{(Pd)}$.

We wish to thank T.B. Shashkina who helped to prepare this paper for publication.

References

(1) E.Z. KURMAEV, V.P. BELASH, S.A. NEMNONOV, and M.F. SOROKINA, Izv. Akad. Nauk SSSR, Ser. fiz. 36, 310 (1972).

(2) S.A. NEMNONOV, E.Z. KURMAEV, V.P. BELASH, and V.A. FOMICHEV, Proc. Internat. Symp. X-Ray Spectra and Electronic Structure of Matter, Munich 1972 (in the press).

(3) O.B. SOKOLOV and YU.A. BABANOV, phys. stat. sol. (b) 51, 469 (1972).

(4) YU.A. BABANOV, O.B. SOKOLOV, and M.F. SOROKINA, phys. stat. sol. (b) 52, 155 (1972).

(Received February 27, 1974)

phys. stat. sol. (b) 62, K103 (1974)
Subject classification: 10.2 and 20.1; 22.8.1

Department of Physics, Indian Institute of Technology, Madras

Cu^{++} Centers in Orthorhombic Alkali Sulphates

By
C. RAMASASTRY, C.S. SUNANDANA, and B.S.V.S.R. ACHARYULU

Introduction Normally copper is known to enter alkali halide crystals in the monovalent state, substituting an alkali ion (1 to 4). In irradiated copper-doped samples divalent ions also are found (3, 5). Certain treatments like chlorination also convert Cu^{+} ions to the Cu^{++} state (6). However, in CsCl-type structures copper is found to enter as Cu^{++}, occupying an interstitial face-center position (7). Thus it would be of interest to know in which valence state copper enters ionic crystals with divalent anions like sulphates, oxides, etc. With this in view, a systematic study on copper-doped potassium, rubidium, and cesium sulphate crystals, grown from solution, was undertaken. In this note are reported the new ultraviolet absorption bands observed in these crystals as also other interesting results.

Results Fig. 1 shows the absorption spectra of K_2SO_4:Cu crystals at RT and LNT. (The undoped K_2SO_4 crystal has no absorption right upto 190 nm.) In the doped crystal, a new absorption band is observed at 240 nm with a half-width of about 1.18 eV.

Undoped Rb_2SO_4 showed an absorption band at 290 nm with a half-width of about 0.93 eV, and a rising absorption towards higher energies (Fig. 1, curve a). This absorption is attributed to NO_3^- ions present as background impurity in the starting material (NO_3^- ion absorbs around 300 nm). This suggestion was confirmed from the infrared spectrum taken of such a crystal which showed a sharp band at 720 cm^{-1} characteristic of the NO_3^- ion (ν_4 mode). Interestingly, when copper is added as impurity, the 290 nm absorption band is completely suppressed and two new bands appear at 250 and 202 nm (Fig. 1, curves b, c).

The absorption spectrum of undoped Cs_2SO_4 crystal (Fig. 2, curve a) showed, in addition to the edge absorption, a band at 215 nm with a width of 0.55 eV which was attributed to a background impurity like Cd^{+} because of the enhancement of this band in Cd-doped Cs_2SO_4 (Fig. 2b, curve b) (8). In the absorption spectra of copper-

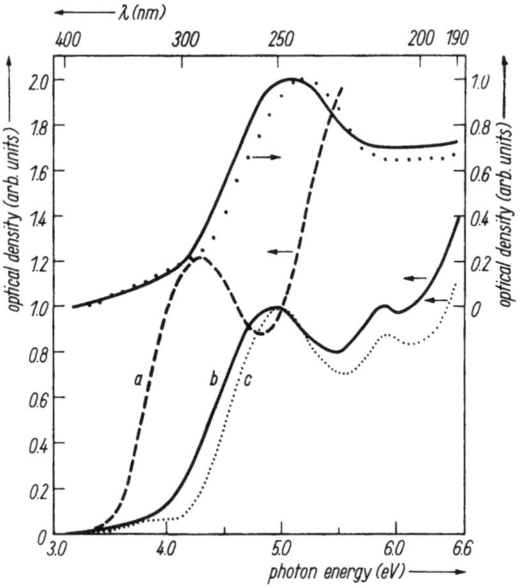

Fig. 1. Optical absorption spectra of K_2SO_4:Cu (—— 300 K, ····· 77 K),
(a) Rb_2SO_4 (undoped, 77 °K),
(b) Rb_2SO_4:Cu (300 K), and
(c) Rb_2SO_4:Cu (77 K)

doped crystal (Fig. 2, curves c, d) a new absorption band is clearly seen at 257 nm with a half-width of 1.14 eV, in addition to the impurity band at 215 nm.

The peak positions and half-widths of the new band observed in these sulphates at RT and LNT are given in Table 1.

Discussion From the ESR data (9, 10) it is clear that copper enters these sulphate lattices in the divalent state, Cu^{++}. Thus the new optical band of K_2SO_4:Cu is directly attributable to the Cu^{++} ion. In the case of Rb_2SO_4, the new band observed at 250 nm is attributed to Cu^{++}. However, one has to explain the suppression of the band at 290 nm due to the NO_3^- ion present in the undoped crystal and the appearance of a band at 202 nm. The best way to do this is to say that NO_3^- loses its electron to Cu^{++} forming Cu^+ centers. The new band appearing at 202 nm in this crystal may be due to these Cu^+ ions. However, it must be remembered that the charge-transfer spectrum of the NO_3^-

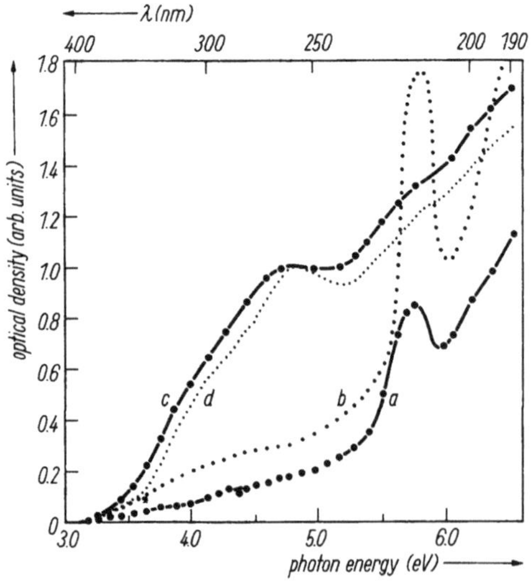

Fig. 2. Optical absorption spectra of
(a) Cs_2SO_4 (undoped, 77 K),
(b) Cs_2SO_4:Cd (77 K), (c) Cs_2SO_4:Cu (300 K),
(d) Cs_2SO_4:Cu (77 K)

Table 1

Peak positions (nm) and half-widths (eV) of Cu^{++} absorption bands in some alkali sulphate crystals

host crystal	RT		LNT	
K_2SO_4	245	(1.20)	240	(1.18)
Rb_2SO_4	253	(1.10)	250	(0.93)
Cs_2SO_4	260	(1.16)	257	(1.14)

ion has two maxima at 303 and 202 nm, with relative intensities in the ratio of 1:10000 (11). Hence the band observed may be the 202 nm band of NO_3^-, with a highly reduced intensity so that the corresponding 303 nm band might not be detectable. Such a suppression of NO_3^- bands is also reported earlier in the case of Rb_2SO_4 doped with cadmium. In this case the presence of Cd^+ formed due to trapping of the electron from NO_3^- by Cd^{++} could be detected by both optical and ESR absorption. In the former, a band at 254 nm characteristic of Cd^+ and in the latter, the low-field hyperfine component attributable to Cd^+ (I = 1/2) were observed (8).

In the case of Cs_2SO_4, the new band observed at 257 nm is attributable directly to the Cu^{++} ions. The longer wavelength shoulder around 295 nm may be one of the Cd^+ bands whose presence in the undoped crystal is known (as mentioned earlier). Another possibility is that some Cu^{++} centers trap electrons and become Cu^+ to give absorption in this region.

That the prominant new band is due to Cu^{++} can also be seen from the fact that the

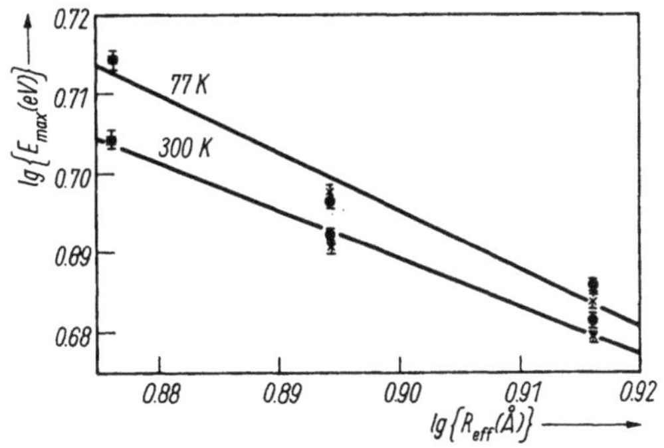

Fig. 3. Mollwo-Ivey type relations for Cu^{++} band in K_2SO_4, Rb_2SO_4, and Cs_2O_4 crystals at 300 and 77 K. The dots and crosses refer to data on samples I and II respectively

band maxima appear to follow a Mollwo-Ivey type relation of the form (Fig. 3)

$$E_{max} = 1.15\, R_{eff}^{-0.66}\ , \text{(RT)}$$

$$E_{max} = 1.08\, R_{eff}^{-0.72}\ , \text{(LNT)}$$

where E_{max} is the peak position (eV) and R_{eff} is the effective lattice parameter given by $R_{eff} = (abc)^{1/3}$, a, b, and c being the lattice constants of the orthorhombic unit cell in Å. From the data of Krätzig et al. (3) we found that the Cu^{++} band maxima in potassium halides also follow a similar relation

$$E_{max} = 3.27\, R^{-3.47}\ .$$

The ground state electron configuration of free Cu^{++} ($3d^9$) gives an inverted doublet $^2D_{5/2,3/2}$ while the first excited state ($3d^8 4s$) gives eight terms - ^4F, ^2F, ^2D, ^4P, ^2P, ^2G, ^4D, and ^2S in order of increasing energy. These eight terms span an energy of about 10 eV while ^4F is 7.54 eV from the ground level (12). The Cu^{++} ion occupies the α and β alkali ion sites in the orthorhombic lattices of the alkali sulphate crystals, and the crystal field of the SO_4^{--} ions contracts the energy levels thus making possible some of the transitions that give rise to absorption bands in the near uv (which for the free ion would be observed only in the far uv).

The width and intensity of the Cu^{++} band in the present case do not seem to be changing very much with temperature. As such, the band may be due to a normally allowed transition like $^2D \rightarrow {}^2P$, the transition energy being reduced from around 8.3 to 5.0 eV. But the large half-width of the band ($\approx$ 1 eV) suggests that it may be the result of more than one transition, for example, $^2D \rightarrow {}^2F$ and $^2D \rightarrow {}^2D$, both being allowed. The nearly O_h symmetry of the oxygen cage around the Cu^{++} ion may not be able to resolve this structure at LNT.

Acknowledgement

The authors are grateful to Dr. S.B.S. Sastry for helpful discussions.

References

(1) A.M. MACMAHON, Z. Phys. 52, 336 (1928).

(2) R. OGGIONI and P. SCARAMELLI, phys. stat. sol. 9, 411 (1965).

(3) E. KRÄTZIG, T. TIMUSK, and W. MARTIENSSEN, phys. stat. sol. 10, 709 (1965).

(4) K. FUSSGAENGER, phys. stat. sol. 34, 157 (1969).

(5) Y.V.G.S. MURTI et al., J. Phys. C 4, 1606 (1971); 5, 2827 (1972).

(6) F. URBACH et al., J. Phys. Chem. Solids 3, 153 (1957).

(7) N. KURODA and A. KAWAMORI, J. Phys. Chem. Solids 32, 1233 (1971).

(8) B.S.V.S.R. ACHARYULU, Ph.D. Thesis, I.I.T., Madras 1972.

(9) M. CHANDRAPAL et al., Curr. Sci. (India) 38, 564 (1969).

(10) B.V.R. CHOWDARI and P. VENKATESWARLU, Proc. Nucl. Phys. and Solid State Phys. (India), 1968 (p. 369).

(11) M.J. BLANDAMER, Chem. Rev. 70, 59 (1970).

(12) C.E. MOORE, Atomic Energy Levels, N.B.S. Circ. 467, Vol. II (1952).

(Received February 27, 1974)

phys. stat. sol. (b) 62, K109 (1974)
Subject classification: 2 and 4; 12.1; 21.2

Commissariat à l'Energie Atomique, Centre d'Etudes de Limeil, Villeneuve-Saint-Georges

A Theoretical Calculation of the Effect of Pressure on the Pair Potential for Liquid Sodium Metal

By
L. DAGENS

The effective interionic pair potentials for uncompressed and compressed liquid sodium metal have be recently derived by Waseda and Suzuki (1) and Waseda and Ohtani (2) (to be referred as to WO) from the observed structural data. WO found that a fairly small pressure (5 kbar) which induces only a 10% change in the density[1)] has a large effect on the pair potential. The pressure effect may be briefly described as follows. (i) The long-range oscillation wavelength is reduced by pressure, although more than expected from the electron theory of metals. (ii) The pair potential is made more repulsive in the range including the first minimum and the first maximum. (iii) The first node r_o is decreased, but its relative change is small (only 1%).

We present here and discuss the density dependence of the liquid sodium metal pair potential as given by an electron theory of simple metals. We use the method of the auxiliary neutral atom (3, 4). It is a first principle method which has given accurate binding energies for the alkali metals (5). The main step is the calculation of the rigid neutral atom valence density n(r). n(r) is defined such that the self-consistent valence density ϱ_{val} of the metal may be written as a superposition $\sum_{ions} n$ of the neutral atom densities centred at the occupied sites. n(r) is structure independent but is still dependent on the mean valence electron density $\varrho_o = k_F^3/3\pi^2$. n(r) have been calculated as described in (4) for four values of k_F given in Table 1 (a graph of n(r) for sodium metal is given in (4)). No explicit use of any pseudopotential is made: n(r) is self-consistently calculated directly from the true Hartree-Fock-Slater ionic potential. Many electron effects are described by the Hartree-Fock-Slater scheme, with the Gaspard-Kohn-Sham (GKS) $\varrho^{1/3}$ exchange potential.

1) According to the values given in their Fig. 1 caption. There is a contradiction between these figures and the k_F values of Table 1 of WO.

Table 1

Some characteristics of the theoretical density dependent pair potential of liquid sodium metal and comparison with experiment. (A) Neutral atom method results, (B) "observed" results (1, 2), (C) pseudo-potential result (9). R_a is the atomic sphere radius, $\vec{k}_F$ the Fermi wavevector, r_o the first node of $\phi(r)$, r_{min} the position of the first minimum in $\phi(r)$, ϕ_{min} is $\phi(r_{min})$, r_1 the position of the second node and ϕ_{max} is the value of the first repulsive maximum. All quantities are given in atomic units

	R_a	k_F	r_o	r_{min}	ϕ_{min}	r_1	$\phi_{max} \times 10^4$
(A)	3.3	0.582	6.6	7.5	-0.00057	9.5	0.5
	3.931	0.4882	6.45	7.4	-0.00094	9.9	0.8
	4.05	0.474	6.4	7.4	-0.00103	10.0	0.8
	4.46	0.430	6.25	7.3	-0.00144	10.1	1.1
(B)	3.91	0.491	6.15	7.1	-0.00084	8.2	4.8
	4.05	0.474	6.07	7.1	-0.00092	8.7	2.8
(C)	3.931	0.4882	6.3	7.3	-0.00125	10.0	0.83

The ion-electron-ion contribution to the pair potential is then calculated from n(r) by means of the approximate relation between n(r) and $\phi_{bs}(q)$ derived in (6). Let $\tilde{n}(r)$ be a pseudo-density equal to n(r) in the range $r \geqq R_m$ and as smooth as possible when $r < R_m$, such that $\int \tilde{n}(r)\, d\vec{r} = Z$ (Z = 1 for Na). Then $\phi(r)$ is given by

$$\phi(r) = \frac{2Z^2}{\pi} \int_0^\infty \left(1 - \phi_{bs}^N(q)\right) j_o(qr)\, dq \tag{1}$$

in the range $r > 2R_m$. ϕ_{bs}^N is the normalized band structure characteristic defined by

$$-\frac{4\pi Z^2}{q^2} \phi_{bs}^N = \frac{\tilde{n}(q)^2\, \tilde{\varepsilon}(q)}{2\,\Omega_o\, \pi_o(q)} . \tag{2}$$

$\pi_o(q)$ is the electron gas static screening function. $\tilde{\varepsilon}(q)$ is the proton-electron dielectric function which must be calculated here with the Gaspar-Kohn-Sham effective interaction $X(q) = -\pi e^2/k_F^2$. As shown in (6) equation (2) is nearly exact when the underlying ionic pseudo-potential is nearly local. This is expected to be true for Na. The form used for $\tilde{n}(r)$ in the range $r < R_m$ is given by formula (32) of reference (5), with $R_m = 0.75\, R_a$ (R_a is the atomic sphere radius).

Fig. 1. Theoretical sodium metal pair potential $\Phi(r)$ for four densities: (a) $R_a = 3.3$; (b) $R_a = 3.931$ (solid sodium metal at $T = 0\ ^oK$); (c) $R_a = 4.05$ (liquid sodium at $T = 100\ ^oC$); (d) $R_a = 4.46$. All quantities are given in atomic units

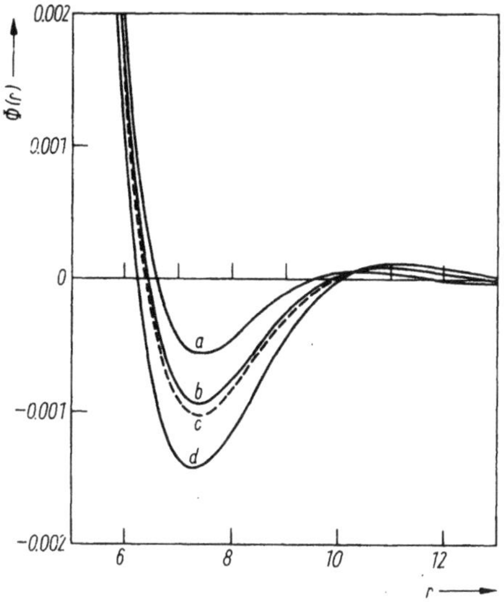

The resulting pair potentials $\Phi(r)$ are shown in Fig. 1 for four different densities. Some interesting characteristics are also displayed in Table 1, together with the corresponding observed values taken from WO.

We first comment the general behaviour of the calculated $\Phi(r)$. The pair potential is strongly density dependent. All the nodes r_i of the pair potential except the first one (r_o) increase with R_a and correspond to Friedel oscillations. The first node r_o increases with pressure. This "repulsive core diameter" (7) is related to the ionic pseudo-potential core radius R_M as shown by Shaw and Heine (8). Its variation with R_a may be approximately described by the formula

$$r_o = 8.0 - 0.39\, R_a \quad \text{at. units} \tag{3}$$

in the range $3.2 < R_a < 4.8$. The repulsive part of $\Phi(r)$ can be fitted in the range $5 < r \lessapprox r_o$ by

$$\Phi(r) = a\left[\left(\frac{r_o}{r}\right)^6 - 1\right] \tag{4}$$

with

$$a = 2.76\text{x}10^{-4}\, R_a^2 - 1.7\text{x}10^{-5}\, R_a^3 \quad \text{at units.}$$

Another important characteristic which can be read on Fig. 1 is that the first minimum Φ_{min} is deepened when the density is decreased. The overall effect of a pressure increase is thus to make the pair potential less attractive in the whole first oscillation range.

We can support this numerical results by a general argument along the lines proposed by Shaw and Heine (8). The integral of $\Phi(r)$ over the total space is

$$\int \Phi \, d\vec{r} = \lim_{q \to 0} \left[\frac{4\pi e^2}{q^2} + \Omega_o \Phi_{bs}(q) \right], \tag{5}$$

i.e. (equation (7) of reference (8), except a factor Ω_o)

$$\int \Phi \, d\vec{r} = -\frac{\pi^2 \hbar^2}{m k_F} + X(0) + \frac{4\pi e^2}{3} R_M^2, \tag{6}$$

where R_M is the model radius of the underlying local pseudo-potential considered by Shaw and Heine (8). X(0) is the q = 0 limit of the exchange and correlation effective function. Its Hartree-Fock and GKS value is $-\pi e^2/k_F^2$. The derivative of (6) with respect to k_F is then positive when k_F is smaller than $2/\pi$: $\int \Phi \, d\vec{r}$ increases with pressure as long as R_a remains larger than about 3 at. units, and Φ becomes thus more repulsive in the mean.

We compare now our theoretical results with the Born-Green analysis results of WO. The overall agreement is qualitatively good. According to Fig. 1 in WO, their pair potential is less attractive in liquid sodium metal under pressure. The nodes r_i (except the first one, r_o) of the WO pair potential change as expected, at least qualitatively. The agreement for Φ_{min} and its change with pressure is quite satisfactory, as shown in Table 1. On the other hand the "observed" hard sphere diameter r_o (6.15 at. units for R_a = 4.05) is significantly lower than the theoretical one (r_o = 6.4) and change in the opposite way with pressure. There is also a large disagreement between the values of the first maximum Φ_{max}. The WO result is much larger than our theoretical Φ_{max}. According to our calculation, $\Phi(r)$ cannot be represented by the Friedel asymptotic form $A_{as} \cos/r^3$ for r values less than 30 at. units. Shyu et al. (9) have obtained the same result. The high value of A_{as} found by Waseda and Suzuki (1) and by WO is puzzling and difficult to reconcile with the well-known theoretical expression which relates A_{as} to the back scattering form factor $W(2k_F)$ (see Harrison (10) (p. 45).[2)] It is worth noting that according to Bal-

2) When applied to the pair potential of Waseda and Suzuki, this theoretical expression (reference (10)), leads to $W(2k_F)$ = 0.10 at. units, in contradiction with the whole theoretical or experimental evidence (11).

lentine and Jones (12) the $\cos/r^3$ behaviour is not obtained for Na when a Percus-Yevick analysis is made.

Our conclusion is that there is a fair agreement between the Born-Green analysis pair potentials of WO and the theoretical ones in the most significant range ($r \gtrsim 10$ at. units). On the other hand, the quantitative discrepancy for the long-range oscillation amplitude is difficult to explain in terms of the electron theory of metals and more work is needed in this respect. According to some authors (12, 13), this behaviour may be an artefact of the Born-Green analysis itself.

References

(1) Y. WASEDA and K. SUZUKI, phys. stat. sol. (b) 49, 643 (1972).

(2) Y. WASEDA and M. OHTANI, phys. stat. sol. (b) 52, K37 (1972).

(3) M.A. BALL, J. Phys. C 2, 1248 (1969).

(4) L. DAGENS, J. Phys. C 5, 2333 (1972).

(5) L. DAGENS, J. de Physique, to be published.

(6) L. DAGENS, J. Phys. C 5, L213 (1972).

(7) N.W. ASHCROFT and J. LEKNER, Phys. Rev. 145, 83 (1966).

(8) R.W. SHAW and V. HEINE, Phys. Rev. B 5, 1646 (1972).

(9) W. SHYU, J.H. WEHLING, and M.R. CORDES, Phys. Rev. B 4, 1802 (1971).

(10) W.A. HARRISON, Pseudo-Potentials in the Theory of Metals, W.A. Benjamin, Inc., New York (p. 45).

(11) M.L. COHEN and V. HEINE, Solid State Phys. 24, 37 (1970).

(12) L.E. BALLENTINE and J.C. JONES, The Properties of Liquid Metals, Ed. S. TAKENCHI, 1972 (p. 51).

(13) W.S. HOWELLS and J.E. ENDERBY, J. Phys. C 5, 1277 (1972).

(Received November 5, 1973)

phys. stat. sol. (b) 62, K115 (1974)

Subject classification: 14.4.1; 20.1; 22.8.2

Institute of Solid State Physics, Bulgarian Academy of Sciences, Sofia

Resonant Interference in Raman Spectrum of KD_2PO_4

By

I.T. SAVATINOVA and J. UZUNOVA

The hydrogen bonded ferroelectrics isomorphous to KH_2PO_4 (KDP) manifest some striking features of anharmonic mode interactions in Raman spectra. Such phenomena can cause a strong distortion of the scattering line-shapes. The case of interference of a broad energy continuum with single discrete states may result in broad and asymmetric bands. These characteristic anomalies are usually analyzed by means of Fano's quantum mechanical formalism (1, 2) developed for Auger processes.

In the case of an approximately constant continuum we may write the scattering intensity as

$$I = Ap \frac{(q + \varepsilon)^2}{1 + \varepsilon^2} . \tag{1}$$

In this expression the parameter q is defined by the transition matrix elements from the initial state to the final states of the interacting scattering processes and by the transition rate W between the final states. The parameter ε is given by

$$\varepsilon = \frac{\omega - \omega_o}{\Gamma/2} ,$$

where ω_o is the frequency of the single-phonon state that has been slightly perturbed by an admixture of states from the continuum. The line halfwidth $\Gamma/2$ is equal to $\pi |W|^2$. For Stokes scattering the value of p is

$$p = \frac{\hbar \omega}{kT} (\bar{n} + 1) ,$$

where n is the Bose population factor. A is a normalization constant.

Using these ideas without any detailed analyses Scott and Wilson (3) have recently interpreted the E spectra of KDP above Θ in the region 0 to 100 cm^{-1}. Wall and She (4) have explained the anomalous line-shape of the B_1 symmetry mode at 470 cm^{-1} associated with the internal vibrations of the $(PO_4)^{3-}$ ion in KDP. They

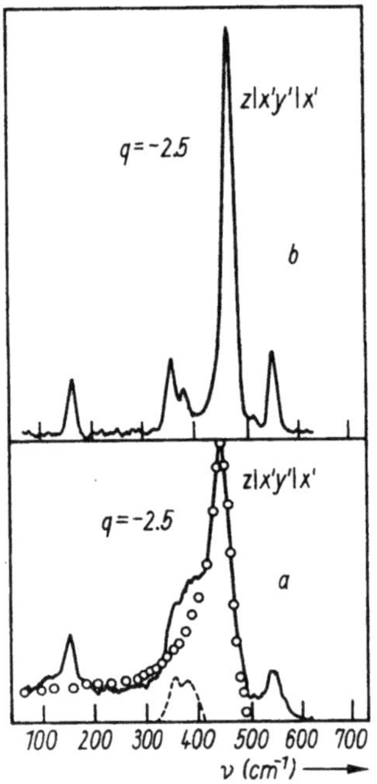

Fig. 1. a) The (x' y') polarization of DKDP at 300 °K. b) The (x'y') polarization of DKDP at 170 °K

suggest that the continuum is associated with proton motions, which may become active in (x'y') polarisation because of the proton disordering.

In this note we analyze the analogous very asymmetric peak at 457 cm^{-1} in (x'y') polarization of the deuterated analog KD_2PO_4 (DKDP). The spectra are obtained using a 1 W Ar laser and a FEU-79 photomultiplier with photon counting equipment. The crystals are placed in a glass cryostate.

Fig. 1a shows the (x'y') Raman spectrum of DKDP at room temperature. The band at 457 cm^{-1} corresponds to the mode at 470 cm^{-1} in KDP shifted towards lower frequencies owing to the deuteration. The left wing of the band is strongly distorted, and an anomalous background continuum extends to the exciting line. Below the temperature of the phase transition (Θ = 203 °K) the continuum practically disappears (Fig. 1b).

A least-squares fit at room temperature was made using equation (1) and q = = -2.5. The lattice mode at 151 cm^{-1} was ignored, since it does not show any interference with the continuum (Fig. 1a, circles). Some traces of the strong Raman lines at 357 cm^{-1} (A_1) and 384 cm^{-1} (B_2) appear in the spectrum because of imperfections of the orientations of crystals. A proposed form of the traces is shown in Fig. 1a (dashed profile) based on estimates of relative intensities and band-widths. After its superposition on our least-squares fit, a good reproduction of the experimental curve was obtained.

Similarly to the KDP case (3, 4), the value of q does not vary with temperature in the paraelectric phase, however, it is two time greater than the same parameter in the B_1 spectrum of KDP (q = -1,3). Since the line-widths of both bands (470 cm^{-1} in KDP and 457 cm^{-1} in DKDP) are equal, we may conclude that the ratio for scattering into the discrete state versus scattering into the continuum in the spectrum of DKDP is approximately twice greater.

Since such interaction phenomena are present in DKDP as well as in KDP it is clear that the continuum is not due to proton motion. More probably it is connected with the disordered state in crystals of the KDP type above θ , only indirectly related to protons because of their disordering role. In a completely ordered crystal the lattice momentum k for the phonons is zero. In a disordered crystal this selection rule weakens to an amount which depends on the degree of disorder in its structure. The disorder-induced Raman spectrum is proportional to the phonon density of states. This leads to the appearance of a continuum overlapping the usual discrete one-phonon spectrum.

References

(1) U. FANO, Phys. Rev. 124, 1866 (1961).

(2) D.L. ROUSSEAU and S.P.S. PORTO, Phys. Rev. Letters 20, 1354 (1968).

(3) J.F. SCOTT and C.M. WILSON, Solid State Commun. 10, 597 (1972).

(4) L.S. WALL and C.Y. SHE, Optics Commun. 5, 123 (1972).

(Received February 15, 1974)

phys. stat. sol. (b) 62, K119 (1974)
Subject classification: 13.5.3

V.I. Lenin State University, Kishinev

Ellipsoidal Optical Polaron

By

E.P. POKATILOV and L.V. TARAKANOVA

Many cubic crystals, parallel with a weak phonon anisotropy, have multi-valley conduction bands with isoenergetic surfaces that have the form of an ellipsoid (1). The problem of the polaron states has been considered under such conditions assuming small ($\alpha \ll 1$) (2) and large values ($\alpha \gg 1$) (3) of the electron-phonon interaction constant $\alpha = \frac{1}{2} e^2 c(2m/\hbar^3 \omega)^{1/2}$. However, the energy and effective masses in (2) contain only a term linear in α. Since in ionic crystals α often has a value larger than one it makes sense to calculate the energy and effective masses taking also into account the terms proportional to α^2. In the limit of strong electron-phonon interaction this problem has been calculated numerically (3), therefore it was interesting to obtain analytic expressions being sufficiently accurate.

The solution of the problem of the optical polaron with ellipsoidal dispersion law of an electron is adduced below, taking into account the mentioned definitions.

To obtain the functional, suitable for all values of the coupling constant, we use Feynman's variational principle (4) which at T (absolute temperature) $\longrightarrow 0$ can be put down in the following form:

$$E = E_o - \frac{1}{\lambda} \langle S - S' \rangle_{S'} ; \qquad \lambda = \frac{1}{kT} , \tag{1}$$

with

$$S = \int_0^\lambda L(\tau) d\tau ; \qquad S' = \int_0^\lambda L'(\tau) d\tau ,$$

where E is a variational functional of the polaron ground state, E_o is the trial system functional, S and L are the exact and S' and L' are trial actions and Lagrange functions of the system, respectively, $\langle \ \rangle$ is the symbol of the functional average (4), k is Boltzmann's constant.

The exact Lagrangian in a linear in $q_{\varkappa_j}$ approximation, corresponding to an anisotropic mass carrier, interacting with lattice oscillations, is

$$L = -\sum_i \frac{m_i \dot{r}_i^2}{2\hbar^2} - \frac{1}{2}\sum_{\varkappa,j}(\dot{q}^2_{\varkappa_j} + \omega^2 q^2_{\varkappa_j}) + \sum_{\varkappa,j}\gamma_{\varkappa_j}(\vec{r})q_{\varkappa_j}. \tag{2}$$

Here $\vec{r}$ is the radius vector of the electron, $q_{\varkappa_j}$ is the normal lattice oscillator coordinate, ω is the optical-phonon wave frequency (without dispersion), $\gamma_{\varkappa_j}(\vec{r})$ is the force of the electron acting on the $(\varkappa, j)$ lattice oscillator, m_i are the electron effective mass components, $i = 1; 2; 3$, $\hbar$ is Planck's constant divided by 2π.

The trial Lagrangian is chosen in a quadratic form:

$$L_o = -\sum_i \frac{m_i \dot{r}_i^2}{2\hbar^2} - \sum_i \frac{M_{if}\dot{R}_i^2}{2\hbar^2} - \sum_i \frac{M_{if}\omega_{if}(R_i - r_i)^2}{2}, \tag{3}$$

where $\vec{R}$ is the fictitious particle radius vector, M_{if} are its mass components, ω_f are the frequencies of the oscillator imitating the electron-phonon coupling. The quantities M_f and ω_f are variational parameters.

In the limit of small coupling constant, we obtain the solution for equation (1), using equations (2) and (3), integrating and minimizing

$$E = -\alpha\hbar\omega\left(\frac{m_{\parallel}}{m_{\perp}}\right)^{1/3} \varphi_o\left(\frac{m_{\parallel}}{m_{\perp}}\right) - \alpha^2\hbar\omega\,\varphi_1\left(\frac{m_{\parallel}}{m_{\perp}}\right). \tag{4}$$

Here we have introduced the following designations:

$$\varphi_o(x) = \begin{cases} \dfrac{1}{2\sqrt{1-x}} \ln\left|\dfrac{1+\sqrt{1-x}}{1-\sqrt{1-x}}\right| & \dfrac{m_{\parallel}}{m_{\perp}} < 1 \\ \dfrac{1}{\sqrt{x-1}} \operatorname{arctg}\sqrt{x-1} & \dfrac{m_{\parallel}}{m_{\perp}} > 1, \end{cases}$$

$$\varphi_1(x) = \frac{1}{54}\frac{x^{2/3}}{(1-x)^2}\left[(2+x^2)\varphi_o^2(x) - 2(2+x)\varphi_o(x) + 3\right].$$

In the same limit of small α, the formulae for polaron effective mass components received by the Krivoglaz-Pekar method (5) have the following form:

$$M_{\perp} = m_{\perp}\left[1 + \alpha\varphi_{\perp}\left(\frac{m_{\parallel}}{m_{\perp}}\right) + \alpha^2\varphi'_{\perp}\left(\frac{m_{\parallel}}{m_{\perp}}\right)\right], \tag{5a}$$

$$M_{\parallel} = m_{\parallel}\left[1 + \alpha\varphi_{\parallel}\left(\frac{m_{\parallel}}{m_{\perp}}\right) + \alpha^2\varphi'_{\parallel}\left(\frac{m_{\parallel}}{m_{\perp}}\right)\right], \tag{5b}$$

where

$$\varphi_{\perp}(x) = \frac{x^{1/3}}{4(x-1)}\left[x\,\varphi_o(x) - 1\right],$$

$$\varphi_{\parallel}(x) = \frac{x^{1/3}}{2(x-1)}\left[1 - \varphi_o(x)\right],$$

$$\varphi'_{\perp(\parallel)}(x) = \frac{x^{2/3}}{(x-1)^3}\left[A_{\perp(\parallel)}(x)\varphi_o^2(x) + B_{\perp(\parallel)}(x)\varphi_o(x) + C_{\perp(\parallel)}(x)\right],$$

$$A_{\perp}(x) = \frac{145x^3 - x^2 + 288x}{2592}; \quad B_{\perp}(x) = -\frac{386x^2 + 382x + 96}{2592}; \quad C_{\perp}(x) = \frac{433x - 1}{2596};$$

$$A_{\parallel}(x) = \frac{-72x^2 + x - 145}{648}; \quad B_{\parallel}(x) = \frac{48x^2 + 46x + 338}{648}; \quad C_{\parallel}(x) = \frac{x - 217}{648}.$$

In equation (4) and (5) the terms linear in α coincide with the results of paper (2) if $m_{\perp}$ is introduced into α as in reference (2), instead of the geometrically averaged mass $(m_{\parallel} m_{\perp}^2)^{1/3}$. The additional quadratic terms slightly reduce the energy and raise the mass values (for $\alpha = 1$ the corrections are about 12% for energy and 15% for mass).

In the limit $m_{\parallel}/m_{\perp} \rightarrow 1$ our results give the well-known Feynman expressions (4) in the second order in α.

In the strong coupling limit we obtain:

$$E = -\alpha^2 \hbar \frac{\omega}{\pi}\left(\frac{m_{\parallel}}{m_{\perp}}\right)^{2/3} \frac{\delta_1^2}{1-\delta_1}\left(2\frac{m_{\parallel}}{m_{\perp}} + 1 - \delta_1\right) F_o^2\left(\frac{m_{\parallel}}{m_{\perp}};\delta_1\right) - \hbar\omega\,\frac{2\frac{m_{\parallel}}{m_{\perp}} + 1 - \delta_1}{1-\delta_1}\ln 2 -$$

$$- \frac{\hbar\omega}{4}\frac{\delta_1}{\sqrt{1-\delta_1}} F_o\left(\frac{m_{\parallel}}{m_{\perp}};\delta_1\right)\left\{\frac{8\frac{m_{\parallel}}{m_{\perp}}}{F_1(\delta_1)} + \frac{1-\delta_1}{F_2(\delta_1)} - \frac{(1-\delta_1)F_1(\delta_1)}{F_2^2(\delta_1)}\right\}; \quad \frac{m_{\parallel}}{m_{\perp}} < 1, \tag{6a}$$

$$E = -\alpha^2 \hbar \frac{\omega}{\pi}\left(\frac{m_{\parallel}}{m_{\perp}}\right)^{2/3} \frac{\delta_2^2}{1-\delta_2}\left[2\frac{m_{\parallel}}{m_{\perp}}(1-\delta_2) + 1\right] F_o^2\left(\frac{m_{\parallel}}{m_{\perp}};\delta_2\right) - \hbar\omega\left[2\frac{m_{\parallel}}{m_{\perp}}(1-\delta_2) + 1\right]\ln 2 -$$

$$- \frac{\hbar\omega}{4}\frac{\delta_2}{\sqrt{1-\delta_2}} F_o\left(\frac{m_{\parallel}}{m_{\perp}};\delta_2\right)\left\{\frac{8\frac{m_{\parallel}}{m_{\perp}}}{F_1(\delta_2)} + \frac{1}{F_2(\delta_2)} - \frac{F_1(\delta_2)}{F_2^2(\delta_2)}\right\}; \quad \frac{m_{\parallel}}{m_{\perp}} > 1, \tag{6b}$$

where

$$F_o\left(\frac{m_\parallel}{m_\perp};\delta\right) = \begin{cases} \dfrac{1-\delta}{2\delta}\,\dfrac{\ln\left|\dfrac{1+\sqrt{\delta}}{1-\sqrt{\delta}}\right|}{2\dfrac{m_\parallel}{m_\perp}+1-\delta} & \dfrac{m_\parallel}{m_\perp} < 1\,, \\ \dfrac{1-\delta}{\delta}\,\dfrac{\arcsin\sqrt{\delta}}{2\dfrac{m_\parallel}{m_\perp}(1-\delta)+1} & \dfrac{m_\parallel}{m_\perp} > 1\,, \end{cases}$$

$$F_1(\delta) = \begin{cases} -\dfrac{1-\delta}{2\sqrt{\delta}}\ln\left|\dfrac{1+\sqrt{\delta}}{1-\sqrt{\delta}}\right| + 1 & \dfrac{m_\parallel}{m_\perp} < 1, \\ \dfrac{1}{\sqrt{\delta}}\arcsin\sqrt{\delta} - \sqrt{1-\delta} & \dfrac{m_\parallel}{m_\perp} > 1\,, \end{cases}$$

$$F_2(\delta) = \begin{cases} \dfrac{1}{2\sqrt{\delta}}\ln\left|\dfrac{1+\sqrt{\delta}}{1-\sqrt{\delta}}\right| - 1 & \dfrac{m_\parallel}{m_\perp} < 1, \\ -\dfrac{1-\delta}{\sqrt{\delta}}\arcsin\sqrt{\delta} + \sqrt{1-\delta} & \dfrac{m_\parallel}{m_\perp} > 1\,, \end{cases}$$

$$\delta_1 = \frac{3.5}{15\dfrac{m_\parallel}{m_\perp}+4}\left[-\left(3\frac{m_\parallel}{m_\perp}+4\right)+\sqrt{\frac{1}{7}\left[-237\left(\frac{m_\parallel}{m_\perp}\right)^2+388\frac{m_\parallel}{m_\perp}+192\right]}\right],$$

$$\delta_2 = 10\,\frac{\dfrac{m_\parallel}{m_\perp}-1}{11\dfrac{m_\parallel}{m_\perp}+3}\,.$$

The minimization of energy functionals has led to transcendental equations, exact solutions of which we could find only numerically. The approximate solutions δ_1 (for $m_\parallel/m_\perp < 1$) and δ_2 (for $m_\parallel/m_\perp > 1$) are just for ratios $m_\parallel/m_\perp$ close to unity, but when the electron mass ratios are of the order of 10, the energy expressions give an error of about 3%.

The formulae for the polaron effective mass in the case $\alpha \gg 1$ are

$$M_\perp = m_\perp \alpha^4 \frac{4}{\pi^2}\left(\frac{m_\parallel}{m_\perp}\right)^{4/3} F_o^2\left(\frac{m_\parallel}{m_\perp};\delta\right)\frac{1}{\delta}F_1^2(\delta)\exp\left\{\frac{2\frac{m_\parallel}{m_\perp}\frac{\delta}{\sqrt{1-\delta}}F_o\left(\frac{m_\parallel}{m_\perp};\delta\right)}{F_1(\delta)}-1\right\}, \tag{7a}$$

$$M_\parallel = m_\parallel \alpha^4 \frac{16}{\pi^2}\left(\frac{m_\parallel}{m_\perp}\right)^{4/3} F_o^2\left(\frac{m_\parallel}{m_\perp};\delta\right)\frac{1}{\delta}F_2^2(\delta)\exp\left\{\frac{\delta^2 F_o\left(\frac{m_\parallel}{m_\perp};\delta\right)}{F_2(\delta)}-1\right\}. \tag{7b}$$

Our analytical expressions for $M_\perp$ and $M_\parallel$ give numerical values that are slightly smaller than in reference (3). In the limit $m_\parallel/m_\perp \rightarrow 1$ these formulae turn to Feynman's expression for the strong coupling polaron (4).

In conclusion we'll note that the intermediate coupling polaron ground energy and effective mass have been calculated numerically by Pekar et al. in their recent paper (6). The calculations show that a reasonable good agreement is observed between these values and our analytical results for $\alpha < 5$ ($\approx 0.5\%$ for $\alpha = 2$ to 3). For not very large α ($5 < \alpha < 10$), both formulae for the energy, (6a) and (6b), give close values to each other which are, however, larger than those obtained in reference (6). The comparison of results of our paper and (3) with the data given in (6) shows that the strong coupling approximation for mass is realized for larger values of α than for energy.

References

(1) V.G. GORCHEV, M.F. DEIGEN, and S.I. PEKAR, Fiz. tverd. Tela 9, 157 (1967).

(2) S.I. PEKAR, Zh. eksper. teor. Fiz. 55, 1997 (1968).

(3) S.I. PEKAR, V.I. SHEKA, and G.V. DMITRENKO, Zh. eksper. teor. Fiz. 63, 1455 (1972).

(4) R.P. FEYNMAN, Phys. Rev. 97, 665 (1955).

(5) M.A. KRIVOGLAZ and S.I. PEKAR, Izv. Akad. Nauk SSSR. Ser. fiz., 21, 16 (1957).

(6) S.I. PEKAR, L.S. KHAZAN, and V.I. SHEKA, Zh. eksper. teor. Fiz. 65, 1999 (1973).

(Received February 15, 1974)

Pre-printed Titles

Papers to be published

in "physica status solidi (b) - basic research"

Vol. 63, No. 1

Review Article

I.V. SVECHKAREV and A.S. PANFILOV, Kharkov

Effects of Pressure on the Electronic Structure of Transition d-Metals

Original Papers

H. BURKHARD, R. GEICK, P. KÄSTNER, and K.-H. UNKELBACH, Aachen and Würzburg

Lattice Vibrations and Free Carrier Dispersion in PbSe

G. ASCARELLI, West Lafayette, Indiana

The Influence of "Rotational" Lattice Deformation States on the F Center Emission

K. IKEDA, T. NAKAMICHI, Y. YAMADA, and M. YAMAMOTO, Sendai

Kondo Effect in the Transport Properties of the CsCl-Type Compounds $Fe_{1-x}Ti_{1+x}$ (II)

B. LISCHKE und W. RODEWALD, Berlin

Magnetische Bereichsstrukturen in Typ I Dünnschichtsupraleitern (II)

J. ROSSAT-MIGNOD, G. QUEZEL, A. BERTON, and J. CHAUSSY, Grenoble

Crystal Field Determination in Rare-Earth Oxysulfides (III)

F.J. OWENS, Dover, New Jersey

Spin-Lattice Coupling Constants of Divalent Manganese in a Tetragonal Lattice

I. P. BONILLA, A. HOLZ, and H.N. PUTT, Porto Alegre

Theory of the Order Parameter of $NaNO_2$ in Molecular Field Approximation

G. SERFÖZÖ, Budapest

Asymptotic and Pre-Asymptotic Region of Charge Density Oscillation around Impurities in Dilute Copper-Based Alloys

V.K. BASHENOV, R.A. ALARASHI, and M.G. FOIGEL, Odessa

Toward a Dielectric Theory of Chemical Bonding

V.M. FAIN, Chernogolovka

On the Theory of Rate Processes in Condensed Media

S.M. KLETSMAN, I.P. POLIKARPOVA, G.N. TATARINOVA, and A.N. TIMOFEEV, Sverdlovsk

Isotope Effect in Self-Diffusion of Cs in Single Crystals of CsI

G. FISHMAN and C. HERMANN, Paris

Optical Pumping and Transverse Magnetic Field Effect for Excitation above the Band Edge in p-Type Semiconductors

C. WEIGEL, D. PEAK, J.W. CORBETT, G.D. WATKINS, and R.P. MESSMER, Albany and Schenectady, New York

LCAO Calculations for the Boron and Nitrogen Interstitial in the Diamond Lattice

N.J. FINK and R.S. POULSEN, Davis, California

An Exact Solution of a New Superconducting State

N. MEISSNER, E.J. SAVINO, J.R. WILLIS, and R. BULLOUGH, Harwell, Didcot, Berks.

The Dislocation Loop in an Anisotropic Medium and Its Interaction with an Interstitial Atom

J. DEPORTES, D. GIGNOUX, and F. GIVORD, Grenoble

Magnetic Properties of a $TmCo_2$ Single Crystal

J. SINGH and H. BAESSLER, Marburg/Lahn

Solution of the Onsager Ion Recombination Model in One Dimension

H. PUSZKARSKI, Poznań

Effect of Magnetic Surface Reconstruction in Spin-Wave Resonance: Two-fold Resonance of the Surface Mode

A.F. SADREEV, Kazan

State Equation of the Ising Model in the Ladder Approximation

G. LEHMANN and C. HAMANN, Karl-Marx-Stadt

A Model of the Electrical Transport Phenomena in Imperfect Crystals of Copper Phthalocyanine (III)

N.C. HALDER, Tampa, Florida

Temperature Dependence of the Nuclear Spin Relaxation Rate of Liquid In

J. MERTSCHING, Berlin

Alloys with Off-Diagonal Disorder

E. MAJERNÍKOVÁ, Bratislava

Stark Effect of Wannier-Mott Excitons Due to Disorder

H.J. FALGE, A. OTTO, and W. SOHLER, München

Dispersion of Surface and Bulk Phonon-Polaritons on α-Quartz Measured by Attenuated Total Reflection

P.L. VUILLERMOZ, A. LAUGIER, and P. PINARD, Lyon

Influence of Clusters on the Thermal Conductivity of GaAs

Papers to be published
in "physica status solidi (a) - applied research"
Vol. 22, No. 2

Original Papers

E. KURAMOTO, F. MINAMI, and S. TAKEUCHI, Tokyo
Computer Simulation of Motion of Screw Dislocation in Model B.C.C. Lattices (II)

H.-J. KAUFMANN, Kharkov
Zum Mechanismus des Tieftemperaturkriechens von $KCl:Ba^{2+}$ Kristallen

W. HAUBENREISSER, W. LÖSER, C. MATTHECK, K.-H. MÖCKEL, and E. STEINBEISS, Dresden
Co_3O_4 as a Model Substance for Thermistor Effect Switching in Coplanar Thin Film Devices

C. BOUMFORD and A.H. MORRISH, Winnipeg, Manitoba
Magnetic Properties of the Iron Selenide Fe_7Se_8

H. NARAMOTO, K. OZAWA, T. OKADA, and T. SUITA, Tokai-mura and Osaka
Correlated Changes of Flow Stress and Optical Absorption in KBr γ-Rayed at LNT

A.D. KING and J. MOERMAN, Chalk River, Ontario
Calcium Diffusion in Pure and YF_3-Doped Single Crystal CaF_2

T.M. RACHEVA-STAMBOLIEVA, G.A. KRASULIN, YU.D. CHISTYAKOV, I.D. DRAGIEVA, and D.H. DJOGLEV, Sofia and Moscow
Thermodynamic Model and Epitaxial Growth of CdSe Layers

A.I. OSEISKII, V.P. SOLDATOV, V.I. STARTSEV, and V.D. NATSIK, Kharkov
Temperature Dependence and Activation Parameters of Creep in Zn in the Temperature Range 1.5 to 80 $^{\circ}$K

G.J. IRWIN, F. GUIU, and P.L. PRATT, London
The Influence of Orientation on Slip and Strain-Hardening of Molybdenum Single Crystals

S.S. BHANDARI and J. VARMA, Udaipur
Mössbauer Effect of $Co^{57}Cl_2$ Deposited on Metals

D.F. CRABTREE, Huddersfield
Luminescence, Optical Absorption, and ESR of ZnS-Se:Mn

S. GUPTA, K.B. LAL, T.M. SRINIVASAN, and G.N. RAO, Kanpur
Mössbauer Studies of Fe-Be, Fe-Nb, Fe-Ru, Fe-Te, and Fe-Ir Alloys

J.-P. KRUMME, Hamburg
The Structure of the Compensation Wall in Ferrimagnets

D.B. PROWSE and J.D. CHASHION, Clayton, Victoria
Mössbauer Effect Measurements on PdGd Alloys

R.V. TELESNIN, A.G. SHISHKOV, V.E. OSUKHOVSKII, A.S. SIGOV, and L.P. OSUKHOVSKAYA, Moscow
Coercivity Spectrum and Domain Wall Creep in 200 to 1000 Å Permalloy Films

S. POROWSKI, M. KONCZYKOWSKI, and J. CHROBOCZEK, Warsaw
Evidence from Transport Measurements at High Pressures for Donor Ions Occupying Non-Equivalent Lattice Positions in InSb

J.A. BRAGAGNOLO, G.M. STORTI, and K.W. BÖER, Newark, Delaware
Bound-Exciton Effects in the Photocurrent Spectrum of CdS Single Crystals

B. BARBARA, M.F. ROSSIGNOL, H.G. PURWINS, and E. WALKER, Grenoble and Genève
Magnetocrystalline Anisotropy in the Cubic $TbAl_2$ Compound

I.B. PUCHALSKA, Salford
Ordered Arrays of Bubbles in Permalloy

Hj. BERNHARDT and P. GÖRLICH, Jena
Investigation of the Dynamical Coloration Behaviour of Cerium-Oxide-Doped Alkaline Earth Fluoride Crystals

V.M. FRIDKIN, K.A. VERKHOVSKAYA, and B.G. BOCHKOV, Moscow
Optical Absorption Edge in Improper Ferroelectrics in the Phase Transition Region

19. Magnetic Resonance
20. Optical Properties
  20.1 Spectra. Optical Constants (X-Ray Spectra See 20)
  20.2 Lasers
  20.3 Luminescence (X-Ray Spectra See 20)
21. Metals. Alloys
  21.1 Transition Metals (Group Sc to Zn) and Their Alloys
    21.1.1 Fe and Fe Alloys
  21.2 Alkali Metals
  21.3 Alkali-Earth Metals
  21.4 Rare-Earth Metals
  21.5 Actinides
  21.6 Noble Metals
  21.7 Semi-Metals (Semiconducting Alloys See 22.7)
22. Semiconductors and Ionic Crystals
  22.1 Elements
    22.1.1 Germanium
    22.1.2 Silicon
    22.1.3 Group VI Elements
  22.2 III–V Compounds
    22.2.1 Arsenides
    22.2.2 Phosphides
    22.2.3 Antimonides
    22.2.4 Mixed Crystals
  22.3 IV–IV, II–V, and III–VI Compounds
    22.3.1 Mixed Crystals
  22.4 II–VI Compounds
    22.4.1 Sulphides
    22.4.2 Selenides
    22.4.3 Tellurides
    22.4.4 Mixed Crystals
  22.5 Halides
    22.5.1 Silver Halides
    22.5.2 Alkali Halides
    22.5.3 Alkali-Earth Halides
    22.5.4 Mixed Crystals
  22.6 Simple Oxides (CdO, ZnO See 22.4)
    22.6.1 Mixed Crystals
  22.7 Semiconducting Intermetallic Compounds
  22.8 Tertiary and Higher Compounds
    22.8.1 Salts of Oxy-Acids (Carbonates, Phosphates, Silicates Including Mica etc.)
    22.8.2 Spinels, Garnets, Ferrites
  22.9 Organic Semiconductors
23. Solidified Gases

Attention

The preceding classification scheme is closely akin to the scheme used by Physics Abstracts to classify the various fields of solid state physics. A paper will either be classified under one or more of the subsections or, if none of these are relevant, it will just appear under the general section heading (i.e. sections and subsections are equivalent in this sense).

The *subject classification* given at the head of each paper begins with the main subject matter of the paper, and this is followed, if necessary, by figures referring to subsidiary subject matter and the type of solids investigated.

The papers published in this issue are listed in the order of their subject classification at the end of the table of contents.

32913

Manuscripts and letters for physica status solidi (b) — basic research and physica status solidi (a) — applied research

should be sent to

DDR, 102 Berlin, Neue Schönhauser Str. 20

or

Internationaler Buchversand GmbH, 1 Berlin 19, Eschenallee 21.

The object of "physica status solidi (b)" and "physica status solidi (a)" is to publish rapidly important and topical results concerning fundamental research in the field of solid state physics and aspects of application and experimental techniques of fundamental solid state research, respectively. The issues of "physica status solidi (b)" and "physica status solidi (a)" will be published on the 1st and the 16th day of each month, respectively. Either series will contain the following categories:

1. Review articles written by arrangement with the editors. They will be remunerated.
2. Original papers selected from those which have not been submitted for publication elsewhere, after being reviewed. (The latest date for acceptance of contributions is the 8th and the 23rd day of each month for "physica status solidi (b)" and "physica status solidi (a)", respectively, 50 days before the delivery of the journal.)
3. Short notes published in the off-set part of the journal. They are, if necessary after being reviewed, selected by the editorial staff. (The latest date for acceptance of manuscripts for this part is the 12th and the 27th day of each month for "physica status solidi (b)" and "physica status solidi (a)", respectively, 16 days before the delivery of the journal.)
4. Pre-printed titles of papers to be published in the next issue of both series.

Editorial Information

Manuscripts should be submitted preferentially in English. Besides, papers in German, French, and Russian are admitted. The final decision on the incorporation of the papers in the series (b) or (a) is taken by the editorial staff after consultation of the referee. Authors are asked to observe the following rules:

1. Manuscripts have to be written as concisely as possible. As most, 20 and 4 pages (28 lines per page; abstracts, tables, references, figures, and captions of figures included, where 3 figures are counted as about one page) are admitted for original papers and short notes, respectively. All manuscripts should be type-written on one side only. Two copies are required. Badly arranged manuscripts are returned.
2. The author's name, initials, and institution should be given.
3. One abstract of about ten lines in English and one in German, French, or Russian are required for original papers.
4. Because of lack of space conclusions (if any) must not have the character of a summary (i.e. substantially a repetition of the contents of the abstract).
5. Formulae should not be written to small and with pencil. Separate lines for formulae are desirable. Indistinct writing will lead to a return of the manuscript.
6. Drawings should be enclosed on separate sheets of paper, not exceeding the maximum and minimum sizes of 20×30 and $10 \times 15\ cm^2$, respectively; photographs should be submitted, on white paper, only for review articles and original papers. Duplicates are required.
7. Captions for the figures should be submitted on separate sheets of paper, and special instructions with regard to the size of illustrations should be given on these.
8. References should be numbered in order of appearance and should be given, on a separate sheet, in the form used in physica status solidi.
9. Units, symbols, etc. should be used in accordance with the recommendations given by the IUPAP.
10. Each paper should contain a proposal for the subject classification according to the classification scheme appearing on the second and third cover of each issue.

The authors will be sent galley proofs of review articles aod original papers. Proof reading of short notes and pre-printed titles will be done by the editorial staff.

For each original paper aud short note a total of 75 reprints is available free of charge. The number of reprints free of charge per review article is 25. Requests for additional reprints (against payment) should be given together with the submitted manuscript. **The deadline for orders of additional reprints is the 15th and the 30th day of each month** for "physica status solidi (b)" and "physica status solidi (a)", respectively, two weeks before the delivery of the journal.

Bei Fragen zur Produktsicherheit wenden Sie sich bitte an:
If you have any questions regarding product safety,
please contact:

Walter de Gruyter GmbH
Genthiner Straße 13
10785 Berlin
productsafety@degruyterbrill.com